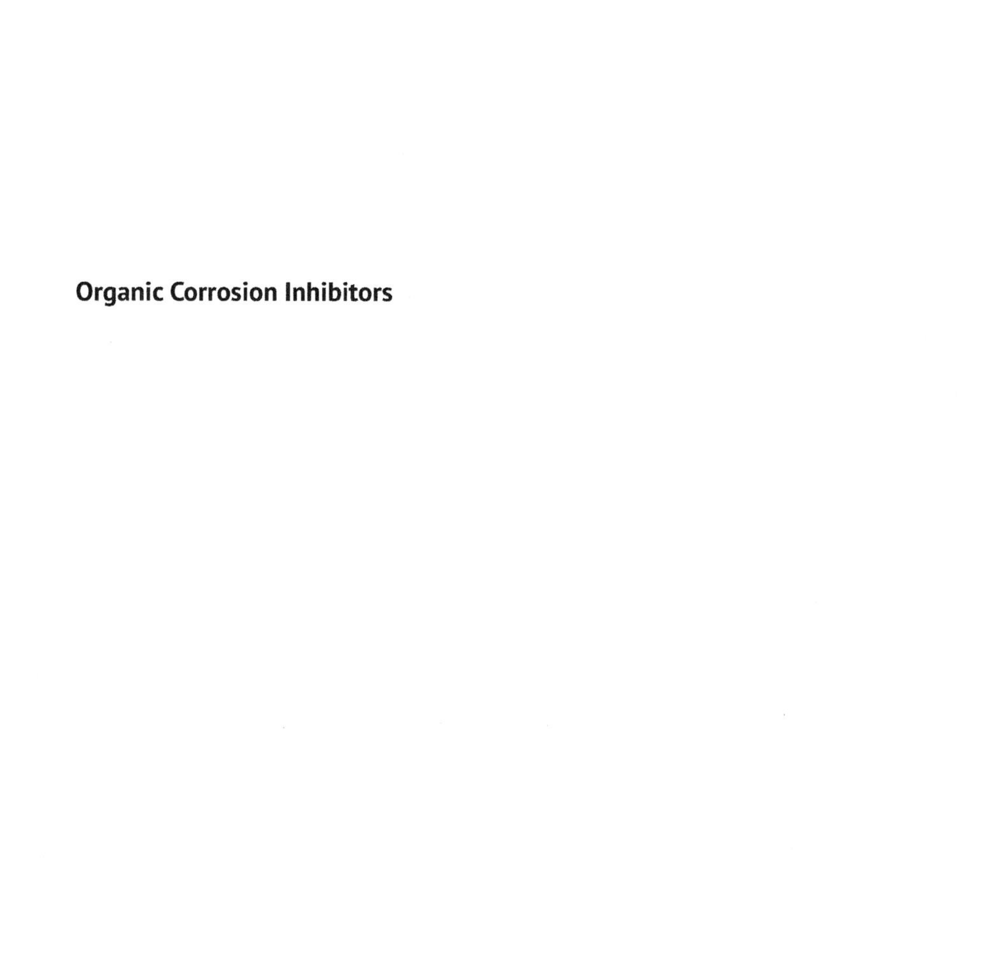

Organic Corrosion Inhibitors

Organic Corrosion Inhibitors

Synthesis, Characterization, Mechanism, and Applications

Edited by

Chandrabhan Verma
King Fahd University of Petroleum and Minerals
Dhahran, Saudi Arabia

Chaudhery Mustansar Hussain
New Jersey Institute of Technology
Newark, NJ, USA

Eno E. Ebenso
University of South Africa
Johannesburg, South Africa

This edition first published 2022

Registered Office
John Wiley & Sons, Inc., 111 River Street, Hoboken, NJ 07030, USA

Editorial Office
111 River Street, Hoboken, NJ 07030, USA

For details of our global editorial offices, customer services, and more information about Wiley products visit us at www.wiley.com.

Wiley also publishes its books in a variety of electronic formats and by print-on-demand. Some content that appears in standard print versions of this book may not be available in other formats.

Library of Congress Cataloging-in-Publication Data

Names: Verma, Chandrabhan, editor. | Hussain, Chaudhery Mustansar, editor. | Ebenso, Eno E., editor.
Title: Organic corrosion inhibitors : synthesis, characterization, mechanism, and applications / edited by Chandrabhan Verma, Chaudhery Mustansar Hussain, Eno E. Ebenso.
Description: First edition. | Hoboken, NJ : Wiley, 2022. | Includes index.
Identifiers: LCCN 2021031915 (print) | LCCN 2021031916 (ebook) | ISBN 9781119794486 (cloth) | ISBN 9781119794493 (adobe pdf) | ISBN 9781119794509 (epub)
Subjects: LCSH: Corrosion and anti-corrosives. | Corrosion and anti-corrosives–Environmental aspects.
Classification: LCC TA462 .O65 2022 (print) | LCC TA462 (ebook) | DDC 620.1/1223–dc23
LC record available at https://lccn.loc.gov/2021031915
LC ebook record available at https://lccn.loc.gov/2021031916

Cover Design and Image: Wiley

Set in 9.5/12.5pt STIXTwoText by Straive, Pondicherry, India

10 9 8 7 6 5 4 3 2 1

Contents

Preface

Corrosion is a highly dangerous phenomenon that causes huge economic and safety problems. Various methods of corrosion monitoring, including cathodic protection, panting and coatings, alloying and dealloying (reduction in metal impurities), surface treatments, and use of corrosion inhibitors have been developed depending upon the nature of metal and environment. Application of organic compounds, especially heterocyclic compounds, is one of the most common, practical, easy, and economic methods of corrosion mitigations. Obviously, these compounds become effective by adsorbing on the metallic surface using electron-rich centers including multiple bonds and polar functional groups. These electron-rich centers act as adsorption sites during their interaction with the metallic surface. Along with acting as adsorption sites, the polar functional groups such as –OH (hydroxyl), $-NH_2$ (amino), –OMe (methoxy), –COOH (carboxyl), $-NO_2$ (nitro), –CN (nitrile), and so on also enhance solubility of organic compounds in polar electrolytes. Present book describes the collection of major advancements in using organic compounds as corrosion inhibitors including their synthesis, characterization, and corrosion inhibition mechanism.

Through this book it can be seen that use of organic compounds serves as one of the most effective, economic, and ease methods of corrosion monitoring. Using previously developed methods, 15% (US $375) to 35% (US $875) of cost of corrosion can be minimized. Different series of organic compounds, including heterocyclic compounds, are effectively used as corrosion inhibitors for different metals and alloys in various environments. Because of the increasing ecological awareness and strict environmental regulations, various classes of environmental-friendly alternatives to the traditional toxic corrosion inhibitors have been developed and being implemented. These series of compounds mostly include carbohydrates, natural polymers and amino acids (AAs), and their derivatives. Corrosion scientists and engineers strongly believe that these environmental-friendly alternatives will be capable to replace, in the near future, the toxic marketable products that are still being used via many worldwide industries.

A book covering the recent developments on using organic compounds as corrosion inhibitors is broadly overdue. It has been addressed by Drs. Verma, Hussain, and Ebenso in this book which attends to fundamental characteristics of organic corrosion inhibitors, their synthesis and characterization, chronological growths, and their industrial applications. The corrosion inhibition using organic compounds, especially heterocyclic compounds, is broad ranging. This book is divided into five sections, where each section contains several chapters. Section 1 "Basics of corrosion and prevention" describes the basic of corrosion, experimental and computational testing of corrosion, and a comparison between organic and inorganic corrosion inhibitors. Section 2 "Heterocyclic and non-heterocyclic corrosion inhibitors" describes the collection of different series of heterocyclic and non-heterocyclic corrosion inhibitors such as amines, imidazole, quinoline, pyridine, indole, and their derivatives. This section also includes organic compounds as corrosion inhibitors for oil and gas industries.

Section 3 "Organic green corrosion inhibitors" entirely focuses on green corrosion inhibitors. This section describes the corrosion inhibition characteristics of carbohydrates, amino acids (AAs), oleochemicals, chemical medicines, ionic liquids (ILs), and their derivatives. Section 4 "Organic compounds based nanomaterials as corrosion inhibitors" describes the corrosion inhibition properties of carbon nanotubes (CNTs: SWCNTs and MWCNTs), graphene oxide (GO), and their composites. In the end, Section 5 "Organic polymers as corrosion inhibitors" gives a description on natural and synthetic polymers as corrosion inhibitors. Overall, this book is written for scholars in academia and industry, working corrosion engineering, materials science students, and applied chemistry. The editors and contributors are well-known researchers, scientists, and true professionals from academia and industry. On behalf of Wiley, we are very thankful to authors of all chapters for their amazing and passionate efforts in making of this book. Special thanks to Prof. M. A. Quraishi, who guided us continuously in drafting of this book. Special thanks to Michael Leventhal (acquisitions editor) and Katrina Maceda (managing editor) for their dedicated support and help during this project. In the end, all thanks to Wiley for publishing the book.

Chandrabhan Verma, PhD
Chaudhery Mustansar Hussain, PhD
Eno E. Ebenso, PhD

About the Editors

Chandrabhan Verma

Chandrabhan Verma is working at the Interdisciplinary Center for Research in Advanced Materials King Fahd University of Petroleum and Minerals (KFUPM), Saudi Arabia. He obtained his PhD in Corrosion Science at the Department of Chemistry, Indian Institute of Technology (Banaras Hindu University) Varanasi, India. He is a member of American Chemical Society (ACS). His research is mainly focused on the synthesis and designing of environmental friendly corrosion inhibitors useful for various industrial applications. Dr. Verma is the author of several research and review articles published in the peer-reviewed international journals of ACS, Elsevier, RSC, Wiley, Springer, etc. He has total citation of more than 4700 with H-index of 39 and i10-index of 91. Dr. Verma received several national and international awards for his academic achievements.

Chaudhery Mustansar Hussain

Chaudhery Mustansar Hussain, PhD, is an adjunct professor and director of labs in the Department of Chemistry & Environmental Sciences at the New Jersey Institute of Technology (NJIT), Newark, New Jersey, USA. His research is focused on the applications of nanotechnology & advanced technologies & materials, analytical chemistry, environmental management, and various industries. Dr. Hussain is the author of numerous papers in peer-reviewed journals, as well as prolific author and editor of several scientific monographs and handbooks in his research areas published with Elsevier, Royal Society of Chemistry, John Wiley & sons, CRC, Springer, etc.

Eno E. Ebenso

Eno E. Ebenso is a research professor at the Institute of Nanotechnology and Water Sustainability in the College of Science, Engineering and Technology, University of South Africa. He has published extensively in local and international peer-reviewed journals of wide readership with over 300 publications (articles in newspapers, plenary/invited lectures, and conference proceedings not included). He currently has an H-Index of 67 and over 10 000 total citations from the Scopus Search Engine of Elsevier Science since 1996. According to the Elsevier SciVal Insights Report (2010–2015), he has a citation impact 10% above world average: second most prolific author in the field of corrosion inhibition worldwide and fifth most downloads of his publications globally in the field of corrosion inhibition. His Google Scholar Citations since 2013 is over 8000 with an H-index of 64 and i10-index of 216. His RESEARCHERID account shows H-index of 44 with total citations of 5779 and average citation per article of 24.78. He is also a B3 NRF Rated Scientist in Chemistry (South African National Research Foundation). INTERPRETATION – B3: Most of the reviewers are convinced that he enjoys considerable international recognition for the high quality and impact of his recent research outputs. He is a member of International Society of Electrochemistry, South African Chemical Institute (M.S.A. Chem. I.), South African Council for Natural Scientific Professions (SACNASP) (Pri. Sci. Nat.), Academy of Science of South Africa (ASSAf), and a fellow of the Royal Society of Chemistry, UK (FRSC).

List of Contributors

Ekemini D. Akpan
Department of Chemistry
School of Chemical and Physical Sciences and Material Science Innovation & Modelling (MaSIM) Research Focus Area
Faculty of Natural and Agricultural Sciences
North-West University
Mmabatho, South Africa

Mustafa R. Al-Hadeethi
Department of Chemistry
College of Education
Kirkuk University
Kirkuk, Iraq

F. A. Ansari
Department of Applied Sciences
Faculty of Engineering
Jahangirabad Institute of Technology
Barabanki, India

Jeenat Aslam
Department of Chemistry
College of Science
Taibah University
Yanbu, Al-Madina
Saudi Arabia

Ruby Aslam
Corrosion Research Laboratory,
Department of Applied Chemistry
Faculty of Engineering and Technology
Aligarh Muslim University
Aligarh
Uttar Pradesh
India

Megha Basik
Corrosion Research Laboratory
Department of Applied Chemistry
Faculty of Engineering and Technology
Aligarh Muslim University
Aligarh
Uttar Pradesh
India

Abdelkarim Chaouiki
Laboratory of Applied Chemistry and Environment
ENSA
University Ibn Zohr
Agadir
Morocco

Dheeraj Singh Chauhan
Center of Research Excellence in Corrosion
Research Institute
King Fahd University of Petroleum and Minerals
Dhahran
Saudi Arabia

Modern National Chemicals
Second Industrial City
Dammam
Saudi Arabia

Omar Dagdag
Laboratory of Industrial Technologies and Services (LITS)
Department of Process Engineering
Height School of Technology
Sidi Mohammed Ben Abdallah University
Fez
Morocco

Amit Kumar Dewangan
Department of Chemistry
Government Digvijay Autonomous Postgraduate College
Rajnandgaon
Chhattisgarh
India

Yeestdev Dewangan
Department of Chemistry
Government Digvijay Autonomous Postgraduate College
Rajnandgaon
Chhattisgarh
India

Eno E. Ebenso
Institute for Nanotechnology and Water Sustainability
College of Science
Engineering and Technology
University of South Africa
Johannesburg
South Africa

M. El Gouri
Laboratory of Industrial Technologies and Services (LITS)
Department of Process Engineering
Height School of Technology
Sidi Mohammed Ben Abdallah University
Fez, Morocco

Lei Guo
School of Material and Chemical Engineering
Tongren University
Tongren
People's Republic of China

Rajesh Haldhar
School of Chemical Engineering
Yeungnam University
Gyeongsan
South Korea

A. El Harfi
Laboratory of Advanced Materials and Process Engineering
Department of Chemistry
Faculty of Sciences
Ibn Tofaïl University
Kenitra, Morocco

Jiyaul Haque
Department of Chemistry
Indian Institute of Technology
Banaras Hindu University
Varanasi
India

Zhongyi He
School of Materials Science and Engineering
East China JiaoTong University
Nanchang
People's Republic of China

Chaudhery Mustansar Hussain
Department of Chemistry and Environmental Science
New Jersey Institute of Technology
Newark
NJ
USA

Savaş Kaya
Health Services Vocational School
Department of Pharmacy
Sivas Cumhuriyet University
Sivas
Turkey

Hassane Lgaz
Department of Architectural Engineering
Hanyang University-ERICA
Ansan
Korea

Han-Seung Lee
Department of Architectural Engineering
Hanyang University-ERICA
Ansan
Korea

Lukman O. Olasunkanmi
Department of Chemistry
Faculty of Science
Obafemi Awolowo University
Ile Ife
Nigeria

Marziya Rizvi
Corrosion Research Laboratory
Department of Mechanical Engineering
Faculty of Engineering
Duzce University
Duzce
Turkey

Sheerin Masroor
Department of Chemistry
A.N. College
Patliputra University
Patna
India

Mohammad Mobin
Corrosion Research Laboratory
Department of Applied Chemistry
Faculty of Engineering and Technology
Aligarh Muslim University
Aligarh
Uttar Pradesh
India

Rachid Salghi
Laboratory of Applied Chemistry and Environment
ENSA
University Ibn Zohr
Agadir
Morocco

Goncagül Serdaroğlu
Faculty of Education
Department of Mathematics and Science Education
Sivas Cumhuriyet University
Sivas
Turkey

Shobha
Department of Physics
Banasthali Vidyapith
Vanasthali
Rajasthan
India

Sudheer
Department of Chemistry
Faculty of Engineering and Technology
SRM-Institute of Science and Technology
Ghaziabad
India

M. A. Quraishi
Interdisciplinary Research Center for Advanced Materials
King Fahd University of Petroleum and Minerals
Dhahran
Saudi Arabia

Taiwo W. Quadri
Department of Chemistry
School of Chemical and Physical Sciences and Material Science Innovation & Modelling (MaSIM) Research Focus Area
Faculty of Natural and Agricultural Sciences
North-West University
Mmabatho, South Africa

Chandrabhan Verma
Interdisciplinary Research Center for Advanced Materials
King Fahd University of Petroleum and Minerals
Dhahran
Saudi Arabia

Dakeshwar Kumar Verma
Department of Chemistry
Government Digvijay Autonomous Postgraduate College
Rajnandgaon
Chhattisgarh
India

Xue Yang
School of Materials Science and Engineering
East China JiaoTong University
Nanchang
People's Republic of China

Saman Zehra
Corrosion Research Laboratory
Department of Applied Chemistry
Faculty of Engineering and Technology
Aligarh Muslim University
Aligarh
Uttar Pradesh
India

Renhui Zhang
School of Materials Science and Engineering
East China JiaoTong University
Nanchang
People's Republic of China

Part I

Basics of Corrosion and Prevention

1

An Overview of Corrosion

Marziya Rizvi

Corrosion Research Laboratory, Department of Mechanical Engineering, Faculty of Engineering, Duzce University, Duzce, Turkey

1 Introduction

1.1 Basics About Corrosion

Corrosion can be scientifically defined in many ways. The term "corrode" is itself obtained from the Latin word "corrodere," i.e. "to gnaw to pieces." The National Association of Corrosion Engineers (NACE) has defined it: "Corrosion is a naturally occurring phenomenon commonly defined as the deterioration of a material (usually a metal) that results from a chemical or electrochemical reaction with its environment" [1]. International Standard Organization explains "corrosion" technically as the "Physio-chemical interaction between a metal and its environment which results in changes in the properties of the metal and which may often lead to impairment of the function of the metal, the environment or the technical system of which these forms a part" [2]. The environment is basically all that present surrounding and in contact with the observed metal/material. The primary factors describing the environment are (i) physical state (gas/liquid/solid); (iii) chemical composition (constituents &concentrations); and (c) the temperature. The corroded metal has obtained a thermodynamic stability in changing to oxides, hydroxides, salts, and carbonates. As per law of entropy, metals post fabrication return to their lowest energy, or natural ore form. Naturally metals are found in their element form or as ores. A lot is incorporated to convert iron ore into steel in the steel factories (Figure 1.1).

Organic Corrosion Inhibitors: Synthesis, Characterization, Mechanism, and Applications, First Edition. Edited by Chandrabhan Verma, Chaudhery Mustansar Hussain, and Eno E. Ebenso.

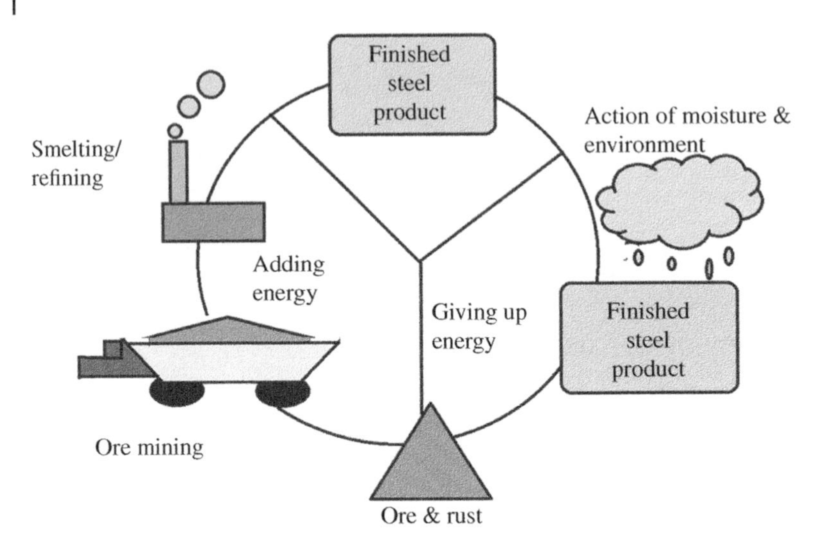

Figure 1.1 Corrosion cycle of steel.

Corrosion is the just reverse of what is known as extractive metallurgy. That implies that the energy utilized to convert an ore into a pure metal is reversed on exposure to environment (oxygen and water). On the exposed metal, oxides, sulfates, and carbonates exist [3–4]. Corrosion science as a subject has been around for many years in the textbooks, and surely its relevance has increased now. Education of corrosion and corrosion mitigation makes the environment safer and more sustainable.

1.2 Economic and Social Aspect of Corrosion

The incurred monetary losses and negative effects on environment geared the current ad on-going researches in the field of corrosion. To sum up the total monetary loss due to corrosion, cost studies have been carried out in several countries. The first significant work on cost of corrosion was presented as a report by Uhlig in 1949, estimating the annual cost of corrosion as US$5.5 billion [5]. However, comprehensively the first study on losses incurred due to corrosion was conducted in the United States in late 1970s. In the year 1978, US$70 billion were wasted, equivalent to approximately 5% of gross national product (GNP) of that year [6]. The US Federal Highway Administration (FHWA) published a breakthrough study back in 2002, estimating the direct corrosion cost associated with USA's industrial sector. The study was conducted by NACE International initiated the study as part of Transportation Equity Act for the 21st Century (TEA-21), having a Congress mandate. The estimated direct cost of corrosion annually is $276 billion, which implies

GNP's 3.1% [7]. This estimation is solely inclusive of the direct costs pertaining to maintenance. Other expenditures after production loss, negative environmental effect, disrupted transports, fatalities, and injuries were computed to be as much as the direct costs. Similarly, some countries conducted corrosion cost studies. These countries were Australia, United Kingdom, Japan, Germany, Kuwait, Finland, India, China, and Sweden. It was inferred that annual corrosion costs was 1–5% of the national GNPs. The recently published material relates the global economic losses due to corrosion, summed up by NACE International in 2016 as $2.5 trillion, which is 3.5% of global GDP [8–10]. The Central Electrochemical Research Institute calculated the cost of corrosion in India by NBS input/output economic model for 2011–2012. The direct cost was US$26.1 billion or 2.4% GDP. The cost avoidable was US$9.3 billion or 35% direct cost of corrosion. The indirect cost was US$39.8 billion or 3.6% of IGDP [11]. NACE International according to the latest global studies estimated Indian cost of corrosion to be GDP's 4.2% [12]. Beyond the cost of corrosion financially are the indirect costs like loss of opportunities and natural resources, potential hazards, etc. A project constructed using building material unable to withstand its environment for the estimated design life, the l resources are being needlessly consumed at later stages for maintenance and repair. Wasting the already depleting natural resources is a direct opposition to the increasing emphasis and demand for sustainable development in order to safeguard for future generations. Along with the wastage of natural resources, weak constructed structures pose threat to lives and well-being. Huge safety concerns have been established in regards with the accidents that might happen in case of corroding structures. A single pipeline that fails, a bridge that collapses, a derailed train compartment due to corroded track, or other accidents is one among numerous that cause enormous indirect losses and huge public outcry. According to the market sector considered, the indirect losses might make up to 5–10 times the direct loss.

1.3 The Corrosion Mechanism

Corrosion occurs by formation of an electrochemical/corrosion cell (Figure 1.2). This particular electrochemical cell comprises of five parts.

a) Anodic zones
b) Cathodic zones
c) Electrical contact between these zones
d) An electrolyte
e) A cathodic reactant

Inside this electrochemical cell, electrons depart from anodic to cathodic sites. The charged particles, ions, move across the conducting solution to balance the

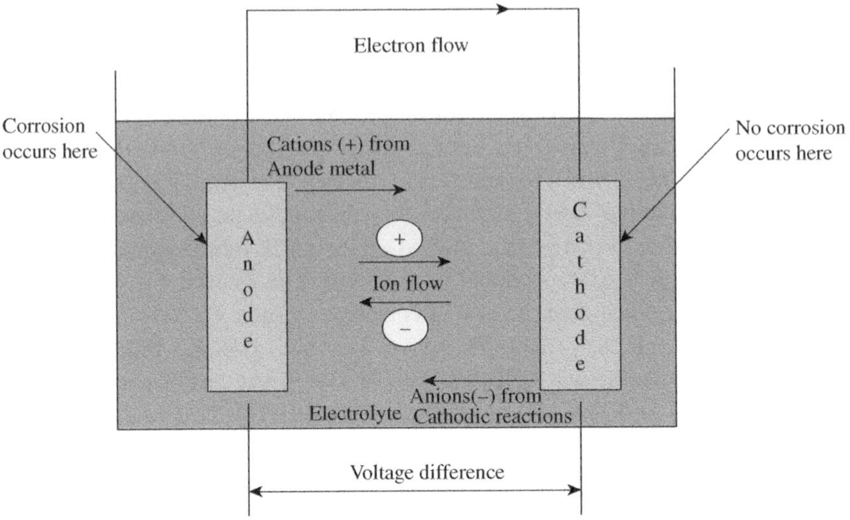

Figure 1.2 An electrochemical cell.

electrons flow. Anions (from cathodic reactions) move toward the anode and cations (from the anode itself) drift toward the cathode. Resultantly, anode corrodes and the cathode does not. There also exists a voltage/potential difference amidst anode and cathode. Numerous discrete micro cells develop on the metal surfaces, due to the constitutional phase difference, from stress variations, coatings, and imperfection levels like dislocations, grain boundaries, kink sites, or from ionic conductivity alterations or compositional changes in the conducting solution. The corrosion process is chemically spontaneous oxidation of the metal on reaction with the cathodic reactant. Every similar cell reaction results from a pair of simultaneous anodic and cathodic reactions going on at identical rates on the surface of metal.

1.3.1 Anodic Reaction

At the anode, the metal corrodes. The anodic reaction is the oxidation of a metal to its ionic form when the electric charge difference exists at the solid–liquid interface. Generally, anodic reaction is an oxidation reaction of a metal to its metal ions, which passes into conductive solution:

$$M \rightarrow M^{n+} + ne^{-} \tag{1.1}$$

where "n" is the metallic valence, e– is the electron, M is metal, and M^{n+} its metalion.

1.3.2 Cathodic Reactions

The cathodic reaction involves the environment and can be represented by the following reaction:

$$R^+ + e^- \rightarrow R^0 \tag{1.2}$$

where R^+ is the positive ion present in the electrolyte, e^- is the metallic electron, and R^0 is the reduced species. Based on the environment, many cathodic reactions and electron consuming reactions are possible. The main reactions are as follows.

The anaerobic acidic aqueous environment

$$2H^+ + 2e^- \rightarrow H2 \tag{1.3}$$

In the anaerobic alkaline aqueous environment

$$2H_2O + 2e^- \rightarrow H_2 + 2OH^- \tag{1.4}$$

In the aerobic acidic aqueous environment

$$2O_2 + 4e^- + 4H^+ \rightarrow 4H_2O \tag{1.5}$$

In the aerobic alkaline aqueous environment

$$2O_2 + 4e^- + 2H_2O \rightarrow 4OH^- \tag{1.6}$$

Some other reactions that are most commonly present in the chemical process are following.

Metal ion reduction

$$M^{3+} + e^- \rightarrow M^{2+} \tag{1.7}$$

Metal ion deposition

$$M^+ + e^- \rightarrow M \tag{1.8}$$

The products of the anodic and cathodic reactions react to form solid corrosion products on the surface of the metal. The Fe^{2+} interacts with OH^- ions as:-

$$Fe^{2+} + 2OH^- \rightarrow Fe(OH)_2 \tag{1.9}$$

$Fe(OH)_2$ is reoxidized to $Fe(OH)_3$, an unstable product, and thus transforms to hydrated ferric oxide commonly called as red rust (Figure 1.3).

$$4Fe(OH)_2 + O_2 + 2H_2O \rightarrow 4Fe(OH)_3 \tag{1.10}$$

$$2Fe(OH)_3 \rightarrow Fe_2O_3 \cdot 3H_2O \tag{1.11}$$

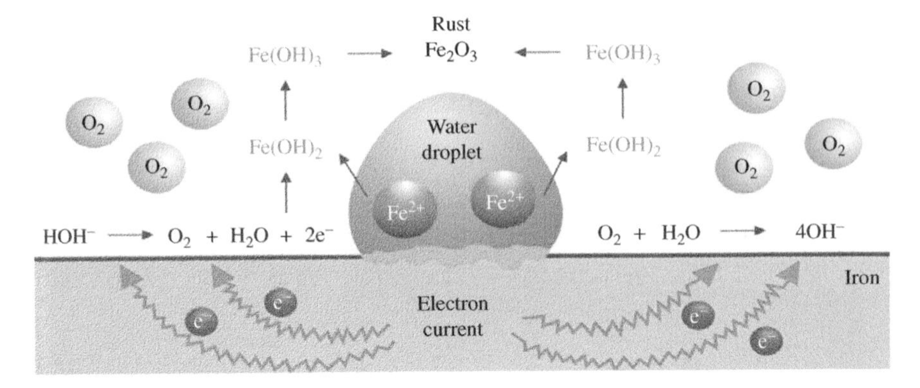

Figure 1.3 Mechanism of rust formation.

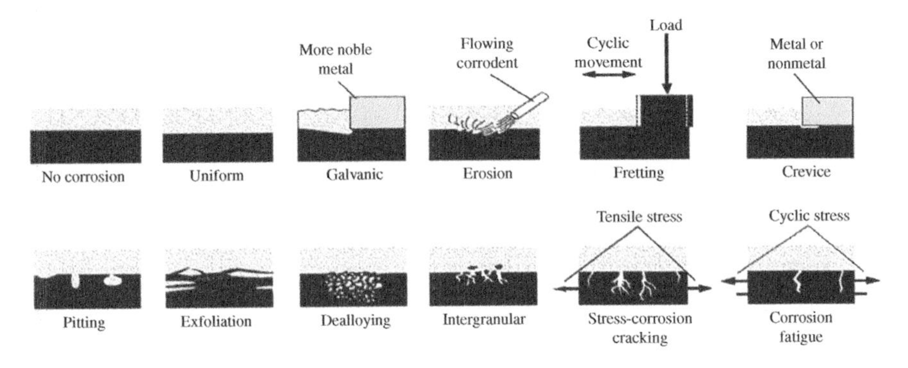

Figure 1.4 Classified forms of corrosion.

1.4 Classification of Corrosion

Seldom is a single class of corrosion discovered in corroding structures. Different metals in contact and contact with different environment hardly allow only one type of corrosion to occur even within a system. Each type of corrosion is caused by their specific reaction mechanisms and has their specific monitoring, prediction, and control methods. Figure 1.4 throws some light on classification of corrosion in a pictorial manner. None of the classifications is a universal standard, even the following classification is an adapted [4, 13].

1.4.1 Uniform Corrosion

This type of corrosion affects a large patch over the metal and causes overall reduction of metallic thickness subject to the fact that metal undergoing corrosion has a uniform composition and metallurgy too. What happens is that anode and cathode do not possess fixed sites; as such there are no sites preferable to corrosion, which occurs here in a uniform fashion. Corrosion rates are easily monitored by electrochemical measuring techniques or gravimetric analysis. A metal suffering from uniform corrosion can be protected using corrosion inhibitors or coatings and also by cathodic protection. Atmospheric corrosion is an example of uniform corrosion. When exposed to dry atmospheres with very less humidity, metals spontaneously tend to form an oxide film. This barrier oxide film acquires a thickness of 2–5 nm [14].

1.4.2 Pitting Corrosion

Pitting corrosion is highly destructive form and a kind of localized attack, which leads to little holes called pits in metal. Small cavities and holes, which are as deep as their diameter, are known as pits. They cause perforations by penetrating into the metal with least loss of weight [15]. Pitting is proportional to the logarithm of electrolyte's concentration of chloride. The prerequisite for pitting to occur is that the electrolyte should be a strong oxidizer for onset of the passive state. The ferric and cupric halide ions are electron acceptors (cathodic reactants), and they do not need oxygen to initiate and propagate pitting. Other propagating factors causing pitting include localized damage chemically and mechanically to a passive oxide film, non-metallic impurities/non-uniformities of metal structure due to nonproportional inhibitor coverage. Pitting, however, can be evaded by reducing aggressiveness of the solution, decreasing the temperature of conductive solution, decreasing Cl– concentration and acidity.

1.4.3 Crevice Corrosion

It is also localized version of corrosion on a microenvironment level but related to a stagnant electrolyte caused by gasket surfaces, lap joints and holes, crevices under bolts, rivet heads, and surface deposits. To evade and limit this type of corrosion, it is suggested to (i) use of welds instead of bolt/rivet joints, (ii) a design to ensure complete draining, (iii) hydrofuging any interstices, which cannot be removed, and (iv) utilizing solely solid and nonporous seals, etc.

1.4.4 Galvanic Corrosion

This type of corrosion is also called "bimetallic corrosion" and "dissimilar metal corrosion," because it occurs due to electrical contact with a more noble metal or maybe a nonmetallic conductor in the electrolyte. The active member of the metallic couple, i.e. less corrosion resistant bears an accelerated corrosion rate, while the noble member is protected by the cathodic effect. The joint between the metals is most corrosion affected. Moving away from this junction, the corrosive attack reduces. As it is already known that each metal has its unique corrosion potential in an electrolyte, the potential difference between the two dissimilar metals causes the less noble metal to corrode. To estimate the corrosion rate conductivity of metal and electrolyte, potential difference of relative anodic and cathodic areas might be accounted [16–18]. To prevent such an attack, the metals should lie close by on the electrochemical series, comparably area of anode should not be too small; if different metals are involved, insulation must be applied; and coatings, paints, and avoiding thread joints are good preventive measures.

1.4.5 Intergranular Corrosion

This corrosion can be referred to as "intercrystalline corrosion"/"interdendritic corrosion" as tensile stress causes it along the grain or crystal boundaries. It might also be known as "intergranular stress corrosion cracking" and "intergranular corrosion cracking." These corrosive attack prefers interdendritic paths. A microstructure examination using a microscope is needed for recognizing this degradation; however, at times, it is recognizable with eyes as in weld decay. The composition's local differences like coring in alloy castings lead to this type of corrosion. The mechanism includes precipitation in grain boundaries like in the case of precipitating chromium carbides in steel. Intermetallic segregation at grain boundaries in aluminum is called "exfoliation." This corrosion type might be prevented and controlled by using mild steel, low carbon type like using post-weld treatment, etc.

1.4.6 Stress-Corrosion Cracking (SCC)

Such a cracking occurs by the simultaneous action of a corrodent and sustained tensile stress. This bars the corrosion-less sections, intercrystalline or trans-crystalline corrosion, which might destroy an alloy without any stress. It is accompanied with hydrogen embrittlement. It might be a conjoint action of a susceptible material, a specific chemical species, and tensile stress. Sedriks and Turnbull reviewed the standard SCC testing [19–20]. Time-consuming techniques, bulky specimens, and expensiveness limit the usage of SCC monitoring techniques. Stress corrosion cracking might be prevented by avoiding chemical species that causes it, controlling hardness and stress, using un-crackable materials specific to environment and temperature/potential control of operation.

1.4.7 Filiform Corrosion

On steel, aluminum, aircraft structures in humidity, flanges, beverage cans, gaskets, and weld zones, this type of corrosion can be detected. Irregular hairlines, sometimes corrosion products filaments present below coatings of paint, rubber, lacquer, tin, silver, enamel, and paper, develop. Material is not lost significantly, but the surface deteriorates. Copper, stainless steel, and titanium alloys are unsusceptible to this attack.

1.4.8 Erosion Corrosion

Rapidly flowing electrolyte and turbulence cause erosion of the metal. The main culprits for turbulence are the pits within a pipeline. Turbulence finally causes a pipeline to have leakage. The velocity of the flowing electrolyte and the physical action of it moving against the surface causes metallic loss at an accelerated rate. Erosion is common occurrence in constriction areas like pump impellers and inlet ends. Erosion can be tackled by less turbulent fluid movement, low velocity of flow, using corrosion resistant pipeline materials, inhibitors, etc.

1.4.9 Fretting Corrosion

A slight oscillatory slip between two surfaces in contact causes fretting corrosion. Bolted/riveted parts are made such that they do not slip or oscillate, which fails in the presence of fluctuation of pressure and vibration. Fretting can be prevented by regular inspection and maintenance of the lubrication.

1.4.10 Exfoliation

At the elongated grain boundaries, the corrosion products present cause the metal to be forced away from the material and form layer-like look, and this is called exfoliation. Also known as lamellar, layered, and stratified corrosion, it proceeds along selected subsurfaces. If the grain boundary attack is severe, it is visible; otherwise a microscope conducts the microstructure examination. Alloys of aluminum are most susceptible to exfoliation. This can be controlled using coatings, heat treatment to control precipitate distribution, and exfoliation-resistant aluminum alloy.

1.4.11 Dealloying

Dealloying is selective corrosion of solid solution of alloy also known as leaching/selective attack/parting. Dealloying can be manifested in various categories like decobaltification (selective leaching of cobalt from cobalt-base alloys), decarburization (selective loss of carbon from the surface layer), dezincification (selective leaching of zinc from zinc-containing alloys), denickelification (selective leaching of nickel from nickel-containing alloys), and graphitic corrosion (gray cast iron in which the metallic constituents are selectively leached). Dealloying might be prevented by selecting more resistant alloys, controlling the selective leaching, sacrificial anode/cathodic protection.

1.4.12 Corrosion Fatigue

Corrosion and cyclic stress when occur simultaneously result in cracks. This is corrosion fatigue. Rapidly fluctuating stress below the tensile strength usually are causative agents. The metallic fatigue strength decreases in corrosive electrolyte. It can be prevented by using high-performance alloys resistant to corrosion fatigue and by using coatings and inhibitors delaying the crack initiation.

1.5 Common Methods of Corrosion Control

Corrosion control is applying the principles of engineering to limit corrosion economically. Each preventive measure bears its own complexities and specificity.

Basically, the idea is to detect the mechanism and causative agents of the degradation and reduce them or completely prevent them from occurring. Let us have a look on some of them as given below.

1.5.1 Materials Selection and Design

There is no all-noble and completely corrosion-resistant metal, but a careful selection might increase the longevity of the metal component. Factors influencing the materials selection are resistance to degradation, test data and design availability, cost, mechanical properties, availability, compatibility with other components, maintainability, life expectancy, appearance, and reliability. Availability, inexpensiveness, and easy fabrication make carbon steel a favorable material for selection [21]. In the petrochemical plants, highly corrosive catalysts and solvents are usually encountered, so stainless steel is best option [22]. Duplex stainless steels are used in pressure vessels, storage tanks, and heat exchangers owing to their good mechanical properties, high resistance to chloride stress corrosion cracking, good erosion and wear resistance, and low thermal expansion [23]. For seawater service, duplex stainless steels of higher molybdenum content (e.g. Zeron 100) have been developed [24]. Appropriate system design is crucial for efficient corrosion control. Numerous factors like materials selection, geometry for drainage, process and construction parameters, avoiding or sealing of crevices, avoidance or electrical separation of dissimilar metals, operating lifetime, and maintenance and inspection requirements are involved.

1.5.2 Coatings

Coatings are generally good option to insulate the metals from exterior aggressive environments. They extend a lengthy protection in wider spectrum of corrosive conditions, atmospheric to aqueous electrolyte solution. Although they provide no structural strength, yet they protect the strength and integrity of a structure. Their function is that of a physical barrier preventing electrolytic attack on metal. Organic coatings like paints, resins, lacquers, and varnishes are the most popular protective coatings. Metallic coating (noble or cathodic and sacrificial or anodic) is also used for corrosion control.

1.5.3 Cathodic Protection (CP)

A metal is completely converted to a cathode to protect it against corrosion. CP is implemented by driving the potential to a negative region/stabilized metal region. Either an external power supply changes the amount of charge on the metal surface or a more reactive metal is converted to a sacrificial anode. The principle involved in CP is to potentially let the metallic article or structure attain corrosion immunity. A stable and unreactive metal is impossible to corrode. This method might be expensive as electricity is consumed, and the extra metals are involved. Cathodic protection can be attained by coupling a given structure (like Fe) with a reactive metal like zinc or magnesium or by impressing a direct current between an inert anode and the metal to be immunized.

1.5.4 Anodic Protection

Based on phenomenon of passivity, anodic protection can control the corrosion in an electrochemical cell. Metal is kept in a passive state; surface is connected as an anode to an inert cathode in the corrosion cell. Anodic protection is used to protect metals that exhibit passivation in environments; when the current density in a corroding structure is much higher than the current density of the same in its passive state over a wide range of potentials, anodic protection can be used. This preventive measure is adopted in aerospace and other critical applications and wherever the cathodic protection is not cost-effective.

1.5.5 Corrosion Inhibitors

The most prevalent, economic, and effective measure against corrosion of metallic surfaces in aggressive media in closed systems is inhibitors [25–30]. Corrosion inhibitors are added in small concentration to effectively reduce the corrosion rate of a metal exposed to corrosive solution. It is similar to a retarding catalyst, which reduces the corrosion rate by increasing or decreasing the reaction of anode and/or cathode, decreasing the reactants' diffusion rate to the metallic surface and decreasing the metallic surface's electrical resistance. Without disrupting the setup, inhibitors can be added in situ. Adsorption theory best explains their functioning in a corrosion cell. Adsorption refers to the adhesion of a chemical species to the superficial single monolayer of the metal without bonding with the bulk of metal. The type of the corrosion medium, the magnitude of the charge at the metal/solution interface, the nature of metal, and the cathodic reaction decide the inhibitor. Three types of environments use inhibitors, namely, recirculating cooling water systems in the pH range of 5–9, primary and secondary production of crude oil and pickling acid solution for the removal of dust and mill scale during the production and fabrication of metals parts, or also for the post-service cleaning. Based on their composition or mechanism of protection, inhibitors are classified as follows.

1.6 Adsorption Type Corrosion Inhibitors

As the name suggests, these organic compounds adsorb on the metal to suppress dissolution and the reduction reaction. Adsorption inhibitors affect both anodic and cathodic process equally or disproportionally.

i) *Vapor-phase corrosion inhibitors*

   Vapor-phase corrosion inhibitors also called volatile corrosion inhibitors. The vapor pressure of these compounds at 20–25°C is usually between 0.1 and

1.0 mm Hg. When kept in the vicinity of the metal to be protected, they sublime and condense in enclosed spaces [31]. For example, phenylthiourea and cyclohexylamine chromate are used for protecting brass. Dicyclohexylamine nitrite protects both ferrous and nonferrous metals/alloys.

ii) *Inorganic inhibitors*
Some metal ions like Pb^{2+}, Mn^{2+}, and Cd^{2+} deposit on the iron surface in acidic environment [32]. Even Br^- and I^- inhibit corrosion in strongly acidic solutions [33]. As_2O_3 and $Sb_2O_3$are also corrosion inhibitors in acidic media. These substances form a metal oxide layer and increase the hydrogen overvoltage to reduce the corrosion rate.

iii) *Organic inhibitors*
It includes large number of organic substances containing N, S, or O atoms in the molecule. Organic inhibitors possess a functional group as the reaction center for the adsorption process. They have heteroatoms like N, S, and O in their structures. The molecular structures majorly influence the extent of inhibition of corrosion.

1.6.1 Anodic Inhibitors

Those substances that reduce the anodic area by acting on the anodic sites and polarizing the anodic reaction are called anodic inhibitors [34]. Anodic inhibitors cause displacements in corrosion potential in positive direction, suppress corrosion current, and reduces corrosion rate. If an anodic inhibitor is not present at a concentration level sufficient to block off all the anodic sites, localized attack such as pitting corrosion can become a serious problem due to the oxidizing nature of the inhibitor, which raises the metal potential and encourages the anodic reaction. Anodic inhibitors are classified as unsafe because they may cause localized corrosion. Examples of anodic inhibitors include orthophosphate, chromate, nitrite, ferricyanide, and silicates.

1.6.2 Cathodic Inhibitors

There are substances that may lessen the cathodic area by polarizing the cathodic reactions on cathodic area [35]. Cathodic inhibitors transfer the corrosion potential in negative direction to retard cathodic reaction and suppress the corrosion rate. Cathodic area is reduced by precipitating the insoluble species on cathodic sites. Cathodic inhibitors do not cause localized corrosion hence safe. They are cathodic poisons/hydrogen-evolution poisons like arsenic and antimony ions. Scavengers like sodium sulfite and hydrazine are filming inhibitors (cathodic precipitates).

1.6.3 Mixed Inhibitors

The formulation contains more than one inhibitor in this case. These inhibitors interfere with anodic, as well as cathodic reactions. These inhibitors are used for multi-metallic substrates and when combined and optimized cathodic/anodic effect is required. The halide ions enhance the action of organic inhibitor in acid solutions.

1.6.4 Green Corrosion Inhibitors

As the chromates/arsenates were restricted, corrosion prevention with ecological green compounds (hazard-free inhibitor formulations) in most oil field applications were designed to effectively meet safety standards and also protect the targeted metal substrates in their service life. In general, most of these efficient corrosion inhibitors are organic compounds containing hetero atoms such as S, N, O, P, and multiple bonds or aromatic rings. The number of lone pairs of electrons and loosely bound π-electrons in these functional groups are determining factors for their activity. These biocompatible substances might be of plants and animal origin. In the past two decades, the researchers sought after the "green" or "eco-friendly" corrosion inhibitors to use cheap, effective compounds at low or "zero" effect on nature. The ever-increasing publications on green corrosion inhibition indicate the interest in exploring the new inhibitors to control the corrosion of various metals.

References

1 Van Delinder, L.S. and deS, A. (1984). *Brasunas, Corrosion basics: An introduction*. Houston, TX: NACE.

2 ISO 8044:1999 (2000). *Corrosion of metals and alloys - Basic terms and definitions*. Brussels: International Organization for Standardization.

3 Veronika, K.B. (2008). Knowledge about metals in the first century. *Korroz. Figy.* 48: 133–137.

4 Fontana, M.G. (1986). *Corrosion Engineering*, 3rde. New York: McGraw-Hill.

5 Uhlig, H.H. (1949). The cost of corrosion in United States. *Chem. Eng. News.* 27: 2764–2767.

6 Bennet, L.H., Kruger, J., Parker, R.I. et al. (1978). *Economic effects of metallic corrosion in the united states*. Washington, DC: National bureau of standards (Special Publication).

7 Koch, G.H., Brongers, M.P.H., Thompson, N.G. et al. (2002). Corrosion costs and preventive strategies in the united states. *Mater. Perform. (Supplement).* 42: 3–11.

8 Wei, H., Wang, Y., Guo, J. et al. (2015). Advanced micro/nanocapsules for self-healing smart anticorrosion coatings. *J. Mater. Chem. A.* 3: 469–480.

9 Haque, J., Srivastava, V., Verma, C., and Quraishi, M.A. (2016). Experimental and quantum chemical analysis of 2-amino-3-((4-((S)-2-amino-2-carboxyethyl)-1H-imidazol-2-yl)thio) propionic acid as new and green corrosion inhibitor for mild steel in 1 M hydrochloric acid solutions. *J. Mol. Liq.* 225: 848–855.

10 Engineering 360 powered by Global Spec, 2016, Annual Global Cost of Corrosion: $2.5 Trillion. http://insights.globalspec.com/article/2340/annual-global-cost-of-corrosion-2- 5 trillion.

11 Bhaskaran, R., Bhalla, L., Rahman, A. et al. (2014). An analysis of the updated cost of corrosion in India. *Mater. Perform.* 53: 56–65.

12 International Measures of Prevention, Application, and Economics of Corrosion Technologies Study (2016). Report No. OAPUS310GKOCH (PP110272)-1 Report prepared by DNV GL U.S.A., Dublin, Ohio and APQC, Houston, TX.

13 Roberge, P.R. (2008). *Corrosion engineering principles and practice*. New York: McGraw-Hill.

14 Leygraf, C., Wallinder, I.O., Tidblad, J., and Graedel, T. (2000). *Atmospheric Corrosion*, 2nde. New York: John Wiley & Sons.

15 Burstein, G.T., Liu, C., Souto, R.M., and Vines, S.P. (2004). Origin of pitting corrosion. *Corros. Eng. Sci. Technol.* 39: 25–30.

16 Deshpande, K.B. (2010). Experimental investigation of galvanic corrosion: comparison between SVET and immersion techniques. *Corros. Sci.* 52: 2819–2826.

17 Tada, E., Sugawara, K., and Kaneko, H. (2003). Distribution of pH during galvanic corrosion of a Zn/steel couple. *Electrochim. Acta.* 49: 1019–1026.

18 Souto, R.M., Gonzalez-Garcia, Y., Bastos, A.C., and Simoes, A.M. (2007). Investigating corrosion processes in the micrometric range: a SVET study of the galvanic corrosion of zinc coupled with iron. *Corros. Sci.* 49: 4568–4580.

19 Sedriks, J. (1990). *Corrosion testing made easy: Stress corrosion cracking test methods*. Houston: NACE.

20 Turnbull, A. (1992). Test methods for environment assisted cracking. *Br. Corros. J.* 27: 271–289.

21 Cherepakhova, G.L., Shreider, A.V., and Charikova, G.P. (1970). Use of galvanized carbon steel tubes in equipment for condensing and cooling. *Chem. Pet. Eng.* 6: 490–492.

22 Martin, J.H. (2006). *Concise encyclopedia of the structure of materials*, 1ste. Elsevier.

23 Erbing Falkland, M.L. (2000). Duplex stainless steels. In: *Uhlig's Corrosion Handbook*, 2ee (ed. R. Winston). New York: Wiley.

24 Grubb, J.F., DeBold, T., and Fritz, J.D. (2005). Corrosion of Wrought Stainless Steels. In: *Corrosion: Materials, ASM Handbook*, vol. 13B (eds. S.D. Cramer and B.S. Covino Jr.). Materials Park, OH: ASM International.

25 Bregman, J.I. (1963). *Corrosion Inhibitors*, 1ste. New York: The MacMillan Co.

26 Eldredge, G.G. and Warner, J.C. (1948). Corrosion inhibitors. In: *H.H. Uhlig's The Corrosion Handbook* (ed. S. Papavinasam), 1021–1032. New York: Wiley.

27 Nathan, C.C. (1973). *Corrosion Inhibitors*. Houston, Texas: National Association of Corrosion Engineers (NACE).

28 Putilova, N., Balezin, S.A., and Barannik, V.P. (1966). *Metallic Corrosion Inhibitors*. London: Pergamon Press.
29 Brooke, M. (1962). *Chemical Inhibitors Checklist*. London: Chemical Engineering, Pergamon Press.
30 Ranney, M.W. (1976). *Inhibitors—Manufacture and Technology*. New Jersey: Noyes Data Corporation.
31 Riggs, O.L. (1973). Theoretical aspects of corrosion inhibitors and inhibition. In: *Corrosion Inhibitors* (ed. C.C. Nathan). Houston, TX: NACE International.
32 Poling, G.W. (1967). Infrared studies of protective films formed by acetylinic corrosion inhibitors. *J. Electrochem. Soc.* 114: 1209–1214.
33 Hausler, R.H. (1979). Corrosion Chemistry. *ACS Symp. Ser.* 89: 263.
34 Bockris, J.O.M. and Khan, S.U.M. (1993). *Surface Electrochemistry: A Molecular Level Approach*. New York: Springer US, Plenum Press.
35 Kirk, R.E. and Othmer, D.F. (1979). *Encyclopedia of Chemical Technology*. Wiley.

2

Methods of Corrosion Monitoring

Sheerin Masroor

Department of Chemistry, A.N. College, Patliputra University, Patna, India

2.1 Introduction

Corrosion is unpredictable phenomenon and is a continuous process. In nature it is observed that every material wants to be in its lowest energy state, for all this to happen metals and alloys like iron, steel, copper or aluminums etc. frequently reacts with components present in environments such as oxygen and water, which leads to the formation of their hydroxides which is so similar to metal ore's composition chemically. The word corrosion is borrowed from the Latin word *Corrodere* that means "to nibble into pieces." There are multiple definitions proposed for the present problem, but most likely accepted are as follows:

a) As per K.E. Heuslerl et.al in 1989, it can be explained as detrimental of the used material that may be physical or mechanical like evaporation, melting, mechanical fracture, and abrasion [1].
b) L.L. Shreir said in 1994 that the term corrosion relates to metals and encompasses all interactions of a metal or alloy in solid or liquid form with its surrounding, irrespective of whether this is beneficial or non-beneficial [2].
c) According to D.A. Jones, it is the destruction of material by chemical reactions between a material and the aggressive environment [3].
d) P.R. Roberge explained corrosion as the destructive intrusion of a material by possible reaction with its environment [4].
e) According to ISO 8044, corrosion is a physicochemical reciprocity in between material and its corresponding environment that causes an alternation in the physical and chemical properties, and further leads to ultimate wreckage of the operation of the used material [5].

Organic Corrosion Inhibitors: Synthesis, Characterization, Mechanism, and Applications,
First Edition. Edited by Chandrabhan Verma, Chaudhery Mustansar Hussain, and Eno E. Ebenso.

f) Later in 2005, M. Fontana demonstrated corrosion as the deterioration of a material because of a reaction with its environment [6].
g) The most recent was given by NACE/ASTM G193, which explains corrosion as the deterioration of a used material (metal) that is consequence of any chemical or electrochemical reaction with its aggressive environment [7].

This explains why utmost materials that are in maximum production and help to build society are metals and hence much susceptible to corrosion [8, 9]. They make strong structures to constitute a great economy. From the very early time, it was presumed that the structures made from metals and its alloys are long lasting for hundreds of years. But they deteriorate as time passes. This takes so much dead full and dangerous form to those large-scale industrial plants, like chemical processing, and electrical power plants shut down as a result of corrosion. All these enhance the problems in economy and lead to losses in many ways.

The percentage of corrosion happened and damage caused by it can be monitored at time before it starts happening by the application of multiple techniques. The main purpose of monitoring corrosion is done because it helps to know the working state of equipment or predicting remaining life of materials and to know locations where defect is occurring, getting good service conditions, specific remedies, and corrosion rates with variables. By knowing all these parameters, we can easily administer corrosion control schemes [10–12]. The key role for corrosion to happen or not can be decided by environments. We can understand the term environment as the integrated surrounding in contact with the metallic structures. Some basic points to keep in mind before describing environments are its physical state (gas, liquid, or solid), chemical composition, constituents, pH, presence of impurities, ions present, temperature, and velocity [13]. So for corrosion to happen, we have to study two components like materials and environments. Further when corrosion is discussed, it is important to think of a combination of a material and an environment. On the other hand, the aggressiveness of an environment cannot be considered without taking metal into consideration. In short, we can surely say that the corrosion performance of the metallic structure can be calculated on to which it is subjected, and the aggressiveness of a surrounding/environment built upon on the material exposed to that environment. Various agencies are engaged to calculate the cost of corrosion in different countries including USA, UK, Australia, Japan, Germany, Finland, Kuwait, Sweden, China, and India. The most common thing among all countries finding is that annually cost of corrosion aligned in between 1 and 5% of the gross national product of each country.

Latest report characterizes direct and indirect corrosion cost of metallic structures in the United States, where total direct (infrastructure – $22.6 billion, utilities – $47.9 billion, transportation – $29.7 billion, production and manufacturing – $17.6 billion, government – $20.1 billion, total – $137.9 billion) and indirect cost (cost of labor attributed to corrosion management activities, cost of the equipment required because of corrosion-related activities, loss of revenue due to disruption in supply of product, and cost of loss of reliability) including total is estimated at $276 billion per year, which comprise of 3.1% of the 1998 US gross domestic product. The data in terms of cost was determined by scrutinizing 26 industrial sectors, where presence of corrosion was expected, and extrapolating the results for a nationwide estimate [14]. From literature, the word monitoring must not be confused with inspection, as the use of electrical methods can be said under monitoring while measurement via nonelectrical methods such as gravimetric analysis comes under inspection or detection [15, 16].

Multiple definitions of corrosion monitoring have been applied since corrosion inhibition came into effect and dominates in Europe after United States. The most accepted definition by authors is that, “It is the organized measurement of the corrosion or deterioration of assets with the aim of assisting the knowledge of corrosion process and getting report for corrosion control.” This clearly explains how we can get important information relating the assistance procured in the operation of a corrosion monitoring program [17]. In another definition Roth well described, “As the estimation of the deterioration of a material which happens through any factor such as chemical reaction, electrochemical, environmental or biological.” This explanation put forth the fact how corrosion reactions and surrounding are interrelated [18]. The most compact definition comes when any technique if used to know or measure the evolvement of corrosion. This definition although is least explanatory [19].

2.2 Methods and Discussion

2.2.1 *Corrosion* Monitoring Techniques

The progression of corrosion precepts, i.e. how lengthily any structures made of metals can be safely operated at specific conditions. Monitoring procedures object to know assertive possibilities in order to elongate the life and forbearance of valuables meantime enhancing defense and diminishing restoration costs. Some key points that are observed during corrosion monitoring are as follows:

a) The failure can be predicted on knowing the deteriorating processes.
b) By correlating the changes taking place and their aftermaths on system corrosively.
c) By getting knowledge of particular corrosion problem and its controlling factors such as temperature, pressure, pH, air flow rate, and many more.

A wide variety of corrosion monitoring techniques have been employed, which are divided into two categories:

a) Destructive methods
   i) Gravimetric analysis
   ii) Potentiodynamic polarization technique
   iii) Electrochemical impedance spectroscopy
   iv) Linear polarization technique
b) Nondestructive methods
   i) Radiography
   ii) Ultrasonic testing
   iii) Eddy current/magnetic flux
   iv) Thermography

a) Destructive Methods
   i) *Gravimetric Analysis*

The feasibility of the process can be reviewed in literature published by NACE, American Society for Testing Materials (ASTM), and other organizations [20, 21].

This method is simplest, inexpensive, and effective method for monitoring the corrosion rate in any suspected system or structure. It is supposed to be accurate and versatile as involves simple measurement. Here the specimen/sample/coupon of material is allowed to expose with environment for a specified duration and then removing the studied sample for further analysis. The basic quantity, which is resolved from corrosion coupons, is loss in weight taking place over the period of exposure to the aggressive surroundings. Expected parameters of single coupon, which have been taken in account for effective corrosion monitoring are presented in Figure 2.1. It provides direct measurement of general corrosion rate [22]. The studied coupons can be exposed to any kind of aggressive environment such as high temperatures, liquid corrosives, different gases, multiple soils, and the atmospheric conditions. The coupons are available in different geometries such as strip, disc, weld, scale, U-bends, C-rings, or stressed (Figure 2.2). However, the most common form of coupon is the metal strip used for equipment surfaces. Coupon samples can be exposed in duplicate/triplicate or multiple batches allowing various numbers of coupons made up of different materials at a specified location.

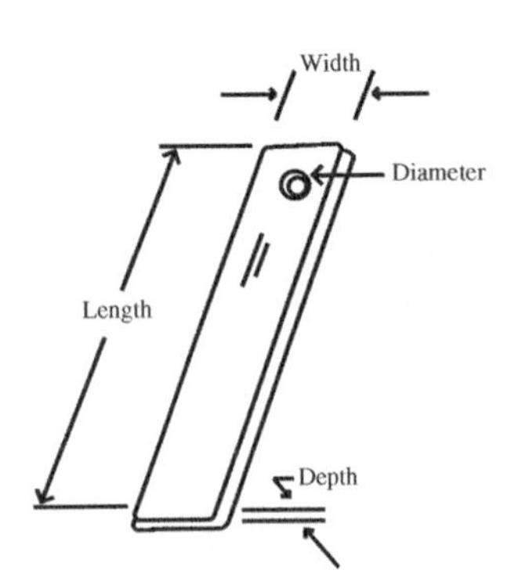

Figure 2.1 Parameters of single coupon.

This analysis was carried out in a thermostated water bath for different time durations ranging from 4 to 12 hours but generally 6 hours can be considered standard as per ASTM designation

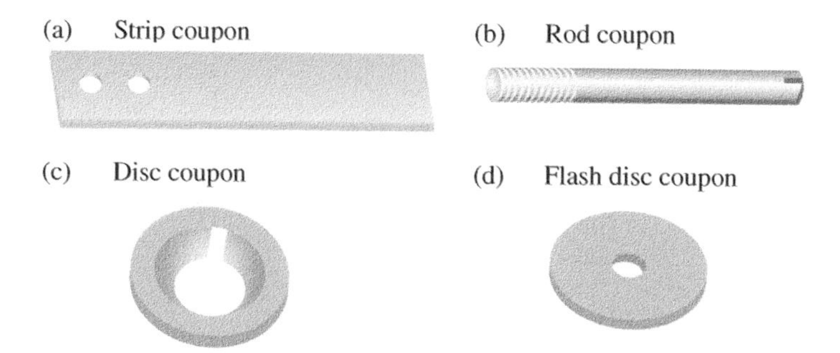

Figure 2.2 Different shapes of metal coupons. (a) Strip/rectangular shape coupon, (b) rod/cylindrical shape coupon, (c) disc shape coupon, (d) flash disk coupon.

G1-90. Here, metal coupons were freshly prepared, which further can be suspended in 250 ml beakers containing 200–250 ml of aggressive/test solutions and allowed to maintained temperature in the range of 20–100°C. The specimens were immersed in triplicate, and average corrosion rate was calculated. The corrosion rate in mpy was calculated using equation:

$$\text{Corrosion Rate}(\text{mpy}) = \frac{534\,\text{W}}{\rho \text{AT}}$$

In the given equation, "W" is weight loss in mg; "ρ" is the density of metal specimen in g/cm^3; "A" is the area of specimen in cm^2, and "t" is exposure time in hours [23, 24].

ii) *Potentiodynamic Polarization*

To carry out these techniques, polarization properties of the metal-surrounding system of interest are measured [25]. The basic theory behind is that polarization curves are acquired by polarizing a working electrode potential comparative to a reference electrode availing external current supplied by way of a counter electrode in a conventional electrochemical cell arrangement. This causes a big problem in getting selection of reference electrode for the measurement of potential. In this investigation, the Tafel constants, i.e. b_a and b_c, are obtained from the slopes of the linear portions present in anodic and cathodic (Figure 2.3) theory, which explains the corrosion mechanism. Further, the corrosion rates can be extracted by extrapolating the linear portions of the obtained curves to intersect at the natural corrosion potential.

These Tafel plots can be made by giving a scan in the range of 250 mV below E_{corr} to 250 mV above E_{corr} with scan rate of 0.1 mV/sec. In the curve, applied potential is present on Y-axis, while logarithm of measured

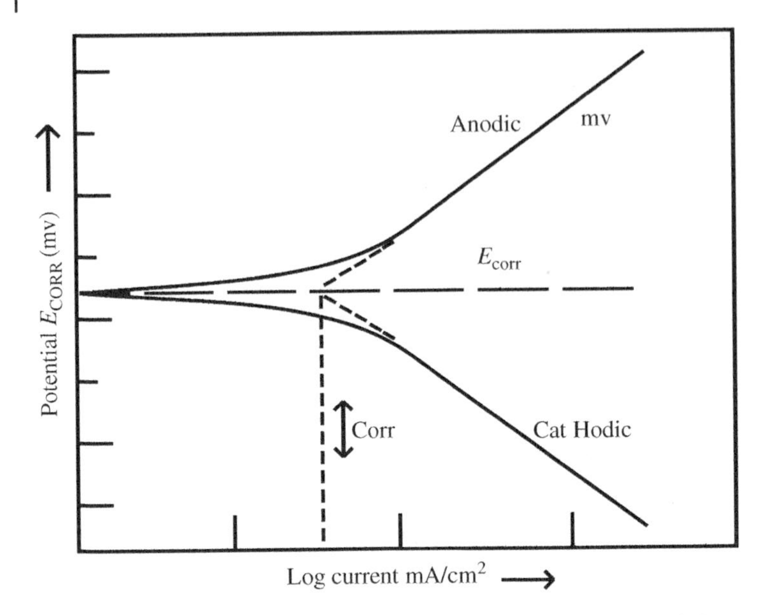

Figure 2.3 Presentation of anodic and cathodic Tafel curves and their extrapolation.

current density along X-axis (Figure 2.3). In next step, a straight line is allowed to cap along the linear portion of the anodic and the cathodic curves and further it is extrapolated to E_{corr}. The intersection point can be named as corrosion current (i_{corr}) [26].

The % *IE* was calculated from the measured I_{corr} values using the relationship:

$$(\%)IE = \left(1 - \frac{i_{corr}}{i^{o}_{corr}}\right) \times 100$$

iii) *Electrochemical Impedance Spectroscopy*

This special technique was designed to dodge severe depreciation of the bared surface of the structure studied and was widely used for examining the corrosion of a working electrode [27]. The monitoring process involves application of frequencies with low amplitude sinusoidal voltage wave to outcome disturbance signals from working electrode. The percentage of corrosion can be analyzed by current response of the frequency or voltages. Specifically, it is generally monitored by giving AC potential to an electrochemical setup and alternatively getting current value via cell.

As we are knowing the concept of electrical resistance, it can be defined as the ability of a circuit element to resist the flow of electrical current. So as per Ohm's law, electrical resistance can be defined as the ratio between voltage, E, and current, I.

$$R = \frac{E}{I}$$

Consider on application of sinusoidal potential excitation, we can get an AC current signal. This obtained current signal can be summed up as sinusoidal functions or Fourier series. The electrochemical impedance is specifically measured applying a modest excitation signal to get cell's response in a pseudo-linear manner as shown in Figure 2.4.

If the excitation signal is expressed as a function of time (t), the equation takes form like,

$$E_t = E_0 \sin(\omega t)$$

Here, in equation, E_t is the obtained potential at time t, E_0 is the amplitude of the signal, and ω is designated as the radial frequency. The interrelationship between radial frequency (ω) who is with unit of radians/second and frequency (f), which is expressed in hertz, is:

$$\omega = 2\pi f$$

For a linear system, the response signal for current, I_t, is shifted in phase (Φ) and has a different amplitude than I_0,

$$I_t = I_0 \sin(\omega t + \Phi)$$

An expression resembling to Ohm's Law permits to determine the impedance of the system as follows:

$$Z = E_t / I_t = Et = E_0 \sin(\omega t) / It = I_0 \sin(\omega t + \Phi) = Z_0 K.$$

where, K is sin (ω t)/sin (ωt+ Φ).

So impedance can be measured in two forms: magnitude (Z_0) and phase shift (Φ). Also in addition, the plot between E_t (X-axis) vs I_t (Y-axis)

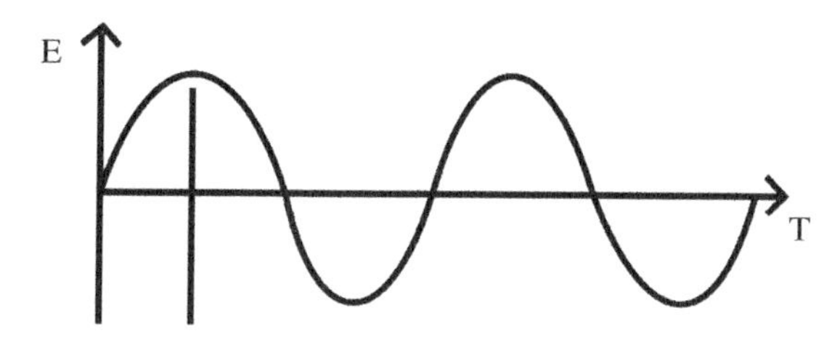

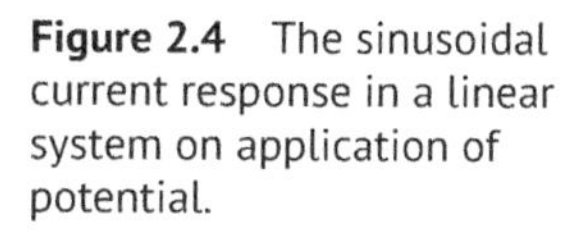
Figure 2.4 The sinusoidal current response in a linear system on application of potential.

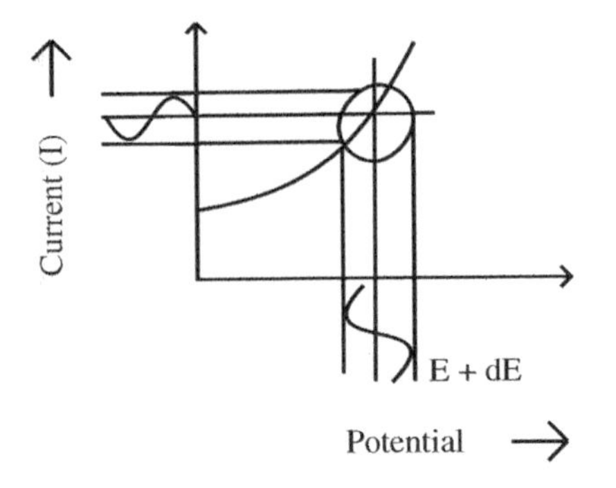

Figure 2.5 The making of Lissajous figure.

was made and oval named as "Lissajous Figure" was obtained as resultant (Figure 2.5). The figure obtained was than analyzed as impedance measurement before the application of modern electrochemical instrumentations.

From Euler's theorem, the expression of impedance can be written in complex form as:

$$\exp(j\Phi) = \cos\Phi + j\sin\Phi$$

So the potential can be best described as:

$$E_t = E_o \exp(j\omega t)$$

While the current response can be recorded as,

$$I_t = I_o \exp(j\omega t - \Phi)$$

Putting these values in impedance formula, the complex equation is formed as,

$$Z(\omega) = E/I = Z_o \exp(j\Phi) = Z_o(\cos\Phi + j\sin\Phi)$$

The complex equation is composed of a real and imaginary components, who are plotted on X-axis and Y-axis in graph forming "Nyquist Plot" (Figure 2.6). The impedance can be best presented as an arrow or vector with dimensions of $|Z|$, whereas the angle made in between this vector and X-axis, commonly known as "Phase Angle."

The most common equivalent circuit, which has been in use to sculpt corrosion of exposed metal in liquiform electrolyte, is called Randles circuit as presented in Figure 2.7.

Where R_Ω is the solution resistance, due to the presence of the electrolyte between the reference and working electrodes, polarization resistance R_p and C_{dl} or CPE_{dl} is the double-layer capacitance or double-layer constant phase element (CPE). Another plot to mark presentation of impedance is the Bode Plot (Figure 2.8), where the impedance is plotted with log frequency on the X-axis and both the absolute values of the impedance ($|Z| = Z_o$) and the phase-shift on the Y-axis. This plot can give frequency information.

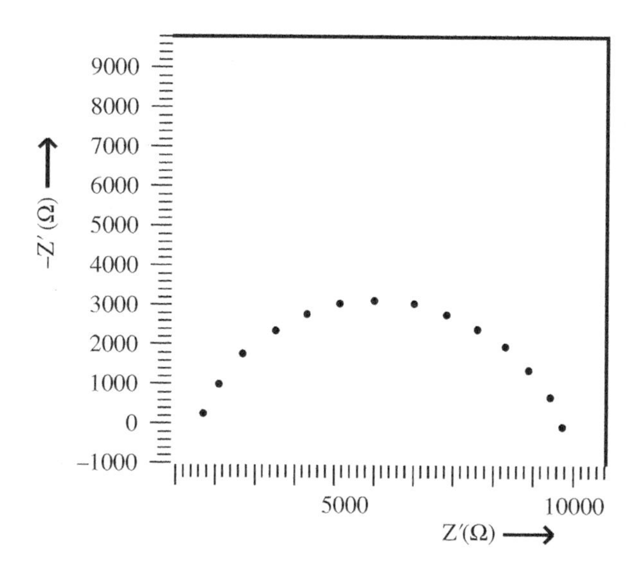

Figure 2.6 Typical Nyquist plots for a Randles equivalent circuit with C_{dl} CPE_{dl} with $N = 0.8$ (red dots).

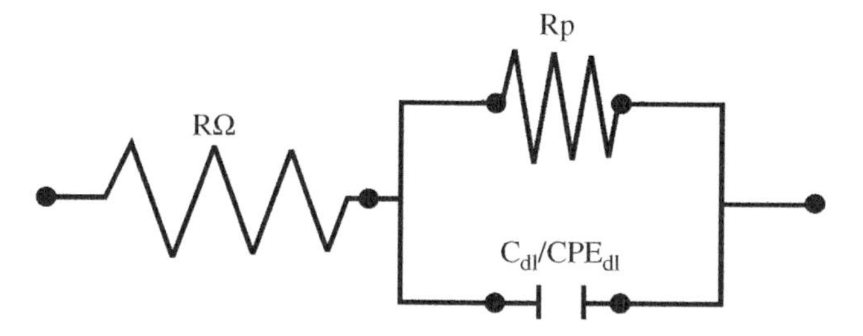

Figure 2.7 A typical Randles plot.

For the recent analysis, the impedance is mainly measured with amplifiers or frequency-response analyzers, which are consider being faster and are also more convenient than impedance bridges. The basic principle involves interpretation of the equivalent resistance and capacitance standards in provision of interfacial aspect. This technique is precise and intermittently in usage for evaluating amalgamate charge transfer criterion to get knowledge of double-layer arrangement. The mathematical value of R_p obtained from electrochemical impedance spectroscopy can be considered more perfect compared to other monitoring techniques. The graph obtained such as Nyquist and Bode gives better understanding of corrosion procedure happening.

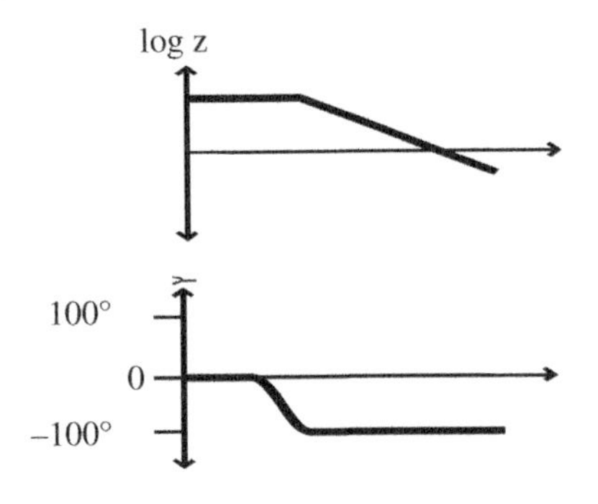

Figure 2.8 Typical Bode plot.

iv) *Linear Polarization Technique*
This electrochemical technique, which commonly also known as linear polarization resistance, is the only corrosion monitoring technique in its type method that permits measurement of corrosion rates directly across real time. So, it is fast and a nonintrusive method that needs an association in between metal reinforcement to get assess of on-going corrosion in structures [28, 29]. One disadvantage of it is that it is only confined to electrolytically conducting liquids. While its response time and data quality are far more superior compared to other corrosion monitoring techniques. It can be harvested to get current–potential (i–E) domain, which can be monitored by applying polarization resistance at a very small voltage differences generally less than 30 mV, above and below its corrosion potential [30]. The obtained current response is linear over narrow range of corrosion potential. So the slope of this current–potential curve is defined as polarization resistance (R_p) whose value is constant. As per Stern and Geary in 1957, current value can be obtained by given equation, where R_p is inversely proportional to the instantaneous corrosion rate, at some conditions [31].

$$\frac{\Delta E}{\Delta I} = R_p = \frac{\beta_a \times \beta_b}{2.303 \times I_{corr} \times (\beta_a + \beta_b)}$$

β_a and β_b are obtained Tafel constants from Tafel plots, polarization measurements for the studied system. In the last phase, the corrosion rate of the structure can be calculated via I_{corr}.

The numerical value of $\Delta E/\Delta I$ is known as the polarization resistance. This variable can be conveniently measured by putting another electrode/auxiliary in the liquid, and in turn connecting it to the working/corroding/test electrode via external power supply. The whole setup is given in circuit provided (Figure 2.9).

Here, R_p is polarization resistance, Rs is solution resistance, while C_e is electrode capacitance. An important advantage of using LPR is that it does not take more than half an hour to provide a conclusion with preliminary apparatus adjustments, balancing of readings, calculating process, and computation of the polarization resistance R_p.

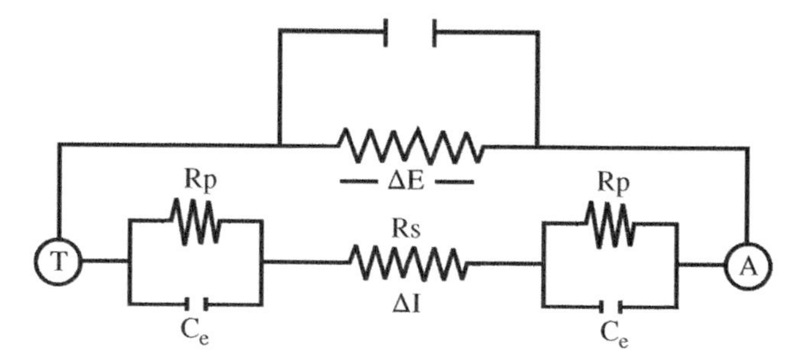

Figure 2.9 Circuit for linear polarization resistance.

b) **Nondestructive**

i) *Gamma Radiography*

Gamma radiography technique is one of the nondestructive methods for thickness loss or testing concrete to obtain information relating concrete quality, and deterioration or defects present in reinforced concrete structures. It is best to report cracks, voids, or any kind of variation present within it [32, 33]. There are two different techniques under this umbrella, tangential radiography technique and double-wall radiography technique as shown in Figure 2.10.

In the tangential method, radiation crosses through the sidewall thickness of the pipe made from metal and the area of the radiograph, which is placed below the tangential position, can be monitored. This specific technique has some important benefits for getting the thickness of the insulation pipes. It also gives the corrosion both internally and externally. In addition, it is having some disadvantage also like it need higher energy and intensity of the provided radiation, which is due to long beam pass via matter. While on the other hand, some effects such as radiography arrangement, power source to film distance, exposure and source beam path thickness, as well as build-up factor and linear atomic absorption coefficient in non-insulated pipes have been quantified in double-wall radiography technique [34–42].

In the tangential technique radiation allowed to passes through the sidewall thickness of the studied pipe and the area of the radiograph, which is located below the tangential position. So the parts of radiograph, which lies behind the tangential location of the pipe, are interpreted only [43].

ii) *Ultrasonic*

This technique helps to measure the thickness of the metal used along with size of defects. The principle involves the determination of thickness by

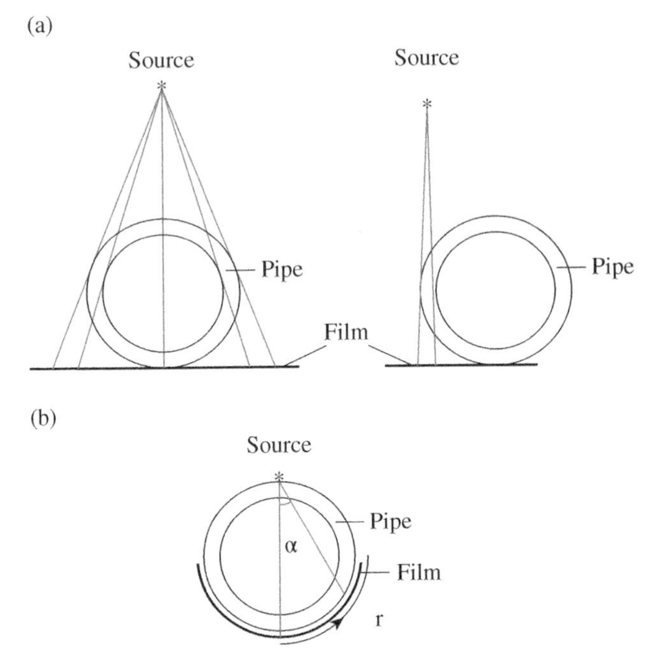

Figure 2.10 Gamma radiography technique for corrosion inspection, (a) tangential radiography technique and (b) double wall radiography technique.

monitoring the amount of time, which ultrasonic wave travels from the transducer via material to the back end of the material, which then reflects back to the transducer. From this, the width of the tested material can be calculated depending upon the speed of the sound passed through tested material. This technique is very efficient for inspecting the condition of vessels, tankers, pipeline in underground. The major advantage of using this technique is that readings can be taken from outside the wall of any operating structures, as they do not need connection on the dual side of the sample structure. This makes easy way to get width of the pipelines where the internal surface cannot be measure. In addition to this, coatings and different kind of linings can also be analyzed. The only known disadvantage of it is the calibration of material to be monitored [44].

iii) *Pulsed Eddy Currents*

This technique applies for knowing wall thickness of the structure during corrosion monitoring generally in refinery units with severe corrosion problems. The principle employs production of pulsed magnetic field to generate eddy currents in the metallic structures. If the metal is specifically steel, hence ferromagnetic, so only topmost exposed layer of structure can be magnetized [45, 46]. The schematic representation is shown in Figure 2.11a and b.

It is showing stages of measurement for eddy currents on the metallic (steel) surface, which is near to the pulse eddy current (PEC) probe. As the time passes, the current passes into the specimen structures showing in stages 2 and 3. At last they reach to farthest surface, which is stage 4 in provided figure. The produced eddy currents induce a voltage signal in the receiver coils of the pulse eddy current probes. The pulse eddy current signals are then displayed in the form of plot, PEC signals vs. time. The free

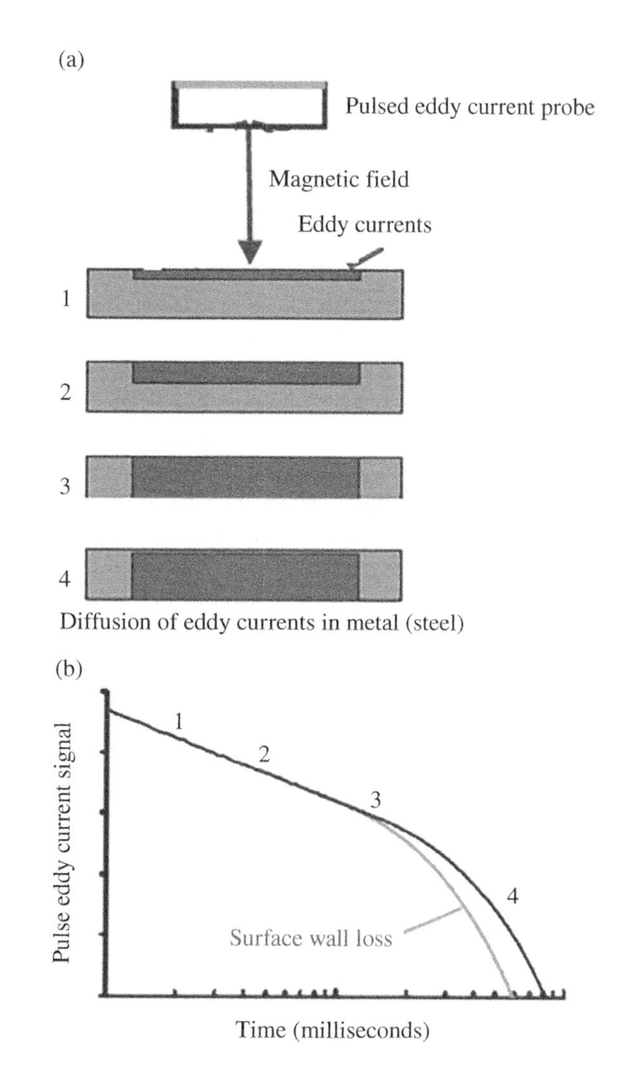

Figure 2.11 Pictorial presentation of (a) generation of pulse eddy current and (b) graph for signal.

expansion of the metal (steel) as experienced by the eddy currents in different stages exposes the strength, which decreases in relatively slow manner. Hence, on reaching to farthest position in structure, the strength dropped suddenly, which can be clearly seen from sharp fall in the PEC signal. At early outset of this, acute decay of the monitored pulse eddy current signal shows wall damage or loss of the structure. The readings of the wall thickness are nearly circular in shape named as "footprints" where eddy currents can flow. The size of these footprints mostly depends on the length in between the probe and the metallic surface along with the dimensions of the probe itself. In all these way, pulse eddy current method is best suited to get wall deterioration or loss in metallic dimensions to get knowledge of corrosion [47].

iv) *Infrared Thermographic Detection*
Out of other routinely detection of deterioration in nondestructive testing methods, one is infrared thermographic detection. This technique seems to be impractical due to many safety reasons and prerequisite to have a two-sided admittance to objects under study [48–50]. The principle associates thermal stimulation of the studied object by channeling with an optical heat source (convective/inductive heat sources) and analyzing the matter surface with an Infrared imager. The obtained data can be sequenced as "IR thermograms" and further recorded on computer for data processing methods like Fourier transform and principal component analysis. The area to study at once can be of dimension 0.2–1.0 m^2. The results obtained can be in terms of binary maps of defects. For bigger areas, the study can be done by applying area-by-area flashing and bringing together multiple infrared thermograms in a panned image. The schematic presentation of infrared thermographic monitoring is shown in Figure 2.12 [51, 52].

Its application has shown versatile inspection of composite materials used in the aerospace, boilers, pipeline jacketing, aluminum airframes, and at many more places [53–59]. It can be seen that this technique can be able to detect material debt up to 10% [60–61].

It has to note also that the screening of metals by using present technique is more burdensome compared to nonmetals. It is due to the fact that metals are examined

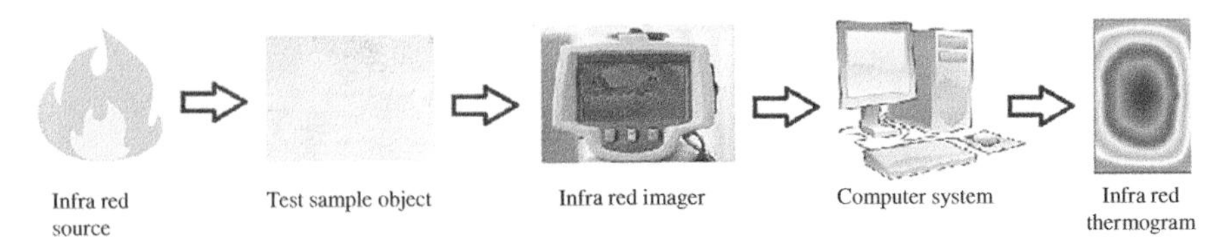

Figure 2.12 Schematic diagram for generating infrared thermogram.

by lower absorption toward optical radiation and higher thermal diffusivity. This leads to the low number of temperature signals with short period. Other known disadvantage of having halogen lamp is the formation of reflected noisy radiation, which can be appeared in both heating and cooling phases. To eliminate these noises, special kind of algorithms are used in infrared thermographic detection [62]. Recently detailed study of this technique was done in 2019 by Doshvarpassand et al., where they showed its application in around 400 literature resources [63].

2.3 Conclusion

Corrosion is the most general obstacle detected in the metallic structure, petrochemical industry, and at oil and gas refineries. This phenomenon of corrosion is occurring due to the metal deterioration and different types of chemical reactions with the pipes. All these problem causes an economic loss at extremely high in multiple industries. The different types of corrosion occur at different position in the same structure. So monitoring and disclosure is the must-to-do places where metal is one of the components. Out of destructive and nondestructive methods of monitoring, coupons and probes like electrical resistance and linear polarized resistance are best to monitor corrosion in pipelines. Electrochemical methods like potentiodynamic polarization and impedance spectroscopy are competent and prudent for corrosion analysis. Specifically electrochemical polarization method has a great potential for corrosion monitoring. It has benefit of being more conscious to and not destroying the assessed metallic surface.

So this cause is the only possible way to monitor deterioration based on the extracted data so that further action of replacement of the pipelines can be done in industries. Corrosion monitoring also offers multiple answers to the problems of whether further corrosion is happening now compared to yesterday. Also by knowing this data, it is important to eliminate the cause of corrosion along with its effects. This can be considered as important asset to altercate corrosion and supposing considerable economic prosperity to the country.

References

1 Heuslerl, K.E., Landolt, D., and Trasatti, S. (1989). Electrochemical corrosion nomenclature. *Pure and Applied Chemistry* 61 (1): 19–22. https://doi.org/10.1351/pac198961010019.

2 Shreir, L.L., Jarman, R.A., and Burstein, G.T. (1994). *Corrosion Volume 1: Metal and Environmental Reactions*, 3rde. Butterworth-Heinemann. ISBN: 0 7506 1077 8 Kindle Edition.

3 Jones, D.A. (1995). *Principles and Prevention of Corrosion*, 2nde. Prentice Hall. ISBN: 978-0133599930.
4 Roberge, P.R. (1999). *Handbook of Corrosion Engineering*, 1ste. McGraw-Hill. ISBN: 0-07-076516-2 or Kindle Edition.
5 ISO 8044:1999 (2000). *Corrosion of Metals and Alloys Đ Basic Terms and Definitions*. Brussels: International Organization for Standardization.
6 Fontana, M.G. (2005). *Corrosion Science and Engineering*, 3rde. Tata McGraw-Hill. ISBN: 978-0070607446.
7 NACE International/ ASTM G193-12d (2012). *Standard Terminology and Acronyms Relating to Corrosion*. West Conshohocken, PA: ASTM International https://doi.org/10.1520/G0193-12D.
8 Chafiq, M., Chaouiki, A., Damej, M. et al. (2020). Bolaamphiphile-class surfactants as corrosion inhibitor model compounds against acid corrosion of mild steel. *Journal of Molecular Liquids* 309: 113070.
9 Lgaz, H., Salghi, R., Masroor, S. et al. (2020). Assessing corrosion inhibition characteristics of hydrazone derivatives on mild steel in HCl: Insights from electronic-scale DFT and atomic-scale molecular dynamics. *Journal of Molecular Liquids* 308: 112998.
10 Zheng, L., Wan, Z., Gao, N., and Zhang, C. (2012). Refining corrosion monitoring technology application and device progress. *Journal of Petroleum Chemical Corrosion and Protection* 2: 47–50.
11 Yu, Z. and Meng, X. (2012). Corrosion monitoring technology and its application in oil and gas field. *Pipeline Technology and Equipment* 2: 48–49.
12 Yang, X., Rao, J., and Wang, Y. (2011). Application of online monitoring technology in petrochemical industry. *Journal of Petroleum Chemical Corrosion and Protection* 3: 40–42.
13 Chapter 1, 2000 ASM International. All Rights Reserved. Corrosion: Understanding the Basics (#06691G).
14 https://corrosiondoctors.org/Principles/Cost.htm#:~:text=Within%20the%20total%20cost%20of,of%20available%20corrosion%20management%20techniques.
15 Rose, J. and Barshinger, J. (1998). Using ultrasonic guided wave mode cutoff for corrosion detection and classification. *Materials Science, 1998 Ultrasonics Symposium. Proceedings (Cat. No. 98CH36102)* 1: 851–854.
16 Pei, J., Yousuf, M., Degertekin, F. et al. (21 April 2009) (1996). Lamb wave tomography and its application in pipe erosion/corrosion monitoring. *Research in Nondestructive Evaluation* 8: 189–197. http://dx.doi.org/10.1007/ BF02433949.
17 Hines, J. (1968). *Industrial Corrosion Monitoring; Committee on Corrosion*. London: Department of Industry H.M.S.O.
18 Rothwell, N. and Tullmin, M. (2000). *The Corrosion Monitoring Handbook*. Kingham, Oxford, UK: Coxmoor Publishing Co.
19 *Glossary, National Corrosion Service*. Teddington, London, UK: National Physical Laboratory (NPL).

20 Standard Guide for Conducting Corrosion Tests in Field Applications; ASTM G4–01, ASTM International, 2005.
21 Field Corrosion Evaluation Using Metallic Test Specimens; RPO497–2004; NACE International: Houston, TX, USA, 2008.
22 Dean, S.W. (1987). *In-Service Monitoring||, ASM Handbook*, vol. 13, 197. ASM International.
23 M. Mobin and S. Masroor, Adsorption and corrosion inhibition behavior of schiff base-based cationic gemini surfactant on mild steel in formic acid, *Journal of Dispersion Science and Technology*, 35:535–543, 2014, 0193-2691 print=1532-2351, https://doi.org/10.1080/01932691.2013.799435.
24 Masroor, S., Mobin, M., Singh, A.K. et al. (2020). Aspartic di-dodecyl ester hydrochloride acid and its ZnO-NPs derivative, as ingenious green corrosion defiance for carbon steel through theoretical and experimental access. *SN Applied Sciences* 2: 144. https://doi.org/10.1007/s42452-019-1515-z.
25 Publication EFC4 (1990). *Guidelines for Electrochemical Corrosion Measurements*. European Federation of Corrosion.
26 Shrier, 1978; Eg&G Application note, 1982.
27 Martin R. Dynamic optimization of chemical additives in a water treatment system. US Patent 6419817, 2000.
28 https://www.alspi.com/lprintro.htm.
29 Rathod, N.G. and Moharana, N.C. (2015). Advanced methods of corrosion monitoring- a review. *IJRET: International Journal of Research in Engineering and Technology* 04 (13) ICISE-2015, Dec- eISSN: 2319-1163, pISSN: 2321-7308.
30 Ropital, F. (2011). *15 - Environmental Degradation in Hydrocarbon Fuel Processing Plant: Issues and Mitigation*, 437–462. Advances in Clean Hydrocarbon Fuel Processing, Science and Technology, Woodhead Publishing Series in Energy https://doi.org/10.1533/9780857093783.5.437.
31 Stern, M. and Geary, A.L. (1957). Electrochemical polarization. *Journal of the Electrochemical Society* 104 (1): 56.
32 Song, H.W. and Saraswathy, V. (2007). Corrosion monitoring of reinforced concrete structures – a review. *International Journal of Electrochemical Science* 2: 1–28.
33 Zaki, A., Chai, H.K., Aggelis, G.D., and Alver, N. (2015). Non destructive evaluation for corrosion monitoring in concrete: a review and capability of acoustic emission technique. *Sensors 15* (8): 19069–19101. ISSN 1424-8220.
34 Burkle, W.S. (1989). Application of tangential radiographic technique for evaluating pipe system erosion/corrosion. *Mater Eval* 47 (10): 1186–1188.
35 Zecherpel U. Corrosion and Deposit Evaluation in Large Diameter Pipes by Radiography. Internal report of the second RCM of the CRP, IAEA, Istanbul, Turkey, March 2004.
36 ASNT (2002). *Nondestructive Testing Handbook—Radiographic Testing*, vol. 4. USA: American Society for Nondestructive Testing, Inc.

37 Krolicki, R.P. (1997). Internal corrosion examination and wall thickness measurement of pipe by radiographic method. *Mater Eval* 35 (2): 32–33.

38 Ekinci, S., Bas, N., Aksu, M. et al. (1998). Corrosion and deposit measurements in pipes by radiographic technique. *Insight* 40 (9): 602–605.

39 Rheinlander, J. and Christiansen, H. (1995). Using film density variations for determination of pipe thickness variation in gamma-ray radiography. *Insight* 37 (9): 691–694.

40 Kajiwara, G. (2000). Examination of the X-ray piping diagnostic system using EGS4 (examination of the film and iron rust). In: *Proceedings of the Second International Workshop on EGS*, 199–208. Japan: Tsukuba.

41 Willems, P., Vaessen, B., Hueck, W., and Ewert, U. (1999). Application of CR for corrosion and wall thickness measurements. *Insight* 41 (10): 635–637.

42 Marstboom, N. (1999). Computed radiography for corrosion and wall thickness measurements. *Insight* 41 (5): 308–309.

43 Edalati, K., Rastkhah, N., Kermani, A. et al. (2006). The use of radiography for thickness measurement and corrosion monitoring in pipes. *International Journal of Pressure Vessels and Piping* 83: 736–741.

44 He, Y. (2016). *Corrosion Monitoring, Reference Module in Materials Science and Materials Engineering*. Hamilton, ON, Canada: Elsevier Inc, Natural Resources Canada https://doi.org/10.1016/B978-0-12-803581-8.03460-3.

45 P.W. van Andel, "Eddy Current Inspection Technique", US patent nr 6,291,992 (September 2001).

46 P.C.N. Crouzen, "Method for inspecting an object of electrically conducting material", US patent nr 6,570,379 (May 2003).

47 Crouzen, P. and Munns, I. (2006). *Pulsed Eddy Current Corrosion Monitoring in Refineries and Oil Production Facilities –Experience at Shell*. Amsterdam, The Netherlands, NDT.net Issue: 2006-11, Publication: 9th European Conference on NDT - September 2006 - Berlin (Germany) (ECNDT 2006), Session: Chemical and Petrochemical: Shell Global Solutions International https://www.ndt.net/search/docs.php3?id=3634.

48 Abdalla, A.N., Farai, M.A., and Samsuri, D. (2019). Challenges in improving the performance of eddy current testing: review. *Measurement and Control* 52 (12): 46–64.

49 IAEA (2005). *Development of Protocols for Corrosion and Deposits Evaluation in Pipes by Radiography*. Vienna: IAEA-TECDOC-1445.

50 Vavilov, V.P. and Chulkov, A.O. (2012). Detecting corrosion in thick metals by active thermal nondestructive testing. *Proceedings SPIE Thermosense-XXXIV* 8354: 117–129.

51 Maldague, X. and Marinetti, S. (1996). Pulse phase infrared thermography. *Journal of Applied Physics* 79 (5): 2694–2698.

52 Rajic, N. (2002). *Principal Component Thermography*. Melbourne, Victoria: Aeronautical and Maritime Research /Laboratory.

53 Marinetti, S. and Vavilov, V. (2010). IR thermographic detection and characterization of hidden corrosion in metals: general analysis. *Corrosion Science* 52 (3): 865–872.

54 Marinetti, S., Bison, P.G., and Grinzato, E. (2002). 3D heat flux effects in the experimental evaluation of corrosion by IR thermography. *Proceedings European Seminar "Quant. IR Thermography – QIRT'02"*: 92–97.

55 Cramer, K.E. and Winfree, W.P. (1998). Thermographic detection and quantitative characterization of corrosion by application of thermal line source. *Proceedings SPIE Thermosense-XX* 3361: 291–297.

56 Prabhu, D.R. and Winfree, W.P. (1993). Neural network based processing of thermal NDE data for corrosion detection. In: *Review of Progress In Quantitative Nondestructive Evaluation*, vol. 12 (eds. D.O. Thompson and D.E. Chimenti), 1260–1265. Plenum Press.

57 Syed, H.I., Winfree, W.P., Cramer, K.E., and Howell, P.A. (1993). *"Thermographic Detection of Corrosion in Aircraft Skin", Review of Progress in Quantitative Nondestructive Evaluation*, 724–729.

58 Alcott, J. (1994). An investigation of nondestructive inspection equipment: detecting hidden corrosion on USAF aircraft. *Materials Evaluation* 5: 64–73.

59 Chulkov, A.O., Vavilov, V.P., and Malakhov, A.S. (2016). A LED-based thermal detector of hidden corrosion flaws. *Russian Journal of Nondestructive Testing* 52 (10): 588–593.

60 Cramer, K.E. and Winfree, W.P. (1998). Thermographic detection and quantitative characterization of corrosion by application of thermal line source. *Proceedings SPIE Thermosense-XX* 3361: 291–297.

61 Prati, J. (2000). Detecting hidden exfoliation corrosion in aircraft wing skin using thermography. *Proceedings SPIE Thermosense-XXII* 4020: 200–209.

62 Simonov, D., Vavilov, V., and Chulkov, A. (2020). Infrared thermographic detector of hidden corrosion. *Sensor Review* 40 (3): 283–289. Emerald Publishing Limited, ISSN 0260-2288, DOI https://doi.org/10.1108/SR-12-2019-0322.

63 Doshvarpassand, S., Wu, C., and Wang, X. (2019). An overview of corrosion defect characterization using active infrared thermography. *Infrared Physics & Technology* 96: 366–389.

3

Computational Methods of Corrosion Monitoring

Hassane Lgaz[1], Abdelkarim Chaouiki[2], Mustafa R. Al-Hadeethi[3], Rachid Salghi[2], and Han-Seung Lee[1]

[1] *Department of Architectural Engineering, Hanyang University-ERICA, Ansan, Korea*
[2] *Laboratory of Applied Chemistry and Environment, ENSA, University Ibn Zohr, Agadir, Morocco*
[3] *Department of Chemistry, College of Education, Kirkuk University, Kirkuk, Iraq*

3.1 Introduction

In recent decades, considerable progress has been made in the development of computational tools. Computer simulations were used as innovative tools to answer complex questions in a way that experiments cannot do. Like many research fields, computational methods were extensively used in corrosion inhibition research [1–4]. Density functional theory (DFT), molecular dynamics (MD) simulations, and Monte Carlo (MC) simulations were extensively applied in many corrosion inhibition studies to discover the underlying inhibition mechanism.

There is a growing body of literature in the field of corrosion science that recognizes the importance of experimental methods for monitoring the corrosion inhibition process of metals and their alloys in aggressive environments [5–7]. It is, actually, the first and the most important step in evaluating the suitability of a corrosion inhibitor. However, questions have been raised about the cost and time taken by these methods. In addition, the relationship between the inhibition efficiency obtained experimentally and the molecular structure of inhibitor compounds cannot be understood from experimental techniques.

In this chapter, efforts were made to discuss the recent progress in using computational methods; especially, quantum chemical (QC) calculations based on DFT, MD, and MC simulations for evaluating the corrosion inhibition performance and adsorption behavior of corrosion inhibitors.

Organic Corrosion Inhibitors: Synthesis, Characterization, Mechanism, and Applications,
First Edition. Edited by Chandrabhan Verma, Chaudhery Mustansar Hussain, and Eno E. Ebenso.

3.2 Quantum Chemical (QC) Calculations-Based DFT Method

QC calculations have emerged in the last decades as an efficient tool for accurate prediction of stability, chemical reactivity, and corrosion inhibition performance of molecules [8–11]. In fact, a good understanding of the underlying reaction mechanisms based on QC analysis of molecular structure of inhibitor molecules is necessary. In this context, DFT has been introduced as a reliable method of examining the molecular structure behavior of corrosion inhibitors. Based on DFT, a series of calculations and vital parameters can be provided at a reasonable computation cost [12]. It can also predict the physicochemical properties of complex molecules in terms of their reactivities, which might be impossible to be evaluated experimentally [13–15]. DFT calculations have been performed at a large variety of theoretical levels to characterize many molecular systems. In general, from QC point of view, successful results would expand the range of chemical compounds that can be used as corrosion inhibitors, which is of great practical importance.

The prediction of corrosion behavior and the examination of electronic propreties of any organic inhibitors can be explored based on DFT analyses with the application of hard and soft acid–base (HSAB) theory [16–18]. It can be noted that HSAB theory offers unique opportunities in the fundamental understanding of structure–property relationships and then adsorption abilities of inhibitor molecules. Of course, Koopmans' theorem [19] is also one of the greatest contributions in theoretical chemistry that has found extensive application in QC calculations.

In this concept of DFT, the ground-state electronic energy is determined completely by the electron density $\rho(\mathrm{r})$ which provides quite accurate results of a molecular system at a theoretical level with no reference to a wave function. In fact, the main goal of the DFT method is to design functionals connecting the electron density with the energy by reducing the complexity of the theoretical system defined by the Schrödinger equation. Nowadays, DFT concept is widely used to extract structural information related to the adsorption of organic inhibitor molecules on the metal surface. Hence, a basic understanding of the theory behind the DFT method is necessary.

3.2.1 Theoretical Framework

The basis of DFT was introduced in1964 by Hohenberg and Kohn as a result of two theorems elegantly demonstrated in [20, 21]. According to these theorems, Hohenberg and Kohn stated that the ground state of a fully interacting system with N electrons is a unique functional of the electronic density $\rho(\mathrm{r})$. So, DFT is a formulation in terms of functional of the density.

For DFT-based simulation, the general expression may then be written as [22]:

$$E_{DFT}\left[\rho\right]=T_S\left[\rho\right]+E_{ne}\left[\rho\right]+J\left[\rho\right]+E_{XC}\left[\rho\right] \tag{3.1}$$

In the above equation, T_s denotes the kinetic energy functional (S indicates that the kinetic energy is obtained from a Slater determinant), E_{ne} and J are the functionals for electron–nuclear attraction and the Coulomb part of the electron–electron repulsion, respectively. E_{xc} is the functional for the exchange correlation. It is important to note that each of these terms is related to electron density.

Of course, the abovementioned theorems were followed by the important work proposed by Kohn and Sham [23], which represented an approximation to define the exchange-correlation energy functional for the DFT method. It can be noted that the exchange-correlation energy functional remains a poorly defined term in the proposed system and is not known. So, the work published by Kohn and Sham (K-S) became the paradigm for solving the problem by replacement of the exchange potential term by a more general exchange-correlation potential. For that, K-S introduced orbitals, also called KS orbitals. As suggested by K-S, the electron density is represented by an auxiliary set of orbitals in which we can determine the electron kinetic energy. In fact, an electronic system simulated with the aid of K-S model-based DFT method is closely linked to the well-known Hartree–Fock (HF) method. In this frame, the formulas used for calculation of kinetic, electron–nuclear, and Coulomb electron–electron energies are identical for both methods.

Concerning the basic functionals used for the description of DFT-based simulations, it has been resulted that DFT can be broken down into many of functional classes. The simplest model is the local density approximation (LDA), which was introduced in 1951 by Slater [24]. In this case, DFT within LDA includes only electron exchange without the correlation. Up to now, LDA was very successful in describing electronic systems and there are similarities between the results expressed by LDA and those described by HF. In the 1980s, the generalized gradient approximations (GGAs) were introduced in order to go beyond a few systematic errors obtained by LDA like overestimations of molecular atomization energies and bond lengths. The prominent idea of GGAs makes use of the spin densities and the spatial variation of the exchange-correlation energy, i.e. the latter taking into account both the density and its gradients. In addition, B3LYP functional was fully proposed and identified in 1998 and it is one of the most popular hybrid functional. In 1994, B3LYP was already been used successfully for the first time in the Gaussian package and it is most successful in terms of overall performance. B3LYP model can be described via the following equation [25]:

$$E_{XC}^{B3LYP}=a_0E_X^{HF}+a_1E_X^{LSD}+a_2E_X^{GGA}+a_3E_C^{LYP} \tag{3.2}$$

Herein, E_X^{HF} represents exchange energy computed with KS orbitals on the basis of HF theoretical level, E_X^{LSD} means the local exchange energy, E_X^{GGA} denotes the gradient corrections to the exchange energy, and E_X^{LYP} represents the gradient corrections for the correlation energy constructed by Lee, Yang, and Parr (LYP) [26].

Another potential problem for computational chemists is how to choose the basis set to express the unknown molecular orbitals in terms of a set of known functions. However, what is not yet understood is the relative importance of any type of basis functions used for representing the electronic wave function. What is important for us to recognize here is that the molecular wave functions can be expressed as the linear combination of atomic orbitals (LCAO).

In principle, many electron systems can be optimized and practically verified using many types of basis functions such as exponential, Gaussian, polynomial, cube functions, wavelets, plane waves, etc. Nevertheless, it is necessary here to note that Gaussian functions are computationally much easier to handle. In the DFT method, STO-3G is a minimal and a simplest basis set that can be used. Briefly, much of the literature [25, 27–31] emphasizes the description of basis sets in the HF and DFT methods, which suggests that increasing the size of the basis set allows a better description of the KS orbitals. For example, 3–21G, 6–31G, 6–311G, and LanL2DZ basis sets are much more likely to be useful in the DFT method for good improvement of QC calculations.

3.2.2 Theoretical Application of DFT in Corrosion Inhibition Studies: Design and Chemical Reactivity Prediction of Inhibitors

As mentioned earlier, DFT-based simulations are widely used in the field of corrosion inhibition to discuss the electronic properties of organic compounds and help in describing the characteristics of inhibitor/surface interactions, which can advance our knowledge about the corrosion inhibition mechanism in its relationship with the structural nature of molecules. In this respect, a variety of QC descriptors such as highest occupied (HO) and lowest unoccupied (LU) molecular orbitals (MO), frontier orbital energies, energy band gap, hardness, electronegativity, Mulliken and Fukui population analyses, electron-donating power, electron-accepting power, chemical potential, hardness, softness, dipole moment and number of electrons transferred, etc., are described and used in the prediction of the characteristic of molecules in terms of chemical reactivity and binding affinities.

On the basis of the DFT method, the results were obtained for all chemical parameters indicated previously at different levels of theory with the aid of reliable open source and commercial software. Within this framework, here, we describe briefly some of the chemical parameters and their mathematical formulas that are the most used in corrosion inhibition studies. More detailed information about all

QC descriptors, and their applications for elucidating and understanding the properties of molecules is presented in the literature [25, 32, 33].

3.2.2.1 HOMO and LUMO Electron Densities

It is well known that HOMO and LUMO distributions (also called FMOs) in a corrosion inhibitor molecule are quite important tools in predicting the chemical reactivities of molecules. These tools are considered for identifying the adsorption points of the molecules which is liable for the interaction with the metal surfaces. As a definition, in an inhibitor molecule, HOMO density is associated with the affinity of electron donation, while LUMO density is related with inclination to receipt an electron. The presence of a wide range of heteroatoms and functional groups in an inhibitor molecule is mainly associated with a higher HOMO density and therefore a higher electron-donating property.

3.2.2.2 HOMO and LUMO Energies

As a consequence of FMOs theory discussed above, HOMO and LUMO were associated with electron-donating and electron-accepting abilities of molecules, respectively. Hence, it can be noted that any inhibitor molecule characterized by high energy of HOMO will be, in most cases, effective in terms of its tendency to transfer the electrons to a metallic surface. On the other hand, low LUMO energy value shows that the molecule is a good electron acceptor.

In addition, Koopmans theorem [34] is a bridge between DFT and MO theory and it can be used in the prediction of ionization potential (IP) and electron affinity (EA) values of molecules. According to this theorem, IP and EA can be expressed via the following equations:

$$IP = -E_{HOMO} \tag{3.3}$$

$$EA = -E_{LUMO} \tag{3.4}$$

Further, the energy difference between LUMO and HOMO called an energy gap (ΔE) is also an essential parameter toward the description of reactivity of a molecule.

$$\Delta E = E_{LUMO} - E_{HOMO} \tag{3.5}$$

A large value of ΔE indicates low reactivity of molecules while a molecule with low value of ΔE can be strongly adsorbed on a metal surface.

3.2.2.3 Electronegativity (η), Chemical Potential (μ), Hardness (η), and Softness (σ) Indices

Parameters such as η, μ, η, and σ are defined as the derivatives of the total electronic energy (E) with respect to the number of electrons (N) at a constant external potential. The η is defined as the negative value of μ. Within the framework of finite differences approaches, these parameters can be expressed in the form of ground-state IP and ground-state EA values of a chemical compound. The theoretical formulas can be expressed as [35]:

$$\mu = -\chi = \left[\frac{\partial E}{\partial N}\right]_{v(r)} = -\left(\frac{IP + EA}{2}\right) \tag{3.6}$$

$$\eta = \frac{1}{2}\left[\frac{\partial^2 E}{\partial N^2}\right]_{v(r)} = \frac{IP - EA}{2} \tag{3.7}$$

$$\sigma = 1/\eta \tag{3.8}$$

3.2.2.4 Electron-Donating Power (ω^-) and Electron-Accepting Power (ω^+)

The new parameters called ω^- and ω^+ were introduced by Gazquez et al. [36] to predict the propensity of chemical species to accept and to donate electrons. Also, these two parameters are related to IP and EA and are mathematically described via the following equations:

$$\omega^+ = \frac{(IP + 3EA)^2}{16(IP - EA)} \tag{3.9}$$

$$\omega^- = \frac{(3IP + EA)^2}{16(IP - EA)} \tag{3.10}$$

Based on these parameters, we can, again, shed light on the inhibition abilities of chemical compounds based on their ability to accept and receive electrons.

3.2.2.5 The Fraction of Electrons Transferred (ΔN)

For the prediction of ΔN, which reflects the tendency of an inhibitor molecule to transfer its electron to a metal surface, the hardness and electronegativity indices have been used. This parameter was computed according to Pearson as per the following equation [35]:

$$\Delta N = \frac{\varnothing_{metal} - \chi_{inh}}{2(\eta_{metal} + \eta_{inh})} \tag{3.11}$$

In this equation, $\varnothing$ is the work function calculated for metal surface and η_{metal} means the global hardness of the metal.

The fraction of electrons transferred provides important insight into the adsorption process and the power of the interaction between the metal surface and inhibitor molecule. Consistent with the literature [37–39], an inhibitor can transfer its electron if $\Delta N > 0$ and vice versa if $\Delta N < 0$.

The application of these parameters in evaluating the adsorption and corrosion inhibition behavior of a large number of molecules is extensively reviewed in the following chapters.

3.2.2.6 Fukui Indices (FIs)

Fukui functions offer unique opportunities to increase the fundamental understanding of the local reactivity and selectivity of chemical species. In other words, Fukui function is a local reactivity index proposed to examine the nucleophilic and electrophilic attack regions of inhibitor molecules. In fact, from this important concept (FIs), we can pinpoint the reactive sites in which the electrophilic or nucleophilic attacks are large or small. As discussed above, HSAB theory provides an important contribution in the prediction and interpretation of many CQ parameters, and the judgment of Fukui functions is also an early attempt in this direction. It is necessary here to clarify exactly what is meant by Fukui function $f(r)$. By definition, $f(r)$ is a first derivative of $\rho(r)$ with respect to number of electrons (N) at a constant external potential $v(r)$.

$$f(r) = \left(\frac{\partial \rho(r)}{\partial N} \right)_{v(r)} \tag{3.12}$$

In addition, FIs were identified with respect to hard or soft reagents by involving the HSAB principle. A simple approximation can be used with the aid of finite difference approximation and Mulliken's population analysis in which FIs were determined as per the following equations [40]:

$$f_K^{+} = q_K(N+1) - q_K(N) \quad (\text{for nucleophilic attack}) \tag{3.13}$$

$$f_K^{-} = q_K(N) - q_K(N-1) \quad (\text{for electrophilic attack}) \tag{3.14}$$

$$f_K^{0} = \frac{q_K(N+1) - q_K(N-1)}{2} \quad (\text{for radical attack}) \tag{3.15}$$

where $q_k(N)$, $q_k(N+1)$, and $q_k(N-1)$ are charge values of neutral, anionic, and cationic forms of atom k, respectively.

3.3 Atomistic Simulations

Microscopic analyses are methods developed to serve as a basis for the investigation and simulation of physical phenomena on a molecular level. As these methods usually allow such a deep analysis, they became essential tools in generating and designing new functional materials. Macroscopic and microscopic characteristics of species constituting a simulation system, i.e. molecules and fine particles, are generated from analyzing the output of simulations.

Two well-known atomistic simulation methods are MC and MD. The advantage of the MD method over the MC method is, besides its ability to analyze thermodynamic equilibrium, it can be used to investigate the dynamic properties of a system in a nonequilibrium state.

3.3.1 Molecular Dynamics (MD) Simulations

The interest in MD simulation is growing due to the rapid development of complex hardware and software that allow simulation of large systems. The concept of MD simulation is based on solving Newton's equations of motion for the atoms in the simulation system using numerical integration [41, 42]. In a given system, its constituents, i.e. atoms and molecules, can move and interact with other constituents in the vicinity. The time evolution and atomic-scale dynamics of this system can be simulated and described by MD. The total energy of the system is mathematically described as a function of all atomic coordinates. A point-like nature of the interactions between atoms in the system are maintained conforming to a given potential energy $E(r_1, r_2, \ldots, r_N)$, where r_j is the vector position of the j-th atom. To obtain atom trajectories, the equation of motion is used to determine the location and velocity vector of each atom at every time-step [43–45].

$$F_i = m_i \times \frac{d^2 r_i}{dt^2} = -\nabla_i E\left(r_1, r_2, \ldots, r_N\right) \tag{3.16}$$

where r is the spatial gradient, E is the system's empirical potential, while r_i and m_i denotes the spatial coordinates and the mass of the i-th atom, respectively.

3.3.1.1 Total Energy Minimization

The aim of the total energy minimization process, also called geometry optimization or structural relaxation, is to find a stable structure, and it is the first essential step in almost all atomistic simulations [1, 2]. By performing this iterative procedure, atoms in a system can reach a lowest energy configuration; their coordinates are adjusted, so the net forces acting on the atoms are zero. In this procedure, forces on atoms as well as the energy of the system are calculated considering only total energies. An effective energy minimization process depends on the used minimization algorithms, which are mostly based on experimentally determined thermodynamic parameters [46, 47].

The simplest minimization algorithm is the steepest descents (SD). It is not a recommended minimization method given the fact that it does not consider previous steps when choosing a search direction. The method of conjugate gradients (CG) is a well-preferred minimization method and can correct the failings of SD. It is generally a reliable and robust minimization algorithm that is preferred for minimizing a large number of simulation systems. Another minimization algorithm is the Quasi-Newton (QN) methods, which seek to build an approximation to the inverse of the Hessian during the minimization. QN methods are often more efficient than Newton's method.

3.3.1.2 Ensemble

The accurate characterization of a system cannot be achieved only by its total energy; the consideration of realistic conditions such as entropy changes, volume, and pressure is essential to produce a useful output. In MD simulation, different thermodynamic potentials, called the ensemble, are used to account for these conditions. The ensemble is a collection of all possible systems that have identical thermodynamic or macroscopic attribute but different microscopic states. An ensemble leaves one parameter variable while it fixes the others [48]. The most common ensembles are the microcanonical ensemble (NVE), canonical (NVT), isobaric–isothermal ensemble (NPT), and grand canonical ensemble (μVT). With these ensembles, the energy can vary during the simulation.

3.3.1.3 Force Fields

In MD simulation, the used potential is a critical factor in determining the reliability of simulations. The set of parameters acting on the nuclei of atoms and mathematical formulas that relate a potential energy (usually described by pair potentials) to a configuration of a molecular system is called Force fields. It is a challenging task in every simulation to choose or create a correct force field for a given system [49]. Force fields are parameterized with experimental data and those from ab initio calculations. In corrosion inhibition, the Condensed-phase Optimized Molecular Potentials for Atomistic Simulation Studies (COMPASS) is the most cited force field. In addition to ab initio calculations, COMPASS was parameterized considering various experimental data including organic compounds made with H, C, N, O, S, P atoms, halogens, and metals [50]. Other used force fields include universal force field (UFF) [51, 52], which is an all-atom potential containing parameters for each atom, and the consistent-valence force field (CVFF) that is a generalized valence force field [53–55]. Because of the high computational cost, three- and more-body interactions are not considered when parameterizing a force field [56].

3.3.1.4 Periodic Boundary Condition

It is impossible to deal with a system composed of 6×10^{23} particles (one-mol-order size) in MD simulation. Fortunately, reasonable results can be obtained by treating a small system of about 100–10000 particles using the periodic boundary condition approach. In the case of a two-dimensional system as that represented in Figure 3.1, the main simulation box is replicated so the simulation region is the central square and the virtual boxes are the surrounding squares. For instance, the

Figure 3.1 Periodic boundary condition.

actual corrosion inhibition process cannot be simulated using an infinite system since it involves thousands of atoms and molecules [50]. To make more realistic simulations with a reduced computational cost, researchers developed several boundary conditions; among them, the periodic boundary conditions, which can make the corrosion inhibition simulation close to an infinite system. It has been used in all MD simulations of corrosion inhibition [57–67].

3.3.2 Monte Carlo (MC) Simulations

In contrast to the MD approach, which deals with both thermodynamic equilibrium and nonequilibrium phenomena, the MC method can only simulate a system under thermodynamic equilibrium [45]. It generates a series of configurations using random trial moves under a certain stochastic law regardless of the equation of motion. The concept of Monte Carlo has been used to develop several methods such as Metropolis MC, kinetic Monte Carlo (kMC), and quantum Monte Carlo which are used, besides atomistic simulations, in many physical and chemical applications. In corrosion inhibition research, Metropolis MC is the widely used approach to simulate the interaction between inhibitor molecules and surface of metals.

3.3.3 Parameters Derived from MD and MC Simulations of Corrosion Inhibition

In the field of corrosion inhibition, atomistic simulations have been used to provide some very relevant and credible insights into the interaction between the chemical compounds used as corrosion inhibitors and metals [68, 69], which cannot be gained from laboratory-based experiments. The comprehensive description of the molecular system on the atomic level can help in the understanding of the adsorption mechanism. Simulation systems differ significantly from work to work; however, most of the atomistic simulations for corrosion inhibition focus on finding a link between the experimental inhibition efficiency and parameters derived from simulations [68, 69]. The choice of suitable surface plans for simulations has been a subject of intense interest. Based on morphology parameters, Cu(111) and Al(111) are so far the most used copper and aluminum plans [3, 70].

In the case of iron plans, for investigating the adsorption of pyrrole, furan, and thiophene on iron surfaces, Guo et al. [71] found that among the three iron plans, Fe(110), Fe(100), and Fe(111), the Fe(110) has a high packed density and it is the most stable plans compared with others. Other parameters derived from MD and MC simulations are reviewed in the following sections.

3.3.3.1 Interaction and Binding Energies

The interaction energy is defined as the required energy for one mole of an inhibitor molecule to be adsorbed on a metal surface [72]. For a simulated system in a vacuum, it can be determined using the following equation [73, 74]:

$$E_{inter} = E_{Total} - \left(E_{Surface} + E_{inh}\right) \tag{3.17}$$

In the presence of a solvent:

$$E_{inter} = E_{Total} - \left(E_{Surface+solvent} + E_{inh}\right) \tag{3.18}$$

where E_{Total}, $E_{Surface}$, $E_{Surface+solution}$, and E_{inh} denote the total energy of the simulated system, surface without solution, surface with solution, and inhibitor molecule alone, respectively.

The binding energy is defined as the negative values of the interaction energy. A large binding energy implies that the inhibitor molecule can be strongly adsorbed over a metal surface [75, 76]:

$$E_{binding} = -E_{inter} \tag{3.19}$$

A representative case where interaction and binding energies are used to discuss the adsorption behavior of corrosion inhibitors have been performed by Lgaz et al. [77]. Authors constructed a simulation cell containing one of the inhibitor's molecule (hydrazone derivatives), water molecules (491 H_2O), and corrosive particles (i.e. 9 Cl^- and 9 H_3O^+) in contact with the Fe(110) surface. Simulations of neutral and protonated molecules have been performed in the presence of solvent using Materials Studio software while the canonical ensemble (NVT), COMPASS force field, and simulation time of 2000ps have been chosen in Forcite module for simulations. In this work, authors found that all inhibitor molecules adopted a flat orientation over iron surface with negative sign of interaction energy, suggesting that all investigated molecules can strongly adsorb on the iron surface.

Interaction and binding energies were found in line with experimentally inhibition efficiency.

In the case of the MC method, some energies such as total energy, adsorption energy, deformation energy, and rigid adsorption energy are obtained as output of the simulation [78–81].

3.3.3.2 Radial Distribution Function

Besides interaction and binding energies, trajectories from MD simulations can be structurally analyzed using a distribution function called the radial distribution function (RDF), which is often written as *g(r)*. In brief, it can be defined as the probability distribution of an atom in a spherical volume with a radius of r in a random system of the same density. The RDF is determined based on the equation proposed by Hansen and McDonald [42].

$$g(r) = \frac{1}{\rho_{Blocal}} \times \frac{1}{N_A} \sum_{i \in A}^{N_A} \sum_{j \in B}^{N_B} \frac{\delta(r_{ij} - r)}{4\pi r^2} \tag{3.20}$$

where $\langle \rho_B \rangle_{local}$ represents the particle density of B average over all shells beside particle A.

The interaction between an inhibitor molecule and metal surface can be judged based on the location of the first peak, which is located at a nearest neighbor distance. It has been shown that a peak located around 1 Å ~ 3.5 Å indicates a small bond length and thus a potential covalent bond, whereas a peak above 3.5 Å is mainly associated with a physical interaction [77].

3.3.3.3 Mean Square Displacement, Diffusion Coefficient, and Fractional Free Volume

The mean square displacement (MSD) has been used by many researchers in different research field as a route to investigate the dynamical aspects of systems. In corrosion inhibition studies, it has been used to calculate the diffusion coefficient of corrosive particles inside a simulated inhibition film [76]. Generally speaking, potent inhibitors are those that could hinder the diffusion of corrosive species, thus preventing the metal from corrosion. Based on this concept, a corrosive particle with a diffusion coefficient higher or like its diffusion in water can easily penetrate the inhibitor film, whereas limiting its diffusion can limit its movement and therefore protecting the metal against corrosion. The following equation is the general formula of the diffusion coefficient [27]:

$$D = \frac{1}{6} \lim_{t \to \infty} \frac{dMSD(t)}{dt}; \text{ where } MSD(t) = \left[\frac{1}{N} \sum_{i=1}^{N} \left| R_i(t) - R_i(0) \right|^2 \right] \tag{3.21}$$

The $R_i(t)$ denotes the position vector of *i*-th particle at time *t*. *N* represents the total number of diffusion particles. The term $|R_i(t) - R_i(0)|^2$ is the ensemble average of the MSD. In MD simulations, the diffusion coefficient is determined by performing a linear fit of the MSD plot using the following equation:

$$D = \frac{m}{6} \tag{3.22}$$

where m is the slope of MSD plot.

In the same context, the free volume inside an inhibitor film can also be determined from MD simulations. It has been confirmed that the diffusion of particles inside an inhibitor film that has large cavities can be very high, while that with lower cavities can hinder their diffusion [74, 76].

The application of atomistic simulations in corrosion inhibition studies is reviewed in Chapter 4 of this book.

Acknowledgments

This research was supported by basic science research program through the National Research Foundation (NRF) of Korea funded by the Ministry of Science, ICT and Future Planning (No. 2015R1A5A1037548).

Suggested Reading

Lewars, E. (2016). Computational chemistry. In: *Introduction to the Theory and Applications of Molecular and Quantum Mechanics*, 3ee, 318. Springer.

Cramer, C.J. (2013). *Essentials of Computational Chemistry: theories and Models*. John Wiley & Sons.

Satoh, A. (2010). *Introduction to Practice of Molecular Simulation: Molecular Dynamics, Monte Carlo, Brownian dynamics, Lattice Boltzmann and Dissipative Particle Dynamics*. Elsevier.

Rapaport, D.C. (2004). *The Art of Molecular Dynamics Simulation*. Cambridge University Press.

Marx, D. and Hutter, J. (2009). *Ab initio Molecular Dynamics: Basic Theory and Advanced Methods*. Cambridge University Press.

References

1 Leimkuhler, B. and Matthews, C. (2016). *Molecular Dynamics*. Springer.

2 Hollingsworth, S.A. and Dror, R.O. (2018). Molecular dynamics simulation for all. *Neuron* 99: 1129–1143.

3 Pareek, S., Jain, D., Hussain, S. et al. (2019). A new insight into corrosion inhibition mechanism of copper in aerated 3.5 wt.% NaCl solution by eco-friendly

Imidazopyrimidine Dye: experimental and theoretical approach. *Chemical Engineering Journal* 358: 725–742.

4 Zhang, X.Y., Kang, Q.X., and Wang, Y. (2018). Theoretical study of N-thiazolyl-2-cyanoacetamide derivatives as corrosion inhibitor for aluminum in alkaline environments. *Computational and Theoretical Chemistry* 1131: 25–32.

5 Yan, Y., Dai, L., Zhang, L.H. et al. (2018). Investigation on the corrosion inhibition of two newly-synthesized thioureas to mild steel in 1 mol/L HCl solution. *Research on Chemical Intermediates* 44: 3437–3454.

6 Chafiq, M., Chaouiki, A., Al-Hadeethi, M.R. et al. (2020). A joint experimental and theoretical investigation of the corrosion inhibition behavior and mechanism of hydrazone derivatives for mild steel in HCl solution. *Colloids and Surfaces A: Physicochemical and Engineering Aspects* 610: 125744.

7 Rbaa, M., Dohare, P., Berisha, A. et al. (2020). New Epoxy sugar based glucose derivatives as eco friendly corrosion inhibitors for the carbon steel in 1.0 M HCl: Experimental and theoretical investigations. *Journal of Alloys and Compounds* 833: 154949.

8 Gouron, A., Le Mapihan, K., Camperos, S. et al. (2018). New insights in self-assembled monolayer of imidazolines on iron oxide investigated by DFT. *Applied Surface Science* 456: 437–444.

9 Belghiti, M.E., Echihi, S., Mahsoune, A. et al. (2018). Piperine derivatives as green corrosion inhibitors on iron surface; DFT, Monte Carlo dynamics study and complexation modes. *Journal of Molecular Liquids* 261: 62–75.

10 Dagdag, O., El Harfi, A., Cherkaoui, O. et al. (2019). Rheological, electrochemical, surface, DFT and molecular dynamics simulation studies on the anticorrosive properties of new epoxy monomer compound for steel in 1 M HCl solution. *RSC Advances* 9: 4454–4462.

11 Chafiq, M., Chaouiki, A., Lgaz, H. et al. (2020). Synthesis and corrosion inhibition evaluation of a new schiff base hydrazone for mild steel corrosion in HCl medium: electrochemical, DFT, and molecular dynamics simulations studies. *Journal of Adhesion Science and Technology* 34 (12): 1283–1314.

12 Khalil, S.M., Ali-Shattle, E.E., and Ali, N.M. (2013). A theoretical study of carbohydrates as corrosion inhibitors of iron. *Zeitschrift für Naturforschung A* 68: 581–586.

13 Saha, S.K. and Banerjee, P. (2015). A theoretical approach to understand the inhibition mechanism of steel corrosion with two aminobenzonitrile inhibitors. *RSC Advances* 5: 71120–71130.

14 Rahmani, R., Boukabcha, N., Chouaih, A. et al. (2018). On the molecular structure, vibrational spectra, HOMO-LUMO, molecular electrostatic potential, UV–Vis, first order hyperpolarizability, and thermodynamic investigations of 3-(4-chlorophenyl)-1-(1yridine-3-yl) prop-2-en-1-one by quantum chemistry calculations. *Journal of Molecular Structure* 1155: 484–495.

15 Chaouiki, A., Lgaz, H., Chung, I.-M. et al. (2018). Understanding corrosion inhibition of mild steel in acid medium by new benzonitriles: Insights from experimental and computational studies. *Journal of Molecular Liquids* 266: 603–616.

16 Peason, R. (1997). *Chemical Hardness: Applications from Molecules to Solids* (ed. R.G. Pearson). Weinheim: Wiley-VCH https://doi.org/10.1002/3527606173.

17 Pearson, R.G. (1963). Hard and soft acids and bases. *Journal of the American Chemical Society* 85: 3533–3539.

18 Kovačević, N. and Kokalj, A. (2011). Analysis of molecular electronic structure of imidazole-and benzimidazole-based inhibitors: a simple recipe for qualitative estimation of chemical hardness. *Corrosion Science* 53: 909–921.

19 Koopmans, T. (1933). Ordering of wave functions and eigenenergies to the individual electrons of an atom. *Physica* 1: 104–113.

20 Hohenberg, P. and Kohn, W. (1964). Inhomogeneous electron gas. *Physical Review* 136: B864.

21 Hohenberg, P. and Kohn, W. (1964). Density functional theory (DFT). *Physical Review* 136: B864.

22 Withnall, R., Chowdhry, B.Z., Bell, S., and Dines, T.J. (2007). Computational chemistry using modern electronic structure methods. *Journal of Chemical Education* 84: 1364.

23 Kohn, W. and Sham, L.J. (1965). Self-consistent equations including exchange and correlation effects. *Physical Review* 140: A1133.

24 Slater, J.C. (1951). A simplification of the Hartree-Fock method. *Physical Review* 81: 385.

25 Obot, I., Macdonald, D., and Gasem, Z. (2015). Density functional theory (DFT) as a powerful tool for designing new organic corrosion inhibitors. Part 1: an overview. *Corrosion Science* 99: 1–30.

26 Lee, C., Yang, W., and Parr, R.G. (1988). Development of the Colle-Salvetti correlation-energy formula into a functional of the electron density. *Physical Review B* 37: 785.

27 Sure, R. and Grimme, S. (2013). Corrected small basis set Hartree-Fock method for large systems. *Journal of Computational Chemistry* 34: 1672–1685.

28 Inada, Y. and Orita, H. (2008). Efficiency of numerical basis sets for predicting the binding energies of hydrogen bonded complexes: evidence of small basis set superposition error compared to Gaussian basis sets. *Journal of Computational Chemistry* 29: 225–232.

29 Cramer, C.J. (2013). *Essentials of Computational Chemistry: Theories and Models*. John Wiley & Sons.

30 Jensen, F. (2017). *Introduction to Computational Chemistry*. John wiley & sons.

31 Frank, J. (October 1999). *Introduction to Computational Chemistry*. Editorial Offices.

32 Young, D. (2004). *Computational Chemistry: A Practical Guide for Applying Techniques to Real World Problems*. John Wiley & Sons.

33 Chong, D.P. (1995). *Recent Advances in Density Functional Methods*. World Scientific.

34 Koopmans, T. (1934). Über die Zuordnung von Wellenfunktionen und Eigenwerten zu den einzelnen Elektronen eines Atoms. *Physica* 1: 104–113.

35 Parr, R.G. and Pearson, R.G. (1983). Absolute hardness: companion parameter to absolute electronegativity. *Journal of the American Chemical Society* 105: 7512–7516.

36 Morell, C., Gázquez, J.L., Vela, A. et al. (2014). Revisiting electroaccepting and electrodonating powers: proposals for local electrophilicity and local nucleophilicity descriptors. *Physical Chemistry Chemical Physics* 16: 26832–26842.

37 Lukovits, I., Kalman, E., and Zucchi, F. (2001). Corrosion inhibitors—correlation between electronic structure and efficiency. *Corrosion* 57: 3–8.

38 Lgaz, H., Chung, I.M., Albayati, M.R. et al. (2020). Improved corrosion resistance of mild steel in acidic solution by hydrazone derivatives: an experimental and computational study. *Arabian Journal of Chemistry* 13: 2934–2954.

39 Lgaz, H., Salghi, R., Masroor, S. et al. (2020). Assessing corrosion inhibition characteristics of hydrazone derivatives on mild steel in HCl: Insights from electronic-scale DFT and atomic-scale molecular dynamics. *Journal of Molecular Liquids* 308: 112998.

40 Perdew, J.P., Burke, K., and Ernzerhof, M. (1996). Generalized gradient approximation made simple. *Physical Review Letters* 77: 3865.

41 Meller, J.A. *Molecular Dynamics*, e LS (2001).

42 Hansson, T., Oostenbrink, C., and van Gunsteren, W. (2002). Molecular dynamics simulations. *Current Opinion in Structural Biology* 12: 190–196.

43 Allen, M.P. (2004). Introduction to molecular dynamics simulation. *Computational Soft Matter: From Synthetic Polymers to Proteins* 23: 1–28.

44 Binder, K., Horbach, J., Kob, W. et al. (2004). Molecular dynamics simulations. *Journal of Physics: Condensed Matter* 16: S429.

45 Rapaport, D.C. (2004). *The Art of Molecular Dynamics Simulation*. Cambridge university press.

46 Mathews, D.H., Sabina, J., Zuker, M., and Turner, D.H. (1999). Expanded sequence dependence of thermodynamic parameters improves prediction of RNA secondary structure. *Journal of molecular biology* 288: 911–940.

47 Zuker, M. (2000). Calculating nucleic acid secondary structure. *Current opinion in structural biology* 10: 303–310.

48 Berendsen, H.J. (1999). Molecular dynamics simulations: The limits and beyond. In: *Computational Molecular Dynamics: Challenges, Methods, Ideas* (eds. M. Griebel, D.E. Keyes, R.M. Nieminen, et al.), 3–36. Springer.

49 Jensen, B. (2016). *Investigation into the Impact of Solid Surfaces in Aqueous Systems*. University of Bergen.

50 Obot, I., Haruna, K., and Saleh, T. (2019). Atomistic simulation: a unique and powerful computational tool for corrosion inhibition research. *Arabian Journal for Science and Engineering* 44: 1–32.

51 Casewit, C.J., Colwell, K.S., and Rappe, A.K. (1992). Application of a universal force field to main group compounds. *Journal of the American Chemical Society* 114: 10046–10053.

52 Rappe, A.K., Casewit, C.J., Colwell, K.S. et al. (1992). UFF, a full periodic table force field for molecular mechanics and molecular dynamics simulations. *Journal of the American Chemical Society* 114: 10024–10035.

53 Hagler, A.T., Huler, E., and Lifson, S. (1974). Energy functions for peptides and proteins. I. Derivation of a consistent force field including the hydrogen bond from amide crystals. *Journal of the American Chemical Society* 96: 5319–5327.

54 Hagler, A.T. and Lifson, S. (1974). Energy functions for peptides and proteins. II. Amide hydrogen bond and calculation of amide crystal properties. *Journal of the American Chemical Society* 96: 5327–5335.

55 Lifson, S., Hagler, A.T., and Dauber, P. (1979). Consistent force field studies of intermolecular forces in hydrogen-bonded crystals. 1. Carboxylic acids, amides, and the C:O.cntdot..cntdot..cntdot.H- hydrogen bonds. *Journal of the American Chemical Society* 101: 5111–5121.

56 Brooks, C.L. (1989). Computer simulation of liquids. *Journal of Solution Chemistry* 18: 99–99.

57 Bahlakeh, G., Ramezanzadeh, M., and Ramezanzadeh, B. (2017). Experimental and theoretical studies of the synergistic inhibition effects between the plant leaves extract (PLE) and zinc salt (ZS) in corrosion control of carbon steel in chloride solution. *Journal of Molecular Liquids* 248: 854–870.

58 Chafai, N., Chafaa, S., Benbouguerra, K. et al. (2017). Synthesis, characterization and the inhibition activity of a new alpha-aminophosphonic derivative on the corrosion of XC48 carbon steel in 0.5 M H2SO4: Experimental and theoretical studies. *Journal of the Taiwan Institute of Chemical Engineers* 70: 331–344.

59 Deyab, M.A., Osman, M.M., Elkholy, A.E., and Heakal, F.E. (2017). Green approach towards corrosion inhibition of carbon steel in produced oilfield water using lemongrass extract. *RSC Advances* 7: 45241–45251.

60 Fouda, A.S., Ismail, M.A., Abousalem, A.S., and Elewady, G.Y. (2017). Experimental and theoretical studies on corrosion inhibition of 4-amidinophenyl-2,2 '-bifuran and its analogues in acidic media. *RSC Advances* 7: 46414–46430.

61 Heakal, F.E., Attia, S.K., Rizk, S.A. et al. (2017). Synthesis, characterization and computational chemical study of novel pyrazole derivatives as anticorrosion and antiscalant agents. *Journal of Molecular Structure* 1147: 714–724.

62 Lgaz, H., Salghi, R., Jodeh, S., and Hammouti, B. (2017). Effect of clozapine on inhibition of mild steel corrosion in 1.0 M HCl medium. *Journal of Molecular Liquids* 225: 271–280.

63 Meng, Y., Ning, W.B., Xu, B. et al. (2017). Inhibition of mild steel corrosion in hydrochloric acid using two novel pyridine Schiff base derivatives: a comparative study of experimental and theoretical results. *RSC Advances* 7: 43014–43029.

64 Roy, P., Saha, S.K., Banerjee, P. et al. (2017). Experimental and theoretical investigation towards anti-corrosive property of glutamic acid and poly-gamma-glutamic acid for mild steel in 1 M HCl: intramolecular synergism due to copolymerization. *Research on Chemical Intermediates* 43: 4423–4444.

65 Sanaei, Z., Bahlakeh, G., and Ramezanzadeh, B. (2017). Active corrosion protection of mild steel by an epoxy ester coating reinforced with hybrid organic/inorganic green inhibitive pigment. *Journal of Alloys and Compounds* 728: 1289–1304.

66 Srivastava, V., Haque, J., Verma, C. et al. (2017). Amino acid based imidazolium zwitterions as novel and green corrosion inhibitors for mild steel: Experimental, DFT and MD studies. *Journal of Molecular Liquids* 244: 340–352.

67 Zhang, C. and Zhao, J.M. (2017). Synergistic inhibition effects of octadecylamine and tetradecyl trimethyl ammonium bromide on carbon steel corrosion in the H2S and CO2 brine solution. *Corrosion Science* 126: 247–254.

68 Quraishi, M.A., Chauhan, D.S., and Saji, V.S. (2020). 3 - Computational methods of inhibitor evaluation. In: *Heterocyclic Organic Corrosion Inhibitors* (eds. M.A. Quraishi, D.S. Chauhan and V.S. Saji), 59–86. Elsevier.

69 Verma, C., Lgaz, H., Verma, D.K. et al. (2018). Molecular dynamics and Monte Carlo simulations as powerful tools for study of interfacial adsorption behavior of corrosion inhibitors in aqueous phase: a review. *Journal of Molecular Liquids* 260: 99–120.

70 Tan, B., Zhang, S., Qiang, Y. et al. (2020). Experimental and theoretical studies on the inhibition properties of three diphenyl disulfide derivatives on copper corrosion in acid medium. *Journal of Molecular Liquids* 298: 111975.

71 Guo, L., Obot, I.B., Zheng, X. et al. (2017). Theoretical insight into an empirical rule about organic corrosion inhibitors containing nitrogen, oxygen, and sulfur atoms. *Applied Surface Science* 406: 301–306.

72 Saha, S.K., Dutta, A., Ghosh, P. et al. (2016). Novel Schiff-base molecules as efficient corrosion inhibitors for mild steel surface in 1 M HCl medium: experimental and theoretical approach. *Physical Chemistry Chemical Physics* 18: 17898–17911.

73 Lgaz, H., Bhat, K.S., Salghi, R. et al. (2017). Insights into corrosion inhibition behavior of three chalcone derivatives for mild steel in hydrochloric acid solution. *Journal of Molecular Liquids* 238: 71–83.

74 Lgaz, H., Salghi, R., Bhat, K.S. et al. (2017). Correlated experimental and theoretical study on inhibition behavior of novel quinoline derivatives for the corrosion of mild steel in hydrochloric acid solution. *Journal of Molecular Liquids* 244: 154–168.

75 Chugh, B., Singh, A.K., Thakur, S. et al. (2019). An Exploration about the Interaction of Mild Steel with Hydrochloric Acid in the Presence of N-(Benzo d thiazole-2-yl)-1-phenylethan-1-imines. *Journal of Physical Chemistry C* 123: 22897–22917.

76 Lgaz, H., Chung, I.M., Salghi, R. et al. (2019). On the understanding of the adsorption of Fenugreek gum on mild steel in an acidic medium: Insights from experimental and computational studies. *Applied Surface Science* 463: 647–658.

77 Lgaz, H., Salghi, R., Masroor, S. et al. (2020). Assessing corrosion inhibition characteristics of hydrazone derivatives on mild steel in HCl: Insights from electronic-scale DFT and atomic-scale molecular dynamics. *Journal of Molecular Liquids* 308: 112998.

78 Khaled, K.F. (2008). Molecular simulation, quantum chemical calculations and electrochemical studies for inhibition of mild steel by triazoles. *Electrochimica Acta* 53: 3484–3492.

79 Khaled, K.F. (2010). Experimental, density function theory calculations and molecular dynamics simulations to investigate the adsorption of some thiourea derivatives on iron surface in nitric acid solutions. *Applied Surface Science* 256: 6753–6763.

80 Khaled, K.F. (2010). Studies of iron corrosion inhibition using chemical, electrochemical and computer simulation techniques. *Electrochimica Acta* 55: 6523–6532.

81 Khaled, K.F. (2010). Electrochemical investigation and modeling of corrosion inhibition of aluminum in molar nitric acid using some sulphur-containing amines. *Corrosion Science* 52: 2905–2916.

4

Organic and Inorganic Corrosion Inhibitors

A Comparison

Goncagül Serdaroğlu[1] *and Savaş Kaya*[2]

[1] *Faculty of Education, Department of Mathematics and Science Education, Sivas Cumhuriyet University, Sivas, Turkey*
[2] *Health Services Vocational School, Department of Pharmacy, Sivas Cumhuriyet University, Sivas, Turkey*

4.1 Introduction

Corrosion is one of the important challenges in the contemporary world, especially, after the introduction of technology into our lives following the industrial revolution, though it has been known and some techniques used to prevent it since the ancient Greek period. Furthermore, following the rapid development of informatics, aviation, and aerospace technology for the last 70 years, it is very important to design the materials used according to suitable environmental conditions, as well as to protect the materials used and to extend their durability periods. In the civilizing world, it is not enough to determine the most effective method, condition, or anticorrosion materials in order to prevent or delay corrosion, but also it is getting important that they are less harmful to the environment, with lower toxicity, and can be obtained from renewable resources, instead of existing natural resources. In this context, organic corrosion inhibitor material design has been very promising in many production areas besides the classical inorganic corrosion inhibitors.

Metal corrosion is known as metal erosion because of chemical or mostly electrochemical reactions when a metal comes into contact with surrounding materials. It is also known that the natural corrosion process is generally an electrochemical phenomenon and in this type of corrosion, it is known that the oxidation process is facilitated by the presence of an electron acceptor known as a suitable depolarizer according to the equation $M \rightarrow M^{n+} + ne^{-}$. In addition,

Organic Corrosion Inhibitors: Synthesis, Characterization, Mechanism, and Applications,
First Edition. Edited by Chandrabhan Verma, Chaudhery Mustansar Hussain, and Eno E. Ebenso.

although there are many types of metal corrosion and classification of them [1], the main types of metal corrosion [2] can be given as follows:

1) Uniform (general) corrosion
2) Galvanic (two-metal) corrosion
3) Thermogalvanic corrosion
4) Crevice corrosion (including deposit corrosion)
5) Pitting corrosion
6) Selective attack, selective leaching (de-alloying)
7) Intergranular corrosion (including exfoliation)
8) Erosion corrosion
9) Cavitation corrosion
10) Fretting corrosion
11) Stress corrosion cracking
12) Corrosion fatigue

In the contemporary world, considering the environmental and economic damages caused by corrosion, it is very important to determine the source of corrosion and to take measures to prevent or at least slow it down: the main ways to slow down and/or preventing corrosion can be summarized as follows.

1) Metal type
2) Protective coating
3) Environmental measures
4) Sacrificial coatings
5) Corrosion inhibitors
6) Design modification

Among these protecting ways, the material selection and design is the simplest way to control the corrosion but in cases where there are not always many options for material selection and micro design, different methods are known to be applied to prevent or slow down corrosion. However, in the liquid phase and atmosphere, the corrosion inhibitors can be often enforced to improve the heat-exchange efficiencies to reduce the corrosion. In this case, the electrochemical methods – anodic, cathodic, or both – have provide a wide profit in terms of the economics and saving the natural sources [3]. Material scientists have commonly exploring the usefulness of the alloys in addition to heat treatment regimes and protective coatings to reduce or prevent the corrosion [3].

Besides, the surface treatments, which are reactive, applied, and biofilm coatings in addition to the anodization, are used to slow down and protect the corrosion. For instance, the use of painting or enamel protects the material from corrosion by creating a corrosion-resistant barrier between the damaging environment and the structural material. In this case, the use of chemicals that form an electrically insulating or chemically impervious coating on exposed metal surfaces to suppress electrochemical reactions make the system less susceptible to scrapes or imperfections in the coating because extra inhibitors can be present wherever the metal is exposed. Nowadays, organic and inorganic corrosion inhibitors, whether natural or artificial, are increasingly important. Because of the great importance of them, in this section, the organic and inorganic corrosion inhibitors are presented considering the molecular structures, main characteristics, and environmental effects.

4.2 Corrosion Inhibitors

Corrosion inhibitors are known to be used in countless fields of both production and commercial areas such as pipelines, cooling systems, refinery units, water treatment, painting, oil refinery units, and so on [4]. A basic classification of corrosion inhibitors is given in Figure 4.1.

4.2.1 Organic Corrosion Inhibitors

Organic corrosion inhibitors reduce and protect metal or alloy metal dissolution in aggressive environment by forming a thin film layer with adsorption process on the target surface. The adsorption of the compounds on the target surface can be physical, chemical, or both ways based on the type and properties of both the organic compound and target surface in addition to the environmental conditions of the corrosion happened in [5, 6]. The kinetic and/or thermodynamic searches on adsorption mechanism of the corrosion inhibition of the organic compounds

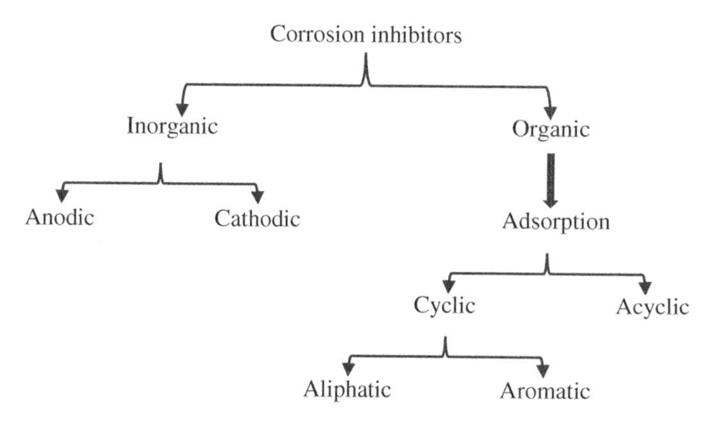

Figure 4.1 A basic classification of the corrosion inhibitors.

have showed that the performance of the corrosion inhibitor have been mostly related to the surface coverage of the compounds depending on the structural properties [7]. It is also well known that heterocyclic compounds containing N, S, O, and P atoms, as well as compounds containing π-bonds and/or polar group that will allow electron delocalization, they provide protection of metals or alloys from corrosion. In the literature, it has been reported in the experimental studies that heterocyclic compounds such as azoles, indoles, and aromatic rings, as well as open-chain organic-based compounds such as epoxy and polymeric systems [8–10], can be successfully used against metal and/or alloy corrosion. In this context, it is important to address the major groups of organic compounds that provide corrosion protection and/or retarding properties.

4.2.1.1 Azoles

Azoles are a class of five-membered aromatic heterocyclic compounds containing at least one nitrogen atom and have an aromatic structure. They are named according to the number of nitrogen atoms in the aromatic ring and their position. In addition, there are other heteroatom-containing azo compounds [11–13] such as O and S in their structure, and the main structures are shown in the Figure 4.2. The aromatic

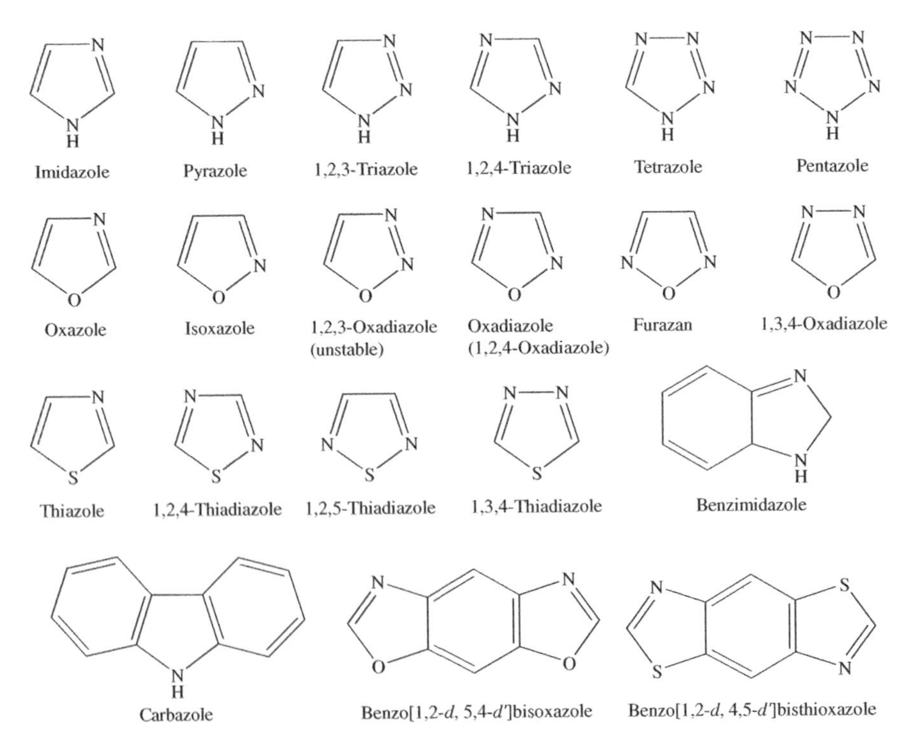

Figure 4.2 The chemical structures of the main azole compounds.

structures of azo compounds and the heteroatoms they contain cause an increase in polarity of these compounds due to electron delocalization in compounds, which make these compounds very useful materials for inhibition of corrosion. Srivastava et al. [14] have recently reported the efficiency of a benzo[d]imidazole derivative green corrosion inhibitor has a very good inhibition capability (>98%) on the carbon steel at optimum dosage conditions and continued the strong inhibition adsorption (>95%) even at 333 K. The inhibition ability of pyridine thiazole compound on copper corrosion in acidic medium has been investigated by a series of experiments techniques (EIS, SEM, AFM), and shown by the EIS technique that the maximum inhibition capacity is reached at 94% at 1 mM concentration [15]. Recently, the inhibition capabilities of a series of imidazole derivatives on the corrosion of mild steel in acidic conditions have been investigated by electrochemical, thermodynamic, and also computational techniques; one of the most important results of this investigation is that the imidazole derivatives provide the anodic protection of mild steel and promote cathodic hydrogen reactions as well [16]. Besides, phenanthroimidazole derivatives [17] and bis-benzothiazoles derivatives [18] have a very high inhibitory potency for mild steel and showed a mixed type inhibitor capacity by TAFEL diagrams, but the priority for the cathode.

4.2.1.2 Azepines

Azepines with seven-membered as a subgroup of heterocyclic organic compounds and their effects are generally well known in medicinal fields [19] (Figure 4.3). Potential use of waste drugs as corrosion inhibitors has received increasing attention in recent years. In 2009, Arslan et al. suggested that drug molecules with quantum chemical methods can show inhibitory properties for mild steel in an acidic environment [20]. It has also been reported that different groups of heterocyclic compounds used in many medical fields are promising as environmentally friendly corrosion inhibitors [21]. For a series of triazoloazepine derivatives, the corrosion and protection constants on steel 45 with a concentration of 1 M have been evaluated and suggested that Cl atom substitutions make the main structure more potent against corrosion among all substituents (H, F, Cl, I, CH_3O, and CH_3 on different position of the molecule) [22]. In addition, the inhibitory characteristics of acetonitrile and secondary amines containing triazoloazepine ring for carbon steel corrosion in HCl and hydrogen sulfate have been investigated and reported that the reciprocal effects of triazoloazepine and p-tolyl groups in the secondary amines provide a maximum synergetic effect during inhibition of steel corrosion in the HCl [23]. In the past, the corrosion inhibition properties of the carbamazepine unused drug on steel in different solvent environments have been examined, and the inhibition efficiency of the carbamazepine has been determined as 85% [24]. Besides, carbamazepine has effective role in both anodic processes and cathodic hydrogen evaluation, and thus the amount of carbamazepine does not affect the corrosion potency [24].

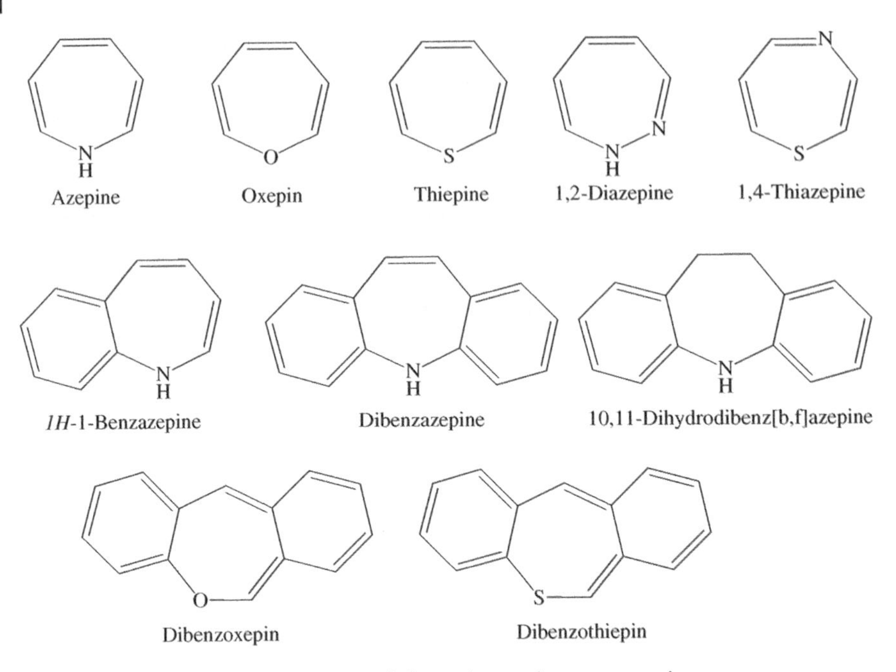

Figure 4.3 The chemical structures of the main azepine compounds.

4.2.1.3 Pyridine and Azines

Pyridine is a heterocyclic organic compound shown by the simple structural formula C_5H_5N and is widely used in many production areas due to its aromatic structure [25] (Figure 4.4). The lone pair of the nitrogen atom in the structure has sp^2 hybridization but does not contribute to the aromaticity of the compound as it lies outside the plane of the ring like sigma bonds and facilitates bond formation with an external system through an electrophilic attack. On the other hand, Azines are an organic compound group with the functional group RR′C = N-N = CRR′, obtained by the condensation of hydrazine with ketones and aldehydes [26]. In the past, many papers have been reported on pyridine and its derivatives and suggested that they have sufficient potency for inhibition of iron, aluminum, carbon steel, N80 steel, mild steel, and zinc in different acidic media [27–34]. Furthermore, Kurmakova and coworkers [35], with empirical and theoretical support, have evaluated the effects of quaternary pyridinium, imidazopyridinium, and imidazoazepinium salts for the inhibition of biocorrosion of steel; they have suggested that the observed inhibition capability of the salts have been largely proportional to both the polarizability and energy gap values of the salts. In addition, a lot of papers also have reported that azine derivatives have been potential corrosion inhibitors for mild steel [36–39], carbon steel [40], iron [41] and

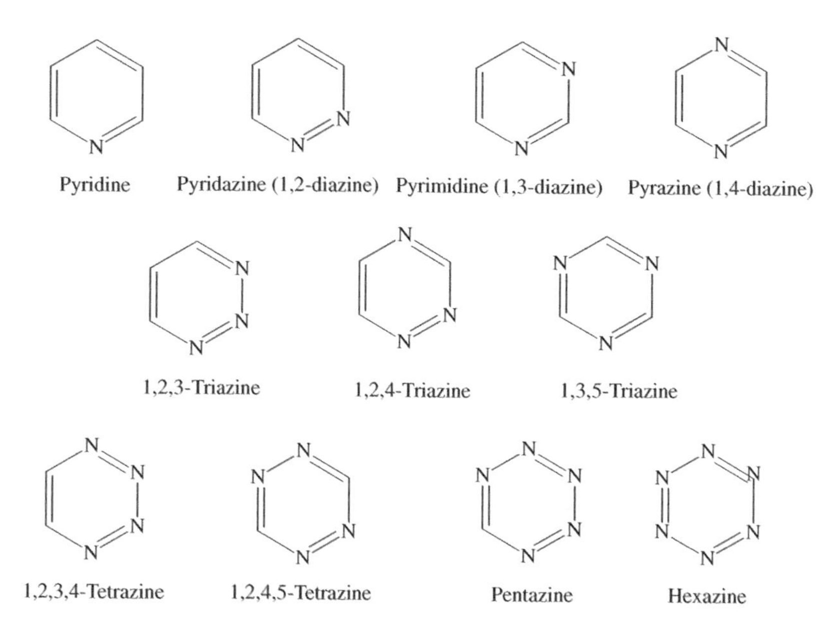

Figure 4.4 The chemical structures of the main pyridine and azine compounds.

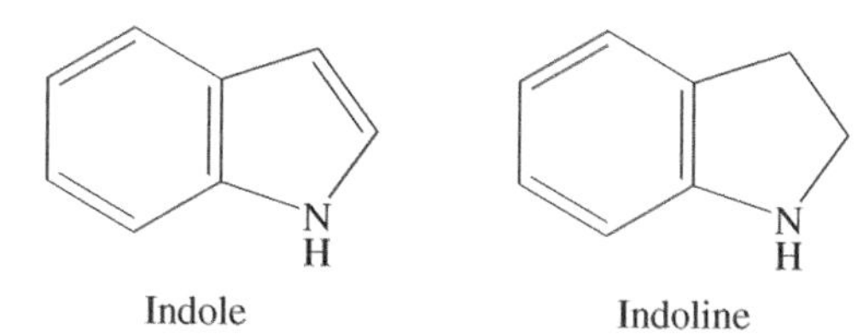

Figure 4.5 The chemical structures of the main indole compounds.

copper [42], an acidic environment. In a recent study, it has been suggested that the newly synthesized polybenzoxazine compound is promising as a highly corrosion-resistant material for mild steel and similar alloys under different conditions, including marine conditions [43].

4.2.1.4 Indoles

Indoles having a formula C8H7N has a bicyclic structure, consisting of a six-membered benzene ring fused to a five-membered pyrrole ring, which provides the usefulness in many production and manufacturing sectors, as well as in the medicinal importance of them [44, 45] (Figure 4.5). In the past, the inhibition capabilities of the indole-based melatonin compound at 30 and 60 °C with a concentration of 10 mM on the steel surface have been determined [46] as 98.3 and 88.6%.In addition, the corrosion efficiencies of 12-(2,3-dioxoindolin-1-yl)-N, N, N-trimethyldodecan-1-ammonium bromide [47] and 1-(2-hydroxyethyl)-2-imidaz olidinone [48] compounds on mild steel have been determined as 95.9% (in 1 M HCl) at 298 K and 85.4% (in 0.5 M HCl) at 293 K, respectively. In addition, the inhibition efficiency of 4-((1H-indol-3-yl) methyl) phenol (IMP) and 4- (di (1H-indol-3-yl) methyl) phenol compounds to protect copper corrosion at 2 mM concentration has been determined [49] to be 99.3 and 97.5%, respectively. In another study, the inhibition potency of new synthesized 3,3-((4-(methylthio)phenyl)methylene)bis(1H-indole) compound on copper corrosion has been examined; it has been suggested to have a good corrosion inhibition capability based on the electrochemical techniques; the adsorption process of the compound on copper surface has been contributed by both physisorption and chemisorption [50].

4.2.1.5 Quinolines

Quinoline is a compound with the formula C9H7N, usually soluble in organic solvents. Although it has little application area with its simple form, its derivatives have a wide range of applications in the literature, which from basic sciences to industrial engineering [51–56] (Figure 4.6). Besides, quinoline derivatives containing highly polarizable groups such as –OH, –OMe, $-NH_2$, and so on in addition to having with both nitrogen atom and π-electronic system, it facilitates interaction with the metal surface, making this group of compounds very important among anticorrosive materials, especially in green corrosion inhibitors [57]. Recently, the anticorrosive behavior or corrosion inhibition potency of "oxoquinolinecarbohydrazide N-phosphonate" [58], "8-hydroxyquinoline" [59–61], "N,N′-((ethane-1,2-diylbis(azanediyl))bis(ethane-2,1-diyl))bis(quinoline-2-carboxamide)" [62], "5-{[(4-dimethylamino-benzylidene)-amino]-methyl}-quinolin-8-ol" [63] derivatives on mild steel in acidic medium has been extensively studied. In a recent work, the inhibition capability of quinoline derivative "5-benzyl-8-propoxyquinoline" [64] on Q235 steel in sulfuric acid medium has been

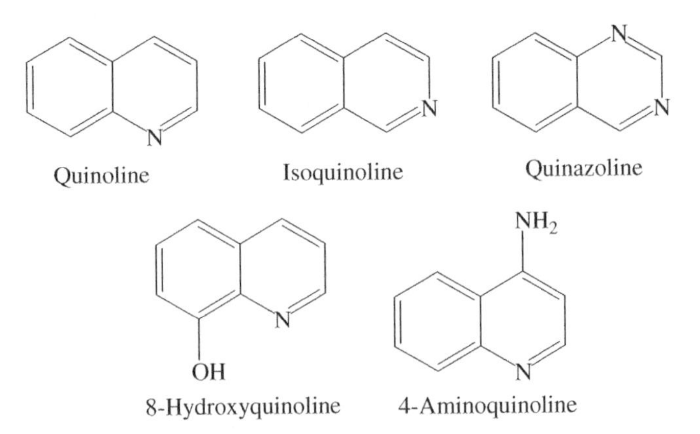

Figure 4.6 The chemical structures of the main quinoline compounds.

determined as 97.7%, and it was shown that the inhibition efficiency has quite high even at high concentrations and temperatures. Furthermore, in another comprehensive study on mild steel in 15% HCl solution, the authors have found that the inhibition potential of quinoline derivative "dibenzylamine-quinoline" [65] has reached its highest value at 363 K with 95.4%, which is quite sufficient to corrosion inhibitor of oil and gas acidification. Mohammadloo and coworkers have investigated the possible usage of "8- hydroxyquinoline" [66] in smart coating applications and have suggested that intelligent corrosion detection can be achieved using 8-HQ as corrosion indicator and inhibitor, based on results obtained from fluorescence microscope.

4.2.1.6 Carboxylic Acid and Biopolymers

As known well, composites and conductive polymers are adsorbed by metal surfaces, suppress the dissolution process, and provide the formation of a protective film for corrosion. In this context, the advantage of polymers and composites over other protective coatings such as paint is that they are environmentally friendly since they do not contain toxic substances [67–70]. In this context, Ateş and Özyılmaz [68] obtained polycarbazole, polycarbazole/nanoclay, and polycarbazole/Zn-nanoparticles film layers by chemical and electrochemical techniques and investigated the corrosion inhibition properties of them on SS304 in saltwater. In their study, they determined that the protection efficiency of the film layers polymerized chemically (PE: 99.81% for PCz, 99.46% for PCz/nanoclay and 99.35%, PCz/Zn-nanoparticle) has been greater than the values obtained by electrochemical methods (PE: 70.68% for PCz, 65.97% for PCz/nanoclay, 66.28%for-PCz/Zn-nanoparticle) [68]. Recently, carbohydrates and derivatives, containing the polar groups such as OH, $-NH_2$, and $-COCH_3$, which facilitate their solubility in electrolyte, have been greatly investigated to use as environmentally friendly corrosion inhibitors for metal and alloys [71]. Vazguez et al. have investigated eight carbohydrates (three commercially obtained and five synthesized) for corrosion inhibition of API 5L X70 steel in acidic medium: the corrosion process is mixed type according to the thermodynamic analysis results and the best inhibition potential has been determined as 87% (for synthesized Methyl-4,6-*O*-Benzylidene-α-D-glucopyranose) at 50 ppm [72]. Also, chitosan as a linear aminopolysaccharide is a copolymer of D-glucosamine and *N*-acetyl-D-glucosamine and is obtained by deacetylation of chitin. Due to the abundance of −OH and −NH2 polar groups in the chitosan structure, they are easy to be adsorbed by the metal surface and are reported to be a good corrosion inhibitor [72–75]. In addition, among green corrosion inhibition research, plant extracts "multi-phytoconstituents from dioscoreaseptemloba" [76] on carbon steel in acidic solution and seven natural polymers for AZ31 Mg-alloy [77] in the saline media have been also explored. Polymeric corrosion inhibitors with recently reported environmentally friendly inhibition properties are given in Figure 4.7.

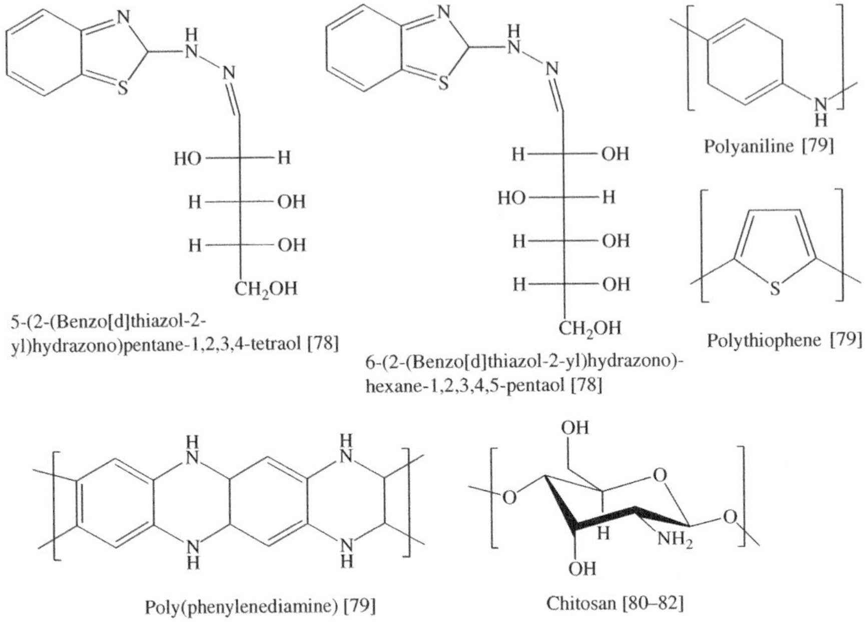

5-(2-(Benzo[d]thiazol-2-yl)hydrazono)pentane-1,2,3,4-tetraol [78]

6-(2-(Benzo[d]thiazol-2-yl)hydrazono)-hexane-1,2,3,4,5-pentaol [78]

Polyaniline [79]

Polythiophene [79]

Poly(phenylenediamine) [79]

Chitosan [80–82]

Figure 4.7 The chemical structures of the green corrosion inhibitors. From Refs. [78–82].

4.2.1.7 Inorganic Corrosion Inhibitors

In addition to organic corrosion inhibitors, inorganic molecules and inorganic salts are also considered as inhibitor in the prevention of corrosion of metal surfaces. Many inorganic complexes, ions, and salts were successfully used against the corrosion of metal surfaces in different corrosive environments [83–85]. Inorganic corrosion inhibitors prevent the corrosion via reaction of anodic and cathodic parts of the system. On the other hand, organic corrosion inhibitors prevent the corrosion process adsorbing on metal surfaces. The most widely used inorganic corrosion inhibitors are the salts of zinc, copper, nickel, arsenic, and additional metals. It should be noted that arsenic compounds are widely considered compared to others. The mentioned arsenic compounds scrape at the cathode cell of metal surfaces when they are mixed in the corrosive medium. It is important to note that the plating decreases the percentage of hydrogen ion interchange. The reason of this situation is the formation of iron sulfide. The reaction between iron sulfide and acid is known as a dynamical process. In the literature, some advantages and disadvantages regarding the using of inorganic corrosion inhibitors are reported. The advantages of them are that they can be used for a long time at high temperatures. Additionally, compared to organic corrosion inhibitors, they are cheaper. As disadvantage, it can be noted that they lose speedily their abilities to connect in the acid solutions that are stronger than 17% hydrochloric acid [86]. Inorganic corrosion inhibitors are classified as anodic and cathodic inhibitors.

4.2.1.8 Anodic Inhibitors

Anodic inhibitors are also known as passivation inhibitors. They cause a reducing anodic reaction. Namely, they support the metal surfaces blocking the anode reaction. In addition, they form a film adsorbed on metal surface. Usually, these inhibitors form the mentioned cohesive and insoluble film reacting with corrosion product initially formed. The corrosion inhibitors and the corrosion potentials of the metals studied affect the anodic reaction [87]. As a result of the reaction with the metal ions (M^{n+}) on anode of corrosion inhibitors, insoluble and impermeable metallic ions hydroxide films occur. If concentrations of inhibitor molecules reach to sufficient height, the cathodic current density becomes higher than the critical anodic current density. Consequently, the metal is passivated. In anodic inhibitors, it is quite important that concentrations of inhibitor molecules should be high in the solution considered. If concentration of inhibitor is low, the film formed cannot cover the entire metal surface. This situation causes a localized corrosion [2]. Nitrates, molybdates, sodium chromates, phosphates, hydroxides, and silicates are the examples of anodic corrosion inhibitors.

4.2.1.9 Cathodic Inhibitors

In the course of corrosion process, the cathodic corrosion inhibitors prevent the occurrence of the cathodic reaction of the metal surfaces. These mentioned inhibitors having some metal ions form insoluble compounds that precipitate in cathodic sites. Here, a compact and adherent film restricting the diffusion of reducible species in these areas settles down on metal surface. The oxygen diffusion and electrons conductive in these areas provide that these inhibitors have a high cathodic inhibition. Magnesium, zinc, and nickel ions can be given as example for cathodic inhibitors because they form the insoluble hydroxides as ($Mg(OH)_2$, $Zn(OH)_2$, $Ni(OH)_2$ reacting with the hydroxide ions of water. The formed insoluble hydroxides are deposited on the cathodic sites of the metal surfaces to protect them. As other examples of cathodic inhibitors, the oxides and salts of antimony, arsenic, and bismuth, which are deposited on the cathode region in acid solutions, can be presented. It is well-known that these inhibitors minimize the release of hydrogen ions [88]. In the current literature, many studies regarding the performances against the corrosion of metal surfaces of inorganic corrosion inhibitors are available [84, 89–91].

References

1 Fontana, M.G. (1986). *Corrosion Engineering*, 3e. Boston, MA: McGraw-Hill.
2 Bardal, E. (2004). *Corrosion and Protection*. London: Springer.
3 Olen, R.R. and Locke, C.E. (1981). *Anodic Protection: Theory and Practice in the Prevention of Corrosion*. New York: Springer.
4 Dariva, C.G. and Galio, A.F. (2014). *Corrosion Inhibitors: Principles, Mechanisms and Applications*. M. Aliofkhazraei, IntechOpen.

5 Onyeachu, I. and Solomon, M.M. (2020). *Journal of Molecular Liquids* 313: 113536.
6 Huang, H. and Bu, F. (2020). *Corrosion Science* 165: 108413.
7 Quraishi, M., Chauhan, D., and Saji, V. (2020). *Heterocyclic Organic Corrosion Inhibitors*. Elsevier.
8 Gladkikh, N., Makarychev, Y., Petrunin, M. et al. (2020). *Progress in Organic Coatings* 138: 105386.
9 Lashgari, S.M., Yari, H., Mahdavian, M. et al. (2021). *Corrosion Science* 178: 109099.
10 Goffin, B., Banthia, N., and Yonemitsu, N. (2020). *Construction and Building Materials* 263: 120162.
11 Bustos-Terrones, V., Serratos, I.N., Vargas, R. et al. (2021). *Materials Science and Engineering B* 263: 114844.
12 Akbarzadeh, S., Ramezanzadeh, M., Ramezanzadeh, B., and Bahlakeh, G. (2020). *Journal of Molecular Liquids* 319: 114312.
13 Ma, T., Tana, B., Xua, Y. et al. (2020). *Colloids and Surfaces A* 599: 124872.
14 Srivastava, V., Salman, M., Chauhan, D.S. et al. (2021). *Journal of Molecular Liquids* https://doi.org/10.1016/j.molliq.2020.115010.
15 Farahati, R., Behzadi, H., Mousavi-Khoshdel, S.M., and Ghaffarinejad, A. (2020). *Journal of Molecular Structure* 1205: 127658.
16 Ouakki, M., Galai, M., Rbaa, M. et al. (2020). *Journal of Molecular Liquids* 319: 114063.
17 Lv, Y.-L., Kong, F.-Y., Zhou, L. et al. (2021). *Journal of Molecular Structure* 1228: 129746.
18 Suhasaria, A., Murmu, M., Satpati, S. et al. (2020). *Journal of Molecular Liquids* 313: 113537.
19 Proctor, G. and Redpath, J. (1996). *MONOCYCLIC AZEPINES: The Syntheses and Chemical Properties of the Monocyclic Azepines*. New York: Wiley.
20 Arslan, T., Kandemirli, F., Ebenso, E.E. et al. (2009). *Corrosion Science* 51: 35–47.
21 Gece, G. (2011). *Corrosion Science* 53: 3873–3898.
22 Demchenko, A.M., Nazarenko, K.G., Makei, A.P., and Prikhod, S.V. (2004). *Russian Journal of Applied Chemistry* 77 (5): 790–793.
23 Bondar, O.S., Prykhodko, S.V., Kurmakova, I.M., and Humenyuk, O.L. (2011). *Materials Science* 47 (3): 90–93.
24 Vaszilcsina, N., Ordodi, V., Borza, A. et al. (2012). *International Journal of Pharmaceutics* 431: 241–244.
25 Krygowski, T.M., Szatyłowicz, H., and Zachara, J.E. (2005). *The Journal of Organic Chemistry* 70: 8859–8865.
26 Chourasiya, S.S., Kathuria, D., Wani, A.A., and Bharatam, P.V. (2019). *Organic & Biomolecular Chemistry* 17: 8486–8521.
27 El-Maksoud, S.A.A. and Fouda, A.S. (2005). *Materials Chemistry and Physics* 93: 84–90.

28 Öğretir, C., Mihci, B., and Bereket, G. (1999). *Journal of Molecular Structure (THEOCHEM)* 488: 223–231.
29 Lashkari, M. and Arshadi, M.R. (2004). *Chemical Physics* 299: 131–137.
30 Ansari, K.R., Quraishi, M.A., and Singh, A. (2015). *Measurement* 76: 136–147.
31 Kosari, A., Moayed, M.H., Davoodi, A. et al. (2014). *Corrosion Science* 78: 138–150.
32 Abdelshafi, N.S. (2020). *Protection of Metals and Physical Chemistry of Surfaces* 56 (5): 1066–1080.
33 Berisha, A. (2020). *Electrochemistry* 1: 188–199.
34 Javadian, S., Ahmadpour, Z., and Yousef, A. (2020). *Progress in Organic Coatings* 147: 105678.
35 Kurmakova, I.M., Bondar, O.S., and Demchenko, N.R. (2016). *Materials Science* 51 (5): 17–23.
36 Ebenso, E.E., Kabanda, M.M., Murulana, L.C. et al. (2012). *Industrial and Engineering Chemistry Research* 51: 12940–12958.
37 Kagatikar, S., Sunil, D., Kumari, P., and Shetty, P. (2020). *Journal of Bio- and Tribo-Corrosion* **6**: 136.
38 Alya, K.I., Mohamed, M.G., Younisc, O. et al. (2020). *Progress in Organic Coatings* 138: 105385.
39 Bouklah, M., Attayibat, A., Kertit, S. et al. (2005). *Applied Surface Science* 242: 399–406.
40 Khamaysa, O.M.A., Selatnia, I., Zeghache, H. et al. (2020). *Journal of Molecular Liquids* 315: 113805.
41 Chetouani, A., Aouniti, A., Hammouti, B. et al. (2003). *Corrosion Science* 45: 1675–1684.
42 Abderrahim, K., Selatnia, I., Sid, A., and Mosset, P. (2018). *Chemical Physics Letters* 707: 117–128.
43 Manoj, M., Kumaravel, A., Mangalam, R. et al. (2020). *Journal of Coating Technology and Research* 17 (4): 921–935.
44 Sundberg, R.J. (1996). *Indoles*. London: Academic Press Limited.
45 Neto, J.S.S. and Zeni, G. (2020). *Organic & Biomolecular Chemistry* 18: 4906–4915.
46 Ituen, E., Mkpenie, V., Moses, E., and Obot, I. (2019). *Bioelectrochemistry* 129: 42–53.
47 Abdellaouia, O., Skallia, M.K., Haoudia, A. et al. (2021). *Chem* 9 (1): 044–056.
48 Keleşoglu, A., Yıldız, R., and Dehri, I. (2019). *Journal of Adhesion Science and Technology* 33 (18): 2010–2030.
49 Feng, L., Zhang, S., Feng, Y. et al. (2020). *Chemical Engineering Journal* 394: 124909.
50 Feng, Y., Feng, L., Suna, Y. et al. (2020). *Crystal Research and Technology* 9 (1): 584–593.
51 Nainwal, L.M., Tasneem, S., Akhtar, W. et al. (2019). *European Journal of Medicinal Chemistry* 164: 121–170.

52 Zhang, J., Zheng, C., Zhang, M. et al. (2020). *Nano Research* 13 (11): 3082–3087.
53 Li, X., Liu, Y., Chen, X. et al. (2020). *Green Chemistry* https://doi.org/10.1039/C9Gc4445K.
54 Zou, L.-H., Zhu, H., Zhu, S. et al. (2019). *The Journal of Organic Chemistry* 84 (19): 12301–12313.
55 Su, T., Zhu, J., Sun, R. et al. (2019). *European Journal of Medicinal Chemistry* 178: 154–167.
56 Qin, Q.-P., Wang, Z.-F., Huang, X.-L. et al. (2019). *European Journal of Medicinal Chemistry* 184: 111751.
57 Verma, C., Quraishi, M.A., and Ebenso, E.E. (2020). *Surfaces and Interfaces* 21: 100634.
58 Fernandes, C.M., Faro, L.V., Pina, V.G.S.S. et al. (2020). *Surfaces and Interfaces* 21: 100773.
59 Douche, D., Elmsellem, H., Anouar, E.H. et al. (2020). *Journal of Molecular Liquids* 308: 113042.
60 Rbaaa, M., Ouakkib, M., Galaic, M. et al. (2020). *Colloids and Surfaces A* 602: 125094.
61 Erazua, E.A. and Radeleke, B.B. (2019). *Journal of Applied Sciences and Environmental Management* 23 (10): 1819–1824.
62 Hassan, N., Ramadanb, A.M., Khalil, S. et al. (2020). *Colloids and Surfaces A* 607: 125454.
63 Fakhry, H., El Faydy, M., Benhiba, F. et al. (2021). *Colloids and Surfaces A: Physicochemical and Engineering Aspects* 610: 125746. https://doi.org/10.1016/j.colsurfa.2020.125746.
64 Chen, S., Chen, S., and Li, W. (2019). *International Journal of Electrochemical Science* 14: 11419–11428.
65 Li, Y., Wang, D., and Zhang, L. (2019). *RSC Advances* **9**: 26464–26475.
66 Mohammadloo, H.E., Mirabedini, S.M., and Pezeshk-Fallah, H. (2019). *Progress in Organic Coatings* 137: 105339.
67 Ates, M. (2016). *Journal of Adhesion Science and Technology* 30 (14): 1510–1536.
68 Ates, M. and Özyılmaz, A.T. (2015). *Progress in Organic Coatings* 84: 50–58.
69 Khatoon, H., Iqbal, S., and Ahmad, S. (2019). *New Journal of Chemistry* 43: 10278–10290.
70 Bekkar, F., Bettahar, F., Moreno, I. et al. (2020). *Polymers* 12: 2227.
71 Rbaa, M., Fardioui, M., Verma, C. et al. (2020). *International Journal of Biological Macromolecules* 155: 645–655.
72 Vázquez, A.E., Robles, M.A.C., Silva, G.E.N. et al. (2019). *International Journal of Electrochemical Science* 14: 9206–9220.
73 Mouaden, K.E.L., Chauhan, D.S., Quraishi, M.A., and Bazzi, L. (2020). *Sustainable Chemistry and Pharmacy* 15: 100213.

74 Macedoa, R.G.M.A., Marquesa, N.N., Tonholob, J., and Balaban, R.C. (2019). *Carbohydrate Polymers* 205: 371–376.

75 Farhadian, A., Varfolomeev, M.A., Shaabani, A. et al. (2020). *Carbohydrate Polymers* 236: 116035.

76 Emoria, W., Zhang, R.-H., Okafor, P.C. et al. (2020). *Colloids and Surfaces A* 590: 124534.

77 Umoren, S.A., Solomon, M.M., Madhankumar, A., and Obot, I.B. (2020). *Carbohydrate Polymers* 230: 115466.

78 Shaw, P., Obot, I.B., and Yadav, M. (2019). *Materials Chemistry Frontiers* **3**: 931–940.

79 Umoren, S.A. and Solomon, M.M. (2019). *Progress in Materials Science* 104: 380–450.

80 Ashassi-Sorkhabi, H. and Kazempou, A. (2020). *Carbohydrate Polymers* 237: 116110.

81 Chauhan, D.S., Quraishi, M.A., Sorour, A.A. et al. (2019). *RSC Advances* 9: 14990–15003.

82 Ansari, K.R., Chauhan, D.S., Quraishi, M.A. et al. (2020). *International Journal of Biological Macromolecules* 144: 305–315.

83 Sayin, K. and Karakaş, D. (2013). *Corrosion Science* 77: 37–45.

84 Saei, E., Ramezanzadeh, B., Amini, R., and Kalajahi, M.S. (2017). *Corrosion Science* 127: 186–200.

85 Ahmed, A.S., Ghanem, W.A., and Hussein, G.A.G. (2020). *Archives of Metallurgy and Materials*: 639–651.

86 Tamalmani, K. and Husin, H. (2020). *Applied Sciences* 10 (10): 3389.

87 Gentil, V. (2003). *Corrosão*, 4ªe. Rio de Janeiro: LTC.

88 Dariva, C.G. and Galio, A.F. (2014). Corrosion inhibitors–principles, mechanisms and applications. *Developments in corrosion protection* 16: 365–378.

89 Ramezanzadeh, M., Bahlakeh, G., and Ramezanzadeh, B. (2019). *Journal of Molecular Liquids* 290: 111212.

90 Uhlig, H.H. and King, P.F. (1959). *Journal of the Electrochemical Society* 106 (1): 1.

91 Nezhad, A.N., Davoodi, A., Zahrani, E.M., and Arefinia, R. (2020). *Surface and Coatings Technology*: 125946.

Part II

Heterocyclic and Non-Heterocyclic Corrosion Inhibitors

5

Amines as Corrosion Inhibitors

A Review

Chandrabhan Verma[1], *M. A. Quraishi*[1], *Eno E. Ebenso*[2], *and Chaudhery Mustansar Hussain*[3]

[1] *Interdisciplinary Research Center for Advanced Materials, King Fahd University of Petroleum and Minerals, Dhahran, Saudi Arabia*
[2] *Institute of Nanotechnology and Water Sustainability, College of Science, Engineering and Technology, University of South Africa, Johannesburg, South Africa*
[3] *Department of Chemistry and Environmental Science, New Jersey Institute of Technology, Newark, NJ, USA*

5.1 Introduction

5.1.1 Corrosion: Basics and Its Inhibition

Corrosion is a universal problem that leads to breakdown of metallic materials by their reaction with the components of environment. Corrosion results into huge safety and monitory losses as estimated annual global cost of corrosion is about US$ 2.5 trillion that equivalent to 3.5–5% of GDPs of different counties [1, 2]. Several methods of corrosion mitigation are developed and adopted by the corrosion scientists and engineers depending upon the nature of environment. Among the implemented methods, use of corrosion inhibitors is one of the best techniques of corrosion monitoring [3–5]. Organic compounds, especially heterocyclic compounds, inhibit corrosion by adsorbing and forming a protective film. Heteroatoms directly participate in the adsorption process as the heteroatoms with unshared electron pair form coordination bonding with the d-orbital of surface metallic atoms. Therefore, it is expected that heteroatom with lower electronegativity forms stronger and effective coordination bonding with the metallic surface compared to the heteroatom with higher electronegativity. The inhibition protectiveness of the most common heteroatoms followed the order: P > S > N > O. Adsorption organic compounds containing heteroatoms may follow either physisorption or

Organic Corrosion Inhibitors: Synthesis, Characterization, Mechanism, and Applications,
First Edition. Edited by Chandrabhan Verma, Chaudhery Mustansar Hussain, and Eno E. Ebenso.

chemisorption or a combination of both. Most of the organic compounds adsorb on metallic surface using their physiochemisorption mechanism of adsorption [6, 7].

Adsorption and corrosion inhibition effect of the organic compounds depends upon numerous factors. One of the most significant factors is temperature. Generally, increase in temperature results into corresponding increase in corrosion rate, i.e. decrease in protection effectiveness [8, 9]. Increase in temperature causes subsequent increase in the kinetic energy of inhibitor molecules that decrease their attraction force or adsorption with/on metallic surface. By considering metal–inhibitor interactions as a chemical reaction between two species, it can be generalized that every 10 °C rise in temperature increase corrosion rate by two times. In highly acidic and basic electrolytic media, increase in temperature may result into acid- or base-catalyzed rearrangement or fragmentation that can also adversely affect the inhibition effect of organic compounds [10]. Another important parameter is inhibitor concentration. Inhibition efficiency of compounds increases with increasing their concentration as the effective surface coverage increases in the same sequence. However, after certain concentration, when inhibitor molecules occupy all the active sites, further increase in inhibitor concentration did not cause any appreciable change in the inhibition performance of the inhibitors.

Molecular electronic structure and molecular size effects are other important factors that determine the corrosion inhibition effectiveness of the organic compounds [11, 12]. Obviously, a compound with high molecular size is expected to behave as superior corrosion inhibitor compared to the compound having lower molecular size. Generally, organic compounds with high molecular size provide superior surface coverage and thereby behave as effective corrosion inhibitors. However, too big molecular sizes can result into the decrease in protection efficiency as solubility of compounds decrease with rise in molecular size. Molecular electron effect is another very important aspect of designing of potential corrosion inhibitors. Generally, an organic compound with planar geometry behaves as better corrosion inhibitors than that of the compounds with vertical geometry. Organic compounds having aromatic rings and extensive conjugation in the form of polar functional groups behave as effective corrosion inhibitors. However, compared to the electron–withdrawing substituents, electron releasing substituents showed better protection effectiveness. Therefore, it is recommended that an effective corrosion inhibitor must have planar geometry and aromatic/heteroaromatic ring(s) with extensive conjugation in the form of polar electron releasing functional groups and multiple bonds.

5.1.2 Amines as Corrosion Inhibitors

Aliphatic and aromatic amines act as effective corrosion inhibitors for different metals and alloys in several electrolytes. Amine-based compounds can be classified into different classes based on their chemical nature and presence of the functional groups. The compounds containing $-NH_2$ group can be regarded as amino compounds, and its derivatives can be treated as amines. Some of the common amine derivatives are 1^0-, 2^0- and 3^0-amines, aniline and its derivatives, Schiff's bases (SBs), and sulfonamides, etc. Amino acids (AAs) and its derivatives containing $-NH_2$ group are also widely used corrosion inhibitors. Urea, thiourea, and their derivatives are also treated as corrosion inhibitors [2, 13–15]. A brief description of these corrosion inhibitors is given below.

5.1.2.1 1^0-, 2^0-, and 3^0-Aliphatic Amines as Corrosion Inhibitors

1^0-, 2^0- and 3^0-aliphatic amines are extensively used as corrosion inhibitors for different metals and alloys in various electrolytes. Generally, this type of compounds become effective by adsorbing on the metallic surface in which polar amine group acts as adsorption center and remaining part(s) of the molecules behave as hydrophobic. Adsorption of these compounds results into the formation of hydrophobic film that avoids the aggressive attack of acidic solution. Obviously, increase in the carbon chain length (hydrophobicity) causes subsequent increase in their inhibition performance. However, increase in the hydrophobicity can result into subsequent decrease in their solubility, especially in the polar electrolytes. Therefore, for an amine compound to be effective corrosion inhibitor, it is essential that it must have a proper combination of hydrophobicity and hydrophilicity. Damborenea and coworkers [16] described the corrosion inhibition effect of four aliphatic amines differing in the nature of aliphatic/hydrophobic hydrocarbon chains for mild steel in 2M HCl solution. Gravimetric and polarization techniques were used to demonstrate the corrosion inhibition effectiveness of the tested amine derivatives. Using both methods, it was derived that inhibition efficiency of the amine derivatives increases on increasing the number of carbon atoms, i.e. hydrophobicity. Inhibition effectiveness of the tested compounds followed the order: $-C_{12}H_{25} > -C_{10}H_{21} > -C_8H_{17} > -C_6H_{13}$. All investigated amine derivatives become effective by adsorbing on the metallic surface that followed the Frumkin adsorption isotherm model. Increase in the inhibition efficiency on increasing the number of carbon atom was further reported [17]. This study demonstrates the corrosion inhibition effect of four diamine compounds for copper corrosion in 0.1 M H_2SO_4 medium. Inhibition efficiency of the investigated compounds followed the order: $-(CH_2)_8- > -CH_2)_6- > -(CH_2)_2- > -(CH_2)_0-$.

In another study, while studying the corrosion inhibition effect of the diamine compounds, Herrag et al. [18] showed that compound containing $-(CH_2)_3-$ moiety showed highest protection efficiency followed by the compound containing $-(CH_2)_2-$ moiety and finally the compound having $-(CH_2)_6-$. In this case, it is expected that increase in the number of methylene (CH_2) substituent beyond three carbons causes significant increase in the hydrophobic character that comes out in the form of low protection efficiency. Effect of substituents and hydrophobic chain lengths are widely investigated. A summary of some major reports is described in Table 5.1. It is important to mention that presence of polar substituent(s) increases the inhibition efficiency by behaving as adsorption center

Table 5.1 Effect of substituents on the corrosion inhibition efficiency of some common aliphatic and aromatic amine derivatives.

S No	Basic structure of inhibitor	Metal/electrolyte	References
1	R-NH_2 R= $-C_6H_{13}$, $-C_8H_{17}$, $-C_{10}H_{21}$, $-C_{12}H_{25}$	Fe/ 2 M HCl	[16]
2	$H_2N—(CH_2)_n—NH_2$ $n = 0, 2, 6, 8$	Cu/ 0.1 M H_2SO_4	[17]
3	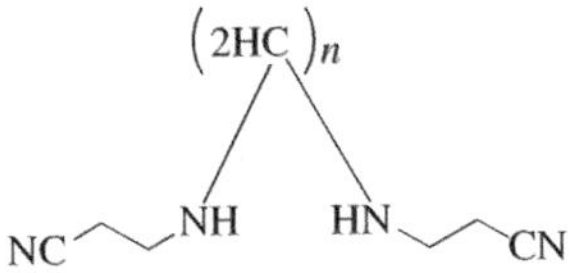 $(2HC)_n$ NC NH HN CN $n = 2, 3, 6$ $n = 2, 3, 6$	Fe/ 1M HCl	[18]
4	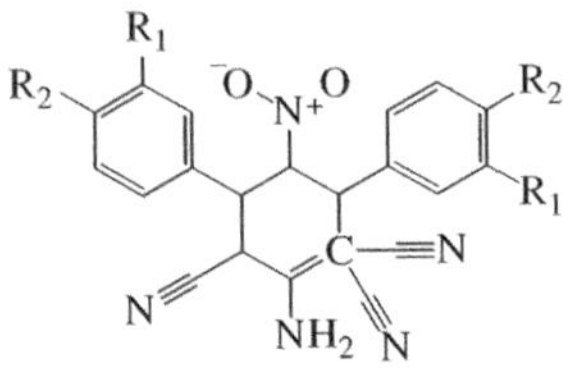 R_1 R_2 ^-O N^+ O R_2 R_1 C≡N N≡ NH_2 N R_1+R_2 = –H + –H; –H + –OCH_3; –OH + –OCH_3 R_1+R_2 = -H+-H; -H+-OCH_3; -OH+-OCH_3	Fe/ 1 M HCl	[19]
5	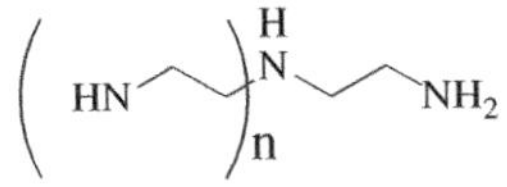 HN H N NH_2 n n= 0, 1, 3 $n = 0, 1, 3$	Fe/ Sea water	[20]
6	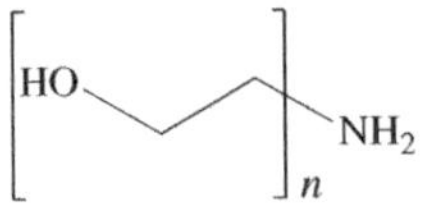 HO NH_2 n $n = 0, 1,2$ $n = 0, 1,2$	Al/ 1 M H_3PO_4	[21]

or increasing the solubility of compounds in the polar electrolytes. Presence of electron donating substituents and increase in the length of hydrophobic carbon chain increase the inhibition efficiency of these compounds. Generally, these compounds become effective by adsorbing on the metallic surface following through Langmuir adsorption isotherm model. Polarization study showed most of the amine derivatives act as mixed-type corrosion inhibitors as their adsorbing obeyed the Langmuir adsorption isotherm model. EIS study showed that most of the amines acted as interface-type corrosion inhibitors as they increase the value of charge transfer process by adsorbing at the interface of metal and electrolytes. Adsorption of these compounds on metallic surface can be accessed through SEM, EDX, AFM, XRD, FT-IR, and UV-vis spectroscopic studies.

5.1.2.2 Amides and Thio-Amides as Corrosion Inhibitors

Amide and thio-amide can be regarded as amine derivatives, and they contain carbonyl (>C=O) and thio-carbonyl (>C=S) functional group at the alpha-position of the $-NH_2$ group. These functional groups are presented as $-CONH_2$ and $-CSNH_2$, respectively. Literature study shows that amide and thio-amides are widely used as corrosion inhibitors for various metals and alloys in various electrolytes. Effect of substituents has also been investigated widely on the corrosion inhibition property of amide and thio-amide groups containing compounds. Fouda and coworkers [22] studied the corrosion inhibition effect of some aryl-methylene cyanothioacetamide derivatives for copper corrosion in 2M HNO_3. This study was carried out using potentiodynamic polarization method. The study was conducted to demonstrate the effect of various substituents on the corrosion inhibition effect of the tested series of compounds. It was observed that study represents a good example of substituents, and corrosion inhibition efficiencies of evaluated compounds were according to their order of electron-donating ability. Inhibition efficiencies of the tested compounds followed the order: $-OCH_3 > -CH_3 > -Cl > -Br > -NO_2$.

Similar observation was further observed while studying the corrosion inhibition effect of substituted phenyl phthalimide derivatives for carbon steel in 0.5M H_2SO_4 using weight loss and potentiodynamic polarization methods [23]. Studies showed that compound containing –OMe (methoxy) substituent showed the highest protection efficiency among the tested compounds, whereas compound containing electron-withdrawing $-NO_2$ (nitro) substituent showed the lowest protection effectiveness. Inhibition efficiencies of various compounds followed the order: $-OCH_3 > -CH_3 > -H > -Cl > -NO_2$. Potentiodynamic polarization study showed that all studied compounds acted as mixed-type corrosion inhibitors, and they adversely affect the anodic as well as cathodic reactions. It is important to mention that all studied compounds become effective by adsorbing on the metallic surface that follow Freundlich adsorption isotherm model. Effect of substituents on the corrosion inhibition effect of amide and thio-amides have also been studied in various other reports, and the summary of some common studies is presented in Table 5.2. It can be clearly seen that presence of electron-donating substituents increase the inhibition efficiency and vice versa. Obviously, they become effective by adsorbing on the metallic surface following mainly through Langmuir adsorption isotherm. Adsorption of amide and thio-amides on metallic surface is generally supported using various surface investigation techniques. Polarization study showed that most of the investigated compounds acted as mixed-type corrosion inhibitors, and they adversely affect the anodic and cathodic Tafel reactions.

5.1.2.3 Schiff Bases as Corrosion Inhibitors

Schiff bases can be regarded as derivatives of amines and aliphatic or aromatic aldehydes. They are commonly symbolized as SBs. SBs are widely used as corrosion inhibitors for different metals and alloys in various electrolytes. Effect of substituents is also extensively investigated on SBs. It is observed that electron-donating substituents increase corrosion inhibition performance of the SBs, while converse is true for electron-withdrawing substituents. Soltani et al. [29] investigated the corrosion inhibition effect of some double Schiff bases (SBs) for mild steel in 2M hydrochloric acid solution. This study was conducted through various methods. All the investigated compounds exhibited reasonable good protection efficiency at relatively lower concentration. Except one SB, remaining three SBs were similar in their chemical structure and differ in the nature of substituents. Study showed that SB containing –OMe (methoxy) substituent showed better corrosion inhibition protection followed by SB containing $-CH_3$ (methyl) substituent and finally the SB without a substituent. Inhibition efficiencies of the tested compounds followed the order: SB-3 ($-OCH_3$) > SB-2 ($-CH_3$) > SB-1 (–H).

In another study [30], corrosion inhibition effect of three SBs differing in the nature of substituents was investigated for mild steel in 1M hydrochloric acid. This study was performed using various electrochemical and surface investigation methods. Outcomes of the study showed that SB containing –OMe substituent (MeO-Salen) showed the highest protection efficiency followed by the SB without any substituent (Salen) and finally by the SB having electron-withdrawing $-NO_2$ substituent (NO_2-Salen). Polarization study showed that investigated SBs behaved as anodic type corrosion inhibitors as they mainly affect the anodic oxidative dissolution reaction. Study further showed that inhibition efficiencies of the tested compounds increase on increasing the concentration of evaluated SBs. All SBs become effective by adsorbing on the metallic surface following through Langmuir adsorption isotherm model. Adsorption mechanism of corrosion inhibition was supported by SEM analyses where significant smoothness in the surface morphology of the metal surface was observed in the presence of investigated SBs. Summary of some common reports on corrosion inhibition effect of SBs is presented in Table 5.3. Inspection of the Table 5.3 showed that electron-donating substituents increase corrosion inhibition performance of the compounds,

Table 5.2 Chemical structures of some common amides and thio-amides used as corrosion inhibitors.

S No	Structure of inhibitor	Metal/electrolyte	References
1	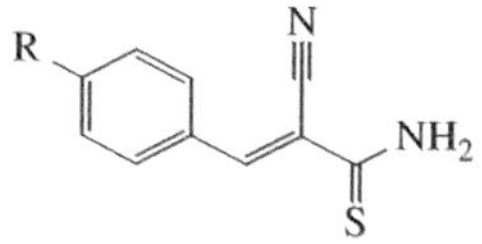	Cu/ 2M HNO_3	[22]
2	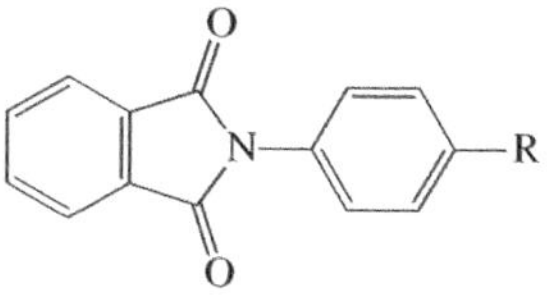	Fe/ 0.5M HCl	[23]
3	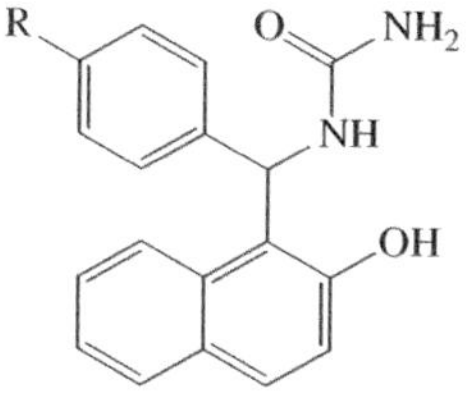	Fe/ 0.5 M H_2SO_4	[24]
4	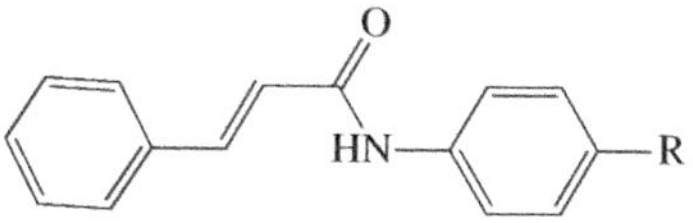	Pb/ 0.1M HCl	[2]
5	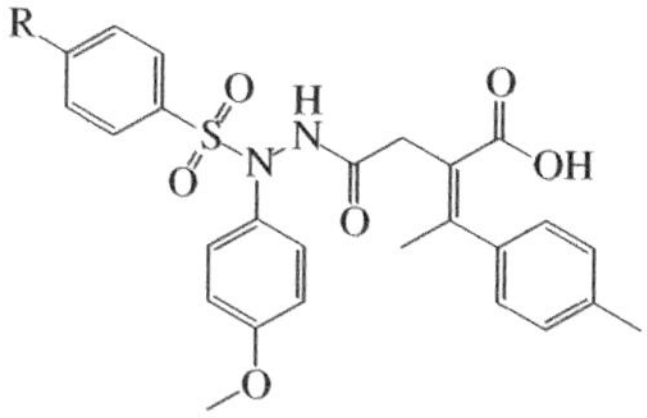	Fe/ 1M HCl	[25]
6	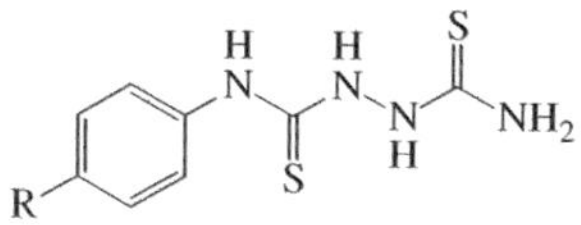	Cu/ 3.5 NaOH	[2]
7	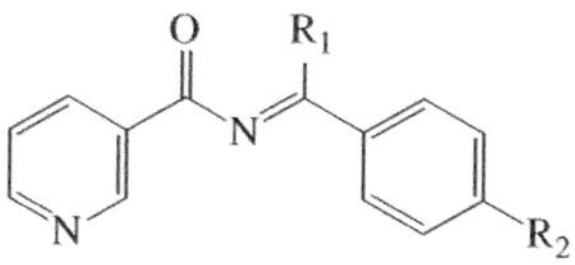	Fe/ 0.5 M HCl	[26]
8	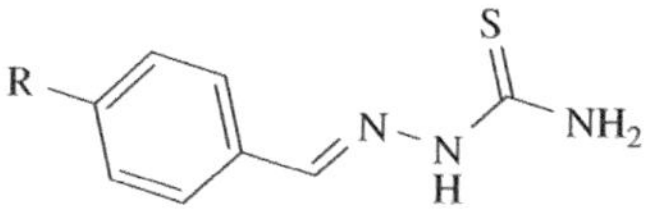	$Fe_{78}B_{13}Si_9$ glassy alloy/ 0.2 M Na_2SO_4	[27]
9	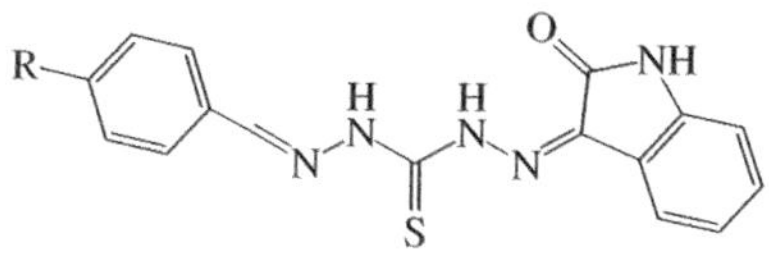	Fe/ 20% H_2SO_4	[28]

Table 5.3 Chemical structures of some common Schiff bases used as corrosion inhibitors.

S No	Structure of inhibitor	Metal/electrolyte	References
1	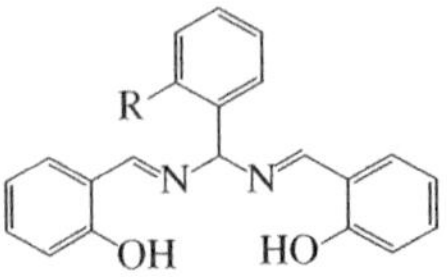	Fe/2M HCl	[29]
3	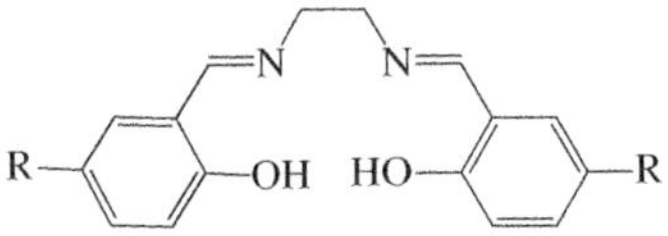	Fe/1M HCl	[30]
4	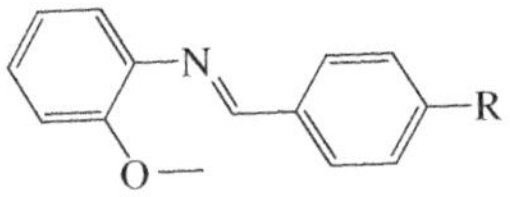	Al/1M HCl	[31]
5	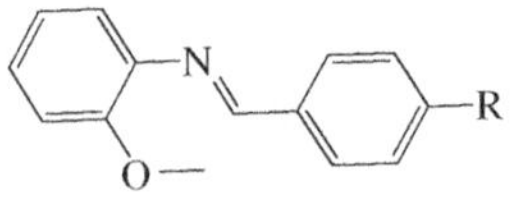	Fe/1M HCl	[32]
7	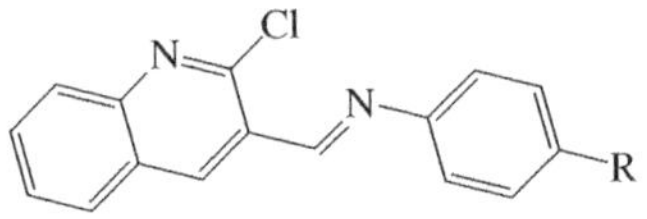	Fe/1M HCl	[33]
8	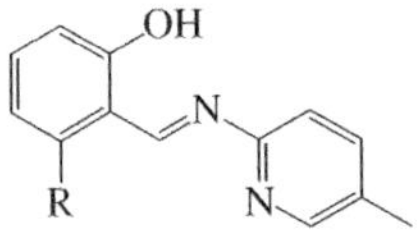	Al/0.1M HCl	[34]
9	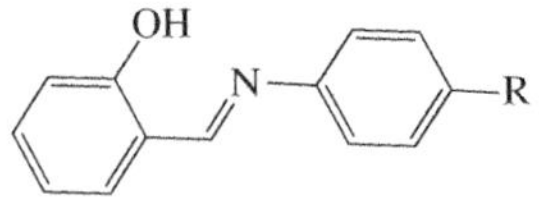	Fe/1M HCl	[35]
10	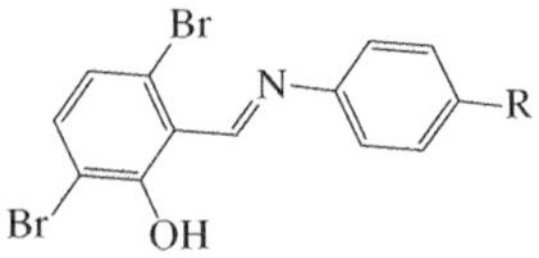	Al/0.5M HCl	[36]
13	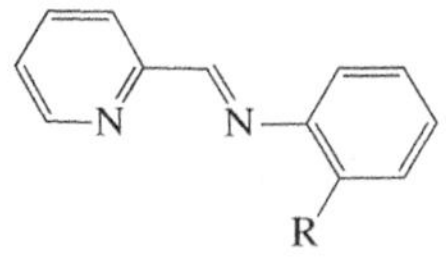	Fe/1M HCl	[37]

Table 5.3 (Continued)

S No	Structure of inhibitor	Metal/electrolyte	References
15	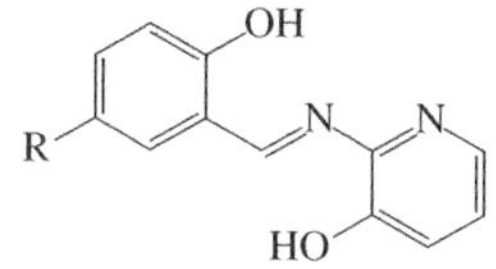	Fe/1M HCl	[38]
17	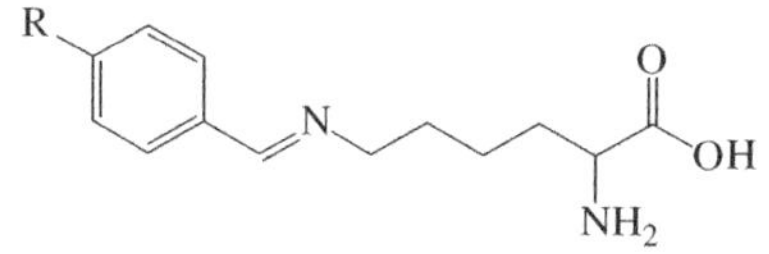	Fe/1M HCl	[39]
18	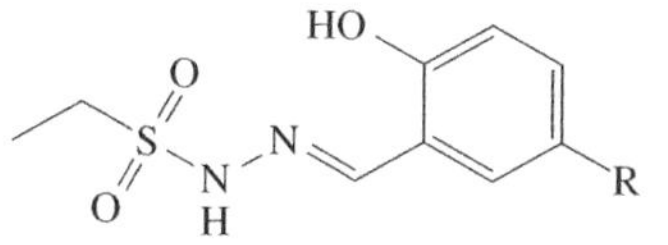	Al/0.1M HCl	[40]

whereas electron-withdrawing substituents exert just inverse effect. These compounds mainly behaves as mixed-type corrosion inhibitors as they adversely affect anodic, as well as cathodic, Tafel reactions. Through EIS analysis, it is observed that SBs behave as interface-type corrosion inhibitors and impose a barrier for charge transfer process. The SBs become effective by adsorbing at the interface of metal and electrolyte following mainly through Langmuir adsorption isotherm model. Adsorption mechanism of corrosion inhibition is extensively supported by surface investigation methods.

5.1.2.4 Amine-Based Drugs and Dyes as Corrosion Inhibitors

Obviously, drugs and dyes are highly complex molecules, and generally, they contain a proper combination of hydrophilicity and hydrophobicity. Because of their bio-tolerability and non-bioaccumulation, drugs and their derivatives can be regarded as green substitute to the traditional toxic corrosion inhibitors [41–43]. However, mostly drug molecules are synthesized using multistep reactions (MSRs) and generally associated with high synthesis and purification costs. Therefore, their use as corrosion inhibitors is a highly expensive task. Nevertheless, recent use of expired drugs as corrosion inhibitors is gaining particular attention. Alongside, dyes are also highly complex molecules and are suitable candidates to be used as corrosion inhibitors literature study showed that various dyes, especially, azo-based dyes, are highly used as corrosion inhibitors for different metals and alloys in various electrolytes. Similarly, several drug molecules are also evaluated as environmental friendly corrosion inhibitors in various systems. Fouda et al. [44] investigated the 2-hydroxyacetophenone-aroyl hydrazone derivatives (dyes) for copper in 3M HNO_3 using various techniques. Investigated compounds differ in the nature of substituents present in their molecular structures. Results showed that compound containing $-CH_3$ (methyl) substituent showed the highest inhibition efficiency, whereas compound containing $-NO_2$ (nitro) substituent showed the lowest protection efficiency.

In another study, corrosion inhibition effect of three azo-based compounds was investigated for aluminum in 0.01 NaOH [45]. Results derived from the study showed that compound containing $-OCH_3$ (methoxy) substituent showed the highest protection effectiveness followed by compound without any substituent and finally by the compound having electron-withdrawing –Cl (chloro) substituent. The effect of substituents on the corrosion inhibition effect of the dyes and drugs are also extensively reported in other studies. Summary of such major reports is presented in Table 5.4. Generally, they behave as effective corrosion inhibitors and become effective by adsorbing on the metallic surface following through Langmuir adsorption isotherm model. Polarization study shows that most of the studied dyes and drugs are mixed-type corrosion inhibitors as they retard both anodic oxidative dissolution and cathodic hydrogen evolution reactions. EIS study shows that dyes and drugs are mainly interface-type corrosion inhibitors as their

Table 5.4 Chemical structures of some common amine-based drugs and dyes used as corrosion inhibitors.

S No	Basic structure of inhibitor	Metal/electrolyte	References
1	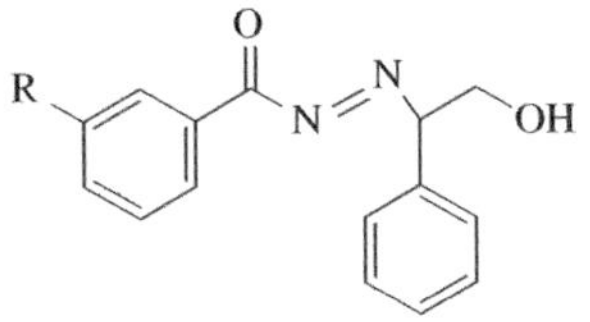	Cu/ 3N HNO_3	[44]
2.	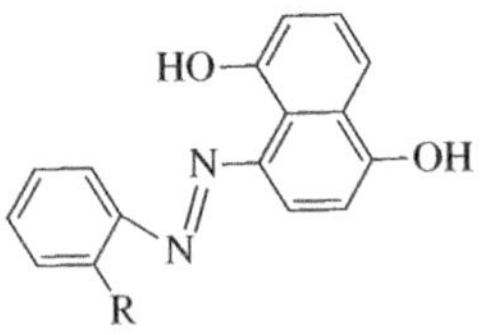	Al/ 0.01 NaOH	[45]
3	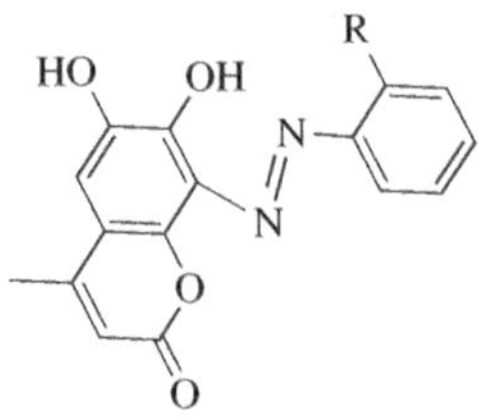	Fe/ 2M HCl	[46]

Table 5.4 (Continued)

S No	Basic structure of inhibitor	Metal/electrolyte	References
4	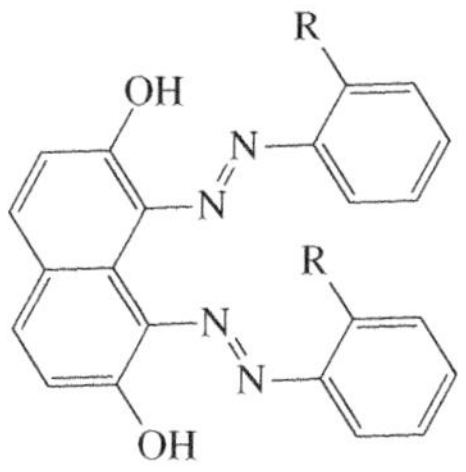 R, OH, N, N, R, N, N, OH	Al/ 2M HCl	[47]
5	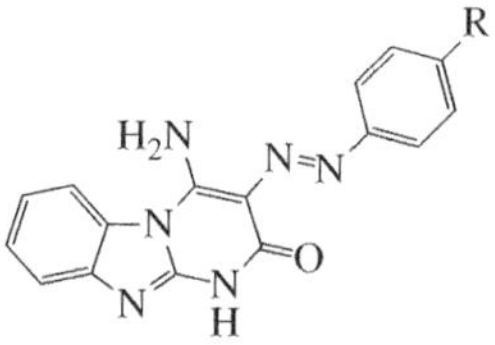 R, H_2N, N=N, N, O, N, N, H	Fe/ 15% HCl	[48]
6	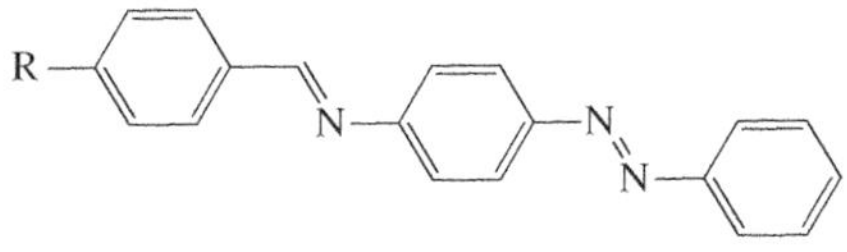 R, N, N, N	Fe/ 1M H_2SO_4	[49]
7	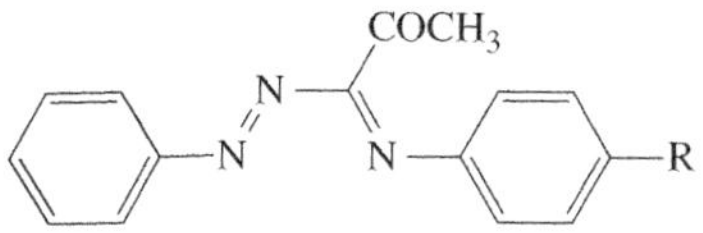 $COCH_3$, N, N, N, R	Fe/ 1M HCl	[50]
8	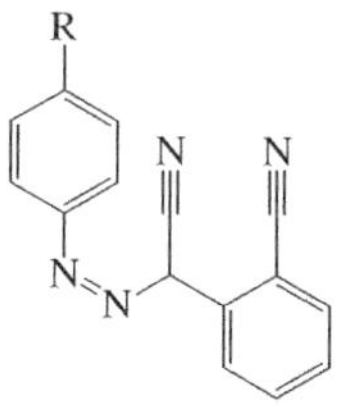 R, N, N, N, N	Fe/ 1M HCl	[51]
9	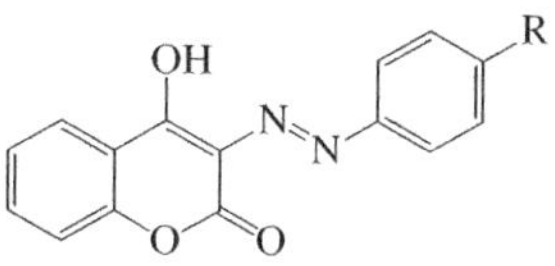 OH, R, N, N, O, O	Cu/ 1M HNO_3	[52]
10	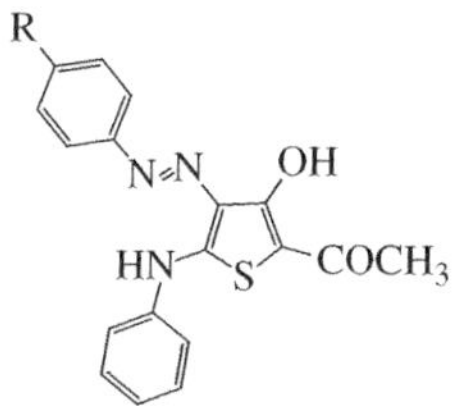 R, N=N, OH, HN, S, $COCH_3$	Fe/ 2M HCl	[53]

(*Continued*)

Table 5.4 (Continued)

S No	Basic structure of inhibitor	Metal/electrolyte	References
11	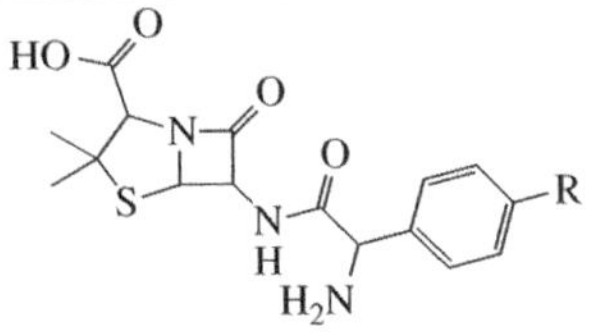 HO, O, O, N, S, O, N, H, H_2N, R	Al/ 2M HCl	[54]
12	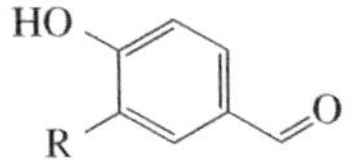 HO, R, O	Fe/ 2M HCl	[55]
13	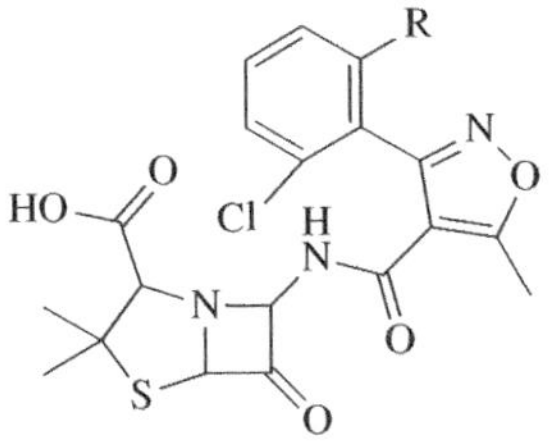 R, N, O, O, HO, Cl, H, N, N, O, S, O	Al 6063/ 0.5 M H3PO4	[56]
14	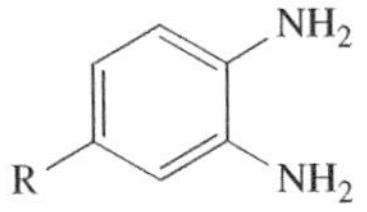 NH_2, R, NH_2	Cu/ 2 M HNO3	[2]

presence affect the charge transfer process of metallic corrosion. Adsorption mechanism of corrosion inhibition is derived through surface investigation methods.

5.1.2.5 Amino Acids and Their Derivatives as Corrosion Inhibitors

AAs are building block of peptide and polypeptide. Because of their natural and biological origin, AAs can be regarded as environmental friendly alternatives to traditional toxic corrosion inhibitors [57–59]. More so, compounds derived from AAs can also be treated as environmental friendly alternatives. Literature study showed that AAs, especially cysteine and their derivatives, are extensively investigated as corrosion inhibitors [60–63]. Most of the AAs and their derivatives acted as mixed-type and interface-type corrosion inhibitors, and their adsorption followed the Langmuir adsorption isotherm model.

5.2 Conclusion and Outlook

Amine-based compounds are widely used as corrosion inhibitors. A proper combination of hydrophilicity and hydrophobicity is essential for designing of effective corrosion inhibitors. Various series of amine derives such as 1°-, 2°- and

3°-amines, amides, and thio-amides, amine-based drugs and dyes, Mannich and Schiff bases, AAs, and their derivatives are widely investigated as corrosion inhibitors for different metals and alloys in various electrolytes. Substituents effect is extensively investigated for above series of amine derivatives, and it was observed that presence of electron-donating substituents enhances the corrosion inhibition effect of the amine derivatives and converse is true for electron-withdrawing substituents. These compounds become by adsorbing on the metallic surface following mostly through Langmuir adsorption isotherm model. EIS and PDP studies show that amine derivatives act as interface- and mixed-type corrosion inhibitors, respectively. Adsorption mechanism of corrosion inhibition by amine derivatives was supported by various surface investigation techniques including SEM, AFM, EDX, XRD, FT-IR, and UV-vis. In majority of the studies, experimental results were supported by the computational studies mainly through density functional theory and molecular dynamics simulations.

Important Websites

https://www.sciencedirect.com/science/article/abs/pii/S0013468696002502
https://www.sciencedirect.com/science/article/abs/pii/0010938X9390085U
https://patents.google.com/patent/US4636256A/en

References

1. Al-Qudsi, S.S. (2010). *Arab Demography and Health Provision, Genetic Disorders Among Arab Populations*, 37–63. Springer.
2. Verma, C., Olasunkanmi, L., Ebenso, E.E., and Quraishi, M. (2018). Substituents effect on corrosion inhibition performance of organic compounds in aggressive ionic solutions: a review. *Journal of Molecular Liquids* 251: 100–118.
3. Dariva, C.G. and Galio, A.F. (2014). Corrosion inhibitors–principles, mechanisms and applications. *Developments in corrosion protection* 16: 365–378.
4. Goyal, M., Kumar, S., Bahadur, I. et al. (2018). Organic corrosion inhibitors for industrial cleaning of ferrous and non-ferrous metals in acidic solutions: a review. *Journal of Molecular Liquids* 256: 565–573.
5. Arthur, D.E., Jonathan, A., Ameh, P.O., and Anya, C. (2013). A review on the assessment of polymeric materials used as corrosion inhibitor of metals and alloys. *International Journal of Industrial Chemistry* 4: 2.
6. Bentiss, F., Lebrini, M., and Lagrenée, M. (2005). Thermodynamic characterization of metal dissolution and inhibitor adsorption processes in mild steel/2, 5-bis (n-thienyl)-1, 3, 4-thiadiazoles/hydrochloric acid system. *Corrosion Science* 47: 2915–2931.
7. Ahamad, I., Prasad, R., and Quraishi, M. (2010). Experimental and quantum chemical characterization of the adsorption of some Schiff base compounds of

phthaloyl thiocarbohydrazide on the mild steel in acid solutions. *Materials Chemistry and Physics* 124: 1155–1165.

8 Yin, Z.F., Feng, Y., Zhao, W. et al. (2009). Effect of temperature on CO_2 corrosion of carbon steel. *Surface and Interface Analysis* 41: 517–523.

9 Xiang, Y., Wang, Z., Li, Z., and Ni, W. (2013). Effect of temperature on corrosion behaviour of X70 steel in high pressure CO2/SO2/O2/H2O environments. *Corrosion Engineering, Science and Technology* 48: 121–129.

10 Verma, C., Olasunkanmi, L.O., Ebenso, E.E. et al. (2016). Adsorption behavior of glucosamine-based, pyrimidine-fused heterocycles as green corrosion inhibitors for mild steel: experimental and theoretical studies. *The Journal of Physical Chemistry C* 120: 11598–11611.

11 Ebenso, E., Ekpe, U., Ita, B. et al. (1999). Effect of molecular structure on the efficiency of amides and thiosemicarbazones used for corrosion inhibition of mild steel in hydrochloric acid. *Materials Chemistry and Physics* 60: 79–90.

12 Muralidharan, S., Quraishi, M., and Iyer, S. (1995). The effect of molecular structure on hydrogen permeation and the corrosion inhibition of mild steel in acidic solutions. *Corrosion Science* 37: 1739–1750.

13 Braun, R.D., Lopez, E.E., and Vollmer, D.P. (1993). Low molecular weight straight-chain amines as corrosion inhibitors. *Corrosion Science* 34: 1251–1257.

14 Khaled, K., Babić-Samardžija, K., and Hackerman, N. (2005). Theoretical study of the structural effects of polymethylene amines on corrosion inhibition of iron in acid solutions. *Electrochimica Acta* 50: 2515–2520.

15 Kaesche, H. and Hackerman, N. (1958). Corrosion inhibition by organic amines. *Journal of the Electrochemical Society* 105: 191.

16 De Damborenea, J., Bastidas, J., and Vazquez, A. (1997). Adsorption and inhibitive properties of four primary aliphatic amines on mild steel in 2 M hydrochloric acid. *Electrochimica Acta* 42: 455–459.

17 Yurt, A. and Bereket, G.z. (2011). Combined electrochemical and quantum chemical study of some diamine derivatives as corrosion inhibitors for copper. *Industrial & Engineering Chemistry Research* 50: 8073–8079.

18 Herrag, L., Bouklah, M., Patel, N. et al. (2012). Experimental and theoretical study for corrosion inhibition of mild steel 1 M HCl solution by some new diaminopropanenitrile compounds. *Research on Chemical Intermediates* 38: 1669–1690.

19 Verma, C., Quraishi, M., and Singh, A. (2015). 2-Amino-5-nitro-4, 6-diarylcyclohex-1-ene-1, 3, 3-tricarbonitriles as new and effective corrosion inhibitors for mild steel in 1 M HCl: Experimental and theoretical studies. *Journal of Molecular Liquids* 212: 804–812.

20 Migahed, M., Attia, A., and Habib, R. (2015). Study on the efficiency of some amine derivatives as corrosion and scale inhibitors in cooling water systems. *RSC Advances* 5: 57254–57262.

21 Fouda, A., Abdallah, M., Ahmed, I., and Eissa, M. (2012). Corrosion inhibition of aluminum in 1 M H3PO4 solutions by ethanolamines. *Arabian Journal of Chemistry* 5: 297–307.

22 Fouda, A., Mohamed, A., and Mostafa, H. (1998). Inhibition of corrosion of copper in nitric acid solution by some arylmethylene cyanothioacetamide derivatives. *Journal de Chimie Physique et de Physico-Chimie Biologique* 95: 45–55.

23 Zaafarany, I. (2009). Phenyl phthalimide as corrosion inhibitor for corrosion of C-Steel in sulphuric acid solution. *Portugaliae Electrochimica Acta* 27: 631–643.

24 Bahrami, M., Hosseini, S., and Pilvar, P. (2010). Experimental and theoretical investigation of organic compounds as inhibitors for mild steel corrosion in sulfuric acid medium. *Corrosion Science* 52: 2793–2803.

25 Khalifa, O.R. and Abdallah, S.M. (2011). Corrosion inhibition of some organic compounds on low carbon steel in hydrochloric acid solution. *Portugaliae Electrochimica Acta* 29: 47–56.

26 Chakravarthy, M., Mohana, K., and Kumar, C.P. (2014). Corrosion inhibition effect and adsorption behaviour of nicotinamide derivatives on mild steel in hydrochloric acid solution. *International Journal of Industrial Chemistry* 5: 19.

27 Arab, S. and Emran, K. (2008). Structure effect of some thiosemicarbazone derivatives on the corrosion inhibition of Fe78B13Si9 glassy alloy in Na2SO4 solution. *Materials Letters* 62: 1022–1032.

28 Ansari, K., Quraishi, M., and Singh, A. (2015). Isatin derivatives as a non-toxic corrosion inhibitor for mild steel in 20% H2SO4. *Corrosion Science* 95: 62–70.

29 Soltani, N., Behpour, M., Ghoreishi, S., and Naeimi, H. (2010). Corrosion inhibition of mild steel in hydrochloric acid solution by some double Schiff bases. *Corrosion Science* 52: 1351–1361.

30 Bayol, E., Gürten, T., Gürten, A.A., and Erbil, M. (2008). Interactions of some Schiff base compounds with mild steel surface in hydrochloric acid solution. *Materials Chemistry and Physics* 112: 624–630.

31 Ashassi-Sorkhabi, H., Shabani, B., Aligholipour, B., and Seifzadeh, D. (2006). The effect of some Schiff bases on the corrosion of aluminum in hydrochloric acid solution. *Applied Surface Science* 252: 4039–4047.

32 Ashassi-Sorkhabi, H., Shaabani, B., and Seifzadeh, D. (2005). Effect of some pyrimidinic Shciff bases on the corrosion of mild steel in hydrochloric acid solution. *Electrochimica Acta* 50: 3446–3452.

33 Prabhu, R., Venkatesha, T., Shanbhag, A. et al. (2008). Inhibition effects of some Schiff's bases on the corrosion of mild steel in hydrochloric acid solution. *Corrosion Science* 50: 3356–3362.

34 Yurt, A., Ulutas, S., and Dal, H. (2006). Electrochemical and theoretical investigation on the corrosion of aluminium in acidic solution containing some Schiff bases. *Applied Surface Science* 253: 919–925.

35 Hegazy, M., Hasan, A.M., Emara, M. et al. (2012). Evaluating four synthesized Schiff bases as corrosion inhibitors on the carbon steel in 1 M hydrochloric acid. *Corrosion Science* 65: 67–76.

36 Fouda, A., Elewady, G., El-Askalany, A., and Shalabi, K. (2010). Inhibition of aluminum corrosion in hydrochloric acid media by three schiff base compounds. *Zaštita Materijala* 51: 205–219.

37 Sukul, D., Pal, A., Saha, S.K. et al. (2018). Newly synthesized quercetin derivatives as corrosion inhibitors for mild steel in 1 M HCl: combined experimental and theoretical investigation. *Physical Chemistry Chemical Physics* 20: 6562–6574.

38 Zaafarany, I. (2014). Corrosion inhibition of 1018 carbon steel in hydrochloric acid using Schiff base compounds. *International Journal of Corrosion and Scale Inhibition* 3: 012–027.

39 Gupta, N.K., Verma, C., Quraishi, M., and Mukherjee, A. (2016). Schiff's bases derived from l-lysine and aromatic aldehydes as green corrosion inhibitors for mild steel: experimental and theoretical studies. *Journal of Molecular Liquids* 215: 47–57.

40 Aytac, A., Özmen, Ü., and Kabasakaloğlu, M. (2005). Investigation of some Schiff bases as acidic corrosion of alloy AA3102. *Materials Chemistry and Physics* 89: 176–181.

41 Gece, G. (2011). Drugs: a review of promising novel corrosion inhibitors. *Corrosion Science* 53: 3873–3898.

42 Abdallah, M. (2002). Rhodanine azosulpha drugs as corrosion inhibitors for corrosion of 304 stainless steel in hydrochloric acid solution. *Corrosion Science* 44: 717–728.

43 Obot, I., Obi-Egbedi, N., and Umoren, S. (2009). Antifungal drugs as corrosion inhibitors for aluminium in 0.1 M HCl. *Corrosion Science* 51: 1868–1875.

44 AS, F., MM, G., and SI, A.E.-R. (2000). 2-Hydroxyacetophenone-aroyl hydrazone derivatives as corrosion inhibitors for copper dissolution in nitric acid solution. *Bulletin of the Korean Chemical Society* 21: 1085–1089.

45 Fouda, A., Abdel-Maksoud, S., and Almetwally, A. (2015). Corrosion Inhibition of Tin in Sodium Chloride Solutions Using Propaneitrile Derivatives. *Chemical Science* 4: 161–175.

46 Abdallah, M., Fouda, A., Shama, S., and Afifi, E. (2008). Azodyes as corrosion inhibitors for dissolution of c-steel in hydrochloric acid solution. *African Journal of Pure and Applied Chemistry* 2: 083–091.

47 Mabrouk, E., Shokry, H., and Abu, A.-N.K. (2011). Inhibition of aluminum corrosion in acid solution by mono-and bis-azo naphthylamine dyes. Part 1. *Chemistry of Metals and Alloys*: 98–106.

48 Yadav, M., Kumar, S., Bahadur, I., and Ramjugernath, D. (2014). Electrochemical and quantum chemical studies on synthesized phenylazopyrimidone dyes as corrosion inhibitors for mild steel in a 15% HCl solution. *International Journal of Electrochemical Science* 9: 3928–3950.

49 Shihab, M.S. and Al-Doori, H.H. (2014). Experimental and theoretical study of [N-substituted] p-aminoazobenzene derivatives as corrosion inhibitors for mild steel in sulfuric acid solution. *Journal of Molecular Structure* 1076: 658–663.

50 Hamani, H., Douadi, T., Al-Noaimi, M. et al. (2014). Electrochemical and quantum chemical studies of some azomethine compounds as corrosion inhibitors for mild steel in 1 M hydrochloric acid. *Corrosion Science* 88: 234–245.

51 Fouda, A., El-Azaly, A.H., Awad, R., and Ahmed, A. (2014). New benzonitrile azo dyes as corrosion inhibitors for carbon steel in hydrochloric acid solutions. *International Journal of Electrochemical Science* 9: 1117–1131.

52 Fouda, A., Rashwan, S., Kamel, M., and Khalifa, M. (2016). 4-hydroxycoumarin derivatives as corrosion inhibitors for copper in nitric acid solutions. *Journal of Materials and Environmental Science* 7: 2658–2678.

53 El-Haddad, M.N. and Fouda, A. (2013). Corrosion inhibition and adsorption behavior of some azo dye derivatives on carbon steel in acidic medium: synergistic effect of halide ions. *Chemical Engineering Communications* 200: 1366–1393.

54 Abdallah, M. (2004). Antibacterial drugs as corrosion inhibitors for corrosion of aluminium in hydrochloric solution. *Corrosion Science* 46: 1981–1996.

55 Emregül, K. and Hayvalı, M. (2004). Studies on the effect of vanillin and protocatechualdehyde on the corrosion of steel in hydrochloric acid. *Materials Chemistry and Physics* 83: 209–216.

56 Nazeer, A.A., El-Abbasy, H., and Fouda, A. (2013). Adsorption and corrosion inhibition behavior of carbon steel by cefoperazone as eco-friendly inhibitor in HCl. *Journal of Materials Engineering and Performance* 22: 2314–2322.

57 Barouni, K., Bazzi, L., Salghi, R. et al. (2008). Some amino acids as corrosion inhibitors for copper in nitric acid solution. *Materials Letters* 62: 3325–3327.

58 Hluchan, V., Wheeler, B., and Hackerman, N. (1988). Amino acids as corrosion inhibitors in hydrochloric acid solutions. *Materials and Corrosion* 39: 512–517.

59 Gece, G. and Bilgic, S. (2010). A theoretical study on the inhibition efficiencies of some amino acids as corrosion inhibitors of nickel. *Corrosion Science* 52: 3435–3443.

60 Farahati, R., Mousavi-Khoshdel, S.M., Ghaffarinejad, A., and Behzadi, H. (2020). Experimental and computational study of penicillamine drug and cysteine as water-soluble green corrosion inhibitors of mild steel. *Progress in Organic Coatings* 142: 105567.

61 Abd El-Hafez, G.M. and Badawy, W.A. (2013). The use of cysteine, N-acetyl cysteine and methionine as environmentally friendly corrosion inhibitors for Cu–10Al–5Ni alloy in neutral chloride solutions. *Electrochimica Acta* 108: 860–866.

62 Aslam, R., Mobin, M., Zehra, S. et al. (2017). N, N′-Dialkylcystine Gemini and Monomeric N-Alkyl Cysteine Surfactants as Corrosion inhibitors on mild steel corrosion in 1 M HCl solution: a comparative study. *ACS Omega* 2: 5691–5707.

63 Yaocheng, Y., Caihong, Y., Singh, A., and Lin, Y. (2019). Electrochemical study of commercial and synthesized green corrosion inhibitors for N80 steel in acidic liquid. *New Journal of Chemistry* 43: 16058–16070.

6

Imidazole and Its Derivatives as Corrosion Inhibitors

Jeenat Aslam[1], Ruby Aslam[2], and Chandrabhan Verma[3]

[1] *Department of Chemistry, College of Science, Taibah University, Yanbu, Al-Madina, Saudi Arabia*
[2] *Corrosion Research Laboratory, Department of Applied Chemistry, Faculty of Engineering and Technology, Aligarh Muslim University, Aligarh, Uttar Pradesh, India*
[3] *Interdisciplinary Research Center for Advanced Materials, King Fahd University of Petroleum and Minerals, Dhahran, Saudi Arabia*

6.1 Introduction

6.1.1 Corrosion and Its Economic Impact

Metals and alloys are broadly used in the industrial area because it has cost-effective, easy smelting process, and excellent economic and physical properties [1–3]. The critical difficulty of this kind of material is that simple to be rusted in the acid pickling process, causing various industrial and financial losses [4–7]. Corrosion is a spontaneous and natural phenomenon that results in the change of pure metals and their alloys into various stable forms, for example, their oxides, hydroxides, and so on via the electrochemical/chemical reactions with the nearby surroundings [8–11]. Although significant progress in the field of corrosion science and technology, the phenomenon of corrosion remains the main problem for industries all over the world [12–14]. A 2002 federal study, initiated by the National Association of Corrosion Engineers (NACE) [15], it was expected that the problem of corrosion is a most challenging and damaging issue in Sweden, Kuwait, Japan, Germany, Australia, United States, India, and China. In the report, it was expected that owing to the corrosion gross domestic productivity (GDP) losses in these countries is approximately 1–5%. According to the most recent observation of NACE international, the annual global cost of corrosion is about the US$2.5 trillion, equivalent to around 3.4% of the global GDP [16, 17]. Recently, the

Organic Corrosion Inhibitors: Synthesis, Characterization, Mechanism, and Applications, First Edition. Edited by Chandrabhan Verma, Chaudhery Mustansar Hussain, and Eno E. Ebenso.

global corrosion costs in Saudi Arabia and India are US $25 billion and the US $100-billion, respectively [18]. In another study, Lim [19] studied the corrosion-related cost for the Gulf Cooperation Council (GCC). This examination suggested that Saudi Arabia spent $24.8 billion to tackle the corrosion, which was the highest amount spent by any Arab state. More so, the corrosion cost in most of the developed and developing countries is estimated to be raised due to the rise in science and technology, industrialization reasons a raise in the utilization of metallic materials for multipurpose applications [20–22]. The corrosion cost can be decreased up to 15–35% using earlier well-known techniques of corrosion mitigation [23].

6.2 Corrosion Mechanism

It is recalled that corrosion is an undesirable reaction of metallic materials with the constituents of the surroundings [24, 25]. Corrosion is a degenerative and irreversible process that follows the second law of thermodynamics. The numerous mechanisms are possible for the corrosive dissolution of the metals depending upon the nature of metal and environment. However, wet or electrochemical corrosion (for example rusting of iron) can be considered as one of the most common forms of corrosion. According to Cushman (1907) [26], electrochemical corrosion proceeds through the formation of numerous anodic and cathodic areas in which more aerated areas behaved as anode and less aerated areas act as a cathode [27, 28]. The anode is a site on the metallic surface where oxidation (divalent metals) takes place as presented follows:

$$M + H_2O \leftrightarrow MOH_{ads} + H^+ + e^- \tag{6.1}$$

$$MOH_{ads} \leftrightarrow MOH^+_{ads} + e^- \tag{6.2}$$

$$MOH^+_{ads} + H^+ \leftrightarrow M^{2+} + H_2O \tag{6.3}$$

The cathodic reactions vary depending upon the environment as presented below [28]:

$$2H^+ + 2e^- \rightarrow H_2 \uparrow (\text{Acidic environment; Hydrogen evolution}) \tag{6.4}$$

$$O_2 + 4H^+ + 2e^- \rightarrow 2H_2O(\text{Reduction of oxygen in acid solution}) \tag{6.5}$$

$$O_2 + 2H_2O + 4e^- \rightarrow 4OH^-(\text{Reduction of oxygen in neutral solution}) \tag{6.6}$$

$$M^{3+} + e^- \rightarrow M^{2+}(\text{Reduction of metal ions}) \tag{6.7}$$

$$M^{2+} + 2e^- \rightarrow M(\text{Metal deposition}) \tag{6.8}$$

Depending upon the relative area anode and cathode, electrochemical corrosion can be divided into three categories, namely, separable anode/cathode (Sep. A/C), inseparable anode/cathode (Insep. A/C), and interfacial anode/cathode (Interfacial A/C) type. Finally, the products of anodic and cathodic areas react to give corrosion product known as rust [28].

$$Fe^{2+} + H_2O + \frac{1}{2}O_2 \rightarrow Fe(HO)_2 \tag{6.9}$$

$$2Fe(HO)_2 + H_2O + \frac{1}{2}O_2 \rightarrow 2Fe(HO)_3 \tag{6.10}$$

$$2Fe(HO)_3 \rightarrow Fe_2O_3(rust) + 3H_2O \tag{6.11}$$

Electrolytes containing dissolved oxygen are more corrosive than the electrolytes devoid of dissolved oxygen, as dissolved oxygen provides alternative paths for electron consumption (cathodic reaction) [29, 30].

6.3 Corrosion Inhibitors

The development of corrosion inhibitors is the most significant solution for the protection of metals and alloys against corrosion. Numerous studies have been dedicated to this topic [31–34]. The inhibitor molecules are added to the aggressive solution to decrease the degradation of metals. In the present time, a lot of new corrosion inhibitors are discovered. Particularly, the organic molecules act as excellent inhibitors because of the heteroatom present in their arrangements, for example, N, O, P, and S beside the aromatic rings [35–38]. The inhibitors protect the metal corrosion via interacting with the surface of the metal by adsorption through the donor atoms, p-orbitals, electron density, and the electronic arrangement of the molecules [39–44]. These organic compounds form a protective layer and are adsorbed on the metal surfaces and protect the metals from the aggressive solution [45–49]. This type of adsorption can be concluded via the two approaches, namely, physical (electrostatic) or chemical (chemisorptions) adsorptions. The chemisorption contains a chemical reaction among the metal surface and the inhibitor molecules to produce strong electronic bonds. On the contrary, physisorption is caused by van der Waals forces [50–52]. The adsorption of these corrosion inhibitors is changed via numerous aspects, for example, electrolyte behavior, presence of charge magnitude on metal, the electronic arrangement of inhibitor molecules, the behavior of substituents, exposure time, and solution temperature [53]. In recent times, the various study turned their attention to describe the interaction among metal surfaces and inhibitors by quantum chemical tools. Numerous computational modeling tools have been used to relate the molecular properties of the inhibitors with their inhibition efficiencies.

6.4 Corrosion Inhibitors: Imidazole and Its Derivatives

In 1887, A. R. Hantzsch (a German chemist) invented the term imidazole [54]. Imidazole is an organic heterocyclic compound with the chemical formula $C_3H_4N_2$, which is having five-membered heterocyclic rings with two nitrogen atoms at positions one and three. Imidazole is highly water-soluble, and it presents in two equivalent tautomeric forms. Imidazole is categorized as an aromatic compound containing the dipole moment (3.67 D) and 6π electron systems [55]. Owing to its high miscibility, highly nucleophilic, and polar nature (because of the resonance; Scheme 6.1), imidazole can strongly interact with a metallic surface and acts as an effective corrosion inhibitor in aqueous electrolytes.

Imidazole is a significant constituent of numerous industrially and biologically useful organic molecules including amino acids, bovine and vitamin B12, and nucleic acid [56]. Due to the unusual ring structure, imidazole and its derivatives exhibit various pharmacological and biological activities such as anxiolytic, analgesic, antidiabetic, anti-mycobacterial, anticancer, anti-protozoa, anti-HIV, anti-tubercular, antifungal, anti-inflammatory, and antibacterial [57–59] are known examples. Apart from these applications, it also has various applications in the industrial sectors. Imidazole and its derivatives have been widely used as a corrosion inhibitor on different metals and alloys, for example, copper, iron, carbon steel, and zinc. Imidazole and its derivatives have been studied as efficient inhibitors because of their outstanding excited-state intramolecular proton transfer, which increases their adsorption performances into the metal surface. In the present chapter, the authors have discussed the corrosion inhibition behavior and current development of imidazole and its derivatives. Table 6.1 exhibits the effective corrosion inhibition effects of imidazole and its derivatives for various metals and alloys in different electrolytic solutions.

Ramkumar et al. [99] synthesized four imidazoline (IDZ) and four isoxazolines (ISO) heterocyclic changing in the behavior of methoxy (-OCH_3) and aromatic (phenyl and naphthyl) moieties and investigated their anticorrosion behavior for

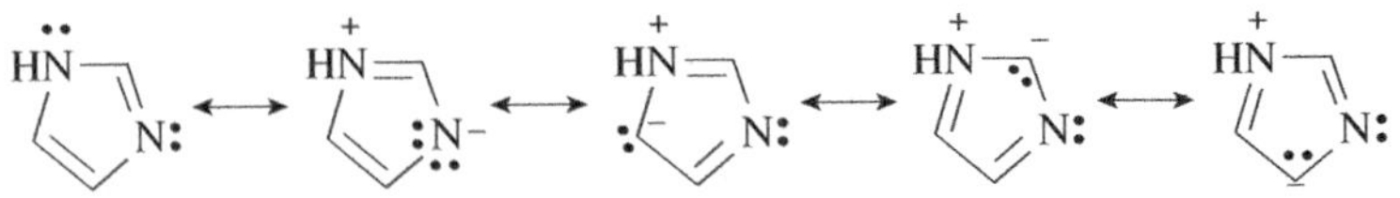

Scheme 6.1 Resonating structures of imidazole.

Table 6.1 Chemical structure of various imidazole derivatives calculated as corrosion inhibitors for different metals and alloys in various electrolytic solutions.

Chemical structures of inhibitors	Metals/electrolytes	References	Chemical structures of inhibitors	Metals/ electrolytes	References
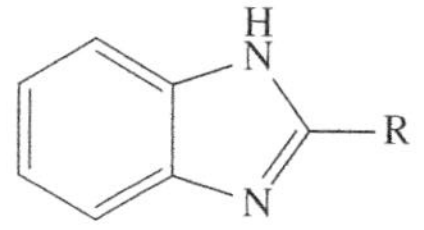 R– = –H, –CH_3, –SH	Fe/ 1M HCl	[60]	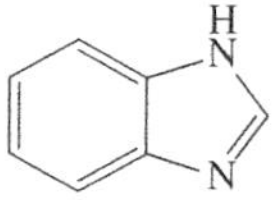 (BI) 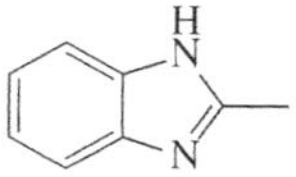 2–CH_3–BI 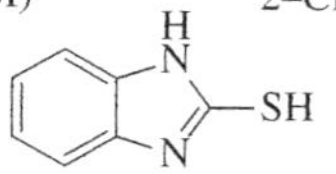 2–SH–BI	Carbon steel/1M HCl and 1M H_2SO_4	[61]
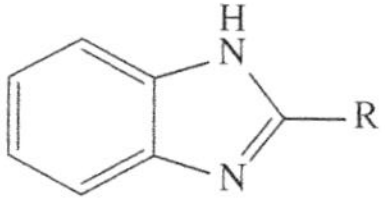 R–= –H, –CH_3, –NH_2	Fe/ 1M H_3PO_4	[62]	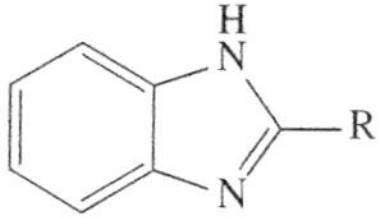 R = –CH_2NH_2, –CH_2Cl, –SCH_3	Fe/ 1M HNO_3	[63]
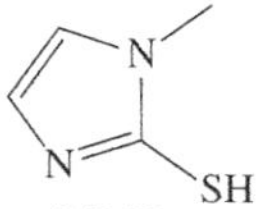 (MMI)	Copper/0.5 M H_2SO_4	[64]	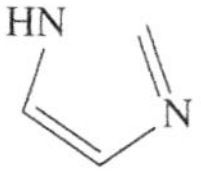 HN N	Iron surface	[65]

(Continued)

Table 6.1 (Continued)

Chemical structures of inhibitors	Metals/electrolytes	References	Chemical structures of inhibitors	Metals/ electrolytes	References
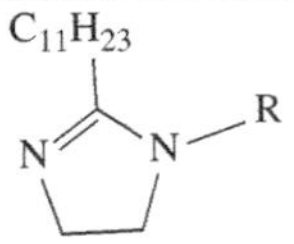 $C_{11}H_{23}$ N N R R = $-CH_2COOH$, $-CH_2CH_2OH$, $CH_2CH_2NH_2$, –H	Fe/ 1M HCl	[66]	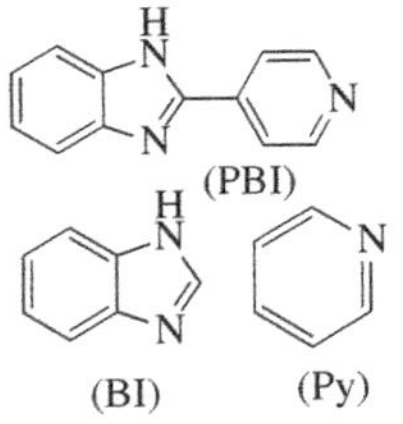 H N N N (PBI) H N N (BI) N (Py)	Mild steel/1M HCl	[67]
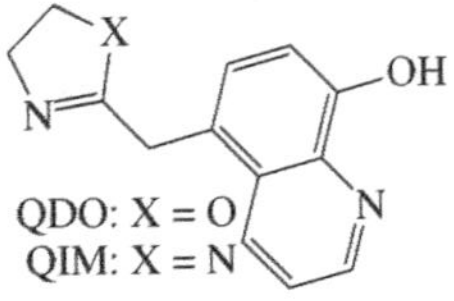 X N OH N QDO: X = O QIM: X = N	Mild steel/1.0M HCl	[68]	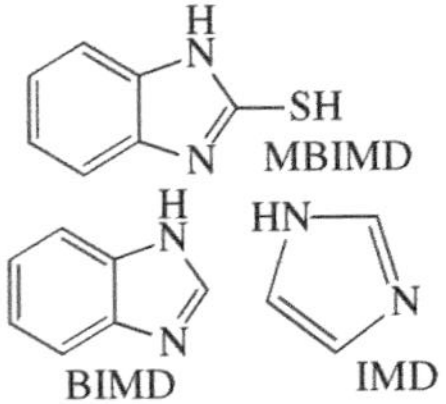 H N N SH MBIMD H N N BIMD HN N IMD	Cu (111) surface	[69]
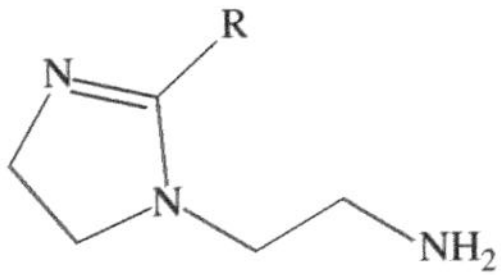 R N N NH_2 R–A: $(CH_2)_6CH_3$; B: $(CH_2)_8CH_3$; C: $(CH_2)_{10}CH_3$;D: $(CH_2)_{12}CH_3$ E: $(CH_2)_{14}CH_3$	Carbon steel/ CO_2	[70]	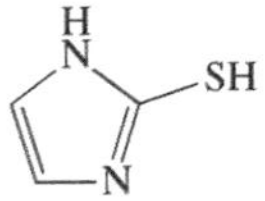 H N SH N	Fe (100) surface	[71]

Inhibitor	Metal/medium	Ref.	Inhibitor	Metal/medium	Ref.
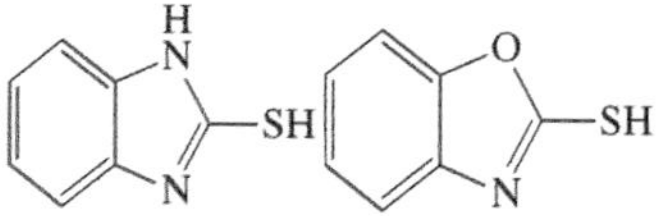	Mild steel/1M HCl	[72]			
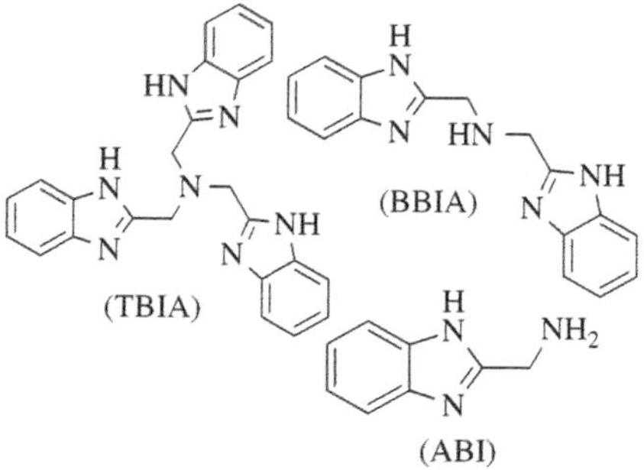 (TBIA) (BBIA) (ABI)	Mild steel/1M HCl	[73, 74]	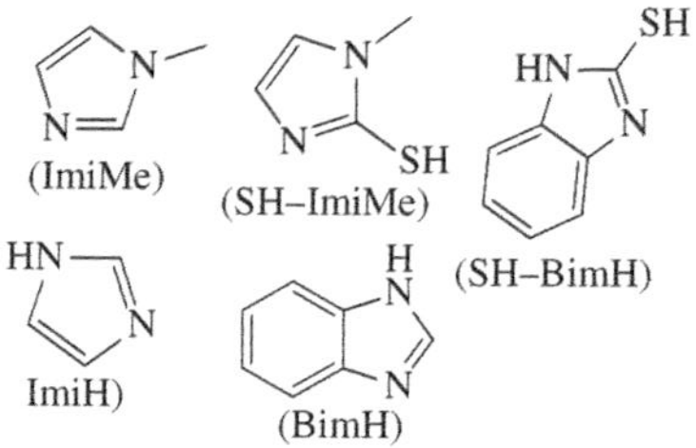 (ImiMe) (SH–ImiMe) (SH–BimH) ImiH) (BimH)	Copper/3wt.% NaCl	[75]
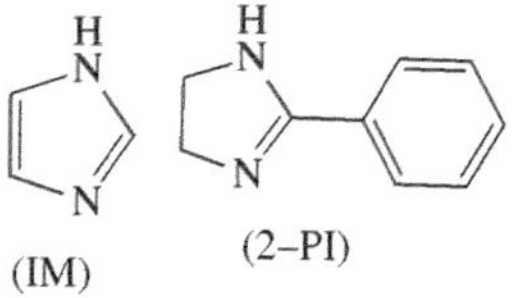 (IM) (2–PI)	AA5052/1.0 M HCl	[76]	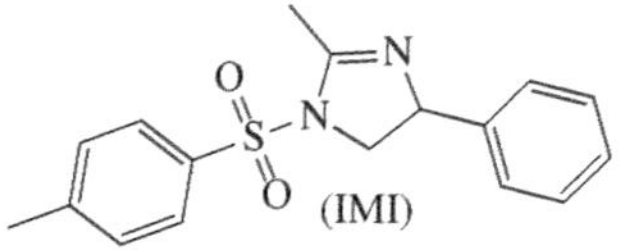 (IMI)	P110 Carbon steel/1.0 M HCl	[77]
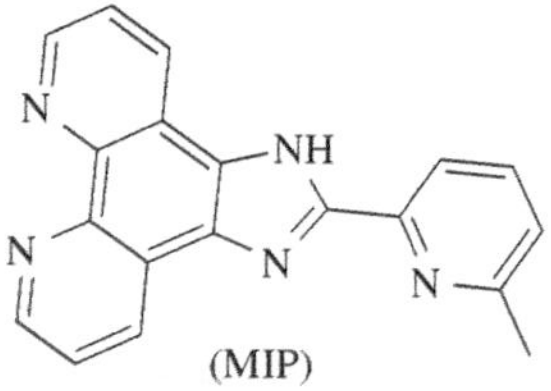 (MIP)	Mild steel/0.5 M H_2SO_4	[78]	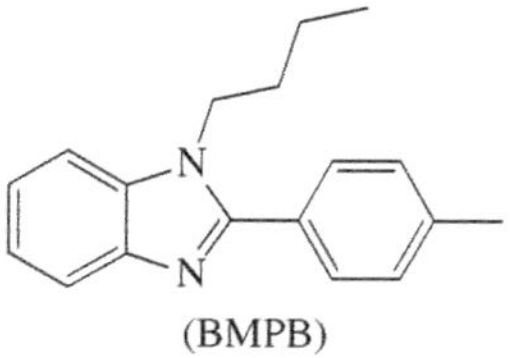 (BMPB)	Mild steel/1M HCl	[79]

(Continued)

Table 6.1 (Continued)

Chemical structures of inhibitors	Metals/electrolytes	References	Chemical structures of inhibitors	Metals/electrolytes	References
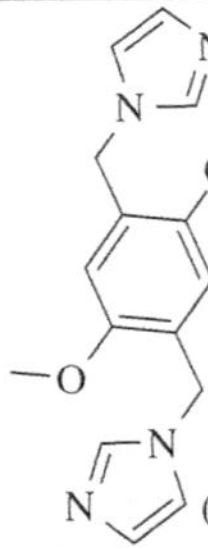 (BIDM) 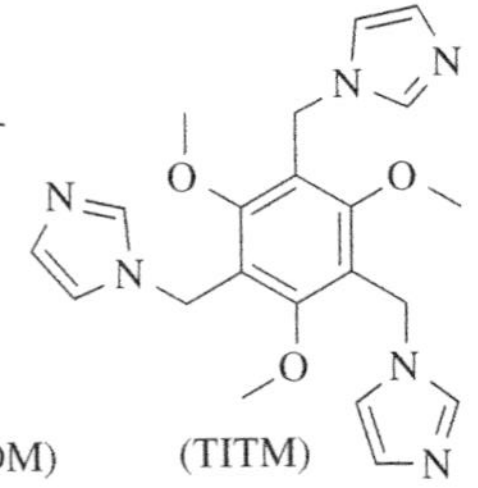 (TITM)	Mild steel/1M HCl	[80]	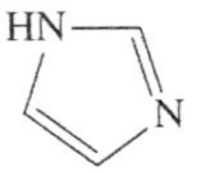 Imidazole 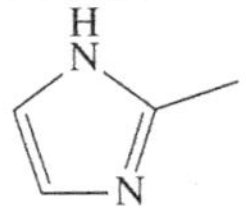 Methyl imidazole	Aluminum/ 0.5 M HCl	[81]
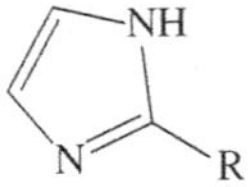 $R = -CH{=}CH_2$; $-CH_2CH{=}CH_2$	Fe/ 1M HCl	[82]	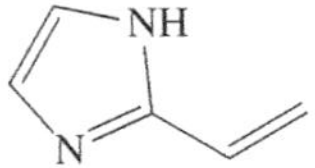 (VI) 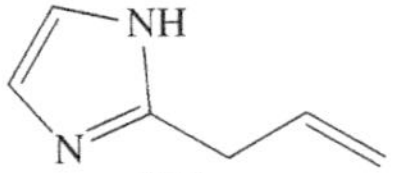 (AI)	Mild steel/1M HCl	[82]
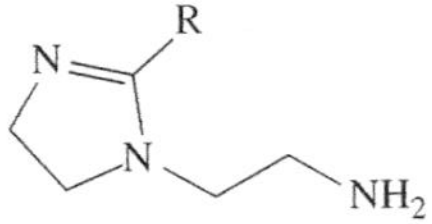 $R = -CF_3$, $-CCl_3$	Fe/ 0.5 M HCl	[83]	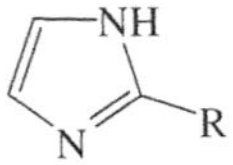 R = –CH3, –Cl	Fe/ 1 M HCl	[84]
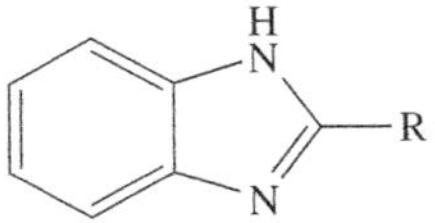 BZ–1:R =– Ph–4–OMe BZ–2:R =– Ph	N80 steel/15% HCl	[85]	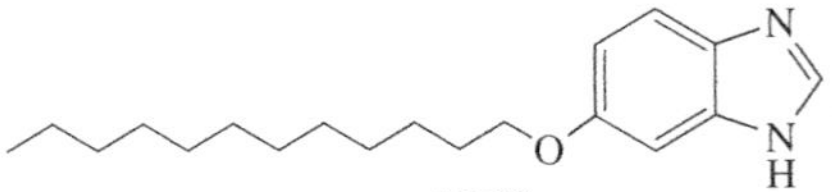 (DBI)	Mild steel/1 M HCl	[76]

Inhibitor	Metal/medium	Ref.	Inhibitor	Metal/medium	Ref.
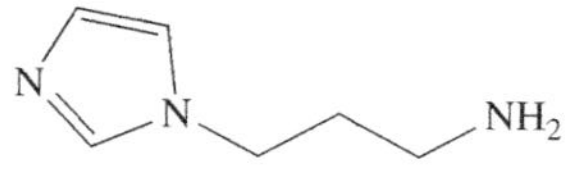 (IMPA)	X70 steel/1 M HCl	[86]	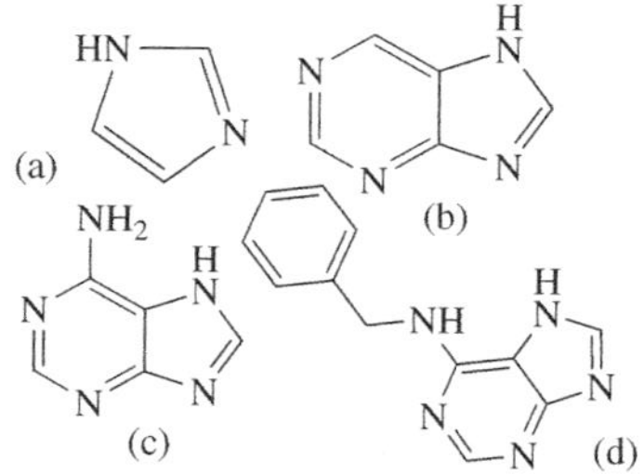 (a) (b) (c) (d)	Cu/ sea water	[87]
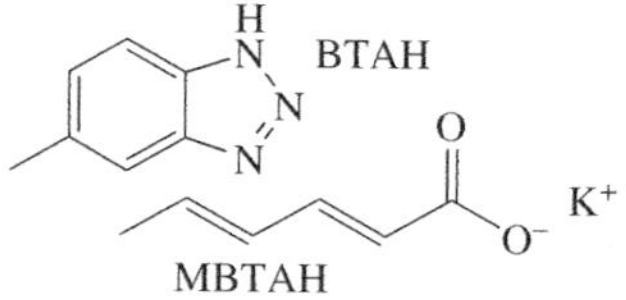 BTAH, MBTAH	Copper/0.01 M H_2SO_4	[88]	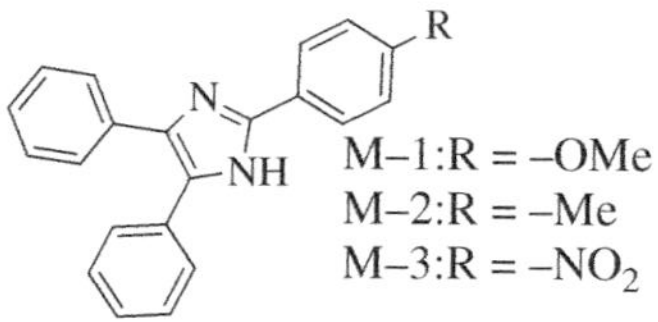 M–1:R = –OMe M–2:R = –Me M–3:R = $–NO_2$	J55 steel/ CO_2-saturated brine solution	[89]
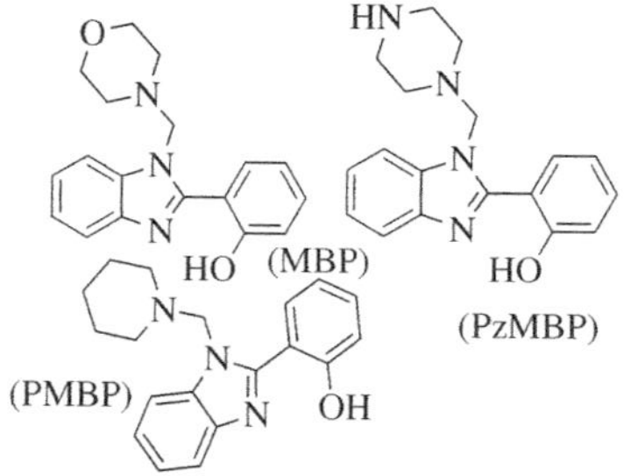 (MBP), (PzMBP), (PMBP)	N80 Steel/15% HCl	[90]	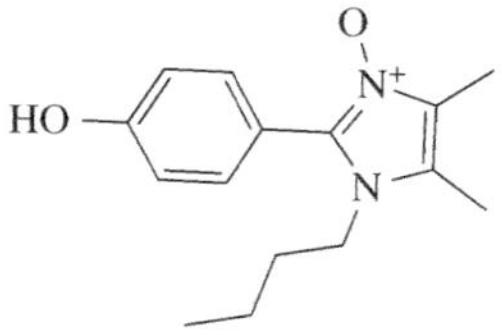	Computational studies only	[91]
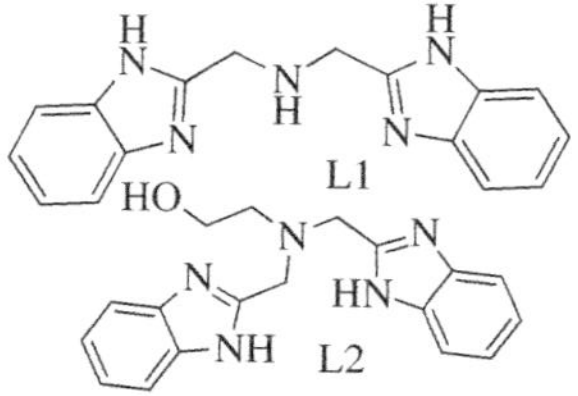 L1, L2	Carbon steel/1 M HCl	[84]	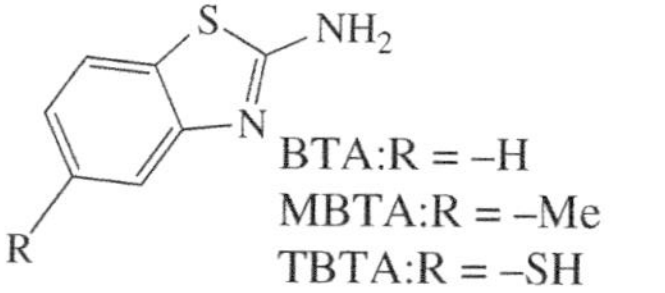 BTA:R = –H MBTA:R = –Me TBTA:R = –SH	Mild steel	[92]

(*Continued*)

Table 6.1 (Continued)

Chemical structures of inhibitors	Metals/electrolytes	References	Chemical structures of inhibitors	Metals/ electrolytes	References
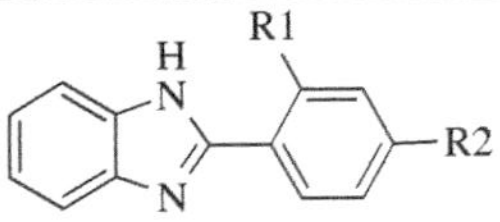 R1 = –H/–NO_2/–NH_2 R2 = –H/–NO_2/–NH_2	Mild steel/1 M HCl	[93]	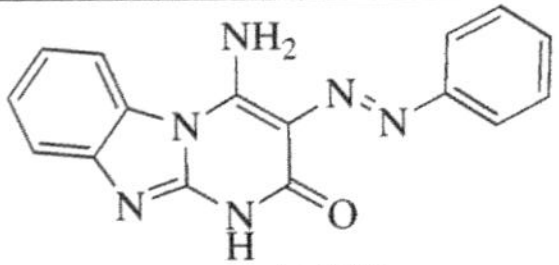 (APIP)	Copper/3.5 wt.% NaCl	[94]
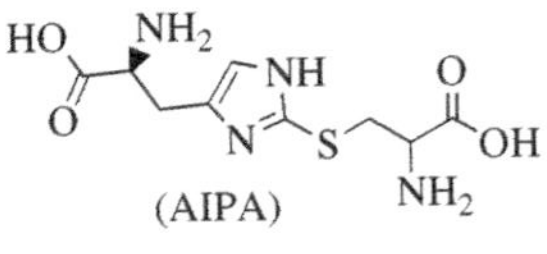 (AIPA)	Mild steel/1 M HCl	[95]	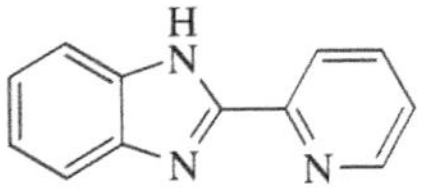 (2PB)	API X60 steel/ CO_2 saturated brine	[96]
(BPMB) 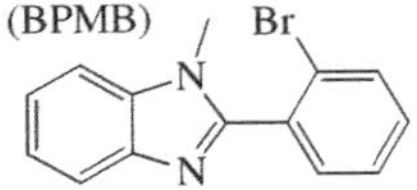  (ME)	Carbon steel/ CO_2-saturated brine	[97]	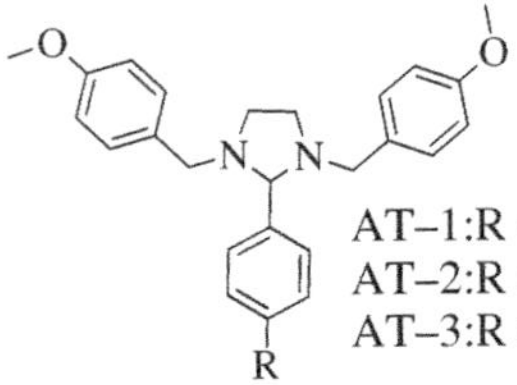 AT–1:R = –Ph–4–OMe AT–2:R = –Ph AT–3:R = –Ph–4–NO_2	J55 steel/ CO_2-saturated Brine	[98]
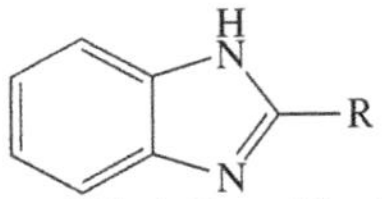 BZ–1:R = –Ph–4–OMe BZ–2:R = –Ph	N80 steel/15% HCl	[85]			

mild steel in 1M HCl. The imidazoline-based compounds act as excellent corrosion inhibitors as compare to isoxazoline based compounds. Both the inhibitors exhibited inhibition effectiveness of more than 85% at 20 mg/l concentration. The findings further demonstrated that inhibitors containing phenyl, methoxy, and naphthyl moieties confirmed maximum protection efficacy compared to the inhibitors without these moieties. PDP measurement showed that the IDZs and ISOs acted as mixed-type corrosion inhibitors and their adsorption mechanism obeyed the Langmuir adsorption isotherm. All the investigational findings were verified by the density functional theory (DFT).

Mihajlovic et al. [87] investigated the anticorrosion behavior of imidazole and its derivatives such as purine, adenine, and 6-benzyl amino purine on copper (Cu) in seawater. The electrochemical analysis explained that these compounds behave as mixed-type corrosion inhibitors. According to the potentiodynamic polarization (PDP) measurements, the inhibition effectiveness at corrosion potential raises in the following order: imidazole <purine <adenine <6-benzyl amino purine, which is in accord with the expectation that the raise of molecular weight and number of heteroatoms in the molecule leads to the raise of inhibition efficacy extent. In a larger possible range, adenine is more effective. The adsorption mechanism of these compounds follows the Langmuir adsorption isotherm. The quantum chemical calculation shows that there is a relationship between the energy gap and inhibition efficiency.

Aljourani et al. [60] reported the anticorrosion behavior of benzimidazole and its derivatives for mild steel in acid solution (1 M HCl). The variation of impedance parameters examined by the change of inhibitor concentration within the range of 50–250 ppm was a sign of their adsorption. The adsorption parameters suggested that these inhibitors delay both the processes (cathodic and anodic) by physisorption and blocking the active sites of corrosion. The adsorption mechanism of these inhibitors on the surface of mild steel obeyed Langmuir's adsorption isotherm. The inhibition efficiency was raised with inhibitor concentration in the order of 2-mercaptobenzimidazole > 2-methylbenzimidazole > benzimidazole, as per the activation energy variation of corrosion.

El-Haddad and Fouda [81] evaluated the anticorrosion effect of imidazole derivatives, namely, imidazole (A) and methyl imidazole (B) on aluminum (Al) in 0.5 M HCl. The findings exhibit that all these imidazole derivatives are excellent corrosion inhibitors. PDP plots confirmed that the imidazole derivatives were of mixed-type inhibitors (Figure 6.1). Tafel slopes (anodic and cathodic) slightly altered in the presence of the imidazole derivatives, which imply that the inhibitors were initially adsorbed on the metallic surface and impeded by merely blocking the reaction sites of the metal surface without influencing the cathodic and anodic reaction mechanism. For imidazole derivatives, the general ground was that the corrosion current density (I_{corr}) reduced and the inhibition efficiency

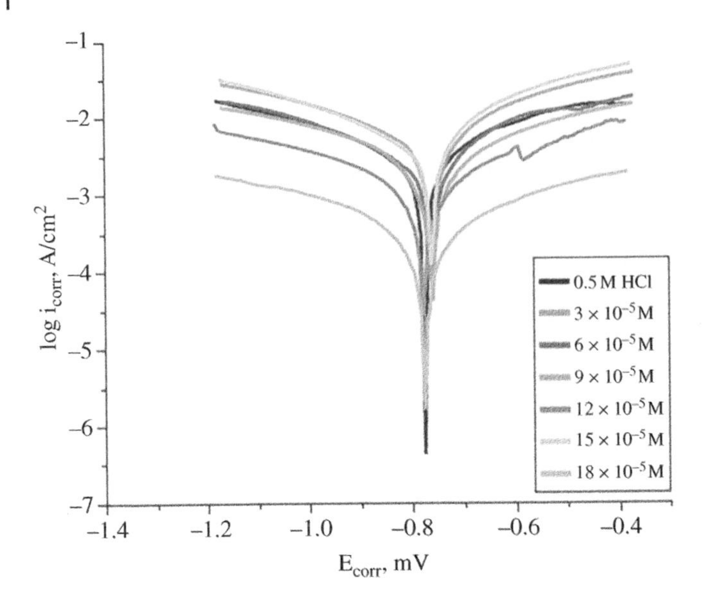

Figure 6.1 PDP plots for the corrosion of Al in 0.5M HCl without and with various concentrations of inhibitor (methyl imidazole) at 30 °C. *Source:* El-Haddad and Fouda [81].

raised with rising inhibitors concentration, and the order of the imidazole derivatives according to their inhibition efficacies are in the following sequence: methyl imidazole > imidazole. The adsorption mechanism of these inhibitors on the Al surface follows the Frumkin adsorption isotherm model. The surface morphology was done by scanning electron microscopy (SEM) to analyze a layer produced on the Al surface. The quantum chemical analysis demonstrates that imidazole derivatives can adsorb on the Al surface via the N atoms, as well as π-electrons in the imidazole ring.

Eldebss et al. [100] investigated various types of novel substituted benzimidazoles embedded with different functional groups that have been prepared from *N*-methyl-2-bromo acetyl benzimidazole. The prepared organic compounds were characterized, and their chemical structures were analyzed by spectral and elemental analysis. In this study, the authors also reported the anticorrosion behavior of few benzimidazole derivatives as corrosion inhibitors.

Ghanbari et al. [62] studied the anticorrosion effect of benzimidazole (BI), 2-methyl benzimidazole (2MBI), and 2-aminobenzimidazole (2ABI) on mild steel in 1 M H_3PO_4 at concentrations between 5×10^{-2} and 10^{-4} M. It was examined that inhibition efficiency raised with raising the concentration of the inhibitor. The adsorption of imidazole derivatives on the mild surface was studied to consider the essential information on the interaction among metal surfaces and inhibitors. Flory–Huggins adsorption isotherm model confirmed that all these inhibitors replaced three or five water molecules on the surface of mild steel. DC polarization and EIS findings explained that the inhibition efficiency of imidazole derivatives followed the order: 2ABI > BI> 2MBI. EIS measurements exhibited that maximum inhibition efficiencies of 2ABI, BI, and 2MBI are 69.5, 62.1, and 51.3%, respectively. The analysis of DC polarization explained that 2ABI, BI, and 2MBI acted as mixed-type inhibitors. The highest inhibition efficiencies of the studied imidazole derivatives were detected at 10^{-4} M concentration.

Zhang et al. [67] studied the anticorrosive behavior of 2-(4-pyridyl)-benzimidazole (PBI) on mild steel in acidic solution (1 M HCl). The inhibition efficacy raised with rising the concentrations of inhibitor but reduced with the rise in acid concentration and temperature. PBI is a mixed-type inhibitor with a predominant effect on cathodic reactions and controls the H^+ ions reduction by merely blocking the reaction sites of the metal surface. For the anodic dissolution, the presence of inhibitor results in the variation of the anodic Tafel slopes. The inhibitor protects the metal from corrosion by the spontaneous adsorption on the metallic surface, and the method deals with Langmuir adsorption isotherm. The theoretical analyses disclose that the PBI adsorption depends on the formation of the coordinative bond between Fe surface and inhibitor molecule.

Zhang et al. [77] synthesized a new imidazoline derivative, 2-methyl-4-phenyl-1-tosyl-4, 5-dihydro-1H-imidazole (IMI), and studied its anticorrosion behavior on P110 carbon steel in acidic solution (1 M HCl). The inhibition effectiveness was raised with the rising IMI inhibitor concentration. The results showed that the IMI acted as a mixed-type inhibitor and follows the Langmuir adsorption isotherm. The surface morphology was performed to examine the surface of metallic samples, illustrating large protection from the acid solution. The Nyquist curves in aggressive solutions without and with inhibitors are given in Figure 6.2. The addition of the IMI inhibitor did not modify the shape of these curves, i.e. the IMI inhibitor managed the activities of the corrosion reaction rather than changed its behavior. The semicircle diameters in the Nyquist curves rise with rising concentrations of IMI inhibitor compared to blank.

Dutta et al. [93] synthesized four novel 2-(substituted phenyl) benzimidazole derivatives (PBI) and studied their corrosion inhibiting effect on mild steel in acidic solution (1M HCl), compared with that of 2-phenylbenzimidazole. The experimental results demonstrated that substitution to the phenyl group attached to benzimidazole brings about obvious raise in inhibition tendency and is related to the property of the substituent group with its position in the phenyl ring. PBI derivatives behave as a mixed-type corrosion inhibitor and maintain almost 90% inhibition efficiency at 1 mM concentration for a long time (96 hours) exposure in acidic solution.

Benali et al. [64] reported the anticorrosion effect of 2-mercapto-1-methylimidazole (MMI) on mild steel in a 5% HCl solution. MMI acted as a mixed type corrosion

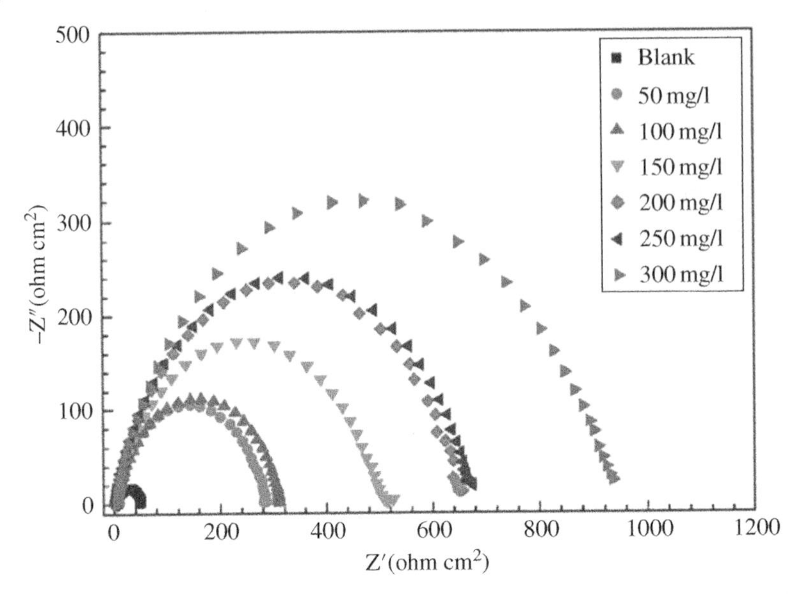

Figure 6.2 Nyquist plots for P110 carbon steel electrode obtained in 1M HCl solution containing various concentrations of IMI inhibitor at 60 °C. *Source:* Zhang et al. [77].

inhibitor and demonstrated 91.18% efficacy at 5×10^{-3} M concentration. The formation of a single semicircle in the Nyquist plots implied that metallic corrosion in acidic solution involves a single charge transfer mechanism. PDP measurements explained that this inhibitor behaved as a cathodic-type corrosion inhibitor and its adsorption mechanism followed the Langmuir adsorption isotherm. The inhibition behavior of various benzimidazole and its derivatives having various other substituents are also studied broadly.

Milosev et al. [75] studied the electrochemical behavior and long-lasting immersion measurements of Cu in 3% NaCl solution in the presence and absence of imidazole (ImiH), benzimidazole (BimH), and its methyl and mercapto derivatives (1-methyl-imidazole (ImiMe), 2-mercapto-1-methyl-imidazole (SH-ImiMe), and 2-mercaptobenzimidazole (SH-BimH)). At 1 mM concentration, the polarization resistance of inhibitors varies noticeably and exhibits the following order: SH-BimH > SH-ImiMe > BimiH > ImiH > ImiMe. These compounds behave as mixed-type corrosion inhibitors with a stronger effect on the anodic side. X-ray photoelectron investigation explained that the variations in inhibition effectiveness are related to variations in composition and development of protective layer formed.

Singh et al. [89] investigated the anticorrosion behavior of three novel imidazole derivatives, namely, 2-(4-methoxyphenyl)-4,5-diphenyl-imidazole (M-1), 4,5-diphenyl-2-(p-tolyl)-imidazole (M-2), and 2-(4-nitrophenyl)-4,5-diphenyl-imidaz

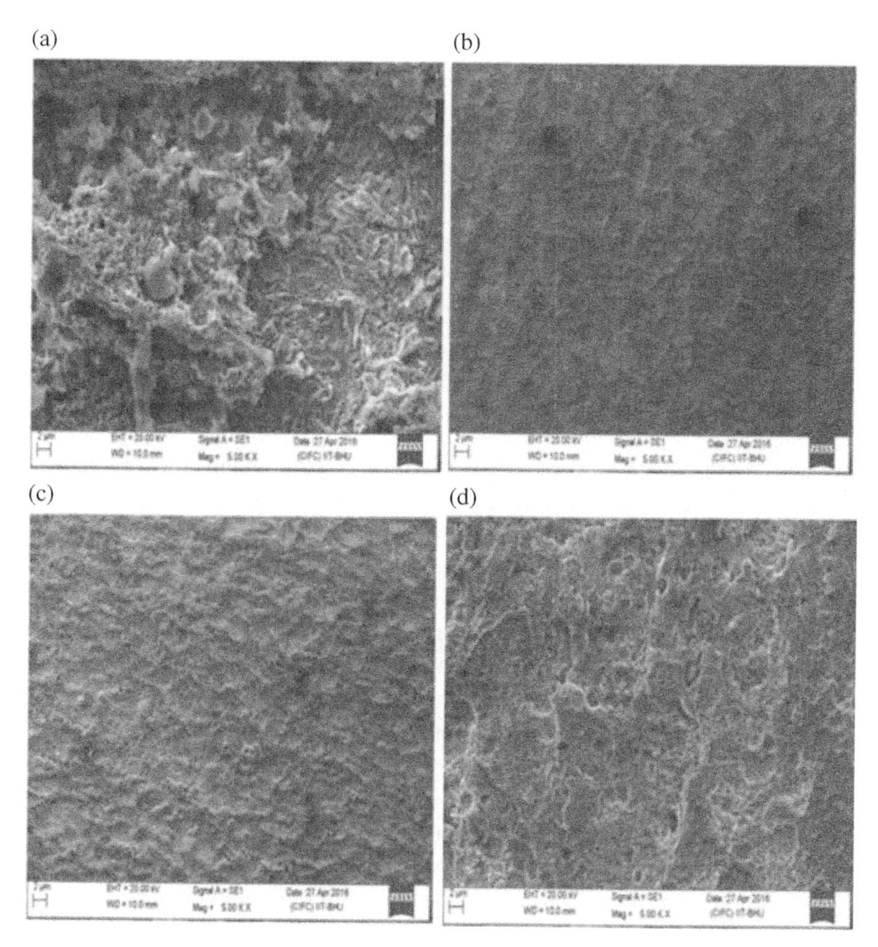

Figure 6.3 SEM micrographs for (a) blank, (b) M-1, (c) M-2, and (d) M-3. *Source:* Singh et al. [89].

ole (M-3) on J55 steel in CO_2-saturated brine solution. M-1 inhibitor explained the highest inhibition efficiency of 93% at 400 mg/l concentration. PDP result exposes that these compounds are mixed-type corrosion inhibitors but suppressing more anodic reactions. The adsorption of all the three imidazole derivatives followed the Langmuir adsorption isotherm. The micrographs of SEM measurement are given in Figure 6.3a–d. Without inhibitor the surface of J55 steel is severely damaged (Figure 6.3a), carrying high corrosion. Though with the presence of inhibitors, the metallic surface attains a smooth morphology (Figure 6.3b–d), i.e. the metal surface is shielded from corrosion. The electrochemical analysis also supports these results.

Table 6.1 shows the abbreviations and chemical structure of imidazole and its derivatives that are used as corrosion inhibitors for different metals and alloys in various electrolytic solutions. Generally, in substituted imidazole molecules, the imidazole ring interrelates with the metal surface as it is extremely electron-rich and behaves as an adsorption center. The presence of substituents plays a significant function in the electron densities delocalization. Typically, electron-donating substituents increase the protection efficiency and the converse is true for electron-withdrawing substituents. The anticorrosion behavior of imidazole and its derivatives are also described in various reports [101–104].

6.5 Computational Studies

Nowadays, computational studies are appeared as a significant tool to theoretically elucidate the corrosion inhibition behavior of heterocyclic organic molecules [105]. The most important significance of computational studies is that the anticorrosion behavior of any compounds or reactivity can be theoretically calculated before their syntheses. This will help in the aiming of efficient corrosion inhibitors without their costly and contaminated laboratory syntheses. The dimension of the molecule responsible for interactions with the surface of metal along with the orientation of inhibitor molecules can be evaluated using these simulation tools. The computational simulations provide numerous theoretical parameters in the term of which global chemical reactivities, as well as corrosion inhibition efficiency of inhibitors, can be explained [105, 106].

The computational modeling based on DFT analysis gives frontier molecular orbital (FMO) illustrations in the term of which molecular active sites responsible for interactions with a metal surface can be depicted. The behavior of substituents on the corrosion inhibition can also be explained using the FMO illustration. Usually, electron-releasing substituents improve the highest occupied molecular orbital (HOMO) or lowest unoccupied molecular orbital (LUMO) FMO contribution, and inhibition efficiency. Tinga et al. [107] investigated the anticorrosion effect of levamisole (LMS) and 4-phenyl imidazole (PIZ) on copper (Cu) in H_2SO_4 solution. At 8 mM concentration, the highest corrosion inhibition efficacies of LMS and PIZ are 99.03 and 95.84%, respectively. All the experimental results found that LMS had superior corrosion inhibition performance than PIZ. LMS is a cathodic corrosion inhibitor, while PIZ belongs to a mixed-type inhibitor. The electrochemical impedance spectroscopy (EIS) analysis shows that the corrosion inhibition effectiveness has a similar tendency when the concentration of inhibitor rises. The adsorption mechanism of these inhibitors follows the Langmuir adsorption isotherm. To further investigate the relationship between the chemical structures of LMS/PIZ and the corrosion inhibition effect, quantum chemical calculations

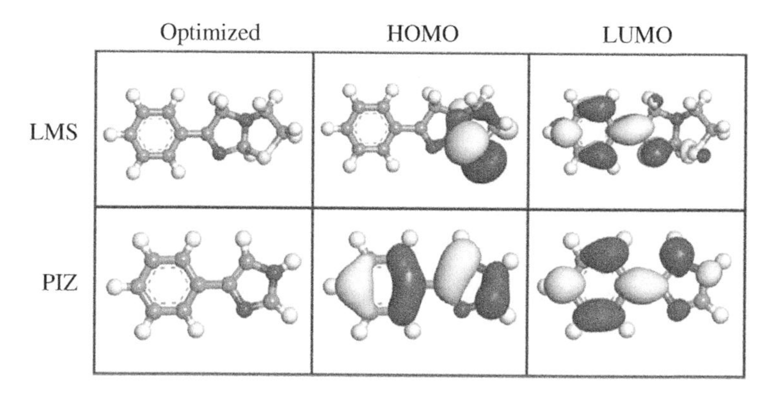

Figure 6.4 Optimized structures of LMS and PIZ (HOMO/LUMO). *Source:* Tinga et al. [107].

were used. The optimized structure of the tested inhibitor based on the FMO theory, LUMO and HOMO are shown in Figure 6.4. Known by the FMO theory, the energy of an electron in the HOMO orbit is high and easy to lose electrons, whereas the LUMO has the lowest energy in all unoccupied orbits and is easy to accept electrons.

Molecular dynamic (MD) and Monte Carlo (MC) simulations are two other useful computational modeling tools that give information about the orientation of corrosion inhibitors on the metal surface. Dohare et al. [108] synthesized three substituted imidazoles, namely, 2-(3-methoxyphenyl)-4,5-diphenyl-1H-imidazole (IM1), 2,4,5-triphenyl-1H-imidazole (IM-2), and 2-(3-nitrophenyl)-4,5-diphenyl-1 Himidazole (IM-3) and studied its corrosion inhibiting behavior by weight loss, EIS, and PDP measurements. The experimental findings demonstrate that methoxy substituted imidazole carried out a good corrosion inhibitor than NO_2-substituted imidazole. These results were confirmed via DFT and MD simulation analysis. IM-1 was found to show the highest inhibition efficiency of 97.5% at 100 mg/l among all the tested imidazoles. PDP measurement explained that all these inhibitors predominantly acted as cathodic inhibitors. The adsorption mechanism of these inhibitors obeyed Langmuir adsorption isotherm. A protective layer formed on the surface of mild steel was verified via SEM and AFM. MD simulation tool explained that the binding energy and interaction energy of the inhibitor molecules on the metal surface followed the order: IM-1> IM-2> IM-3. All these imidazole molecules are adsorbed onto the metal surface via smooth orientation. Figure 6.5 signifies the top view and side view of imidazole inhibitors' adsorption onto the metal surface.

To increase the contact surface area, the inhibitor molecules slowly moved parallel to the iron surface, which could improve the degree of surface coverage, by

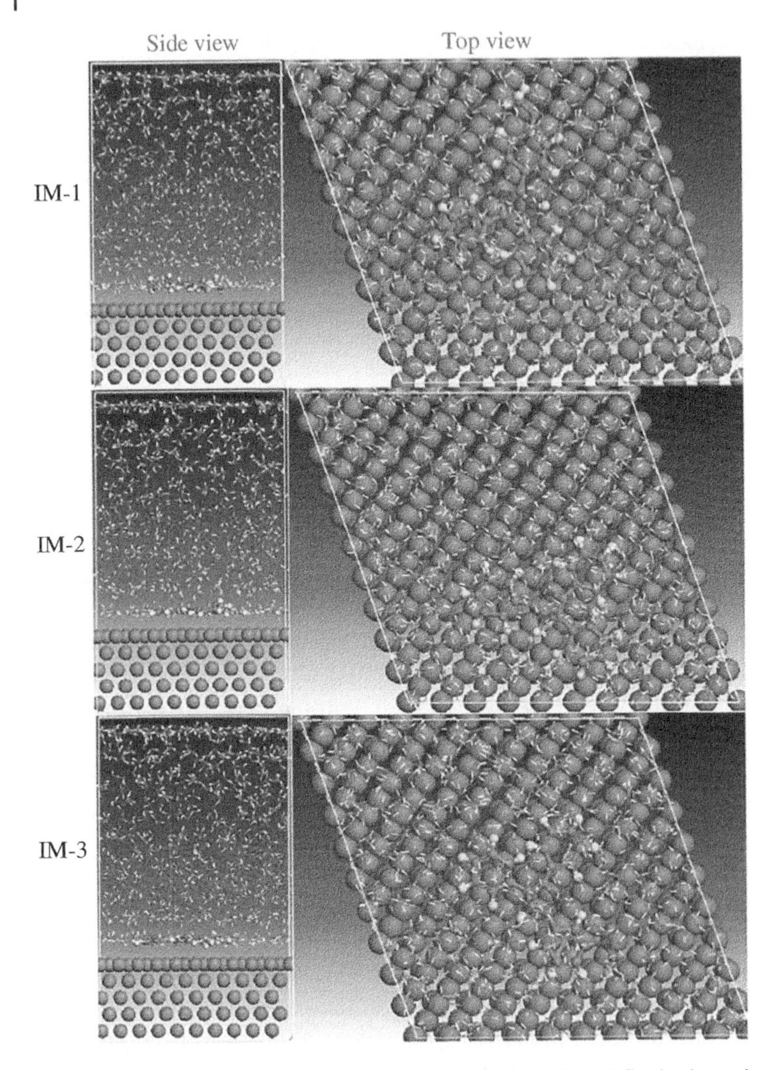

Figure 6.5 Top view and side view of the neutral inhibitors' final adsorption on the Fe (110) surface in aqueous solution. *Source:* Dohare et al. [108].

forming a protective layer. There are important interactions among the tested inhibitors and Fe atoms, generally because of the existence of π bonds in phenyl rings, as well as of N atoms in imidazole rings. In the literature, it is studied that more negative interaction energy values show the maximum adsorption capacity of an inhibitor onto the iron surface [109–112]. Moreover, the higher magnitude of binding energies means easier inhibitor molecule adsorption onto the metal surface and higher stability of the adsorptive system [113]. The lowest IM-3 inhibition efficiency is ascribed to the electrons withdrawing ($-NO_2$) [114], whereas the maximum inhibition efficiency of IM-1 is owing to the electron-donating group ($-OCH_3$) [115]. The orientation analysis of corrosion inhibitors on the metal surface gives essential information concerning the relative effectivity of various inhibitors as a compound with smooth/horizontal orientation will improve corrosion inhibitor that of the compound with vertical orientation. The substituent's effect on the adsorption performance of heterocyclic organic compounds can also be verified using these tools. Usually, in the presence of electron-withdrawing and donating substituent, inhibitor molecules obtain a vertical and flat orientation, respectively. Although studying the corrosion inhibition performance of three heterocyclic organic compounds [116] obtained from imidazole, namely 2-(1,4,5-triphenyl-1H-imidazol-2-yl)phenol (IM-OH), 1,4,5-triphenyl-2-(4 -metyoxyphenyl)-1H-imidazole (IM-OCH_3), and 3-methoxy-4-(1,4,5-triphenyl-1 H-imidazole-2-yl)phenol (IM-H) on mild steel in 0.5M H_2SO_4 medium. The obtained findings show that IM-OH, IM-OCH3, and IM-H behave as excellent inhibitors. EIS results explain that IM-OH, IM-OCH3, and IM-H imparted large resistance and act as mixed-type corrosion inhibitors. The inhibition efficiency rises with the raise of inhibitor concentration to attain 97.7% (IM-OH), 98.9% (IM-OCH_3), and 88.9% (IM-H) at 10^{-3} M.

6.6 Conclusions

In this chapter, we discussed the current development of imidazole and its derivatives as corrosion inhibitors for different metals and alloys in various electrolytic solutions. The inhibition behavior of imidazole and its derivatives was analyzed by various techniques such as chemical, electrochemical, surface, and computational. Most of the imidazole and its derivatives acted as mixed-type inhibitors and their adsorption mechanism followed the Langmuir adsorption isotherm. Imidazole derivatives have excellent biological and industrial activities along with the corrosion inhibition effect. It is observed that most of the imidazole-based corrosion inhibitors are obtained for carbon steel/mild steel in hydrochloric acid solution. Thus, the inhibition effect of imidazole and its derivatives should also be examined for other metals and alloys in various electrolytes. Moreover, their corrosion inhibition effect can also be done by computation studies that are novel, environmentally friendly, and cost-effective techniques.

References

1 Kharbach, Y., Qachchachi, F.Z., Haoudi, A. et al. (2017). Anticorrosion performance of three newly synthesized isatin derivatives on carbon steel in hydrochloric acid pickling environment: electrochemical, surface and theoretical studies. *J. Mol. Liq. 246*: 302.

2 Al Hamzi, A.H., Zarrok, H., Zarrouk, A. et al. (2013). The role of acridin-9(10H)-one in the inhibition of carbon steel corrosion: thermodynamic, electrochemical and DFT studies. *Int. J. Electrochem. Sci. 8* (2): 2586.
3 Hegazy, M.A., El-Etre, A.Y., El-Shafaie, M., and Berry, K.M. (2016). Novel cationic surfactants for corrosion inhibition of carbon steel pipelines in oil and gas wells applications. *J. Mol. Liq. 214*: 347.
4 El Guerraf, A., Titi, A., Cherrak, K. et al. (2018). The synergistic effect of chloride ion and 1,5-diaminonaphthalene on the corrosion inhibition of mild steel in 0.5 M sulfuric acid: experimental and theoretical insights. *Surf. Interfaces 13*: 168.
5 Bashir, S., Sharma, V., Lgaz, H. et al. (2018). The inhibition action of analgin on the corrosion of mild steel in acidic medium: a combined theoretical and experimental approach. *J. Mol. Liq. 263*: 454.
6 El-Hajjaji, F., Belkhmima, R.A., Zerga, B. et al. (2014). Temperature performance of a thionequinoxaline compound as mild steel corrosion inhibitor in hydrochloric acid medium. *Int. J. Electrochem. Sci. 9*: 4721.
7 Salghi, R., Ben Hmamou, D., Benali, O. et al. (2015). Study of the corrosion inhibition effect of pistachio essential oils in 0.5 M H2SO4. *Int. J. Electrochem. Sci. 10*: 8403.
8 Mai, W., Soghrati, S., and Buchheit, R.G. (2016). A phase field model for simulating the pitting corrosion. *Corros. Sci. 110*: 157.
9 Verma, C., Ebenso, E.E., Bahadur, I., and Quraishi, M.A. (2018). An overview on plant extracts as environmental sustainable and green corrosion inhibitors for metals and alloys in aggressive corrosive media. *J. Mol. Liq. 266*: 577.
10 Nazeer, A.A. and Madkour, M. (2018). Potential use of smart coatings for corrosion protection of metals and alloys: a review. *J. Mol. Liq. 253*: 11.
11 Kıcır, N., Tansuğ, G., Erbil, M., and Tüken, T. (2016). Investigation of ammonium (2, 4-dimethylphenyl)-dithiocarbamate as a new, effective corrosion inhibitor for mild steel. *Corros. Sci. 105*: 88.
12 Lavanya, K., Saranya, J., and Chitra, S. (2018). Recent reviews on quinoline derivative as corrosion inhibitors. *Corros. Rev. 36* (4): 365.
13 Bammou, L., Belkhaouda, M., Salghi, R. et al. (2014). Corrosion inhibition of steel in sulfuric acidic solution by the chenopodium ambrosioides extracts. *J. Assoc. Arab Univ. Basic Appl. Sci. 16* (1): 83.
14 Daoud, D., Douadi, T., Issaadi, S., and Chafaa, S. (2014). Adsorption and corrosion inhibition of new synthesized thiophene Schiff base on mild steel X52 in HCl and H2SO4 solutions. *Corros. Sci. 79*: 50.
15 Mishra, A., Aslam, J., Verma, C. et al. (2020). Imidazoles as highly effective heterocyclic corrosion inhibitors for metals and alloys in aqueous electrolytes: a review. *J. Taiwan Inst. Chem. Eng. 114*: 341–358a.
16 http://insights.globalspec.com/article/2340/annual-global-cost-of-corrosion-2-5-trillion

17 http://www.mintek.co.za/2011/11/15/the-high-cost-of-corrosion

18 http://impact.nace.org/documents/appendix-a.pdf

19 Lim, H.L. (2012). Assessing level and effectiveness of corrosion education in the UAE. *Int. J. Corros.* 2012 (7): 1–10. Article ID 785701. doi: https://doi.org/10.1155/2012/785701.

20 Goyal, M., Kumar, S., Bahadur, I. et al. (2018). Organic corrosion inhibitors for industrial cleaning of ferrous and non-ferrous metals in acidic solutions: a review. *J. Mol. Liq. 256*: 565–573.

21 Verma, C., Ebenso, E.E., and Quraishi, M. (2017). Ionic liquids as green and sustainable corrosion inhibitors for metals and alloys: an overview. *J. Mol. Liq. 233*: 403–414.

22 Verma, C., Olasunkanmi, L., Ebenso, E.E., and Quraishi, M. (2018). Substituents effect on corrosion inhibition performance of organic compounds in aggressive ionic solutions: a review. *J. Mol. Liq. 251*: 100–118.

23 Koch, G. (2017). Cost of corrosion. In: *Trends in Oil and Gas Corrosion Research and Technologies*, 1ee (ed. A.M. El-Sherik), 3–30. Boston: Woodhead Publishing, Elsevier.

24 Singh, R. and Dahotre, N.B. (2007). Corrosion degradation and prevention by surface modification of biometallic materials. *J. Mater. Sci. Mater. Med. 18*: 725–751.

25 Upadhyay, D., Panchal, M.A., Dubey, R., and Srivastava, V. (2006). Corrosion of alloys used in dentistry: a review. *Mat. Sci. Eng.: A* 432: 1–11.

26 Corrosion of iron: 4,700 w (1907). *Electrochemical and Metallurgical Industry*, vol. 5, 363. Gives in condensed form papers by Walker and Cushman. See also editorial, p. 343.

27 (a) Noor, E.A. and Al-Moubaraki, A.H. (2008). *Mater. Chem. Phys. 110*: 145–154;(b) Abdallah, M., Al-Tass, H.M., AL Jahdaly, B.A., and Fouda, A.S. (2016). *J. Mol. Liq. 216*: 590–597;(c)Mostafa, H.A., Abd EI-Maksoud, S.A., and Moussa, M.N.H. (2001). *Port. Electrochim. Acta* 19: 109–120.

28 Cushman, A.S. and Gardner, H.A. (1910). *The Corrosion and Preservation of Iron and Steel*. McGraw-Hill.

29 Hiromoto, S., Tsai, A.-P., Sumita, M., and Hanawa, T. (2000). Effects of surface finishing and dissolved oxygen on the polarization behavior of Zr65Al7. 5Ni10Cu17. 5 amorphous alloy in phosphate buffered solution. *Corros. Sci. 42*: 2167–2185.

30 Venkatraman, M.S., Cole, I.S., and Emmanuel, B. (2011). Model for corrosion of metals covered with thin electrolyte layers: pseudo-steady state diffusion of oxygen. *Electrochim. Acta 56*: 7171–7179.

31 Karthik, R., Muthukrishan, P., Chen, S.M. et al. (2015). Anti-corrosion inhibition of mild steel in 1M hydrochloric acid solution by using tiliacoraaccuminata leaves extract. *Int. J. Electrochem. Sci. 10*: 3707.

32 Ghibate, R., Sabry, F., Kharbach, Y. et al. (2015). Valuation of surfactant phosphonates synthesized in the protection of metal surfaces against corrosion of mild steel in 0.5M H2SO4 media. *Int. J. Eng. Res. Appl.* 5 (12): 22.

33 Fouda, A.S., Shalabi, K., and Elmogazy, H. (2014). Corrosion inhibition of α-brass in HNO3 by indole and 2-oxyindole. *J. Mater. Environ. Sci.* 5 (6): 1691.

34 Ouakki, M., Rbaa, M., Galai, M. et al. (2018). Experimental and quantum chemical investigation of imidazole derivatives as corrosion inhibitors on mild steel in 1.0 M hydrochloric acid. *J. Bio-Tribo. Corros.* 4: 35.

35 Petersen, A., Rodrigues, S.R., Dalmoro, V. et al. (2017). Anthocyanins as a corrosion inhibitor for 2024-T3 aluminum alloys: a study of electrochemical behavior. *Int. J. Corros. Scale Inhib.* 6 (3): 29.

36 Negm, N.A., El Hashash, M.A., Abd-Elaal, A. et al. (2018). Amide type nonionic surfactants: synthesis and corrosion inhibition evaluation against carbon steel corrosion in acidic medium. *J. Mol. Liq.* 256: 574.

37 Ghazoui, A., Zarrouk, A., Bencaht, N. et al. (2014). New possibility of mild steel corrosion inhibition by organic heterocyclic compound. *J. Chem. Pharm. Res.* 6 (2): 704.

38 Bedair, M.A., El-Sabbah, M.M.B., Fouda, A.S., and Elaryian, H.M. (2017). Synthesis, electrochemical and quantum chemical studies of some prepared surfactants based on azodye and Schiff base as corrosion inhibitors for steel in acid medium. *Corros. Sci.* 128: 54.

39 Verma, C., Olasunkanmi, L.O., Ebenso, E.E. et al. (2016). Adsorption behavior of glucosamine-based, pyrimidine- fused heterocycles as green corrosion inhibitors for mild steel: experimental and theoretical studies. *J. Phys. Chem. C* 120 (21): 11598.

40 Shainy, K.M., Ammal, P.R., Unni, K.N. et al. (2016). Surface interaction and corrosion inhibition of mild steel in hydrochloric acid using pyoverdine, an eco-friendly biomolecule. *J. Bio-Tribo. Corros.* 2: 20.

41 Jiang, L., Qiang, Y., Lei, Z. et al. (2018). Excellent corrosion inhibition performance of novel quinoline derivatives on mild steel in HCl media: experimental and computational investigations. *J. Mol. Liq.* 255: 53.

42 Alaoui, K., El Kacimi, Y., Galai, M. et al. (2016). Poly(1-phenylethene): as a novel corrosion inhibitor for carbon steel/hydrochloric acid interface. *Anal. Bioanal. Electrochem.* 8 (7): 830.

43 Galai, M., Rbaa, M., El Kacimi, Y. et al. (2017). Anti-corrosion properties of some triphenylimidazole substituted compounds in corrosion inhibition of carbon steel in 1.0 M hydrochloric acid solution. *Anal. Bioanal. Electrochem.* 9: 80.

44 Abd El-Raouf, M., Khamis, E.A., Abou Kana, M.T.H., and Negm, N.A. (2018). Electrochemical and quantum chemical evaluation of new bis (coumarins) derivatives as corrosion inhibitors for carbon steel corrosion in 0.5 M H2SO4. *J. Mol. Liq.* 255: 341.

45 Shaban, A., Felhosi, I., and Telegdi, J. (2017). Laboratory assessment of inhibition efficiency and mechanism of inhibitor blend (P22SU) on mild steel corrosion in high chloride containing water. *Int. J. Corros. Scale Inhib.* 6 (3): 262.

46 He, X., Jiang, Y., Li, C. et al. (2014). Inhibition properties and adsorption behavior of imidazole and 2-phenyl- 2-imidazoline on AA5052 in 1.0 M HCl solution. *Corros. Sci. 83*: 124.

47 Ansari, F.A., Verma, C., Siddiqui, Y.S. et al. (2018). Volatile corrosion inhibitors for ferrous and non-ferrous metals and alloys: a review. *Int. J. Corros. Scale Inhib.* 7 (2): 126.

48 Qiao, L., Wang, Y., Wang, W. et al. (2014). The preparation and corrosion performance of self-assembled monolayers of stearic acid and MgO layer on pure magnesium. *Mater. Trans. 55* (8): 1337.

49 Ruan, L., Zhang, Z., Huang, X. et al. (2017). Evaluation of corrosion inhibition of two schiff bases selfassembled films on carbon steel in 0.5 M HCl. *Int. J. Electrochem. Sci. 12*: 103.

50 Bousskri, A., Anejjar, A., Messali, M. et al. (2015). Corrosion inhibition of carbon steel in aggressive acidic media with 1-(2-(4-chlorophenyl)-2-oxoethyl) pyridazinium bromide. *J. Mol. Liq. 211*: 1000.

51 Arellanes-Lozada, P., Olivares-Xometl, O., Likhanova, N.V. et al. (2018). Adsorption and performance of ammonium-based ionic liquids as corrosion inhibitors of steel. *J. Mol. Liq. 265*: 151.

52 Umoren, S.A. and Solomon, M.M. (2015). Effect of halide ions on the corrosion inhibition efficiency of different 4 organic species: a review. *J. Ind. Eng. Chem. 21*: 81.

53 Verma, C., Olasunkanmi, L.O., Ebenso, E.E., and Quraishi, M.A. (2018). Adsorption characteristics of green 5-arylaminomethylene pyrimidine2,4,6-triones on mild steel surface in acidic medium: experimental and computational approach. *Results Phys. 8*: 657.

54 Hantzsch, A. and Weber, J. (1887). Ueber verbindungen des thiazols (pyridins der thiophenreihe). *Ber. Dtsch. Chem. Ges. 20*: 3118–3132.

55 Christen, D., Griffiths, J.H., and Sheridan, J. (1981). The microwave spectrum of imidazole; complete structure and the electron distribution from nuclear quadrupole coupling tensors and dipole moment orientation. *Zeitschrift für Naturforschung A. 36*: 1378–1385.

56 Alaoui, K., Dkhireche, N., Ebn Touhami, M. et al. (2020). Review of application of imidazole and imidazole derivatives as corrosion inhibitors of metals. In: *New Challenges and Industrial Applications for Corrosion Prevention and Control*, 1ee (eds. Y. El Kacimi, S. Kaya and R. Touir), 101–131. IGI Global.

57 Joule, J.A. and Mills, K. (2008). *Heterocyclic Chemistry*. Wiley.

58 Katritzky, A.R., Ramsden, C.A., Joule, J.A., and Zhdankin, V.V. (2010). *Handbook of Heterocyclic Chemistry*. Elsevier.

59 Gribble, G.W. and Joule, J. (2009). *Progress in Heterocyclic Chemistry*. Elsevier.

60 Aljourani, J., Raeissi, K., and Golozar, M. (2009). Benzimidazole and its derivatives as corrosion inhibitors for mild steel in 1M HCl solution. *Corros. Sci. 51*: 1836–1843.

61 Aljourani, J., Golozar, M., and Raeissi, K. (2010). The inhibition of carbon steel corrosion in hydrochloric and sulfuric acid media using some benzimidazole derivatives. *Mater. Chem. Phys. 121*: 320–325.

62 Ghanbari, A., Attar, M., and Mahdavian, M. (2010). Corrosion inhibition performance of three imidazole derivatives on mild steel in 1 M phosphoric acid. *Mater. Chem. Phys. 124*: 1205–1209.

63 Khaled, K. (2011). Experimental and computational investigations of corrosion and corrosion inhibition of iron in acid solutions. *J. Appl. Electrochem. 41*: 277–287.

64 Benali, O., Larabi, L., and Harek, Y. (2010). Influence of the 2-Mercapto-1-Methyl Imidazole (MMI) on the corrosion inhibition of mild steel in 5% HCl. *J. Saud. Chem. Soc. 14*: 231–235.

65 Mendes, J.O., da Silva, E.C., and Rocha, A.B. (2012). On the nature of inhibition performance of imidazole on iron surface. *Corros. Sci. 57*: 254–259.

66 Zhang, J., Qiao, G., Hu, S. et al. (2011). Theoretical evaluation of corrosion inhibition performance of imidazoline compounds with different hydrophilic groups. *Corros. Sci. 53*: 147–152.

67 Zhang, F., Tang, Y., Cao, Z. et al. (2012). Performance and theoretical study on corrosion inhibition of 2-(4-pyridyl)-benzimidazole for mild steel in hydrochloric acid. *Corros. Sci. 61*: 1–9.

68 Rbaa, M. and Lakhrissi, B. (2019). Novel oxazole and imidazole based on 8-hydroxyquinoline as a corrosion inhibition of mild steel in HCl Solution: insights from Experimental and Computational Studies. *Surf. Interfaces 15*: 43–59.

69 Sun, S., Geng, Y., Tian, L. et al. (2012). Density functional theory study of imidazole, benzimidazole and 2-mercaptobenzimidazole adsorption onto clean Cu (1 1 1) surface. *Corros. Sci. 63*: 140–147.

70 Zhang, J., Liu, J., Yu, W. et al. (2010). Molecular modeling of the inhibitionmechanismof 1-(2-aminoethyl)-2-alkyl-imidazoline. *Corros. Sci. 52*: 2059–2065.

71 Radilla, J., Negron Silva, G.E., Palomar Pardave, M. et al. (2013). DFT study of the adsorption of the corrosion inhibitor 2-mercaptoimidazole onto Fe (1 0 0) surface. *Electrochim. Acta 112*: 577–586.

72 Mahdavian, M. and Ashhari, S. (2010). Corrosion inhibition performance of 2-mercaptobenzimidazole and 2-mercaptobenzoxazole compounds for protection of mild steel in hydrochloric acid solution. *Electrochim. Acta 55*: 1720–1724.

73 Tang, Y., Zhang, F., Hu, S. et al. (2013). Novel benzimidazole derivatives as corrosion inhibitors of mild steel in the acidic media. Part I: gravimetric, electrochemical, SEM and XPS studies. *Corros Sci. 74*: 271–282.

74 Cao, Z., Tang, Y., Cang, H. et al. (2014). Novel benzimidazole derivatives as corrosion inhibitors of mild steel in the acidic media. Part II: theoretical studies. *Corros. Sci. 83*: 292–298.

75 Milosev, I., Kovacevic, N., Kovac, J., and Kokalj, A. (2015). The roles of mercapto, benzene and methyl groups in the corrosion inhibition of imidazoles on copper: I. Experimental characterization. *Corros. Sci. 98*: 107–118.

76 Zhang, D., Tang, Y., Qi, S. et al. (2016). The inhibition performance of longchain alkyl-substituted benzimidazole derivatives for corrosion of mild steel in HCl. *Corros. Sci. 102*: 517–522.

77 Zhang, L., He, Y., Zhou, Y. et al. (2015). A novel imidazoline derivative as corrosion inhibitor for P110 carbon steel in hydrochloric acid environment. *Petroleum. 1*: 237–243.

78 Obi-Egbedi, N., Obot, I.B., and Eseola, A. (2014). Synthesis, characterization and corrosion inhibition efficiency of 2-(6-methylpyridin-2-yl)-1H-imidazo [4, 5-f][1, 10] phenanthroline on mild steel in sulphuric acid. *Arab. J. Chem. 7*: 197–207.

79 Xu, B., Gong, W., Zhang, K. et al. (2015). Theoretical prediction and experimental study of 1-butyl-2-(4-methylphenyl) benzimidazole as a novel corrosion inhibitor for mild steel in hydrochloric acid. *J. Taiwan Inst. Chem. Eng. 51*: 193–200.

80 Gopi, D., Sherif, E.S.M., Surendiran, M. et al. (2014). Experimental and theoretical investigations on the inhibition of mild steel corrosion in the ground water medium using newly synthesised bipodal and tripodal imidazole derivatives. *Mater. Chem. Phys. 147*: 572–582.

81 El-Haddad, M.N. and Fouda, A. (2015). Electroanalytical, quantum and surface characterization studies on imidazole derivatives as corrosion inhibitors for aluminum in acidic media. *J. Mol. Liq. 209*: 480–486.

82 Obot, I., Umoren, S., Gasem, Z. et al. (2015). Theoretical prediction and electrochemical evaluation of vinylimidazole and allylimidazole as corrosion inhibitors or mild steel in 1M HCl. *J. Ind. Eng. Chem. 21*: 1328–1339.

83 Zhang, K., Xu, B., Yang, W. et al. (2015). Halogen-substituted imidazoline derivatives as corrosion inhibitors for mild steel in hydrochloric acid solution. *Corros. Sci. 90*: 284–295.

84 Gutierrez, E., Rodríguez, J.A., Cruz-Borbolla, J. et al. (2016). Development of a predictive model for corrosion inhibition of carbon steel by imidazole and benzimidazole derivatives. *Corros. Sci. 108*: 23–35.

85 Yaocheng, Y., Caihong, Y., Singh, A., and Lin, Y. (2019). Electrochemical study of commercial and synthesized green corrosion inhibitors for N80 steel in acidic liquid. *New J. Chem. 43*: 16058–16070.

86 Eduok, U., Faye, O., and Szpunar, J. (2016). Corrosion inhibition of X70 sheets by a film-forming imidazole derivative at acidic pH. *RSC Adv. 6*: 108777–108790.

87 Mihajlović, M.B.P., Radovanović, M.B., Tasić, Ž.Z., and Antonijević, M.M. (2017). Imidazole based compounds as copper corrosion inhibitors in seawater. *J. Mol. Liq. 225*: 127–136.

88 Tasic, Z.Z., Mihajlovic, M.B.P., and Antonijevic, M.M. (2016). The influence of chloride ions on the anti-corrosion ability of binary inhibitor system of 5-methyl-1H-benzotriazole and potassium sorbate in sulfuric acid solution. *J. Mol. Liq. 222*: 1–7.

89 Singh, A., Ansari, K., Kumar, A. et al. (2017). Electrochemical, surface and quantum chemical studies of novel imidazole derivatives as corrosion inhibitors for J55 steel in sweet corrosive environment. *J. Alloys Compd. 712*: 121–133.

90 Yadav, M., Kumar, S., Purkait, T. et al. (2016). Electrochemical, thermodynamic and quantum chemical studies of synthesized benzimidazole derivatives as corrosion inhibitors for N80 steel in hydrochloric acid. *J. Mol. Liq. 213*: 122–138.

91 Benzon, K., Mary, Y.S., Varghese, H.T. et al. (2017). Spectroscopic DFT. molecular dynamics and molecular docking study of 1-butyl-2-(4-hydroxyphenyl)-4, 5-dimethyl-imidazole 3-oxide. *J. Mol. Struct. 1134*: 330–344.

92 Behzadi, H. and Forghani, A. (2017). Correlation between electronic parameters and corrosion inhibition of benzothiazole derivatives-NMR parameters as important and neglected descriptors. *J. Mol. Struct. 1131*: 163–170.

93 Dutta, A., Saha, S.K., Adhikari, U. et al. (2017). Effect of substitution on corrosion inhibition properties of 2-(substituted phenyl) benzimidazole derivatives on mild steel in 1 M HCl solution: a combined experimental and theoretical approach. *Corros. Sci. 123*: 256–266.

94 Pareek, S., Jain, D., Hussain, S. et al. (2019). A new insight into corrosion inhibition mechanism of copper in aerated 3.5wt.% NaCl solution by eco-friendly Imidazopyrimidine Dye: experimental and theoretical approach. *Chem. Eng. J. 358*: 725–742.

95 Haque, J., Srivastava, V., Verma, C., and Quraishi, M.A. (2017). Experimental and quantum chemical analysis of 2-amino-3-((4-((S)-2-amino-2-carboxyethyl)-1H-imidazol-2-yl) thio) propionic acid as new and green corrosion inhibitor for mild steel in 1M hydrochloric acid solution. *J. Mol. Liq. 225*: 848–855.

96 Onyeachu, I.B., Obot, I.B., Sorour, A.A., and Abdul-Rashid, M.I. (2019). Green corrosion inhibitor for oilfield application I: electrochemical assessment of 2-(2-pyridyl) benzimidazole for API X60 steel under sweet environment in NACE brine ID196. *Corros. Sci. 150*: 183–193.

97 Obot, I.B., Onyeachu, I.B., and Umoren, S.A. (2019). Alternative corrosion inhibitor formulation for carbon steel in CO2-saturated brine solution under high turbulent flow condition for use in oil and gas transportation pipelines. *Corros. Sci.* 159: 108140.

98 Singh, A., Ansari, K., Lin, Y. et al. (2019). Corrosion inhibition performance of imidazolidine derivatives for J55 pipeline steel in acidic oilfield formation water: electrochemical, surface and theoretical studies. *J. Taiwan Inst. Chem. Eng.* 95: 341–356.

99 Ramkumar, S., Nalini, D., Quraishi, M.A. et al. (2020). Anti-corrosive property of bioinspired environmental benign imidazole and isoxazoline heterocyclics: A cumulative studies of experimental and DFT methods. *J. Heterocyclic Chem. 57*: 103–119.

100 Eldebss, T.M., Farag, A.M., and Shamy, A.Y. (2019). Synthesis of some Benzimidazole-based heterocycles and their application as copper corrosion inhibitors. *J. Heterocyclic Chem. 56*: 371–390.

101 Ozbay, S., Yanardag, T., Dincer, S., and Aksut, A. (2014). Benzimidazole Schiff bases as corrosion inhibitors for copper and brass. *Europ. Internat. J. Sci. Tech. 3*: 1–6.

102 Singh, A., Ansari, K.R., Quraishi, M.A., and Lgaz, H. (2019). Effect of electron donating functional groups on corrosion inhibition of J55 steel in a sweet corrosive environment: experimental, density functional theory, and molecular dynamic simulation. *Materials 12*: 17.

103 Moreira, R.R., Soares, T.F., and Ribeiro, J. (2014). Electrochemical investigation of corrosion on AISI 316 stainless steel and AISI 1010 carbon steel: study of the behaviour of imidazole and benzimidazole as corrosion inhibitors. *Adv. Chem. Eng. Sci. 4*: 503.

104 Yang, D., Ye, Y., Su, Y. et al. (2019). Functionalization of citric acid-based carbon dots by imidazole toward novel green corrosion inhibitor for carbon steel. *J. Clean. Prod. 229*: 180–192.

105 Obot, I., Macdonald, D., and Gasem, Z. (2015). Density functional theory (DFT) as a powerful tool for designing new organic corrosion inhibitors. Part 1: an overview. *Corros. Sci. 99*: 1–30.

106 Verma, C., Lgaz, H., Verma, D. et al. (2018). Molecular dynamics and Monte Carlo simulations as powerful tools for study of interfacial adsorption behavior of corrosion inhibitors in aqueous phase: a review. *J. Mol. Liq. 260*: 99–120.

107 Tinga, Y., Shengtao, Z., Li, F. et al. (2020). Investigation of imidazole derivatives as corrosion inhibitors of copper in sulfuric acid: combination of experimental and theoretical researches. *J. Taiwan Inst. Chem. Eng. 106*: 118–129.

108 Dohare, P., Quraishi, M.A., Lgaz, H., and Salghi, R. (2019). Electrochemical DFT and MD simulation study of substituted Imidazoles as novel corrosion inhibitors for mild steel. *Port. Electrochim. Acta 37* (4): 217–239.

109 Salarvanda, Z., Amirnasr, M., Talebianb, M. et al. (2017). Enhanced corrosion resistance of mild steel in 1 M HCl solution by trace amount of 2-phenyl-benzothiazole derivatives: Experimental, quantum chemical calculations and molecular dynamics (MD) simulation studies. *Corros. Sci. 114*: 133–145.

110 Kaya, S., Guo, L., Kaya, C. et al. (2016). Quantum chemical and molecular dynamic simulation studies for the prediction of inhibition efficiencies of some piperidine derivatives on the corrosion of iron. *J. Taiwan Inst. Chem. Eng. 65*: 522.

111 Saha, S.K., Murmu, M., Murmu, N.C., and Banerjee, P. (2016). Evaluating electronic structure of quinazolinone and pyrimidinone molecules for its corrosion inhibition effectiveness on target specific mild steel in the acidic medium: a combined DFT and MD simulation study. *J. Mol. Liq. 224*: 629.

112 Yan, Y., Wang, X., Zhang, Y. et al. (2013). Theoretical evaluation of inhibition performance of purine corrosion inhibitors. *Mol. Simul. 39*: 1034.

113 Singh, A., Lin, Y., Quraishi, M.A. et al. (2015). Porphyrins as corrosion inhibitors for N80 Steel in 3.5% NaCl solution: electrochemical, quantum chemical. QSAR and QSAR and Monte Carlo simulations studies. *Molecules 20*: 15122–15146.

114 Hasanov, R., Sadıkoğlu, M., and Bilgiç, S. (2007). Electrochemical and quantum chemical studies of some Schiff bases on the corrosion of steel in H2SO4 solution. *Appl. Surf. Sci. 253*: 3913.

115 Fouda, A.S., El-Ewady, Y.A., Abo-El-Enien, O.M., and Agizah, F.A. (2008). Cinnamoylmalononitriles as corrosion inhibitors for mild steel in hydrochloric acid solution. *Anti-Corros Method Mater. 55*: 317.

116 Ouakki, M., Galai, M., Rbaa, M. et al. (2020). Investigation of imidazole derivatives as corrosion inhibitors for mild steel in sulfuric acidic environment: experimental and theoretical studies. *Ionics 26*: 5251–5272.

7

Pyridine and Its Derivatives as Corrosion Inhibitors

Chandrabhan Verma[1], M. A. Quraishi[1], and Chaudhery Mustansar Hussain[2]

[1] *Interdisciplinary Research Center for Advanced Materials, King Fahd University of Petroleum and Minerals, Dhahran, Saudi Arabia*
[2] *Department of Chemistry and Environmental Science, New Jersey Institute of Technology, Newark, NJ, USA*

7.1 Introduction

Corrosion causes huge safety, economic, and environment-related problems, especially in the petroleum-based industries. According to the recent estimation of NACE (National Association of Corrosion Engineers), global cost of corrosion is about US \$2.5 trillion, which constitutes about 3.4% of world's GDP [1]. Further, the cost of corrosion is expected to increase because of the increasing industrialization and consumption of metallic materials for industrial, as well as house hold, applications. In view of high cost of corrosion, corrosion scientists and engineers developed and used various corrosion monitoring methods including coatings, paintings, galvanization, and by using corrosion inhibitors. It is well documented that cost of the corrosion can be reduced up to 15% (US \$3745 billion) to 35% (US \$8745 billion) by applying previously developed methods of corrosion mitigation [1]. Use of organic compounds, especially heterocyclic compounds, is termed as one of the best practices of corrosion mitigation because of their ease of synthesis, high inhibition performance, cost-effectivity, and ease of application [2, 3].

It is important to mention that organic compounds containing heterocyclic ring(s), polar substituents, and multiple bonds strongly interact with the metallic surface and form inhibitive film through their adsorption. Adsorption of organic compounds on metallic surface may follow physisorption (electrostatic interactions), chemisorption (charge-sharing), or mixed (physiochemisorption) mechanism. Through their adsorption, organic corrosion inhibitors build a

Organic Corrosion Inhibitors: Synthesis, Characterization, Mechanism, and Applications,
First Edition. Edited by Chandrabhan Verma, Chaudhery Mustansar Hussain, and Eno E. Ebenso.

barrier between metallic surface and the corrosive environment. Literature study shows the adsorption of most of the organic corrosion inhibitors mainly followed physiochemisorption mechanism. Adsorption behavior of organic corrosion inhibitors affect by numerous factors including the chemical and electronic structures of the inhibitor molecules, nature of the substituents, aggressiveness of the corrosive environment, and surrounding temperature [4, 5]. Organic compounds containing polar functional groups such as –CN (nitrile), –COOH (carboxyl), –NH_2 (amino), –$COOC_2H_5$ (ester), –NO_2 (nitro), –O– (ether), –$CONH_2$ (amide), –OMe (methoxy), >C=O (carbonyl), >C=N– (imine), –COCl (acid chloride), and –N=N– (azo) act as effective corrosion inhibitors as these substituents enhance the corrosion inhibition performance of the organic compounds by increasing the solubility in the polar electrolytes, as well as by behaving as adsorption centers.

7.1.1 Pyridine and Its Derivatives as Corrosion Inhibitors

Pyridine (C_5H_5H) is an important nitrogenous-based heteroaromatic compound that acquires various biological and industrial applications [6]. Pyridine is a building block of various medicines, vitamins, and useful agrochemicals [7, 8]. Pyridine acquires dipole moment value of 2.2 Debye [9, 10]. Because of the presence of a unshared electron pair, pyridine is relatively more basic compared to the benzene; therefore, it is expected that pyridine and its derivatives are more effective corrosion inhibitors than benzene and its derivatives, respectively [11–22]. It is also important to mention that reactivity and therefore corrosion inhibition effect of pyridine derivatives would be greater than that of the pyridine itself. Along with pyridine, 2-aminopyridine and 2,6-diaminopyridine derivatives are widely used as corrosion inhibitors. Optimized, HOMO, and LUMO molecular orbital pictures of the pyridine, 2-aminopyridine, and 2,6-diaminopyridine derived through Gaussian 09 software is shown in Figure 7.1. It can be clearly seen that HOMO and LUMO frontier electron densities are distributed over the entire segments of the molecules. This observation reveals that entire parts of the pyridine, 2-aminopyridine and 2,6-diaminopyridine involve in charge sharing with the metallic surface. It can also be seen that HOMO and LUMO also distributed over the amino substituent(s) of 2-aminopyridine and 2,6-diaminopyridine indicating that amino substituent also participated in the charge sharing. Along with amino (–NH_2), –NHMe,–OMe, –OH, –SH, and –NMe_2 substituted pyridine derivatives at second and/or sixth positions are also expected to behave as effective corrosion inhibitors.

It is important to notice that a lesser value of energy band gap ($\Delta E = E_{LUMO} - E_{HOMO}$) is consistent with the high protection effectiveness and vice versa. In above compounds, it can be clearly seen that 2,6-diaminopyridine has smallest value of ΔE; therefore, it is expected the best corrosion inhibitor followed by 2-aminopyridine and finally pyridine.

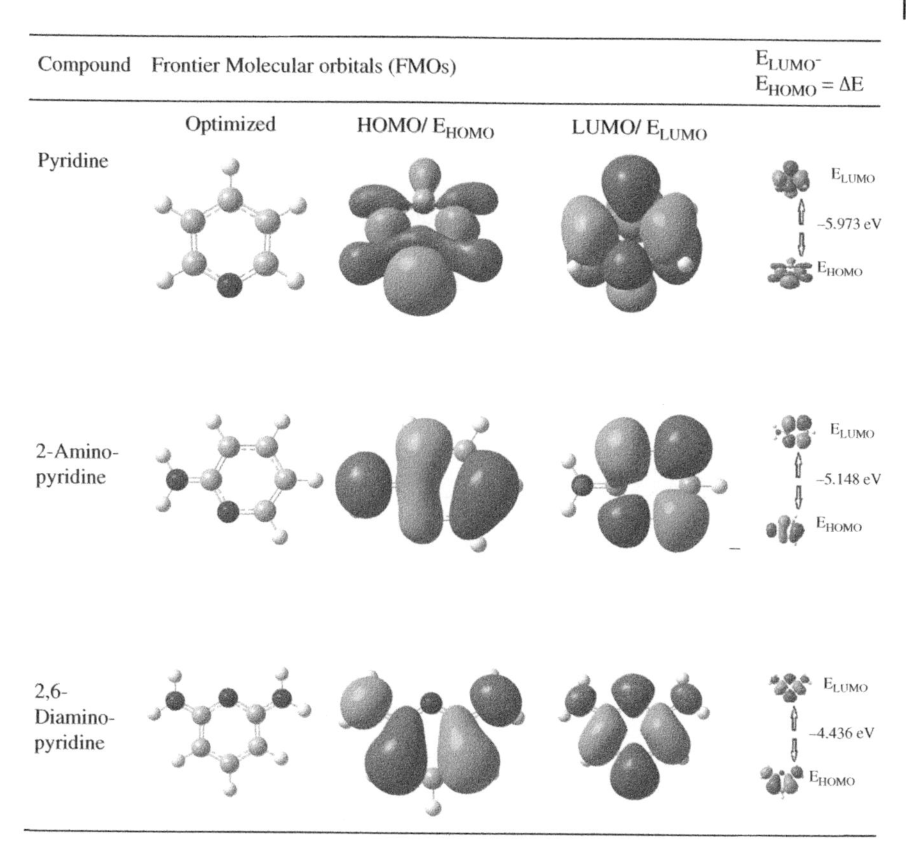

Figure 7.1 FMOs of benzene, pyridine, 2-aminopyridine, and 2,6-diaminopyridine.

7.1.2 Literature Survey

7.1.2.1 Substituted Pyridine as Corrosion Inhibitors

Pyridine and its derivatives are widely utilized as corrosion inhibitors. Because of the presence of addition nitrogen as the place of carbon atom benzene, pyridine is expected to more basic and superior corrosion inhibitors compared to the benzene. However, as pyridine nitrogen is less basic compared to the pyperidine's nitrogen, pyperidine and its derivatives are expected as better corrosion inhibitors compared to the pyridine and its derivatives [23]. It is important to mention that pyridine nitrogen is sp^2-hybridized, whereas pyperidine nitrogen is sp^3-hybridized. In a study, Awad [24] demonstrated that increase in the number of methyl group from one to three causes subsequent increase in the inhibition performance of the methyl substituted pyridine. This is attributed to increase in the electron density at the donor sites because of the electron-donating ability of methyl substituent.

Increase in the inhibition performance of the pyridine derivatives compared to the pyridine in the presence of other electron-donating substituents and lowering in the presence of electron-withdrawing substituents are greatly reported [25–27]. Study suggests that presence of–SH (thiol) and/or –OH (hydroxyl) substituent at second or sixth position significantly improves the inhibition performance of the pyridine derivatives [28]. Pyridine and its derivatives are extensively used as corrosion inhibitors in various systems [29–42].

Corrosion inhibition performance and other salient properties related with pyridine and its derivatives are presented in Table 7.1 [25–27, 43–58]. Most of the pyridine derivatives act as effective corrosion inhibitors, and they become effective by adsorbing on the metallic surface. Their inhibition performance increases

Table 7.1 A summary of corrosion inhibition effect of some common reports on substituted pyridine for different metals and alloys.

S. No.	Pyridine derivative(s)/ abbreviation	Adsorption behavior	Metal/ electrolytes	References
1	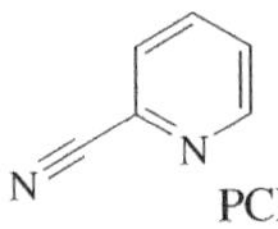 PCN	LAI/ mixed-type	MS/ 0.1M HCl	[25]
2	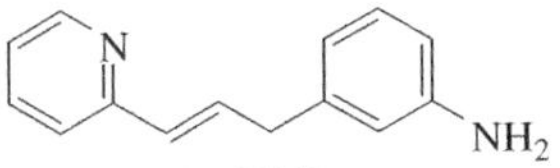 (4AAPA)	LAI// mixed-type	MS/ 1M HCl	[43]
3	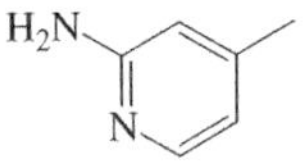 (AMP)	LAI// mixed-type	MS/ 0.5 M HCl	[44]
4	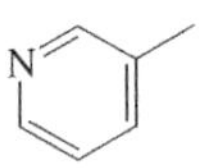 3-MP 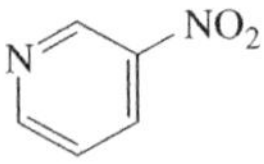 3-NM	--	Aluminum/ 1M KOH	[26]
5	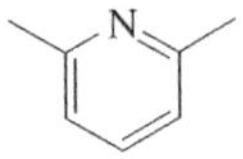 (DMP)	LAI// mixed-type	Aluminum/ water	[27]
6	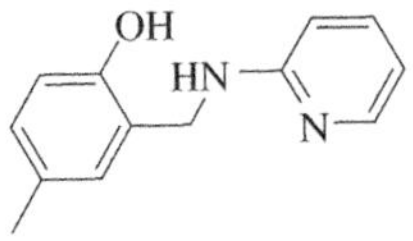	LAI/	MS/1M HCl	[45]

Table 7.1 (Continued)

S. No.	Pyridine derivative(s)/ abbreviation	Adsorption behavior	Metal/ electrolytes	References
7	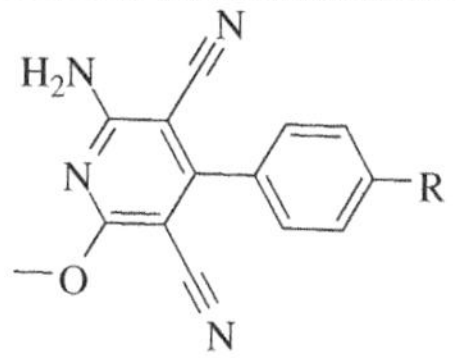 PC–1: R=–H; PC–2:R=–Me PC–3: R=–OMe	LAI// mixed-type	MS/1M HCl	[46]
8	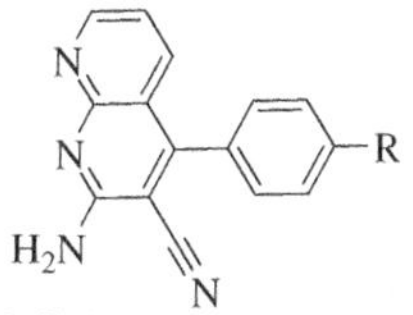 ANC–1: R=–OMe; ANC–2: R=–Me ANC–3:R=–NO_2	LAI// cathodic-type	N80 steel/15% HCl	[47]
9	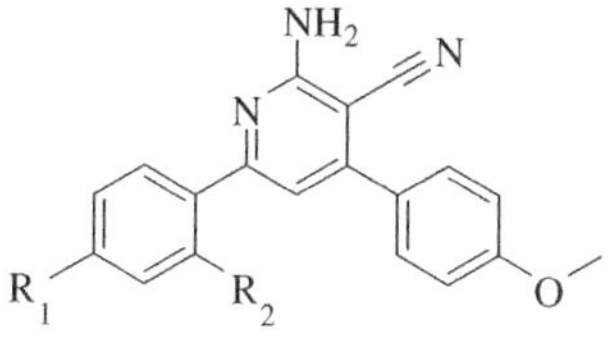 AMP:R_1=R_2=–H ADP:R_1=R_2=–OH	LAI// cathodic-type	N80 steel/15% HCl	[48]
10	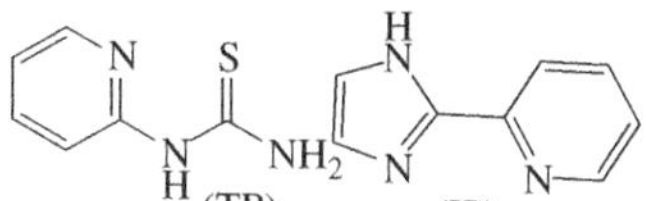 (TP) (IP)	LAI// anodic-type	MS/1M HCl	[49]
11	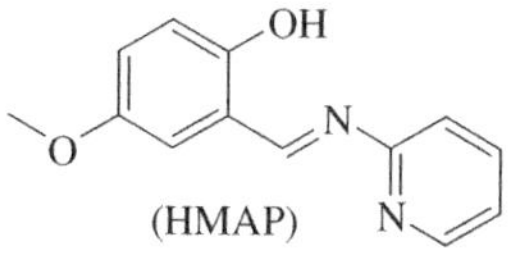 (HMAP)	LAI// anodic-type	MS/1M HCl	[50]
12	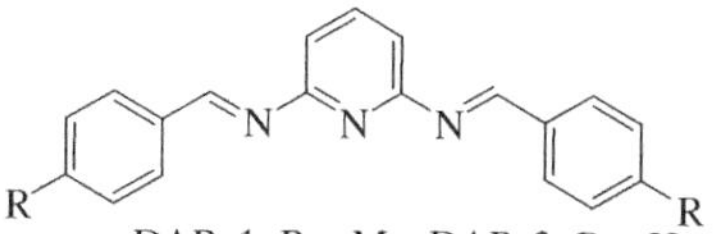 DAP–1: R=–Me; DAP–2: R=–H DAP–3:R=–NO_2	LAI// anodic-type	MS/ 1M HCl	[51]

(*Continued*)

Table 7.1 (Continued)

S. No.	Pyridine derivative(s)/ abbreviation	Adsorption behavior	Metal/ electrolytes	References
13	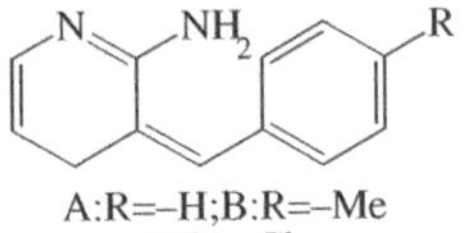 A:R=–H;B:R=–Me C:R=–Cl	Donor–acceptor interactions	Carbon steel/acidic media	[52, 53]
14	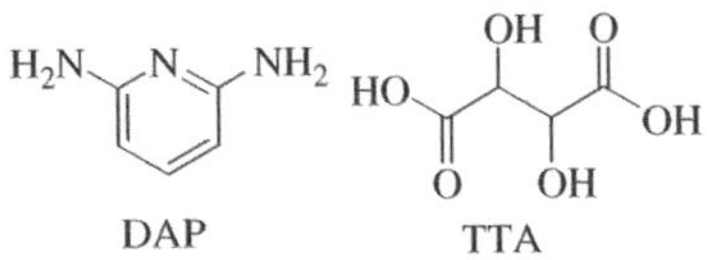 DAP TTA	Mixed-type inhibitors	Mild steel/ 0.5M HCl	[54]
15	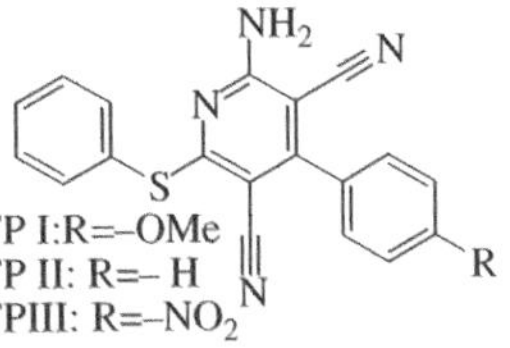 ADTP I:R=–OMe ADTP II: R=– H ADTPIII: R=–NO_2	LAI// mixed-type inhibitors	Mild steel/ 1M HCl	[55]
16	DPPN: R=–H DHPN: R=–OH DMPN: R=–OMe 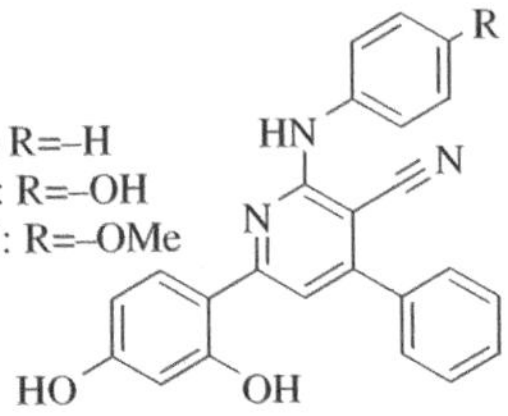	LAI// mixed-type inhibitors	Mild steel/1M HCl	[56]
17	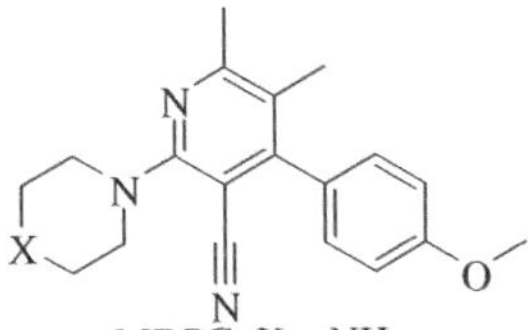 MDPC: X=–NH MDMC: X=O	LAI// mixed-type inhibitors	N80 Steel/15% HCl	[57]
18	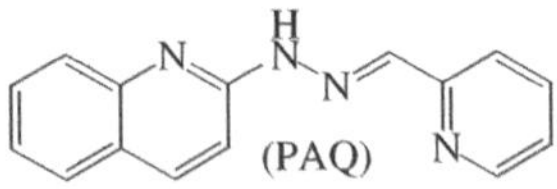 (PAQ)	LAI// mixed-type inhibitor	Q235 steel/ 1M HCl	[23]
19	2PCOX 3PCOX 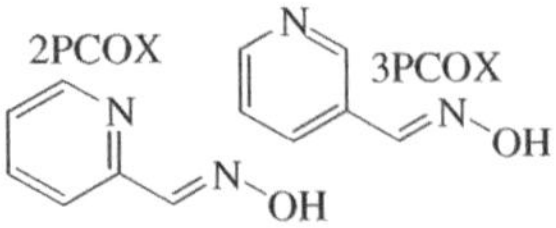	El-Awady & LAI// mixed-type inhibitor	Mild steel/1M HCl	[58]

on increasing their concentration. Potentiodynamic polarization study shows that pyridine derivatives become effective by retarding both anodic and cathodic reactions and behave as mixed-type corrosion inhibitors. Electrochemical study reveals that pyridine derivatives mostly behave as interface-type corrosion inhibitors as they become effective by adsorbing on the surface of metal and electrolyte. In study, Yıldız and coworkers [25] demonstrated the corrosion inhibition effect of 2-pyridinecarbonitrile (PCN) for mild steel in 1M hydrochloric acid solution. The electrochemical and surface investigation techniques were used to demonstrate the corrosion inhibition effectiveness of the PCN. Analysis showed that PCN acts as an effective corrosion inhibitor and exhibited as high as 94.7% inhibition efficiency at 10 mM concentration. Analysis further showed that PCN become effective by adsorbing on the metallic surface following through Langmuir adsorption isotherm model. Polarization investigation showed that PCN acts as mixed-type inhibitor and become effective by retarding both anodic and cathodic Tafel reaction. Pyridine derivatives containing other polar substituents such as –Me, –NH_2, –NO_2, and –OH are also tested for iron-based alloys [43–45] and aluminum [26, 27]. In these compounds, increase in the inhibition performance of the pyridine derivatives is expected due to increase in the number of active sites, increase in molecular size, and increase in the solubility of the inhibitor molecules in polar electrolytes. Computational studies are widely used to support the outcomes of the experimental studies.

7.1.3 Pyridine-Based Schiff Bases (SBs) as Corrosion Inhibitors

SBs are the compounds derived from the reaction of primary amine with aldehydes. It is important to mention that amino substituent present at second and/or sixth position of the pyridine ring can easily react with aliphatic and aromatic aldehyde to form suitable SBs. Pyridine-based SBs are widely used as corrosion inhibitors for different metals and alloys in various electrolytes [59–61]. Most of the pyridine-based SBs become effective by adsorbing on the metallic surface following mostly through Langmuir adsorption isotherm model. Through polarization studies, it is observed that most of the pyridine-based SBs become effective by retarding the anodic, as well as cathodic Tafel reactions and act as mixed-type corrosion inhibitors [62–66]. Adsorption of these compounds can be supported by surface investigation of corroded metallic specimens through SEM, AFM, EDX, FT-IR, XRD, XPS, and UV-vis spectroscopic analyses. Three imidazole-fused pyridine derivatives (PPs) were used as inhibitors by Gupta et al. [62]. Both chemical and electrochemical methods were used to demonstrate the corrosion inhibition effect of the PPs. Study showed that halogen-substituted PPs showed better corrosion protection compared to the without substituted PP. Polarization studies showed that PPs behaved as mixed-type corrosion inhibitors and retards both anodic metallic dissolution and cathodic hydrogen evolution reactions. A summary of corrosion inhibition effect of various other pyridine-based SBs is presented in Table 7.2 [59–75].

7.1.4 Quinoline-Based Compounds as Corrosion Inhibitors

Quinoline is a bicyclic hetero-aromatic compound in which pyridine moiety is fused with benzene ring. Because of the presence of high electron density in the form of five double bonds and unshared electron pair of nitrogen, quinoline and its derivatives are expected to behave as effective corrosion inhibitors. Theoretically, quinoline-based heterocyclic compounds are expected to behave as better corrosion inhibitors than that of the pyridine-based heterocyclic compounds. Similar to the substituted pyridine derivatives, quinoline derivatives substituted at second position is expected to form chelating complexes with the metallic atoms therefore they are supposed to behave as effective corrosion inhibitors. In various studies it is observed that mostly quinoline and quinoline derivatives act as mixed-type corrosion inhibitors. They become effective by adsorbing on the metallic surface following mainly through Langmuir adsorption isotherm model. Various reports are published recently on describing the corrosion inhibition effect of quinoline and its derivatives [76]. One of the most frequently utilized quinoline derivatives is 8-hydroxyquinoline. Summaries of corrosion inhibition effect of quinoline and 8-hydroxyquinoline-based compounds are given in Tables 7.3 and 7.4, respectively.

7.2 Summary and Outlook

From the ongoing discussion, it is clear that pyridine and pyridine-based compounds are widely used as corrosion inhibitors for different metals and alloys. Pyridine derivatives containing either electron donating or withdrawing substituents at second and/or sixth positions show relatively high inhibition efficiency compared to the non-substituted pyridine and pyridine derivatives substituted at other position. This observation suggests that pyridine derivatives substituted at second and/or sixth positions form chelating complexes with the metallic surface atoms. Therefore, they form relatively more stable surface protective film. Pyridine derivatives containing one substituent at either second or sixth positions act as bidentate ligands, and bi-substituted pyridine derivatives (at either at second and sixth positions) act as tri-dentate ligands. Present review articles feature the collection of corrosion inhibition effect of pyridine and its derivatives including quinoline and 8-hydroxyquinoline. Pyridine derivatives mostly behave as mixed-type corrosion inhibitors as they become effective by retarding the anodic and cathodic Tafel reactions. They inhibit corrosion by adsorbing on the metallic surface

Table 7.2 A summary of corrosion inhibition effect of some common reports on substituted pyridine for different metals and alloys.

S. No.	Pyridine-based SB(s)/abbreviation	Adsorption behavior	Metal/electrolyte	References
1	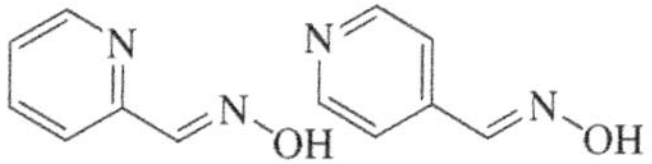 Hp2ylm, Hp3ylm	LAI//anodic-type	copper/ 0.1M HNO_3	[59]
2	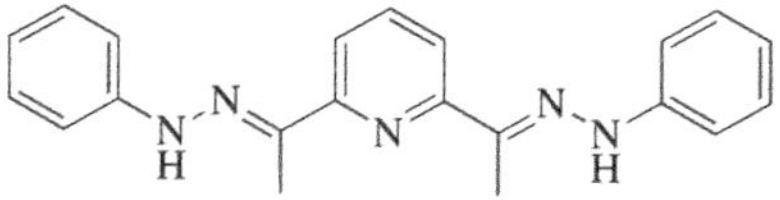 (BPEP)	LAI//mixed-type	Zn/ 1M HCl	[60]
3	(BPEP) 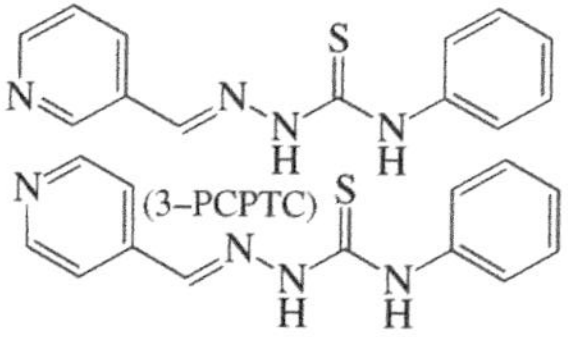 (3–PCPTC) (4–PCPTC)	LAI//mixed-type	MS/ 1M HCl	[61]
4	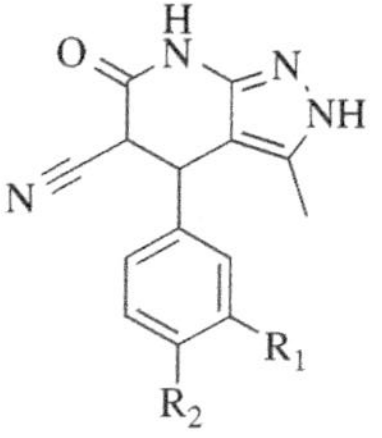 PP1: R_1=–H; R_2=–Br PP2: R_1=–H; R_2=–Cl PP3: R_1=–OPh; R_2=–H	LAI//mixed-type	MS/1M HCl	[62]

(Continued)

Table 7.2 (Continued)

S. No.	Pyridine-based SB(s)/abbreviation	Adsorption behavior	Metal/electrolyte	References
5	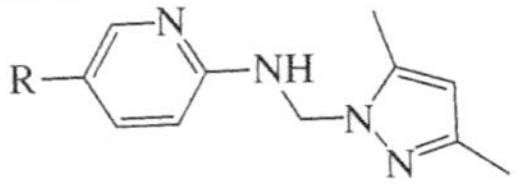 Br–PP:R=–Br; Cl–PP:R=–Cl H–PP:R=–H; Me–PP: R=–Me	Donor–acceptor interactions	--	[63]
6	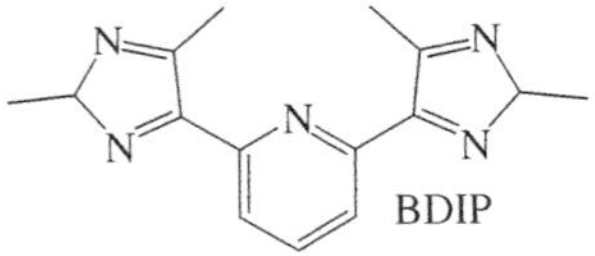 BDIP	LAI//mixed-type	MS/1 M HCl	[64]
7	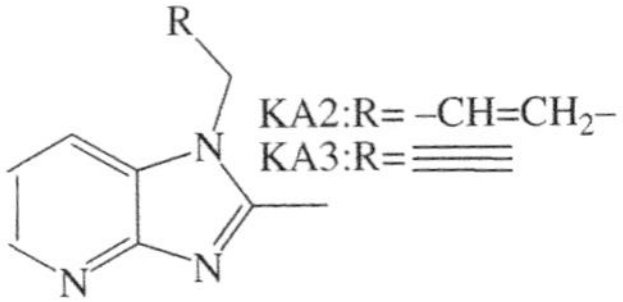 KA2:R= –CH=CH_2– KA3:R=≡	LAI//cathodic-type	MS/1 M HCl	[65]
8	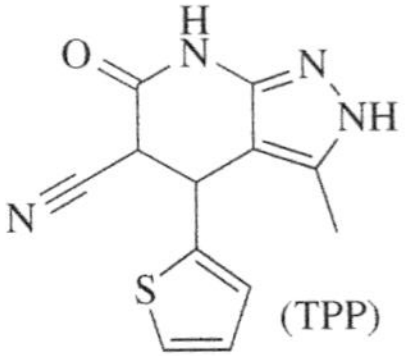 (TPP)	LAI//cathodic-type	MS/0.5M HCl	[66]

9	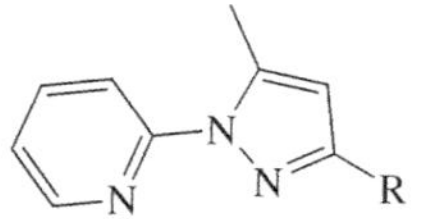 P1: R=–CH_2OH P2: R: –$COOC_2H_5$	Frumkin adsorption isotherm/ cathodic type	Steel/1M HCl	[67]
10	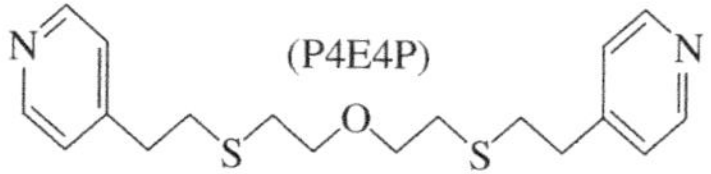 (P4E4P)	LAI//mixed-type	MS/1M HCl	[68]
11	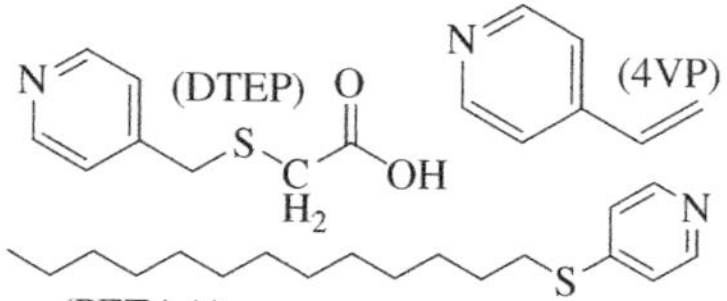 (DTEP) (4VP) (PETAA)	Mixed-type but mainly cathodic	MS/1M HCl	[69]
12	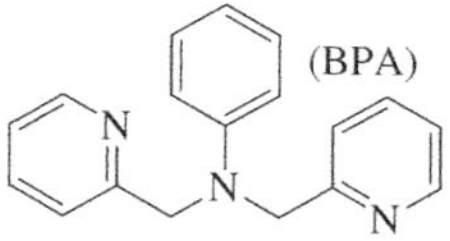 (BPA)	LAI//mixed type inhibitor	Mild steel/1M HCl	[70]
13	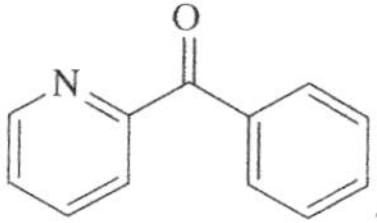 2BP	LAI/	MS/1M HCl	[71]
14	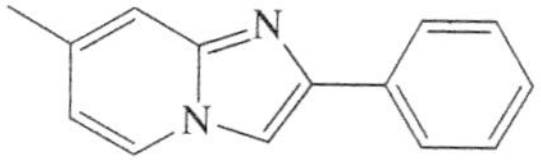 MPP	LAI//mixed-type	Carbon steel/1M HCl	[72]

(Continued)

Table 7.2 (Continued)

S. No.	Pyridine-based SB(s)/abbreviation	Adsorption behavior	Metal/electrolyte	References
15	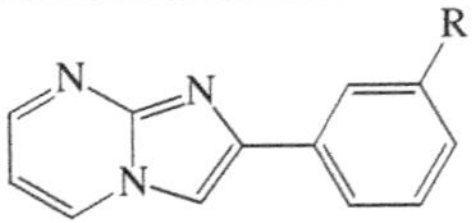 R N N N P1: R=–H; P2: R=–OMe	LAI//mixed-type	C38 Steel/1M HCl	
16	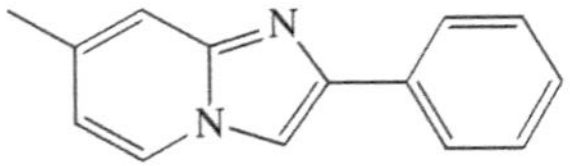 N N PPI PPI	LAI//mixed-type	Carbon steel/2M H_3PO_4	[73]
17	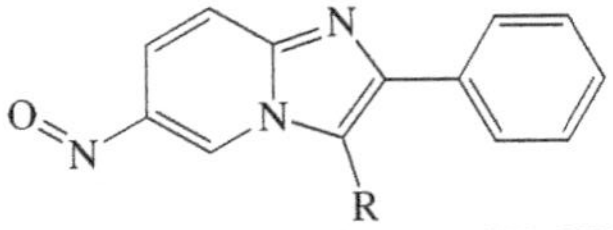 N O N N R C1:R=–CHO; C2: R=–CH_2OH	LAI//anodic-type	Carbon steel/1M HCl	[74]

Table 7.3 Informations about corrosion inhibition using quinoline derivatives.

Quinoline-based compound/abbreviation	Metal and electrolytes	References	Quinoline-based compound/abbreviation	Metal and electrolytes	References
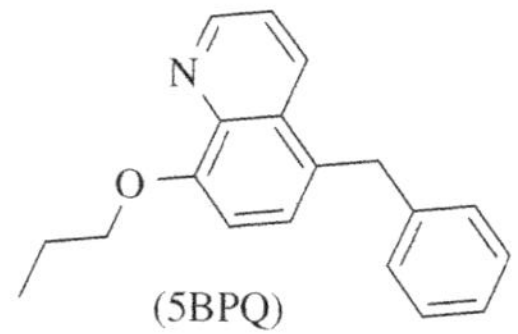 (5BPQ)	Q235 steel/0.5M H_2SO_4	[77]	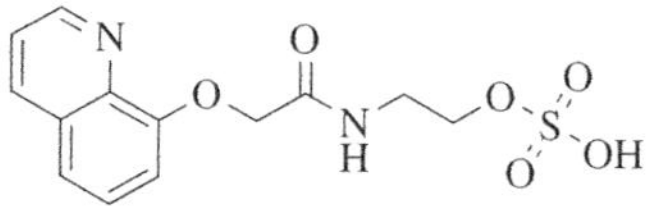 (8HQD)	Al/1M $HClO_4$	[78]
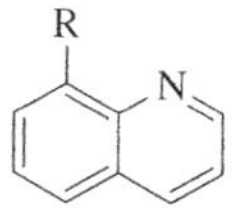 8–AQ: R=–NH_2 8–NQ: R=–NO_2	Al-AA5052 alloy/3% NaCl	[79]	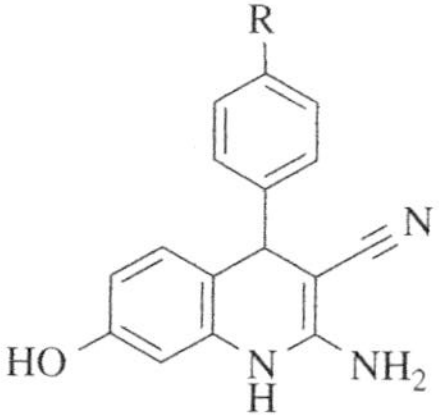 Q1: R=–H Q2: R=–Me Q3: R=–OMe Q4: R=–NMe_2	Fe (110) surface	[80]
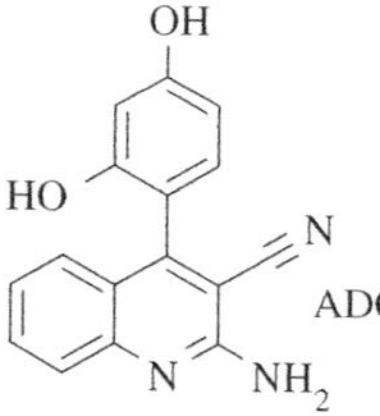 ADQC	SAE 1006 steel/1M HCl	[81]	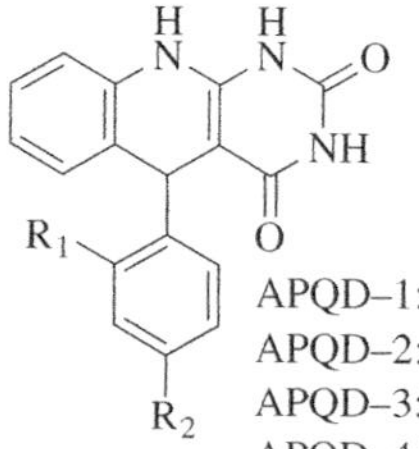 APQD–1:R_1=–H; R_2=–NO_2 APQD–2:R_1=–H; R_2=–H APQD–3:R_1=–H; R_2=–OH APQD–4:R_1=–OH; R_2=–OH	MS/1M HCl	[82]

(Continued)

Table 7.3 (Continued)

Quinoline-based compound/abbreviation	Metal and electrolytes	References	Quinoline-based compound/abbreviation	Metal and electrolytes	References
Q–1: R=–H Q–2: R=–Me Q–3: R=–OMe Q–4: R=–NMe_2 R, N, HO, N H, NH_2	MS/1M HCl	[83]	AAC–1: R=–NO_2 AAC–2: R=–H AAC–3: R=–OH R, N, N, NH_2	MS/1M HCl	[84]
R_2, N, Cl, O, R_1 QA–1:R_1=–Cl; R_2=–H QA–2:R_1=–NO_2; R_2=–H QA–3:R_1=–Cl; R_2=–NO_2	MS/1M HCl	[85]	R, N QL: R=–H QLD: R=–Me QLDA: R=–COOH	MS/0.5 M HCl	[86]
Compound 1: X= $\overset{\ominus}{O}$–Ph Compound 2: X= $\overset{\ominus}{O}$–Ph–p–Me Compound 3: X= $\overset{\ominus}{O}$–Ph–p–NO_2 $X^{\ominus}$, $N^{\oplus}$, C_2H_4OH	Zinc-Mg batteries/26% NH_4Cl	[87, 88]	R_1, N–N, R_2, O, O AQ–1:–R_1=–H; –R_2=––Me AQ–2:R_1=–H; –R_2=––OH AQ–3:R_1=–NO_2; –R_2=––H	N80 steel/15% HCl	[89]

Inhibitor	Metal/medium	Ref.	Inhibitor	Metal/medium	Ref.
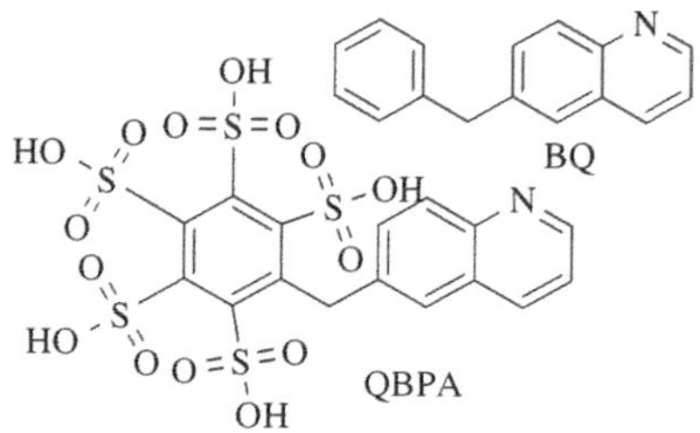 BQ, QBPA	MS/1M HCl	[90]	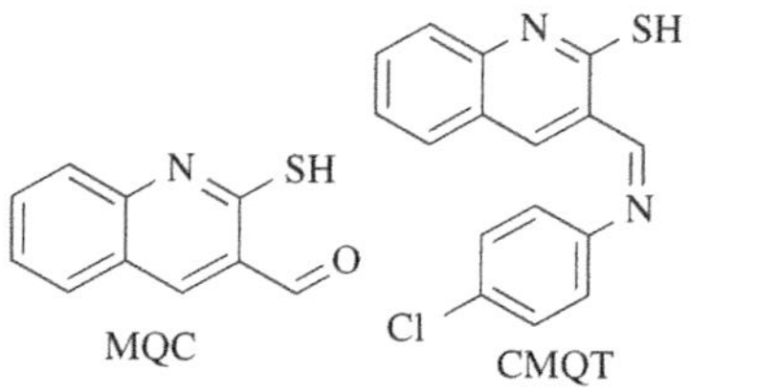 MQC, CMQT	MS/1M HCl	[91]
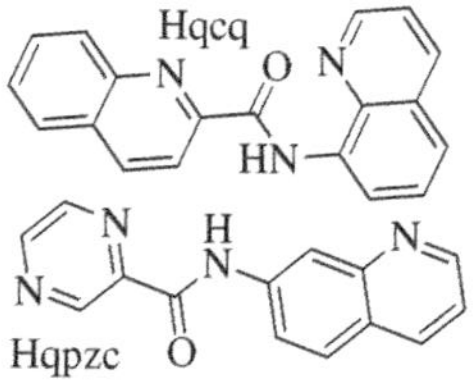 Hqcq, Hqpzc	MS/1M HCl	[92]	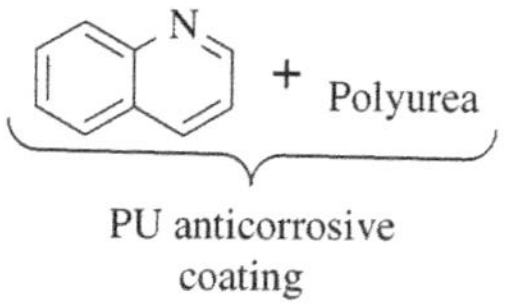 Quinoline + Polyurea (PU anticorrosive coating)	MS/5 wt% HCl	[93, 94]
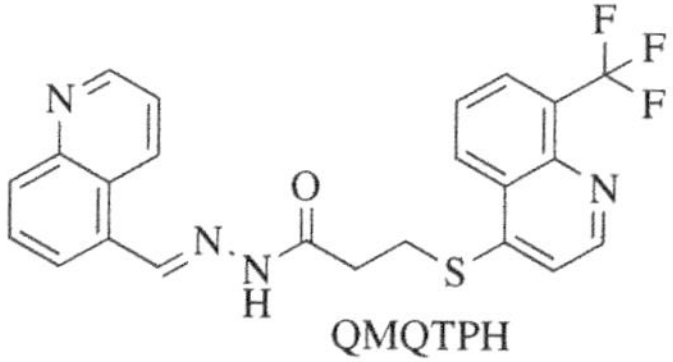 QMQTPH	MS/1 & 2M HCl	[95]	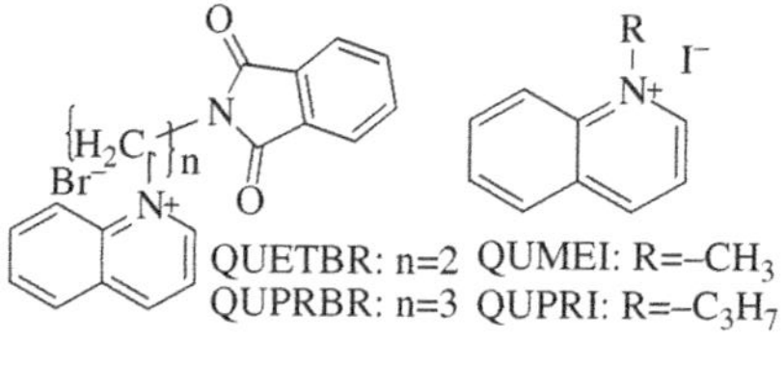 QUETBR: n=2, QUPRBR: n=3; QUMEI: R=$-CH_3$, QUPRI: R=$-C_3H_7$	MS/0.5M H_2SO_4	[96]
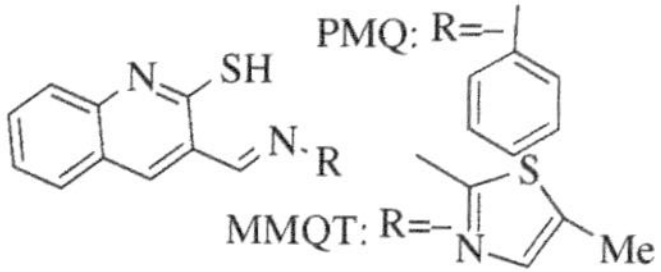 PMQ: R=–(p-tolyl); MMQT: R=–(thiazolyl, Me)	MS surface	[97]	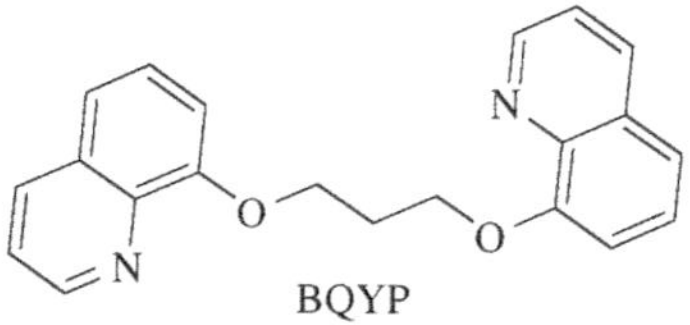 BQYP	MS/2M H_2SO_4	[98]

Table 7.4 Informations about corrosion inhibition using 8-hydroxyquinoline derivatives.

8-Hydroxyquinoline-based compound/abbreviation	Metal and electrolytes	References	Quinoline-based compound/abbreviation	Metal and electrolytes	References
N HO X — NH_2 HMQN: X = NH AMQN: X = O	Carbon steel/1M HCl	[99]	N HO N H O O O EHQP N HO HN O O BHQC	Carbon steel/1M HCl	[100]
HL1: R=–Me HL2: R=–H HL3: R=–NO_2 N HO N N R O	Carbon steel/2M HCl	[101]	N HO H N N R_1 R_2 BIMQ: R_1=R_2=–H; MBMQ; R_1= –H; R_2=–CH_3; CBMQ: R_1=–H; R_2=–Cl; DCBMQ: R_1= R_2=–Cl	Carbon steel in 2M H_3PO_4	[102]
N HO N N O N N (HL)	Carbon steel/ 2M HCl	[103]	R N OH HQMT HN S N H Me TCHQ: R= HN S NH_2	Carbon steel/1M HCl	[104]

Inhibitor	Metal/medium	Ref.	Inhibitor	Metal/medium	Ref.
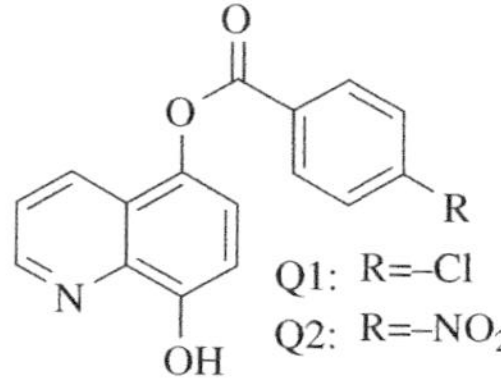 Q1: R=–Cl Q2: R=–NO_2	MS/ 1M HCl	[105]	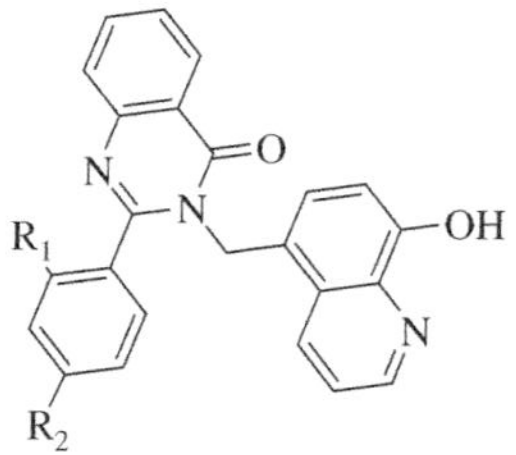 HQ–ZH: R_1=R_2=–H HQ–ZNO_2: R_1=–H; R_2=–NO_2 HQ–OH: R_1=–OH; R_2=–H	MS/1M HCl	[106]
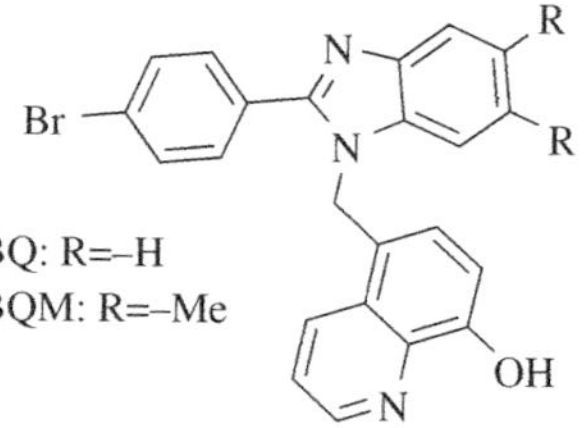 BQ: R=–H BQM: R=–Me	MS/1M HCl	[107]	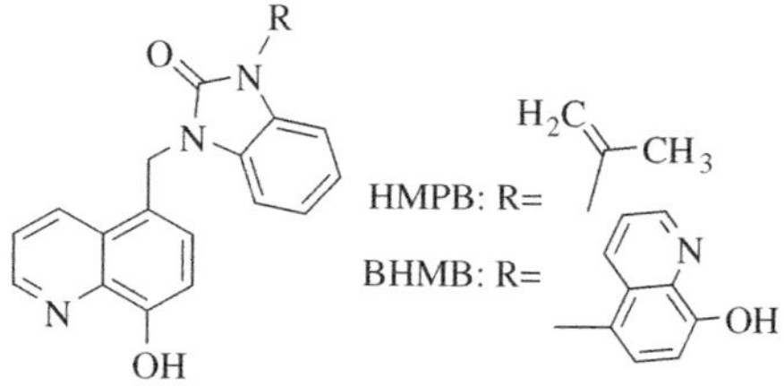 HMPB: R= H_2C=C–CH_3 BHMB: R= (8-hydroxyquinolin-5-yl)	Carbon steel/1M HCl	[108]
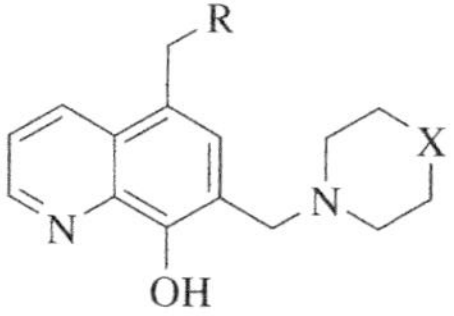 HM1: R=–N_3; X=O HM2: R=–CN; X=O HM3: R=–N_3; X=CH_2	Mild steel/1M HCl	[109]	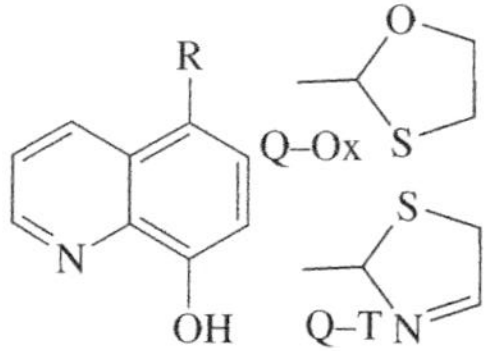 Q–Ox Q–T	MS/1M HCl	[110]

following through Langmuir adsorption isotherm model. Computational modelings are generally used to corroborate the results derived through experimental analyses. Adsorption of the pyridine-based compounds on metallic surface is supported by various surface investigations including SEM, AFM, EDX, RDX, XPS, FR-IR, and UV-vis analyses.

References

1 Verma, C., Ebenso, E.E., and Quraishi, M. (2017). Ionic liquids as green and sustainable corrosion inhibitors for metals and alloys: an overview. *Journal of Molecular Liquids* 233: 403–414.
2 Quraishi, M. and Jamal, D. (2002). Development and testing of all organic volatile corrosion inhibitors. *Corrosion* 58: 387–391.
3 Popoola, L.T. (2019). Organic green corrosion inhibitors (OGCIs): a critical review. *Corrosion Reviews* 37: 71–102.
4 Verma, C., Olasunkanmi, L.O., Ebenso, E.E. et al. (2016). Adsorption behavior of glucosamine-based, pyrimidine-fused heterocycles as green corrosion inhibitors for mild steel: experimental and theoretical studies. *The Journal of Physical Chemistry C* 120: 11598–11611.
5 Verma, C., Quraishi, M.A., Kluza, K. et al. (2017). Corrosion inhibition of mild steel in 1M HCl by D-glucose derivatives of dihydropyrido [2, 3-d, 6, 5-d′] dipyrimidine-2, 4, 6, 8 (1H, 3H, 5H, 7H)-tetraone. *Scientific Reports* 7: 44432.
6 Pal, S. (2018). Pyridine: a useful ligand in transition metal complexes. *Pyridine* 57.
7 Shimizu, S., Watanabe, N., Kataoka, T. et al. (2000). *Pyridine and Pyridine Derivatives*. Ullmann's Encyclopedia of Industrial Chemistry.
8 Balasubramanian, M. and Keay, J.G. (1996). *Pyridines and Their Benzo Derivatives: Applications*. Elsevier.
9 Leis, D. and Curran, B.C. (1945). Electric moments of some γ-substituted Pyridines1. *Journal of the American Chemical Society* 67: 79–81.
10 Brownson, G. and Yarwood, J. (1971). Far-infrared intensity and normal coordinate studies on pyridine-halogen complexes. *Journal of Molecular Structure* 10: 147–153.
11 Hassan, A., Hussein, R., Abou-krisha, M., and Attia, M. (2020). Density functional theory investigation of some Pyridine Dicarboxylic acids derivatives as corrosion inhibitors. *International Journal of Electrochemical Science* 15: 4274–4286.
12 Quraishi, M. (2015). The corrosion inhibition effect of aryl pyrazolo pyridines on copper in hydrochloric acid system: computational and electrochemical studies. *RSC Advances* 5: 41923–41933.
13 Benabdellah, M., Ousslim, A., Hammouti, B. et al. (2007). The effect of poly (vinyl caprolactone-co-vinyl pyridine) and poly (vinyl imidazol-co-vinyl pyridine) on the corrosion of steel in H 3 PO 4 media. *Journal of Applied Electrochemistry* 37: 819–826.

14 Dandia, A., Gupta, S., Singh, P., and Quraishi, M. (2013). Ultrasound-assisted synthesis of pyrazolo [3, 4-b] pyridines as potential corrosion inhibitors for mild steel in 1.0 M HCl. *ACS Sustainable Chemistry & Engineering* 1: 1303–1310.

15 Donya, A., Pakter, M., Shalimova, M., and Lambin, V. (2002). The effect of polar substituents in aniline and pyridine derivatives on the inhibition of steel corrosion in acids. *Protection of Metals* 38: 216–219.

16 Berisha, A. (2020). Experimental, Monte Carlo and molecular dynamic study on corrosion inhibition of mild steel by pyridine derivatives in aqueous perchloric acid. *Electrochemistry* 1: 188–199.

17 Mahmoud, N.F. and El-Sewedy, A. (2018). Multicomponent reactions, solvent-free synthesis of 2-Amino-4-aryl-6-substituted Pyridine-3, 5-dicarbonitrile derivatives, and corrosion inhibitors evaluation. *Journal of Chemistry* 2018: 1–9.

18 Lashgari, M., Arshadi, M., and Parsafar, G.A. (2005). A simple and fast method for comparison of corrosion inhibition powers between pairs of pyridine derivative molecules. *Corrosion* 61: 778–783.

19 Ashry, E., Nemr, A., and Ragab, S. (2012). Quantitative structure activity relationships of some pyridine derivatives as corrosion inhibitors of steel in acidic medium. *Journal of Molecular Modeling* 18: 1173–1188.

20 Chaitra, T., Mohana, K., and Tandon, H. (2016). Study of new thiazole based pyridine derivatives as potential corrosion inhibitors for mild steel: theoretical and experimental approach. *International Journal of Corrosion* 2016: 1–21.

21 Zuo, Y., Li, Z., Chen, L. et al. (2017). Treatment of the rust layer by different pyridine derivatives and its effect on the epoxy-polyvinylbutyral coating directly painted onto the rust mild steel. *International Journal of Electrochemical Science* 12: 11728–11741.

22 Abd El-Maksoud, S. and Fouda, A. (2005). Some pyridine derivatives as corrosion inhibitors for carbon steel in acidic medium. *Materials Chemistry and Physics* 93: 84–90.

23 Chaudhary, R., Namboodhiri, T., Singh, I., and Kumar, A. (1989). Effect of pyridine and its derivatives on corrosion of 0040, 2826MB, and 2605–8–2 metallic glasses in sulphuric acid solution at 25° C. *British Corrosion Journal* 24: 273–278.

24 Awad, H. (2006). The corrosion and inhibition of Zn-Al alloy in acidic media by pyridine and its methyl-containing derivatives. *Anti-Corrosion Methods and Materials* 53: 110–117.

25 Yıldız, R., Döner, A., Doğan, T., and Dehri, İ. (2014). Experimental studies of 2-pyridinecarbonitrile as corrosion inhibitor for mild steel in hydrochloric acid solution. *Corrosion Science* 82: 125–132.

26 Patil, D. and Sharma, A. (2014). Inhibition of corrosion of aluminium in potassium hydroxide solution by pyridine derivatives. *International Scholarly Research Notices* 2014: 1–5.

27 Padash, R., Jamalizadeh, E., and Jafari, A.H. (2017). Adsorption and corrosion inhibition behavior of aluminium by 2, 6-di methyl pyridine in distilled water. *Anti-Corrosion Methods and Materials* 64: 550–554.

28 Han, P., Li, W., Tian, H. et al. (2018). Comparison of inhibition performance of pyridine derivatives containing hydroxyl and sulfhydryl groups: Experimental and theoretical calculations. *Materials Chemistry and Physics* 214: 345–354.

29 Öğretir, C., Mihci, B., and Bereket, G. (1999). Quantum chemical studies of some pyridine derivatives as corrosion inhibitors. *Journal of Molecular Structure: THEOCHEM* 488: 223–231.

30 Xiao-Ci, Y., Hong, Z., Ming-Dao, L. et al. (2000). Quantum chemical study of the inhibition properties of pyridine and its derivatives at an aluminum surface. *Corrosion Science* 42: 645–653.

31 Veloz, M. and Martínez, I. (2006). Effect of some pyridine derivatives on the corrosion behavior of carbon steel in an environment like NACE TM0177. *Corrosion* 62: 283–292.

32 Elmsellem, H., Basbas, N., Chetouani, A. et al. (2014). Quantum chemical studies and corrosion inhibitive properties of mild steel by some pyridine derivatives in 1 N HCl solution. *Portugaliae Electrochimica Acta* 32: 77–108.

33 Ser, C.T., Žuvela, P., and Wong, M.W. (2020). Prediction of corrosion inhibition efficiency of pyridines and quinolines on an iron surface using machine learning-powered quantitative structure-property relationships. *Applied Surface Science* 512: 145612.

34 Hazani, N.N., Dzulkifli, N.N., Ghazali, S.A.I.S.M. et al. (2018). Synthesis, characterisation and effect of temperature on corrosion inhibition by thiosemicarbazone derivatives and its tin (IV) complexes. *Malaysian Journal of Analytical Sciences* 22: 758–767.

35 Tang, J., Hu, Y., Han, Z. et al. (2018). Experimental and theoretical study on the synergistic inhibition effect of pyridine derivatives and sulfur-containing compounds on the corrosion of carbon steel in CO2-saturated 3.5 wt.% NaCl solution. *Molecules* 23: 3270.

36 Lashkari, M. and Arshadi, M. (2004). DFT studies of pyridine corrosion inhibitors in electrical double layer: solvent, substrate, and electric field effects. *Chemical Physics* 299: 131–137.

37 Kuprin, V., Ivanova, M., and Skrypnik, Y.G. (1999). Peculiarities of adsorption and inhibitory action of some pyridine derivatives on steels. *Materials Science* 35: 811–817.

38 Kliskic, M., Radosevic, J., and Gudic, S. (1997). Pyridine and its derivatives as inhibitors of aluminium corrosion in chloride solution. *Journal of Applied Electrochemistry* 27: 947–952.

39 Jamalizadeh, E., Jafari, A., and Hosseini, S. (2008). Semi-empirical and ab initio quantum chemical characterisation of pyridine derivatives as HCl inhibitors of aluminium surface. *Journal of Molecular Structure: THEOCHEM* 870: 23–30.

40 Gurudatt, D.M. and Mohana, K.N. (2014). Synthesis of new pyridine based 1, 3, 4-oxadiazole derivatives and their corrosion inhibition performance on mild steel in 0.5 M hydrochloric acid. *Industrial & Engineering Chemistry Research* 53: 2092–2105.

41 Sampat, S. and Vora, J. (1974). Corrosion inhibition of 3s aluminium in trichloroacetic acid by methyl pyridines. *Corrosion Science* 14: 591–595.

42 Zhang, W., Li, H.J., Wang, Y. et al. (2018). Adsorption and corrosion inhibition properties of pyridine-2-aldehyde-2-quinolylhydrazone for Q235 steel in acid medium: electrochemical, thermodynamic, and surface studies. *Materials and Corrosion* 69: 1638–1648.

43 Karthik, R., Vimaladevi, G., Chen, S.-M. et al. (2015). Corrosion inhibition and adsorption behavior of 4-amino acetophenone pyridine 2-aldehyde in 1 M hydrochloric acid. *International Journal of Electrochemical Science* 10: 4666–4681.

44 Mert, B.D., Yüce, A.O., Kardaş, G., and Yazıcı, B. (2014). Inhibition effect of 2-amino-4-methylpyridine on mild steel corrosion: experimental and theoretical investigation. *Corrosion Science* 85: 287–295.

45 Al-Amiery, A. and Shaker, L. (2020). Corrosion inhibition of mild steel using novel pyridine derivative in 1 M hydrochloric acid. *Koroze a ochrana materialu* 64: 59–64.

46 Ansari, K., Quraishi, M., and Singh, A. (2015). Corrosion inhibition of mild steel in hydrochloric acid by some pyridine derivatives: an experimental and quantum chemical study. *Journal of Industrial and Engineering Chemistry* 25: 89–98.

47 Ansari, K. and Quraishi, M. (2015). Experimental and computational studies of naphthyridine derivatives as corrosion inhibitor for N80 steel in 15% hydrochloric acid. *Physica E: Low-dimensional Systems and Nanostructures* 69: 322–331.

48 Ansari, K., Quraishi, M., and Singh, A. (2015). Pyridine derivatives as corrosion inhibitors for N80 steel in 15% HCl: electrochemical, surface and quantum chemical studies. *Measurement* 76: 136–147.

49 Zhang, W., Li, H.-J., Wang, Y. et al. (2018). Gravimetric, electrochemical and surface studies on the anticorrosive properties of 1-(2-pyridyl)-2-thiourea and 2-(imidazol-2-yl)-pyridine for mild steel in hydrochloric acid. *New Journal of Chemistry* 42: 12649–12665.

50 Al-Amiery, A., Salman, T.A., Alazawi, K.F. et al. (2020). Quantum chemical elucidation on corrosion inhibition efficiency of Schiff base: DFT investigations supported by weight loss and SEM techniques. *International Journal of Low Carbon Technologies* 15: 202–209.

51 Dohare, P., Quraishi, M., and Obot, I. (2018). A combined electrochemical and theoretical study of pyridine-based Schiff bases as novel corrosion inhibitors for mild steel in hydrochloric acid medium. *Journal of Chemical Sciences* 130: 8.

52 Ju, H., Li, X., Cao, N. et al. (2018). Schiff-base derivatives as corrosion inhibitors for carbon steel materials in acid media: quantum chemical calculations. *Corrosion Engineering, Science and Technology* 53: 36–43.

53 Ashassi-Sorkhabi, H., Shaabani, B., and Seifzadeh, D. (2005). Corrosion inhibition of mild steel by some Schiff base compounds in hydrochloric acid. *Applied Surface Science* 239: 154–164.

54 Qiang, Y., Guo, L., Zhang, S. et al. (2016). Synergistic effect of tartaric acid with 2, 6-diaminopyridine on the corrosion inhibition of mild steel in 0.5 M HCl. *Scientific Reports* 6: 33305.

55 Quraishi, M.A. (2014). 2-Amino-3, 5-dicarbonitrile-6-thio-pyridines: new and effective corrosion inhibitors for mild steel in 1 M HCl. *Industrial & Engineering Chemistry Research* 53: 2851–2859.

56 Verma, C., Olasunkanmi, L.O., Quadri, T.W. et al. (2018). Gravimetric, electrochemical, surface morphology, DFT, and Monte Carlo simulation studies on three N-substituted 2-aminopyridine derivatives as corrosion inhibitors of mild steel in acidic medium. *The Journal of Physical Chemistry C* 122: 11870–11882.

57 Yadav, M. and Kumar, S. (2014). Experimental, thermodynamic and quantum chemical studies on adsorption and corrosion inhibition performance of synthesized pyridine derivatives on N80 steel in HCl solution. *Surface and Interface Analysis* 46: 254–268.

58 Thomas, K.J. (2017). Electro analytical and gravimetrical investigations on corrosion inhibition properties of pyridine-carbaldehyde derivatives on carbon steel. *Chemical Science Review and Letters* 6: 2300–2308.

59 Varghese, C., Thomas, K., Raphael, V., and Shaju, K. (2019). Corrosion inhibition capacity of two heterocyclic oximes on copper in nitric acid: electrochemical, quantum chemical and surface morphological investigations. *Current Chemistry Letters* 8: 1–12.

60 Abdallah, M., Ahmed, S., Altass, H. et al. (2019). Competent inhibitor for the corrosion of zinc in hydrochloric acid based on 2, 6-bis-[1-(2-phenylhydrazono) ethyl] pyridine. *Chemical Engineering Communications* 206: 137–148.

61 Meng, Y., Ning, W., Xu, B. et al. (2017). Inhibition of mild steel corrosion in hydrochloric acid using two novel pyridine Schiff base derivatives: a comparative study of experimental and theoretical results. *RSC Advances* 7: 43014–43029.

62 Gupta, S., Dandia, A., Singh, P., and Qureishi, M. (2015). Green synthesis of pyrazolo [3, 4-b] pyridine derivatives by ultrasonic technique and their application as corrosion inhibitor for mild steel in acid medium. *Journal of Materials and Environmental Science* 6: 168–177.

63 El Adnani, Z., Benjelloun, A., Benzakour, M. et al. (2014). DFT-based QSAR study of substituted pyridine-pyrazole derivatives as corrosion inhibitors in molar hydrochloric acid. *International Journal of Electrochemical Science* 9: 4732–4746.

64 El Khattabi, O., Zerga, B., Sfaira, M. et al. (2012). On the adsorption properties of an imidazole-pyridine derivative as corrosion inhibitor of mild steel in 1 M HCL. *Der Pharma Chemica* 4: 1759–1768.

65 Bouayad, K., Rodi, Y.K., Elmsellem, H. et al. (2016). Density-functional theory and experimental evaluation of inhibition mechanism of Novel Imidazo [4, 5-b] pyridine derivatives. *Journal of Taibah University for Science* 10: 139–147.

66 Singh, P., Quraishi, M., Gupta, S., and Dandia, A. (2016). Investigation of the corrosion inhibition effect of 3-methyl-6-oxo-4-(thiophen-2-yl)-4, 5, 6, 7-tetrahydro-2H-pyrazolo [3, 4-b] pyridine-5-carbonitrile (TPP) on mild steel in hydrochloric acid. *Journal of Taibah University for Science* 10: 139–147.

67 Tebbji, K., Oudda, H., Hammouti, B. et al. (2005). Inhibition effect of two organic compounds pyridine–pyrazole type in acidic corrosion of steel. *Colloids and Surfaces A: Physicochemical and Engineering Aspects* 259: 143–149.

68 Khadiri, A., Ousslim, A., Bekkouche, K. et al. (2018). 4-(2-(2-(2-(2-(Pyridine-4-yl) ethylthio) ethoxy) ethylthio) ethyl) pyridine as new corrosion inhibitor for mild steel in 1.0 M HCl solution: experimental and theoretical studies. *Journal of Bio-and Tribo-Corrosion* 4: 64.

69 Krim, O., Elidrissi, A., Hammouti, B. et al. (2009). Synthesis, characterization, and comparative study of pyridine derivatives as corrosion inhibitors of mild steel in HCl medium. *Chemical Engineering Communications* 196: 1536–1546.

70 Xu, B., Ji, Y., Zhang, X. et al. (2016). Experimental and theoretical evaluation of N, N-Bis (2-pyridylmethyl) aniline as a novel corrosion inhibitor for mild steel in hydrochloric acid. *Journal of the Taiwan Institute of Chemical Engineers* 59: 526–535.

71 James, A. and Oforka, N. (2014). 2-Benzoyl pyridine: an inhibitor for mild steel corrosion in hydrochloric acid solution. *Advances in Applied Science Research* 5: 1–6.

72 Ghazoui, A., Saddik, R., Benchat, N. et al. (2012). Comparative study of pyridine and pyrimidine derivatives as corrosion inhibitors of C38 steel in molar HCl. *International Journal of Electrochemical Science* 7: 7080–7097.

73 Hmamou, D.B., Salghi, R., Zarrouk, A. et al. (2013). Electrochemical and gravimetric evaluation of 7-methyl-2-phenylimidazo [1, 2-α] pyridine of carbon steel corrosion in phosphoric acid solution. *International Journal of Electrochemical Science* 8: 11526–11545.

74 Ech-chihbi, E., Nahlé, A., Salim, R. et al. (2019). An investigation into quantum chemistry and experimental evaluation of imidazopyridine derivatives as corrosion inhibitors for C-steel in acidic media. *Journal of Bio-and Tribo-Corrosion* 5: 24.

75 Anejjar, A., Zarrouk, A., Salghi, R. et al. (2013). Computational and experimental evaluation of the acid corrosion inhibition of carbon steel by 7-methyl-2 phenylimidazo [1, 2-α] pyridine. *International Journal of Electrochemical Science* 8: 5961–5979.

76 Verma, C., Quraishi, M., and Ebenso, E.E. (2020). Quinoline and its derivatives as corrosion inhibitors: a review. *Surfaces and Interfaces* 100634.

77 Chen, S., Chen, S., and Li, W. (2019). Corrosion inhibition effect of a New Quinoline derivative on Q235 steel in H2SO4 solution. *International Journal of Electrochemical Science* 14: 11419–11428.

78 Aly, M.R.E.S., Shokry, H., Sharshar, T., and Amin, M.A. (2016). A newly synthesized sulphated 8-hydroxyquinoline derivative to effectively control aluminum corrosion in perchloric acid: Electrochemical and positron annihilation studies. *Journal of Molecular Liquids* 214: 319–334.

79 Wang, D., Yang, D., Zhang, D. et al. (2015). Electrochemical and DFT studies of quinoline derivatives on corrosion inhibition of AA5052 aluminium alloy in NaCl solution. *Applied Surface Science* 357: 2176–2183.

80 Erdoğan, Ş., Safi, Z.S., Kaya, S. et al. (2017). A computational study on corrosion inhibition performances of novel quinoline derivatives against the corrosion of iron. *Journal of Molecular Structure* 1134: 751–761.

81 Verma, C. and Quraishi, M. (2017). 2-Amino-4-(2, 4-dihydroxyphenyl) quinoline-3-carbonitrile as sustainable corrosion inhibitor for SAE 1006 steel in 1 M HCl: electrochemical and surface investigation. *Journal of the Association of Arab Universities for Basic and Applied Sciences* 23: 29–36.

82 Verma, C., Olasunkanmi, L., Obot, I. et al. (2016). 5-Arylpyrimido-[4, 5-b] quinoline-diones as new and sustainable corrosion inhibitors for mild steel in 1 M HCl: a combined experimental and theoretical approach. *RSC Advances* 6: 15639–15654.

83 Singh, P., Srivastava, V., and Quraishi, M. (2016). Novel quinoline derivatives as green corrosion inhibitors for mild steel in acidic medium: electrochemical, SEM, AFM, and XPS studies. *Journal of Molecular Liquids* 216: 164–173.

84 Verma, C., Quraishi, M., Olasunkanmi, L., and Ebenso, E.E. (2015). L-Proline-promoted synthesis of 2-amino-4-arylquinoline-3-carbonitriles as sustainable corrosion inhibitors for mild steel in 1 M HCl: experimental and computational studies. *RSC Advances* 5: 85417–85430.

85 Lgaz, H., Salghi, R., Bhat, K.S. et al. (2017). Correlated experimental and theoretical study on inhibition behavior of novel quinoline derivatives for the corrosion of mild steel in hydrochloric acid solution. *Journal of Molecular Liquids* 244: 154–168.

86 Ebenso, E.E., Obot, I.B., and Murulana, L. (2010). Quinoline and its derivatives as effective corrosion inhibitors for mild steel in acidic medium. *International Journal of Electrochemical Science* 5: 1574–1586.

87 Zhang, D., Li, L., Cao, L. et al. (2001). Studies of corrosion inhibitors for zinc–manganese batteries: quinoline quaternary ammonium phenolates. *Corrosion Science* 43: 1627–1636.

88 Zhao, J., Duan, H., and Jiang, R. (2015). Synergistic corrosion inhibition effect of quinoline quaternary ammonium salt and Gemini surfactant in H2S and CO2 saturated brine solution. *Corrosion Science* 91: 108–119.

89 Ansari, K., Ramkumar, S., Nalini, D., and Quraishi, M. (2016). Studies on adsorption and corrosion inhibitive properties of quinoline derivatives on N80 steel in 15% hydrochloric acid. *Cogent Chemistry* 2: 1145032.

90 Jiang, L., Qiang, Y., Lei, Z. et al. (2018). Excellent corrosion inhibition performance of novel quinoline derivatives on mild steel in HCl media: experimental and computational investigations. *Journal of Molecular Liquids* 255: 53–63.

91 Mistry, B.M., Sahoo, S.K., and Jauhari, S. (2013). Experimental and theoretical investigation of 2-mercaptoquinoline-3-carbaldehyde and its Schiff base as an inhibitor of mild steel in 1 M HCl. *Journal of Electroanalytical Chemistry* 704: 118–129.

92 Erami, R.S., Amirnasr, M., Meghdadi, S. et al. (2019). Carboxamide derivatives as new corrosion inhibitors for mild steel protection in hydrochloric acid solution. *Corrosion Science* 151: 190–197.

93 Gite, V.V., Tatiya, P.D., Marathe, R.J. et al. (2015). Microencapsulation of quinoline as a corrosion inhibitor in polyurea microcapsules for application in anticorrosive PU coatings. *Progress in Organic Coatings* 83: 11–18.

94 Marathe, R., Tatiya, P., Chaudhari, A. et al. (2015). Neem acetylated polyester polyol – renewable source based smart PU coatings containing quinoline (corrosion inhibitor) encapsulated polyurea microcapsules for enhance anticorrosive property. *Industrial Crops and Products* 77: 239–250.

95 Saliyan, V.R. and Adhikari, A.V. (2008). Quinolin-5-ylmethylene-3-{[8-(trifluoromethyl) quinolin-4-yl] thio} propanohydrazide as an effective inhibitor of mild steel corrosion in HCl solution. *Corrosion Science* 50: 55–61.

96 Mourya, P., Singh, P., Tewari, A. et al. (2015). Relationship between structure and inhibition behaviour of quinolinium salts for mild steel corrosion: experimental and theoretical approach. *Corrosion Science* 95: 71–87.

97 Saha, S.K., Ghosh, P., Hens, A. et al. (2015). Density functional theory and molecular dynamics simulation study on corrosion inhibition performance of mild steel by mercapto-quinoline Schiff base corrosion inhibitor. *Physica E: Low-dimensional Systems and Nanostructures* 66: 332–341.

98 Alamshany, Z.M. and Ganash, A.A. (2019). Synthesis, characterization, and anti-corrosion properties of an 8-hydroxyquinoline derivative. *Heliyon* 5: e02895.

99 Rouifi, Z., Rbaa, M., Benhiba, F. et al. (2020). Preparation and anti-corrosion activity of novel 8-hydroxyquinoline derivative for carbon steel corrosion in HCl molar: computational and experimental analyses. *Journal of Molecular Liquids* 112923.

100 El Faydy, M., Lakhrissi, B., Guenbour, A. et al. (2019). In situ synthesis, electrochemical, surface morphological, UV–visible, DFT and Monte Carlo simulations of novel 5-substituted-8-hydroxyquinoline for corrosion protection of carbon steel in a hydrochloric acid solution. *Journal of Molecular Liquids* 280: 341–359.

101 Eldesoky, A. and Nozha, S. (2017). The adsorption and corrosion inhibition of 8-hydroxy-7-quinolinecarboxaldehyde derivatives on C-steel surface in hydrochloric acid. *Chinese Journal of Chemical Engineering* 25: 1256–1265.

102 El Faydy, M., Lakhrissi, B., Jama, C. et al. (2020). Electrochemical, surface and computational studies on the inhibition performance of some newly synthesized 8-hydroxyquinoline derivatives containing benzimidazole moiety against the corrosion of carbon steel in phosphoric acid environment. *Journal of Materials Research and Technology* 9: 727–748.

103 Abou-Dobara, M., Omar, N., Diab, M. et al. (2019). Polymer complexes. LXXV. Characterization of quinoline polymer complexes as potential bio-active and anti-corrosion agents. *Materials Science and Engineering: C* 103: 109727.

104 El Faydy, M., Benhiba, F., Lakhrissi, B. et al. (2019). The inhibitive impact of both kinds of 5-isothiocyanatomethyl-8-hydroxyquinoline derivatives on the corrosion of carbon steel in acidic electrolyte. *Journal of Molecular Liquids* 295: 111629.

105 Rbaa, M., Benhiba, F., Obot, I. et al. (2019). Two new 8-hydroxyquinoline derivatives as an efficient corrosion inhibitors for mild steel in hydrochloric acid: synthesis, electrochemical, surface morphological, UV–visible and theoretical studies. *Journal of Molecular Liquids* 276: 120–133.

106 Rbaa, M., Galai, M., Benhiba, F. et al. (2019). Synthesis and investigation of quinazoline derivatives based on 8-hydroxyquinoline as corrosion inhibitors for mild steel in acidic environment: experimental and theoretical studies. *Ionics* 25: 3473–3491.

107 Rbaa, M., Lgaz, H., El Kacimi, Y. et al. (2018). Synthesis, characterization and corrosion inhibition studies of novel 8-hydroxyquinoline derivatives on the acidic corrosion of mild steel: experimental and computational studies. *Materials Discovery* 12: 43–54.

108 El Faydy, M., Rbaa, M., Lakhrissi, L. et al. (2019). Corrosion protection of carbon steel by two newly synthesized benzimidazol-2-ones substituted 8-hydroxyquinoline derivatives in 1 M HCl: experimental and theoretical study. *Surfaces and Interfaces* 14: 222–237.

109 Douche, D., Elmsellem, H., Guo, L. et al. (2020). Anti-corrosion performance of 8-hydroxyquinoline derivatives for mild steel in acidic medium: gravimetric, electrochemical, DFT and molecular dynamics simulation investigations. *Journal of Molecular Liquids* 308: 113042.

110 Rbaa, M., Abousalem, A.S., Rouifi, Z. et al. (2020). Synthesis, antibacterial study and corrosion inhibition potential of newly synthesis oxathiolan and triazole derivatives of 8-hydroxyquinoline: experimental and theoretical approach. *Surfaces and Interfaces* 19: 100468.

8

Quinoline and Its Derivatives as Corrosion Inhibitors

Chandrabhan Verma and M. A. Quraishi

Interdisciplinary Research Center for Advanced Materials, King Fahd University of Petroleum and Minerals, Dhahran, Saudi Arabia

8.1 Introduction

Metallic alloys are broadly used as constructional resources in industries and house-hold applications. Nevertheless, pure metals and their alloys are extremely imprudent and readily experience corrosion by the components of atmosphere [1–4]. Corrosion is a decidedly destructive occurrence that causes massive financial, security, and environmental losses. According to the estimation of NACE, the National Association of Corrosion Engineers, both developed and developing countries are badly affected by the corrosion failure problems as corrosion causes the loss of about 3.5% of the world's GDP [5]. Several accidents have been reported because of the corrosion failure. A summary of some major accidents that have been happened because of corrosion failure is given in Figure 8.1.

Consequently, corrosion and its anticipation has fascinated huge desirability, and several methods of its lessening are developed from time-to-time depending ahead the character of metal and environment [6, 7]. One of the most effective methods of corrosion mitigation is the use of synthetic corrosion inhibitors. Synthetic inhibitors may be of organic and inorganic type; however, nowadays use of inorganic corrosion inhibitors is totally restricted because of their high toxicity and ability to accumulate in animal bodies, i.e. bioaccumulation. Unlike to inorganic compounds, most of the organic compounds are relatively safe and environmental friendly alternatives to be used as inhibitors against corrosion. Organic compounds inhibit corrosion by its adsorption through chemical, physical, or physiochemical (mixed-) mechanism. After getting adsorbed, organic compounds

Organic Corrosion Inhibitors: Synthesis, Characterization, Mechanism, and Applications,
First Edition. Edited by Chandrabhan Verma, Chaudhery Mustansar Hussain, and Eno E. Ebenso.

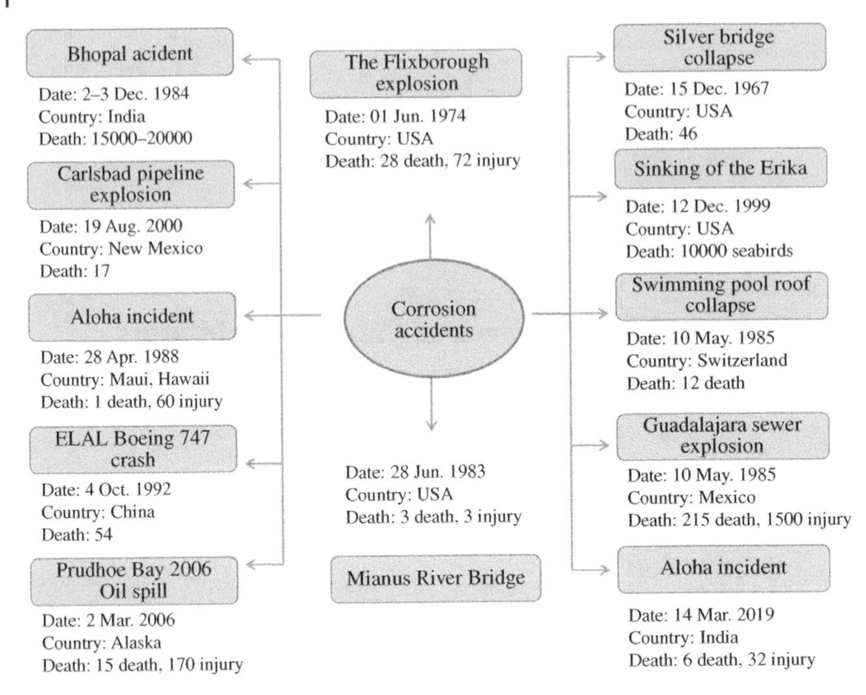

Figure 8.1 Pictorial illustration of some major accidents that have been happened because of corrosion failure.

build a hydrophobic film that avoids the contact of metallic surfaces with their environments. Organic inhibitors may be of nature or synthetic origins (Figure 8.2).

Adsorption of organic compounds depends upon numerous factors including aggressiveness of the corrosive environment, presence of salts and moisture, and atmospheric temperature. Generally, metals feel more susceptibility in the presence of salts and moisture, as well as at the high temperature. Along with several external factors, adsorption as well as corrosion inhibition property of organic compounds depend upon their molecular structures and electronic factors. Nature of substituents greatly alters the adsorption behavior and strength. The effect of substituents on the adsorption behavior of organic compounds can be described by Hammett substituent constant values that can be derived using following Hammett equation [8, 9]:

$$\log \frac{K_R}{K_H} = \rho\sigma \tag{8.1}$$

$$\log \frac{1-\eta\%_R}{1-\eta\%_H} = \rho\sigma \tag{8.2}$$

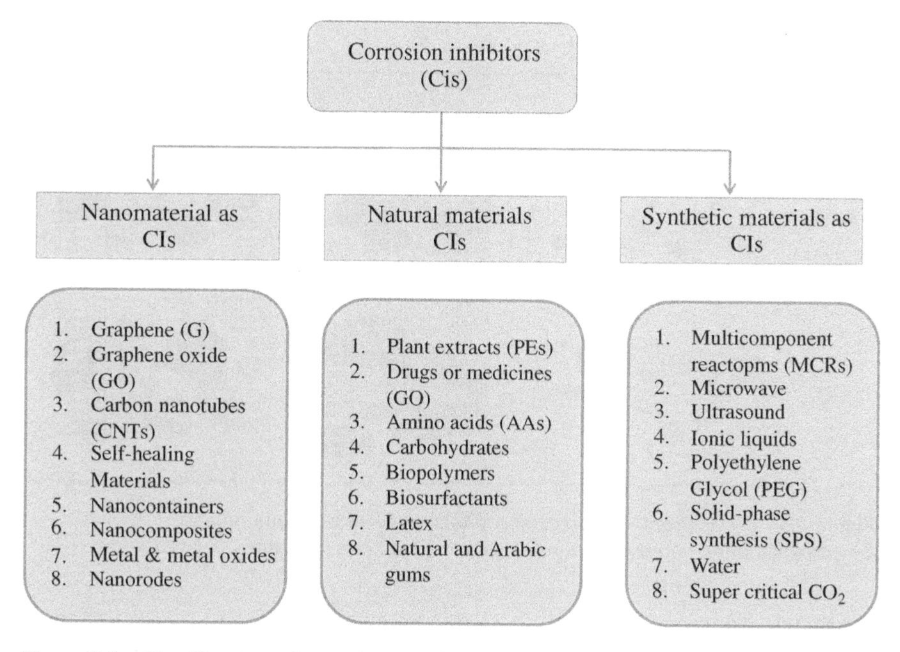

Figure 8.2 Classification of organic corrosion inhibitors. Organic corrosion inhibitors may be nanomaterials and molecular organic compounds of natural and synthetic origin.

$$\log\frac{\eta\%_R}{\eta\%_H} = \log\frac{C_{rH}}{C_{rR}} = \rho\sigma - \log\frac{\theta_R}{\theta_H} \tag{8.3}$$

In the above equations, ρ is the reaction indices and σ is the Hammett substituent constant. $\eta\%$, Cr, and θ are the inhibition efficiency, corrosion rate, and surface coverage, respectively, with (-*R*) and without (-*H*) substituent. Generally, electron-realizing substituents possess negative sign of σ, and they are expected to increase the inhibition efficiency of organic compounds in their presence and converse is true for the electron-withdrawing substituents with positive sign of σ [8].

8.2 Quinoline and Its Derivatives as Corrosion Inhibitors

It is well known as quinoline and its derivatives are highly effective against metallic corrosion which is attributed due to their association with high electron density. The FMOs pictures of quinoline and 8-hydroxyquinoline are presented Figure 9.3. This observation suggests that whole part of the quinoline and 8-hydroxyquinoline molecules participate in the bonding with the metallic surface. It is important to mention that during interaction with metallic surface,

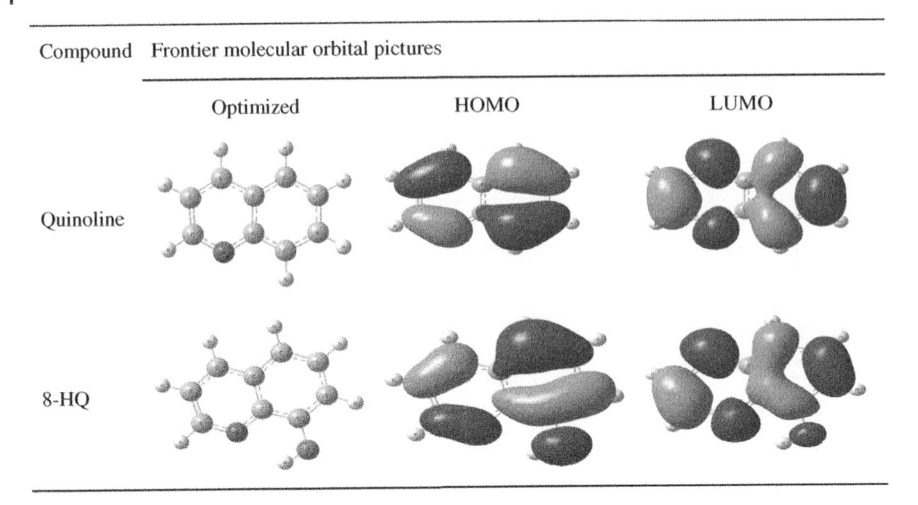

Figure 8.3 FMOs (frontier molecular orbitals) pictures of quinoline and 8-hydroxyquioline derived though Gaussian 09 software package.

HOMO (highest occupied molecular orbital) of inhibitor molecules can be considered as sites for electron donation, and LUMO (lowest unoccupied molecular orbital) can be considered as sites for electron acceptance. Inspection of the Figure 8.3 shows that HOMO and LUMO both are distributed over entire molecules; therefore, quinoline and 8-hydroxyquinoline are greatly involved in bonding with the metallic surface. It is also very interesting to see that HOMO and LUMO are also distributed over the hydroxyl (-OH) functional group present at eighth position of the 8-hydroxyquinoline. Based on this observation, it can be concluded that polar substituents especially electron donating increase the HOMO and LUMO contribution and therefore inhibition efficiency. Therefore, derivatization of the quinoline and 8-hydroxyquinoline enhances their ability of adsorption on metallic surface and inhibition effect against metallic corrosion.

8.2.1 8-Hydroxyquinoline and Its Derivatives as Corrosion Inhibitors

Chemical structures, abbreviation and other relevant information about the 8-hydroquinoline and its derivatives are presented in Table 8.1. It is important to mention that 8-hydroxyquinoline can form chelating complex with the metal surface by involving the unshared electron pairs of nitrogen and oxygen several reports dealing with the anticorrosion effect of 8-hydroxyquinoline and its derivatives are published [31, 32]. Obot and his coworkers reported inhibition effect of 8-hydroxyquinoline for X60 in 15% HCl solution with and without 0.4wt% KI [10].

Table 8.1 Chemical structures, abbreviations, and nature of metal and electrolytes of 8-hydroxyquinoline and its common derivatives evaluated as corrosion inhibitors.

No.	Chemical structure/abbreviation	Nature of inhibitor	Metal/ electrolyte	Ref.s	S No.	Chemical structure/abbreviation	Nature of inhibitor	Metal/ electrolyte	References
1	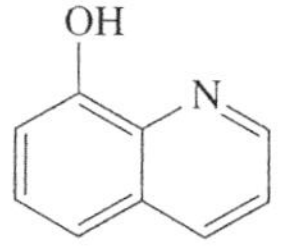 8–HQ	Mixed type	X60 steel/15% HCl	[10]	2	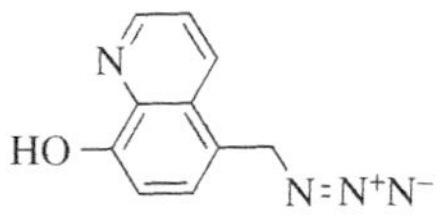 Copper (II) and zinc (II).	Interface type/ coating	Mg-AZ91D alloy/3% NaCl	[11, 12]
4	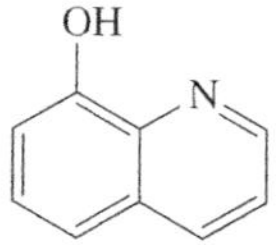 8–HQ	Mixed type/ LAI	Mild steel/1M HCl	[13]	5	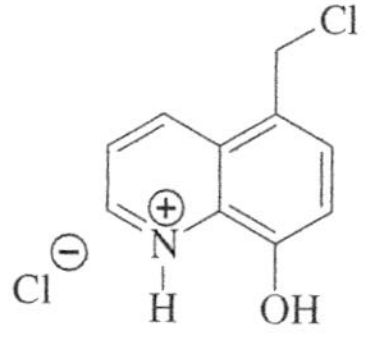 Cl–QH **Cl-QH**	Mixed type/ LAI	XC38 steel/1M HCl	[14]
3	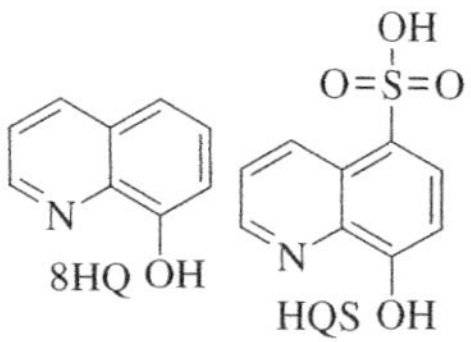 8HQ HQS	HQ mixed & HQS cathodic type	Al 2024-T3/3.5% NaCl 8-	[15]	6	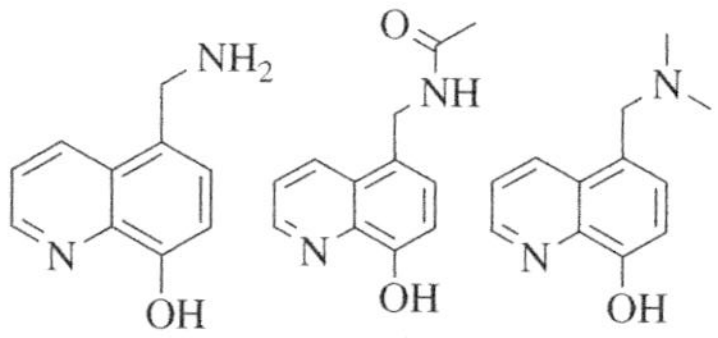 (AMHQ) (ACAMHQ) (DMHQ)	Mixed type/LAI	C40E steel/1M HCl	[16]

(Continued)

Table 8.1 (Continued)

No.	Chemical structure/abbreviation	Nature of inhibitor	Metal/ electrolyte	Ref.s	S No.	Chemical structure/abbreviation	Nature of inhibitor	Metal/ electrolyte	References
9	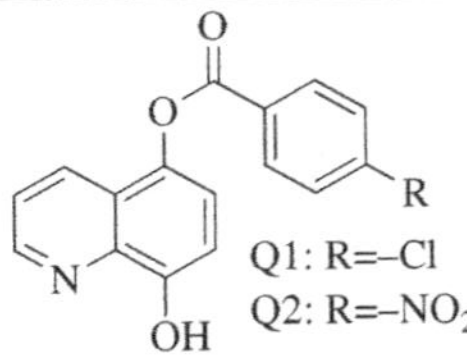 Q1: R=–Cl Q2: R=–NO_2	Mixed type/ LAI	Carbon steel/1M HCl	[17]	7	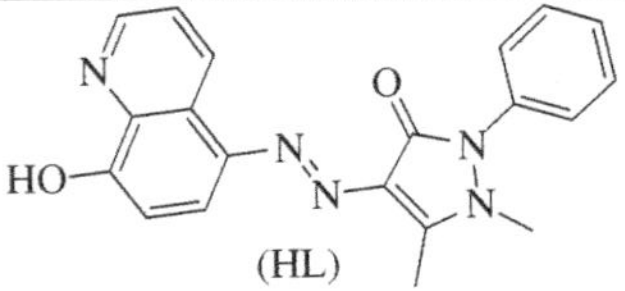 (HL)	Mixed type/LAI	Carbon steel/1M HCl	[18]
12	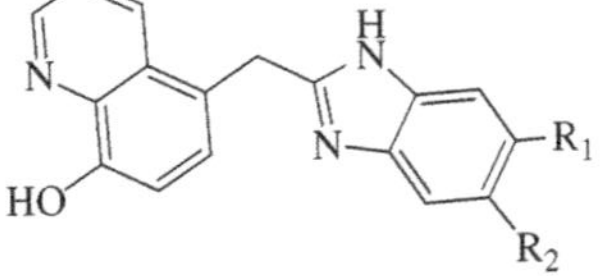 BIMQ:R_1=R_2=–H; MBMQ; R_1= –H; R_2=–CH_3; CBMQ:R_1=–H; R_2=–Cl; DCBMQ: R_1= R_2=–Cl	Mixed type/ LAI	Mild steel /1M HCl	[19]	10	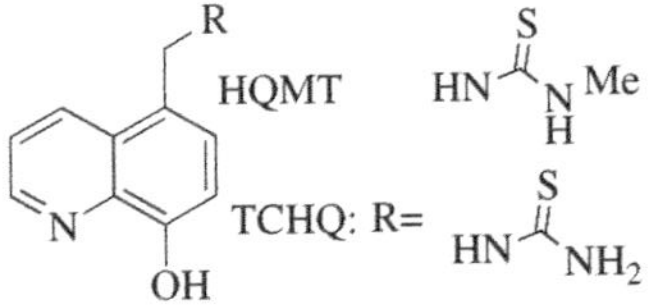 HQMT TCHQ: R=	Mixed type/LAI	Carbon steel/1M HCl	[20]
15	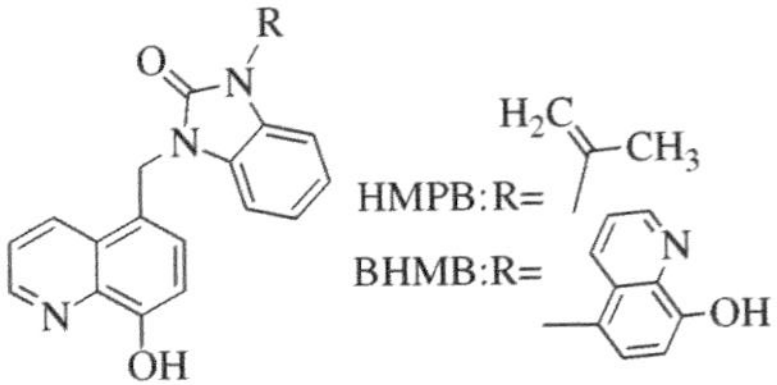 HMPB:R= BHMB:R=	Mixed-type inhibitors	Carbon steel/2M HCl	[21]	8	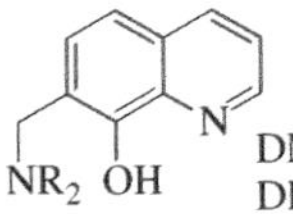 DMQ: R=–Me DPQ: R=–C_3H_7	Mixed type/LAI	Carbon steel/1M HCl	[22]
14	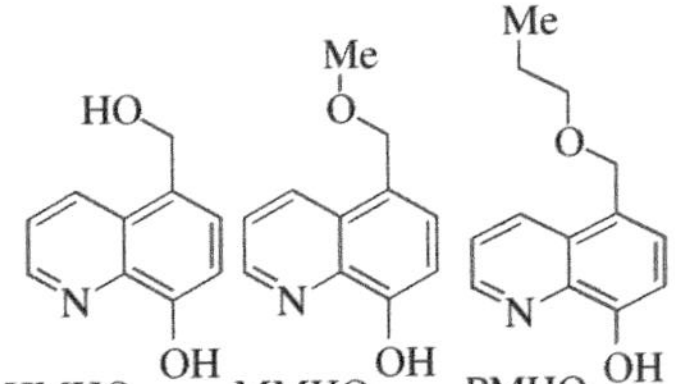 HMHQ MMHQ PMHQ	Mixed type/ LAI	Carbon steel in 2M H_3PO_4	[23]	13	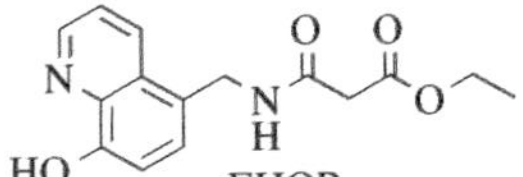 EHQP 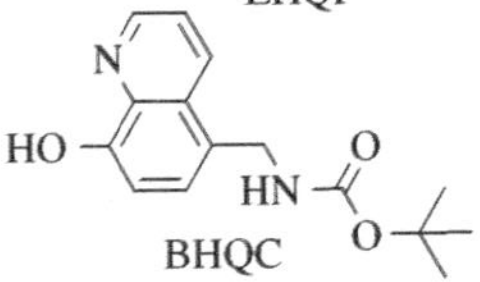 BHQC	Mixed type/LAI	Carbon steel /2M HCl	[24]

11	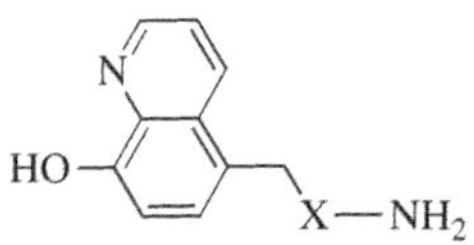 HMQN: X = NH AMQN: X = O	Mixed type/ LAI	Carbon steel/1M HCl	[25]	18	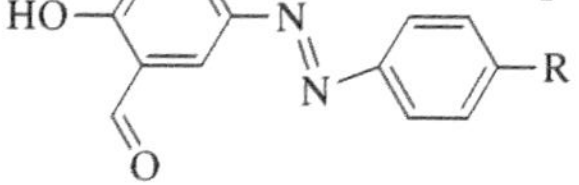 HL1: R=–Me HL2: R=–H HL3: R=–NO_2	Mixed type/LAI	Mild steel /1M HCl	[26]
16	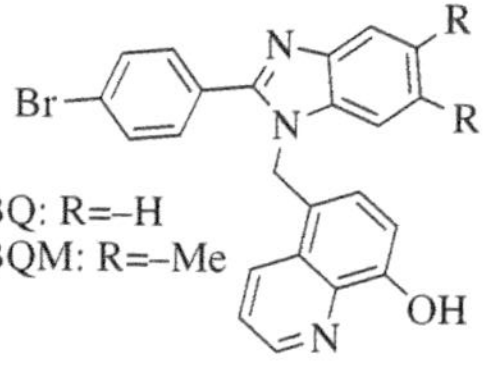 BQ: R=–H BQM: R=–Me	Cathodic type/LAI	Carbon steel/1 M HCl	[27]	17	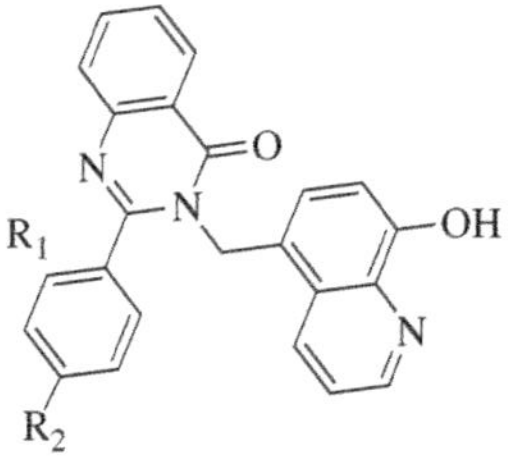 HQ–ZH:R_1=R_2=–H HQ–ZNO_2: R_1=–H; R_2=–NO_2 HQ–OH:R_1=–OH; R_2=–H	Mixed type/LAI	Mild steel/1M HCl	[28]
19	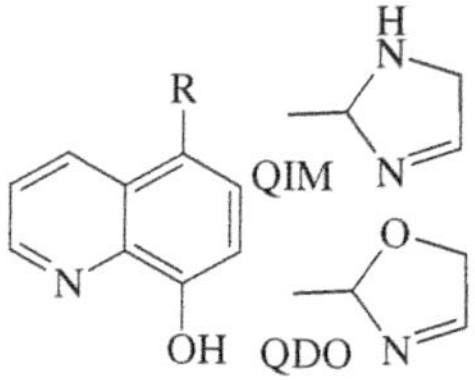 QIM QDO	Mixed type/ LAI	Mild steel / 1M HCl	[29]	20	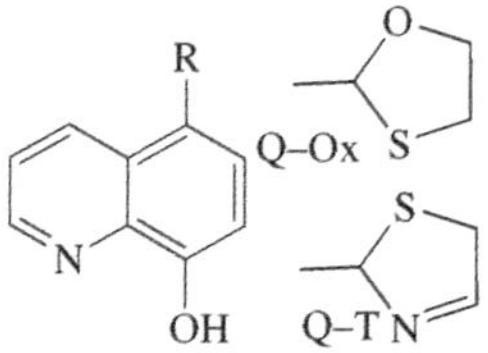 Q–Ox Q–T	Mixed type/LAI	Mild steel/1M HCl	[30]

Results showed that presence of KI synergistically enhances the protection effectiveness of the 8-hydroxyquinoline. It was observed that inhibitor inhibits corrosion by adsorption mechanism. Adsorption of the 8-hydroxyquinoline on X60 surface was monitored by surface analyses through FT-IR, SEM and EDX methods. 8-hydroxyquinoline acted as mixed-type inhibitor as derived through PDP study. Corrosion inhibition by 8-hydroxyquinolie for Mg-AZ91D alloy was reported in 3% NaCl elsewhere [11, 12]. In another study [15], 8-hydroxyquinoline (8HQ) and 8-hydroxy-quinoline-5-sulfonic acid (HQS) are tested as inhibitors for Al alloy (AA2024-T3) in 3.5% NaCl. Both 8HQ and HQS acted as effective corrosion inhibitors and their presence affect the nature of anodic and cathodic reactions. In several other studies corrosion inhibition effect of the quinoline and its derivatives are evaluated using several experimental and computational analyses. Most of such compounds acted as effective corrosion inhibitors and their adsorption followed the Langmuir adsorption isotherm model (LAI). 8-hydroxyquinoline and its derivative act as mixed type corrosion inhibitors. Adsorption mechanism of corrosion inhibition was studied using SEM, EDX, AFM, FT-IR and UV-visible methods. Using computational techniques it can be derived that in most of the studies, 8-hydroxyquinoline and its derivatives interact with metal using donor-acceptor interactions.

8.2.2 Quinoline Derivatives Other Than 8-hydroxyquinoline as Corrosion Inhibitors

Literature study showed that besides 8-hydroxyquinoline and its derivatives, other quinoline derivatives are widely used as corrosion inhibitors, and effect of substituent has also been investigated extensively. A summary of such results is presented in Table 8.2. Generally, it has been found that electron-donating substituents such as –OH, –OMe, and $–NH_2$ increases the inhibition efficiency in their presence and electron withdrawing substituents exert just inverse impact. During studying the anticorrosion property of 2-amino-4-arylquinoline-3-carbonitriles (AACs) for 1M HCl/mild steel system, our research team observed that electron-donating –OH substituent (AAC-3) increases the protection efficiency that of the non-substituted (AAC-2) and nitro-substituted (AAC-1) compounds [36]. Analyses were conducted using weight loss, electrochemical, surface, and DFT methods. Experimental results showed the tested compounds followed the order: AAC-3 (-OH) > AAC-2 (-H) > AAC-1 (-NO2). Results showed that AACs become effective by adsorbing on the metallic surface and their adsorption obeyed the Langmuir adsorption isotherm. EIS study showed that AACs acted as interface-type inhibitors, and their presence enhances the charge transfer resistance values. PDP analyses showed that all tested compounds acted as mixed-type inhibitors with slight cathodic predominance. SEM and EDX analyses were conducted to demonstrate the adsorption

Table 8.2 Chemical structures, abbreviations, and nature of metal and electrolytes of 8quinoline derivatives other than 8-hydroxyquinoline.

S No.	Chemical structure/abbreviation	Nature of inhibitor	Metal/ electrolyte	References	S No.	Chemical structure/abbreviation	Nature of inhibitor	Metal/ electrolyte	References
3	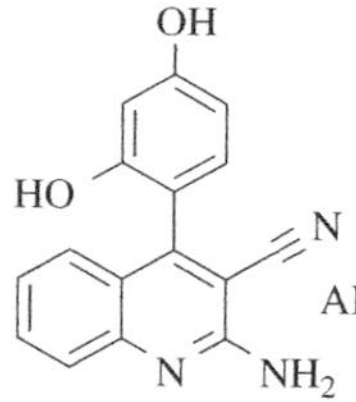 ADQC	Cathodic type	SAE 1006 steel/1M HCl	[33]	1	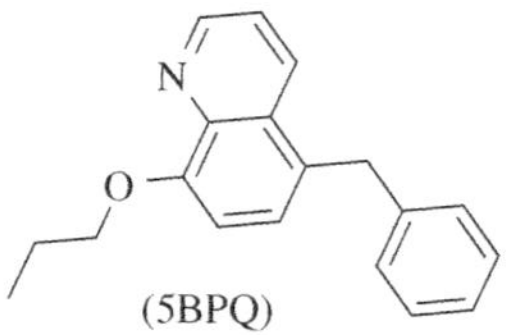 (5BPQ)	Mixed-type inhibitor	Q235 steel/ 0.5M H_2SO_4	[34]
7	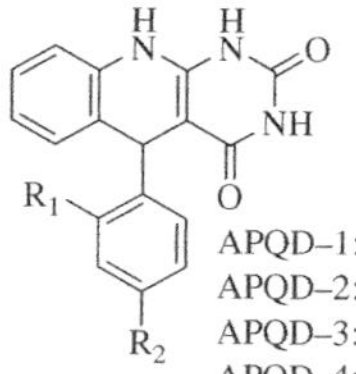 APQD–1: R_1=–H; R_2=–NO_2 APQD–2: R_1=–H; R_2=–H APQD–3: R_1=–H; R_2=–OH APQD–4: R_1=–OH; R_2=–OH	Cathodic type/LAI	Mild steel/1M HCl	[35]	6	AAC–1: R=–NO_2 AAC–2: R=–H AAC–3: R=–OH 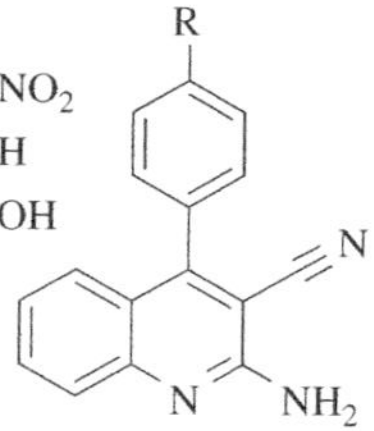	Cathodic type/LAI	Mild steel/1M HCl	[36]
9	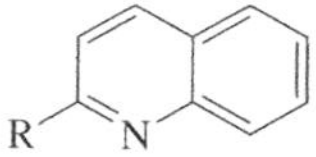 QL: R=–H QLD: R=–Me QLDA: R=–COOH	Mild steel in 0.5M HCl	LAI	[37]	5	Q–1: R=–H Q–2: R=–Me Q–3: R=–OMe Q–4: R=–NMe_2 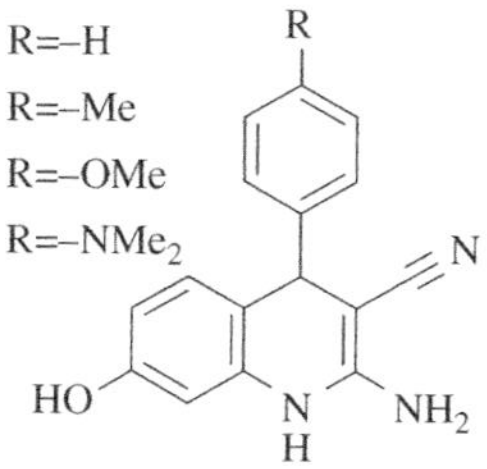	Donor–acceptor	Fe (110) surface	[38]

(Continued)

Table 8.2 (Continued)

S No.	Chemical structure/abbreviation	Nature of inhibitor	Metal/ electrolyte	References	S No.	Chemical structure/abbreviation	Nature of inhibitor	Metal/ electrolyte	References
4	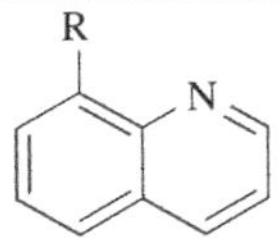 8–AQ: R=$-NH_2$ 8–NQ: R=$-NO_2$	Anodic type/ interface type	Al-AA5052 alloy/ 3% NaCl	[39]	2	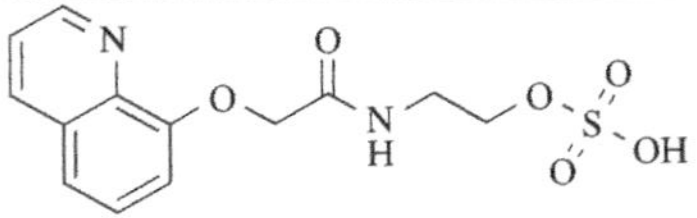 (8HQD)	Mixed-type	Al/ 1M $HClO_4$	[40]
10	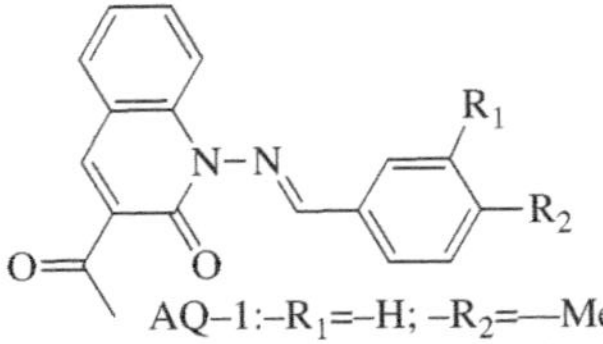 AQ–1:–R_1=–H; –R_2=—Me AQ–2:R_1=–H; –R_2=—OH AQ–3:R_1=–NO_2; –R_2=—H	Cathodic type/LAI	N80 steel/15% HCl	[41]	8	Q–1: R=–H Q–2: R=–Me Q–3: R=–OMe Q–4: R=–NMe_2 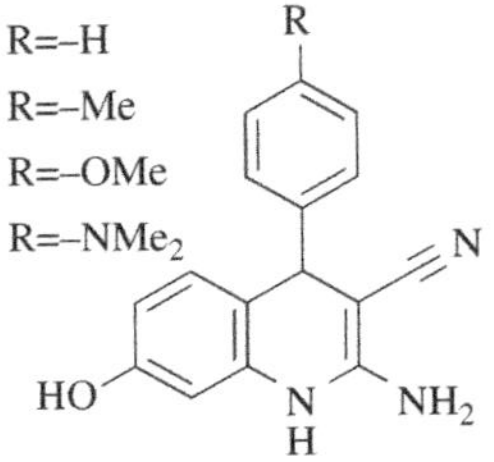	Mixed type/ LAI	Mild steel/1M HCl	[42]
14	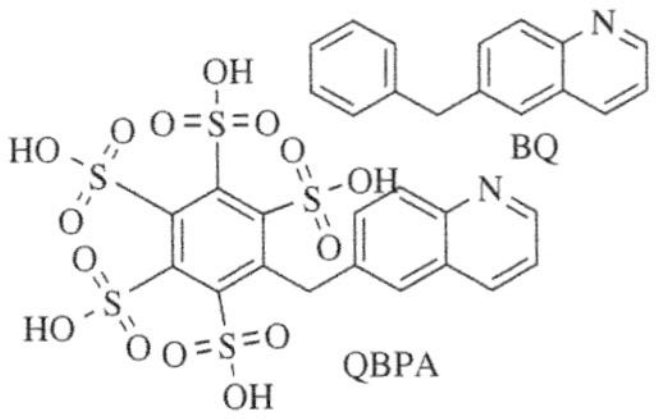 BQ QBPA	Mixed type/ LAI	Mild steel/1M HCl	[43]	11	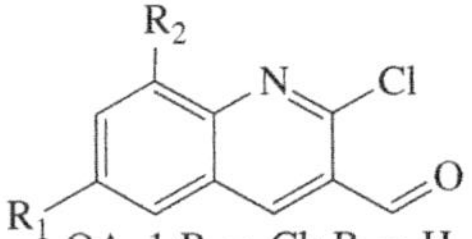 QA–1:R_1=–Cl; R_2=–H QA–2:R_1=–NO_2; R_2=–H QA–3:R_1=–Cl; R_2=–NO_2	Mild steel/1M HCl	Mixed type/ LAI	[44]

19	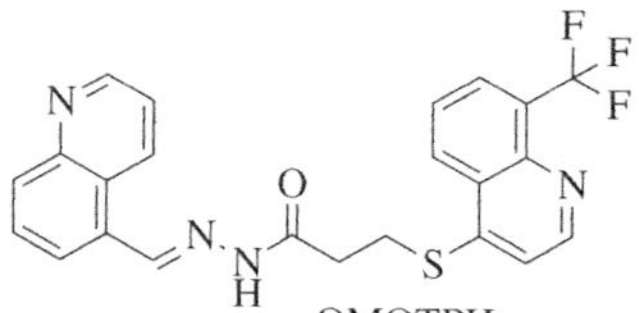 QMQTPH	Anodic type/ LAI	Mild steel/1 & 2M HCl	[45]	**13**	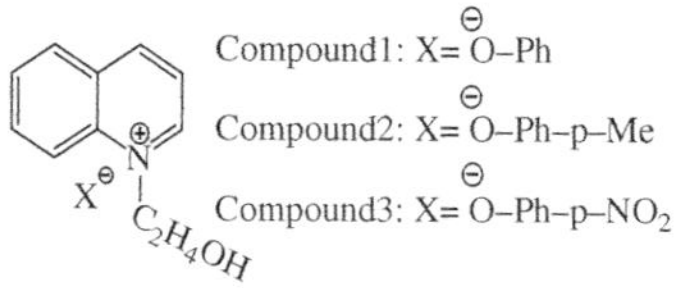 Compound1: X= $\overset{\ominus}{O}$–Ph; Compound2: X= $\overset{\ominus}{O}$–Ph–p–Me; Compound3: X= $\overset{\ominus}{O}$–Ph–p–NO_2	Anodic type	Zinc-Mg batteries/26% NH_4Cl	[46, 47]
18	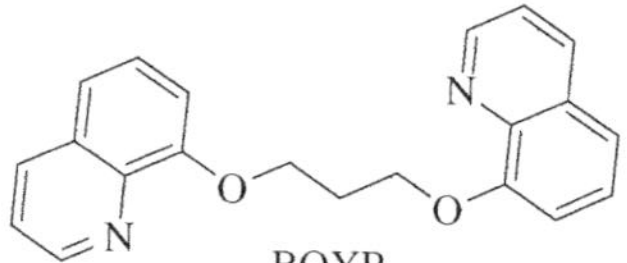 BQYP	Mixed type/ LAI	Mild steel/2M H_2SO_4	[48]	**17**	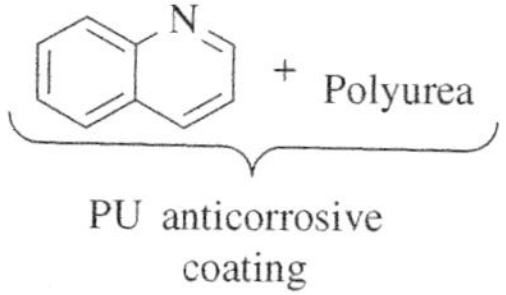 + Polyurea; PU anticorrosive coating	Mild steel /5 wt% HCl	coating	[49, 50]
12	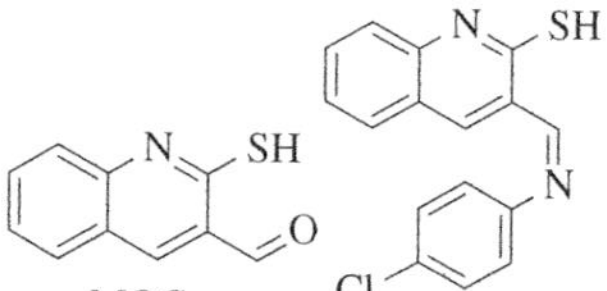 MQC; CMQT	Mild steel/1M HCl	Mixed type/ LAI	[51]	**16**	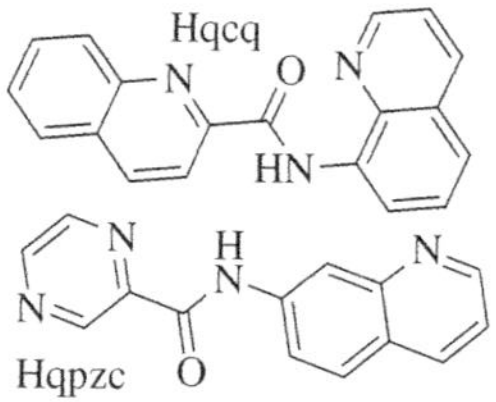 Hqcq; Hqpzc	Mixed type/ LAI	Mild steel/1M HCl	[52]
21	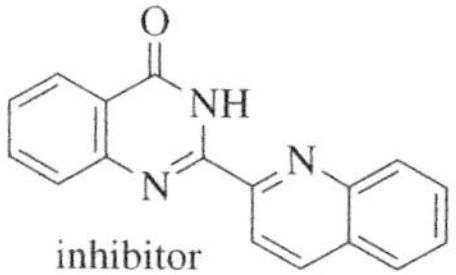 inhibitor	Mixed type/ LAI	Mild steel/1M HCl	[53]	**15**	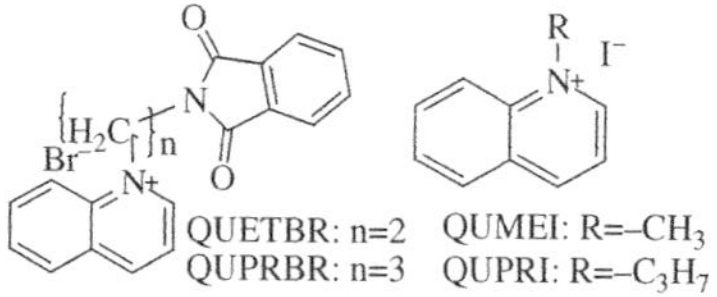 QUETBR: n=2; QUPRBR: n=3; QUMEI: R=–CH_3; QUPRI: R=–C_3H_7	Mixed type/ LAI	Mild steel/0.5M H_2SO_4	[54]

behavior of the tested compounds on metallic surface, and results showed that AACs inhibit corrosion through adsorbing on the metallic surface. DFT study provide most solid clue that how substituents affect the corrosion inhibition effect of tested AAC molecules. It was observed that contribution of HOMO and LUMO was greatly reduced in the presence of electron-withdrawing nitro-substituents and converse was true for electron-donating –OH substituent. Similar observation was further observed by our research team while studying the anticorrosive effect of 5-Arylpyrimido-[4, 5-b] quinoline-diones (APQDs) for mild steel in 1M HCl. Experimental results showed that inhibition efficiencies of tested compounds followed the order: APQD-4 (2-OH)> APQD-3 (-OH) > APQD-2 (-H) > APQD-1 (-NO2). DFT study showed that HOMO and LUMO contribution was increased in the presence of –OH substituent(s) and same was decreased in the presence of nitro-substituents. Most of the quinoline derivatives acted as mixed-type inhibitors and their adsorption followed the Langmuir isotherm model.

8.3 Conclusion and Outlook

From the ongoing discussion, it is clear that quinoline-based heterocyclic compounds are widely used as corrosion inhibitors. Presence of the high electron density in the form of five double bonds and nonbonding electrons of nitrogen, quinoline, and its derivatives offer strong bonding with the metallic surface. Presence of the electron-donating substituents enhances the protection effectiveness and converse is true for electron-withdrawing substituents. Generally, electron-withdrawing substituents decrease HOMO and LUMO contributions and electron-donating substituents increase these contribution. 8-Hydroxyquinoline and its derivatives form the chelating complexes with the metal surface. Formation of chelating complexes is also possible for other quinoline derivatives substituted by polar functional group(s) at eighth position. Most of the quinoline derivatives acted as mixed-type inhibitors and their presence affect the anodic, as well as cathodic Tafel reactions. Adsorption of the quinoline derivatives mostly followed Langmuir adsorption isotherm model. Using DFT methods, it can be derived that quinoline and its derivatives interact with metallic surface mostly using donor–acceptor mechanism. MD simulations showed that quinoline-based compounds acquire flat or nearly flat orientation on the metallic surface and therefore act as effective inhibitors.

Declaration of Interest Statement: The authors declare that they have no known competing financial interests or personal relationships that could have appeared to influence the work reported in this paper.

Conflict of Interest Statement: The authors declare that they have no known conflict of interest.

References

1 Finšgar, M. and Jackson, J. (2014). Application of corrosion inhibitors for steels in acidic media for the oil and gas industry: a review. *Corrosion Science* 86: 17–41.

2 Popoola, L.T., Grema, A.S., Latinwo, G.K. et al. (2013). Corrosion problems during oil and gas production and its mitigation. *International Journal of Industrial Chemistry* 4: 35.

3 Papavinasam, S. (2013). *Corrosion Control in the Oil and Gas Industry*. Elsevier.

4 Palacios, C.A. and Chaudary, V. (1996). Corrosion control in the oil and gas industry using nodal analysis and two-phase flow modeling techniques, SPE Latin America/Caribbean Petroleum Engineering Conference SPE-36127-MS, Society of Petroleum Engineers.

5 Verma, C., Ebenso, E.E., and Quraishi, M. (2017). Ionic liquids as green and sustainable corrosion inhibitors for metals and alloys: an overview. *Journal of Molecular Liquids* 233: 403–414.

6 Sastri, V.S. (2012). *Green Corrosion Inhibitors: Theory and Practice*. Wiley.

7 Eddy, N.O. (2011). *Green Corrosion Chemistry and Engineering: Opportunities and Challenges*. Wiley.

8 Leffler, J.E. and Grunwald, E. (2013). *Rates and Equilibria of Organic Reactions: As Treated by Statistical, Thermodynamic and Extrathermodynamic Methods*. Courier Corporation.

9 Thornton, E.R. (1964). Rates and equilibria of organc reactions, as treated by statistical, thermodynamis and extrathermodynamic methods. *Journal of the American Chemical Society* 86: 1273–1273.

10 Obot, I., Ankah, N., Sorour, A. et al. (2017). 8-Hydroxyquinoline as an alternative green and sustainable acidizing oilfield corrosion inhibitor. *Sustainable Materials and Technologies* 14: 1–10.

11 Shen, S., Zuo, Y., and Zhao, X. (2013). The effects of 8-hydroxyquinoline on corrosion performance of a Mg-rich coating on AZ91D magnesium alloy. *Corrosion Science* 76: 275–283.

12 Zong, Q., Wang, L., Sun, W., and Liu, G. (2014). Active deposition of bis (8-hydroxyquinoline) magnesium coating for enhanced corrosion resistance of AZ91D alloy. *Corrosion Science* 89: 127–136.

13 Rbaa, M., Abousalem, A.S., Touhami, M.E. et al. (2019). Novel Cu (II) and Zn (II) complexes of 8-hydroxyquinoline derivatives as effective corrosion inhibitors for mild steel in 1.0 M HCl solution: computer modeling supported experimental studies. *Journal of Molecular Liquids* 290: 111243.

14 El Faydy, M., Galai, M., El Assyry, A. et al. (2016). Experimental investigation on the corrosion inhibition of carbon steel by 5-(chloromethyl)-8-quinolinol hydrochloride in hydrochloric acid solution. *Journal of Molecular Liquids* 219: 396–404.

15 Li, S.-m., Zhang, H.-r., and Liu, J.-h. (2007). Corrosion behavior of aluminum alloy 2024-T3 by 8-hydroxy-quinoline and its derivative in 3.5% chloride solution. *Transactions of Nonferrous Metals Society of China* 17: 318–325.

16 Lakhrissi, B., Warad, I., Verma, C. et al. (2020). Experimental and computational investigations on the anti-corrosive and adsorption behavior of 7-N, N'-dialkyaminomethyl-8-Hydroxyquinolines on C40E steel surface in acidic medium. *Journal of Colloid and Interface Science* 576: 330–344.

17 El Faydy, M., Galai, M., Touhami, M.E. et al. (2017). Anticorrosion potential of some 5-amino-8-hydroxyquinolines derivatives on carbon steel in hydrochloric acid solution: gravimetric, electrochemical, surface morphological, UV–visible, DFT and Monte Carlo simulations. *Journal of Molecular Liquids* 248: 1014–1027.

18 El Faydy, M., Touir, R., Touhami, M.E. et al. (2018). Corrosion inhibition performance of newly synthesized 5-alkoxymethyl-8-hydroxyquinoline derivatives for carbon steel in 1 M HCl solution: experimental, DFT and Monte Carlo simulation studies. *Physical Chemistry Chemical Physics* 20: 20167–20187.

19 Rbaa, M., Benhiba, F., Obot, I. et al. (2019). Two new 8-hydroxyquinoline derivatives as an efficient corrosion inhibitors for mild steel in hydrochloric acid: synthesis, electrochemical, surface morphological, UV–visible and theoretical studies. *Journal of Molecular Liquids* 276: 120–133.

20 El Faydy, M., Lakhrissi, B., Guenbour, A. et al. (2019). In situ synthesis, electrochemical, surface morphological, UV–visible, DFT and Monte Carlo simulations of novel 5-substituted-8-hydroxyquinoline for corrosion protection of carbon steel in a hydrochloric acid solution. *Journal of Molecular Liquids* 280: 341–359.

21 Abou-Dobara, M., Omar, N., Diab, M. et al. (2019). Polymer complexes. LXXV. Characterization of quinoline polymer complexes as potential bio-active and anti-corrosion agents. *Materials Science and Engineering: C* 103: 109727.

22 Rouifi, Z., Rbaa, M., Benhiba, F. et al. (2020). Preparation and anti-corrosion activity of novel 8-hydroxyquinoline derivative for carbon steel corrosion in HCl molar: Computational and experimental analyses. *Journal of Molecular Liquids* 307: 112923.

23 El Faydy, M., Lakhrissi, B., Jama, C. et al. (2020). Electrochemical, surface and computational studies on the inhibition performance of some newly synthesized 8-hydroxyquinoline derivatives containing benzimidazole moiety against the corrosion of carbon steel in phosphoric acid environment. *Journal of Materials Research and Technology* 9: 727–748.

24 Eldesoky, A. and Nozha, S. (2017). The adsorption and corrosion inhibition of 8-hydroxy-7-quinolinecarboxaldehyde derivatives on C-steel surface in hydrochloric acid. *Chinese Journal of Chemical Engineering* 25: 1256–1265.

25 El Faydy, M., Benhiba, F., Lakhrissi, B. et al. (2019). The inhibitive impact of both kinds of 5-isothiocyanatomethyl-8-hydroxyquinoline derivatives on the corrosion of carbon steel in acidic electrolyte. *Journal of Molecular Liquids* 295: 111629.

26 Rbaa, M., Lgaz, H., El Kacimi, Y. et al. (2018). Synthesis, characterization and corrosion inhibition studies of novel 8-hydroxyquinoline derivatives on the acidic corrosion of mild steel: experimental and computational studies. *Materials Discovery* 12: 43–54.

27 El Faydy, M., Rbaa, M., Lakhrissi, L. et al. (2019). Corrosion protection of carbon steel by two newly synthesized benzimidazol-2-ones substituted 8-hydroxyquinoline derivatives in 1 M HCl: experimental and theoretical study. *Surfaces and Interfaces* 14: 222–237.

28 Rbaa, M., Galai, M., Benhiba, F. et al. (2019). Synthesis and investigation of quinazoline derivatives based on 8-hydroxyquinoline as corrosion inhibitors for mild steel in acidic environment: experimental and theoretical studies. *Ionics* 25: 3473–3491.

29 Rbaa, M. and Lakhrissi, B. (2019). Novel oxazole and imidazole based on 8-hydroxyquinoline as a corrosion inhibition of mild steel in HCl Solution: insights from Experimental and Computational Studies. *Surfaces and Interfaces* 15: 43–59.

30 Rbaa, M., Abousalem, A.S., Rouifi, Z. et al. (2020). Synthesis, antibacterial study and corrosion inhibition potential of newly synthesis oxathiolan and triazole derivatives of 8-hydroxyquinoline: experimental and theoretical approach. *Surfaces and Interfaces* 19: 100468.

31 Ser, C.T., Žuvela, P., and Wong, M.W. (2020). Prediction of corrosion inhibition efficiency of pyridines and quinolines on an iron surface using machine learning-powered quantitative structure-property relationships. *Applied Surface Science* 512: 145612.

32 Popova, A., Christov, M., and Vasilev, A. (2007). Inhibitive properties of quaternary ammonium bromides of N-containing heterocycles on acid mild steel corrosion. Part I: gravimetric and voltammetric results. *Corrosion Science* 49: 3276–3289.

33 Verma, C. and Quraishi, M. (2017). 2-Amino-4-(2, 4-dihydroxyphenyl) quinoline-3-carbonitrile as sustainable corrosion inhibitor for SAE 1006 steel in 1 M HCl: electrochemical and surface investigation. *Journal of the Association of Arab Universities for Basic and Applied Sciences* 23: 29–36.

34 Chen, S., Chen, S., and Li, W. (2019). Corrosion inhibition effect of a New Quinoline derivative on Q235 Steel in H2SO4 solution. *International Journal of Electrochemical Science* 14: 11419–11428.

35 Verma, C., Olasunkanmi, L., Obot, I. et al. (2016). 5-Arylpyrimido-[4, 5-b] quinoline-diones as new and sustainable corrosion inhibitors for mild steel in 1 M HCl: a combined experimental and theoretical approach. *RSC Advances* 6: 15639–15654.

36 Verma, C., Quraishi, M., Olasunkanmi, L., and Ebenso, E.E. (2015). L-Proline-promoted synthesis of 2-amino-4-arylquinoline-3-carbonitriles as sustainable

corrosion inhibitors for mild steel in 1 M HCl: experimental and computational studies. *RSC Advances* 5: 85417–85430.

37 Ebenso, E.E., Obot, I.B., and Murulana, L. (2010). Quinoline and its derivatives as effective corrosion inhibitors for mild steel in acidic medium. *International Journal of Electrochemical Science* 5: 1574–1586.

38 Erdoğan, Ş., Safi, Z.S., Kaya, S. et al. (2017). A computational study on corrosion inhibition performances of novel quinoline derivatives against the corrosion of iron. *Journal of Molecular Structure* 1134: 751–761.

39 Wang, D., Yang, D., Zhang, D. et al. (2015). Electrochemical and DFT studies of quinoline derivatives on corrosion inhibition of AA5052 aluminium alloy in NaCl solution. *Applied Surface Science* 357: 2176–2183.

40 Aly, M.R.E.S., Shokry, H., Sharshar, T., and Amin, M.A. (2016). A newly synthesized sulphated 8-hydroxyquinoline derivative to effectively control aluminum corrosion in perchloric acid: electrochemical and positron annihilation studies. *Journal of Molecular Liquids* 214: 319–334.

41 Ansari, K., Ramkumar, S., Nalini, D., and Quraishi, M. (2016). Studies on adsorption and corrosion inhibitive properties of quinoline derivatives on N80 steel in 15% hydrochloric acid. *Cogent Chemistry* 2: 1145032.

42 Singh, P., Srivastava, V., and Quraishi, M. (2016). Novel quinoline derivatives as green corrosion inhibitors for mild steel in acidic medium: electrochemical, SEM, AFM, and XPS studies. *Journal of Molecular Liquids* 216: 164–173.

43 Jiang, L., Qiang, Y., Lei, Z. et al. (2018). Excellent corrosion inhibition performance of novel quinoline derivatives on mild steel in HCl media: experimental and computational investigations. *Journal of Molecular Liquids* 255: 53–63.

44 Lgaz, H., Salghi, R., Bhat, K.S. et al. (2017). Correlated experimental and theoretical study on inhibition behavior of novel quinoline derivatives for the corrosion of mild steel in hydrochloric acid solution. *Journal of Molecular Liquids* 244: 154–168.

45 Saliyan, V.R. and Adhikari, A.V. (2008). Quinolin-5-ylmethylene-3-{[8-(trifluoromethyl) quinolin-4-yl] thio} propanohydrazide as an effective inhibitor of mild steel corrosion in HCl solution. *Corrosion Science* 50: 55–61.

46 Zhang, D., Li, L., Cao, L. et al. (2001). Studies of corrosion inhibitors for zinc–manganese batteries: quinoline quaternary ammonium phenolates. *Corrosion Science* 43: 1627–1636.

47 Zhao, J., Duan, H., and Jiang, R. (2015). Synergistic corrosion inhibition effect of quinoline quaternary ammonium salt and Gemini surfactant in H2S and CO2 saturated brine solution. *Corrosion Science* 91: 108–119.

48 Alamshany, Z.M. and Ganash, A.A. (2019). Synthesis, characterization, and anti-corrosion properties of an 8-hydroxyquinoline derivative. *Heliyon* 5: e02895.

49 Gite, V.V., Tatiya, P.D., Marathe, R.J. et al. (2015). Microencapsulation of quinoline as a corrosion inhibitor in polyurea microcapsules for application in anticorrosive PU coatings. *Progress in Organic Coatings* 83: 11–18.

50 Marathe, R., Tatiya, P., Chaudhari, A. et al. (2015). Neem acetylated polyester polyol – renewable source based smart PU coatings containing quinoline (corrosion inhibitor) encapsulated polyurea microcapsules for enhance anticorrosive property. *Industrial Crops and Products* 77: 239–250.

51 Mistry, B.M., Sahoo, S.K., and Jauhari, S. (2013). Experimental and theoretical investigation of 2-mercaptoquinoline-3-carbaldehyde and its Schiff base as an inhibitor of mild steel in 1 M HCl. *Journal of Electroanalytical Chemistry* 704: 118–129.

52 Erami, R.S., Amirnasr, M., Meghdadi, S. et al. (2019). Carboxamide derivatives as new corrosion inhibitors for mild steel protection in hydrochloric acid solution. *Corrosion Science* 151: 190–197.

53 Zhang, W., Ma, R., Liu, H. et al. (2016). Electrochemical and surface analysis studies of 2-(quinolin-2-yl) quinazolin-4 (3H)-one as corrosion inhibitor for Q235 steel in hydrochloric acid. *Journal of Molecular Liquids* 222: 671–679.

54 Mourya, P., Singh, P., Tewari, A. et al. (2015). Relationship between structure and inhibition behaviour of quinolinium salts for mild steel corrosion: experimental and theoretical approach. *Corrosion Science* 95: 71–87.

9

Indole and Its Derivatives as Corrosion Inhibitors

Taiwo W. Quadri[1], Lukman O. Olasunkanmi[2], Ekemini D. Akpan[1], and Eno E. Ebenso[3]

[1] *Department of Chemistry, School of Chemical and Physical Sciences and Material Science Innovation & Modelling (MaSIM) Research Focus Area, Faculty of Natural and Agricultural Sciences, North-West University, Mmabatho, South Africa*
[2] *Department of Chemistry, Faculty of Science, Obafemi Awolowo University, Ile Ife, Nigeria*
[3] *Institute for Nanotechnology and Water Sustainability, College of Science, Engineering and Technology, University of South Africa, Johannesburg, South Africa*

9.1 Introduction

Heterocyclic organic compounds have gained wide interest and applications in several fields of scientific research over the years owing to their chemical and biological activities [1, 2]. Heterocycles are cyclic organic compounds consisting of heteroatoms such as nitrogen, phosphorus, selenium, sulfur, or oxygen in their structure. Prominent among these heterocycles is the five-membered ring system known as indoles, which is probably one of the most naturally abundant, widely investigated, and well-reported heterocyclic compound as demonstrated by its numerous applications and meteoric rise in the number of citations [3, 4]. The history of indoles is dated back to 1866 when Adolf von Baeyer [5] reduced oxindole to indole with the use of zinc dust. Indole is a nitrogen-containing heterocycle having a five-membered pyrrole ring fused with benzene ring as shown in Figure 9.1. Indoles have been found to exist in natural biological compounds such as serotonin, melatonin, tryptophan, and in synthetic drugs such as tadalafil, fluvastatin, rizatriptan, and sumatriptan. It is worthy of mention that these four indole-based drugs recorded a large sales output to the tune of US $3.2 billion in 2010 [6]. Besides, several research reports have documented the presence of indole alkaloids in plants. Generally, indole and its derivatives have been

Organic Corrosion Inhibitors: Synthesis, Characterization, Mechanism, and Applications, First Edition. Edited by Chandrabhan Verma, Chaudhery Mustansar Hussain, and Eno E. Ebenso.

investigated and extensively used in biological and medicinal fields as antibacterial, anticancer, anti-inflammatory, anti-HIV, antioxidant, antifungal, antitubercular, antimalarial agents and so on [7–10]. In addition, indole compounds have also found applicability in the development of super-capacitors, textile dyes, fertilizers, insecticides, etc. [11–13]. Apart from these listed wide range of applications, indole molecules have also been explored for their corrosion inhibition properties.

Isatin, an indole derivative, has gained considerable attention among corrosion research experts, which makes it necessary to present some facts about this group of molecules as well. Isatin or 1H-indole-2,3-dione was synthesized for the first time from the reaction of indigo dye with chromic and nitric acids by Erdmann and Laurent in 1840 [14]. Isatin, an aromatic bicyclic heterocyclic compound with both a five and six-membered rings, is characterized with a N atom fixed at position 1 with two carbonyl groups at positions 2 and 3 (Figure 9.1). Isatin is abundant in nature and is known to possess wide biological, pharmaceutical, and medical applications [15–17].

Indole and its numerous derivatives have been synthesized and investigated as potential anticorrosive additives for industrial metals such as copper, zinc, mild steel, aluminum, and brass. These N-containing compounds have been reported to adsorb on the surface of metal, thereby creating a shielding effect on the metallic surface. This chapter presents a comprehensive overview of the application of indole and its derivatives as anticorrosive materials for different metals in diverse corrosive media.

9.2 Synthesis of Indoles and Its Derivatives

Several methods of synthesis have been identified in the preparation of indoles, which have been reported in literature with the advent of newer synthetic routes [6, 18]. About the oldest, conventional, and most reliable method of indole preparation is the Fischer indole synthetic method, which entails aromatic C–H functionalization. In corrosion inhibition studies, much attention is paid to substituted indole derivatives as the presence of substituents significantly impact on

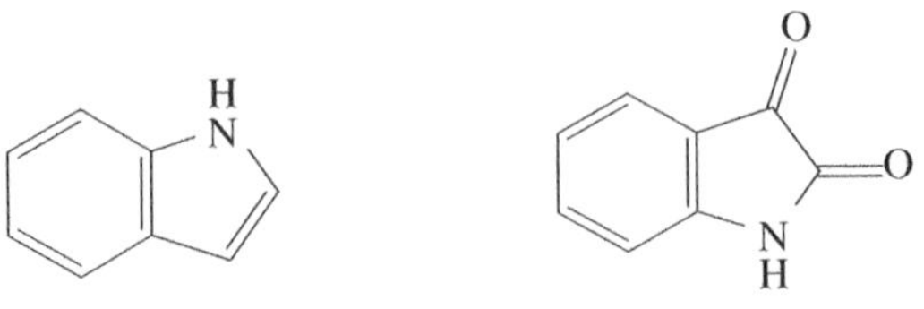

Figure 9.1 Structure of indole and isatin. Structures of indole (left-hand side) and isatin (right-hand side).

the chemical activity of the compound. Some of the indole synthetic methods are enumerated below:

1) Alum-catalyzed synthesis of bis(indolyl) methanes by reaction of 1H-indoles with aldehydes/ketones by ultrasound approach (Scheme 9.1) [19]
2) Oxidative Michael reaction of Baylis–Hillman adducts with indoles by 2-iodoxybenzoic acid under neutral conditions (Scheme 9.2) [20]
3) Refluxing with hydrazine hydrate and condensation with different substituted aromatic aldehydes in methanol (Scheme 9.3) [21]
4) One-pot synthesis reaction between phenylhydrazines and pyruvic acid via microwave irradiation (Scheme 9.4) [22]
5) Hetero-annulation of 2-haloaniline compounds and phenylacetylene in a pot reaction with Pd $(PPh_3)_2Cl_2$ as catalyst (Scheme 9.5) [23]
6) Synthesis of N-substituted indole aldehydes from substituted benzyl halides (Scheme 9.6) [24]
7) Preparation of 3-amino-alylated indoles via a three-component Mannich-type reaction using organocatalyzed solvent-free conditions (Scheme 9.7) [25]

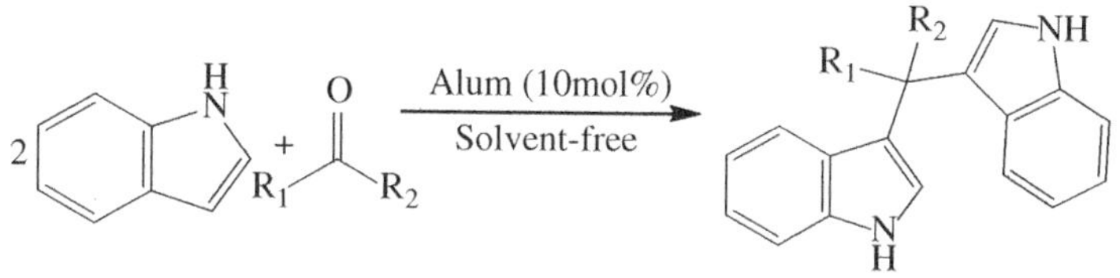

Scheme 9.1 Synthetic scheme of alum catalyzed bis(indolyl)methanes.

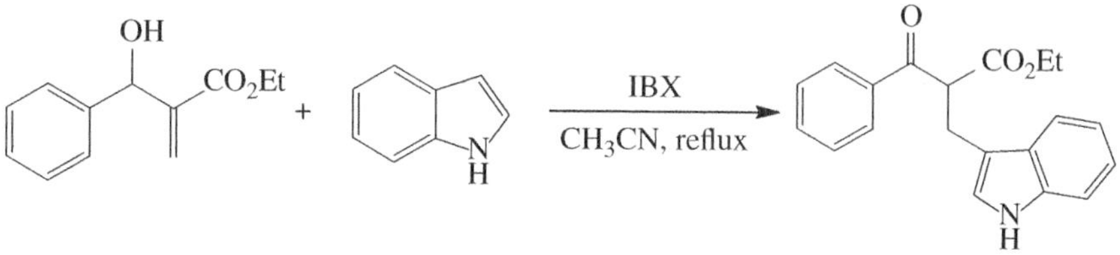

Scheme 9.2 One-pot oxidative Michael reaction of Baylis–Hillman adducts.

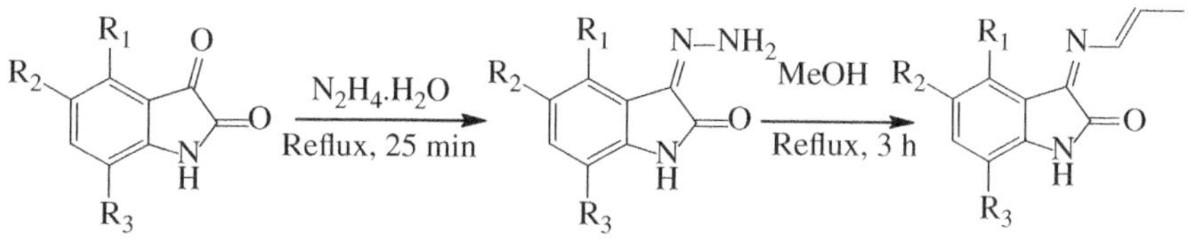

Scheme 9.3 Synthetic route of bis-Schiff bases from isatin.

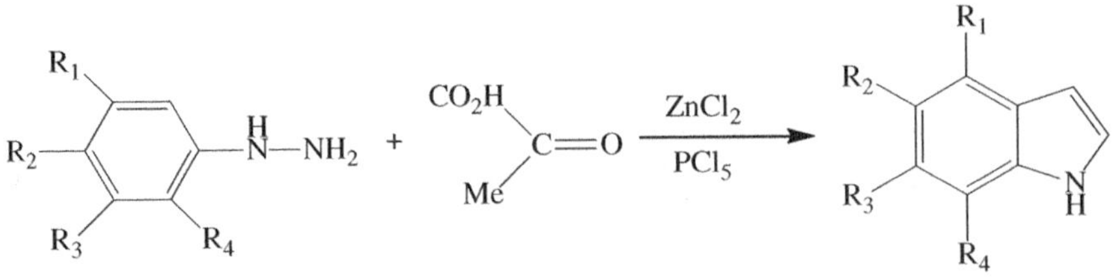

R_1, R_2, R_3, R_4 — N(H)–NH_2 + CO_2H, C=O, Me — $ZnCl_2$ / PCl_5 → R_1, R_2, R_3, R_4, NH

Scheme 9.4 One-pot synthesis of indole-2-carboxylic acids.

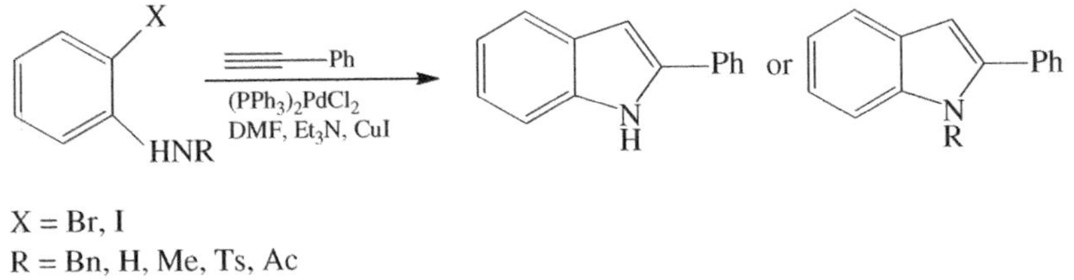

X, HNR — Ph; $(PPh_3)_2PdCl_2$; DMF, Et_3N, CuI → NH, Ph or NR, Ph

X = Br, I
R = Bn, H, Me, Ts, Ac

Scheme 9.5 Synthetic scheme of 2-substituted indoles by heteroannulation.

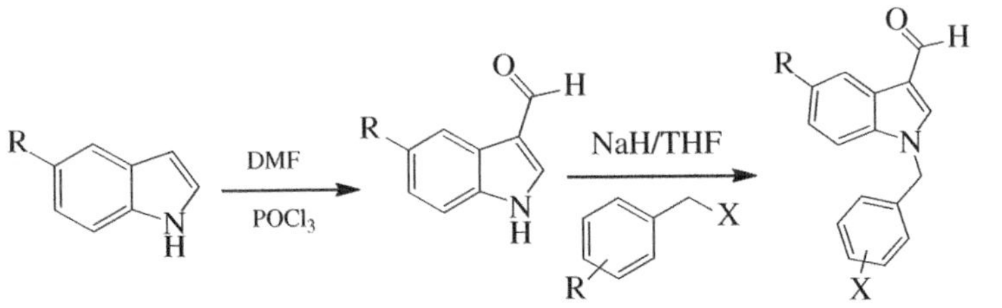

R, NH — DMF / $POCl_3$ → R, O, H, NH — NaH/THF; R, X → R, O, H, N, X

Scheme 9.6 Preparation of *N*-benzyl indole aldehydes from indole.

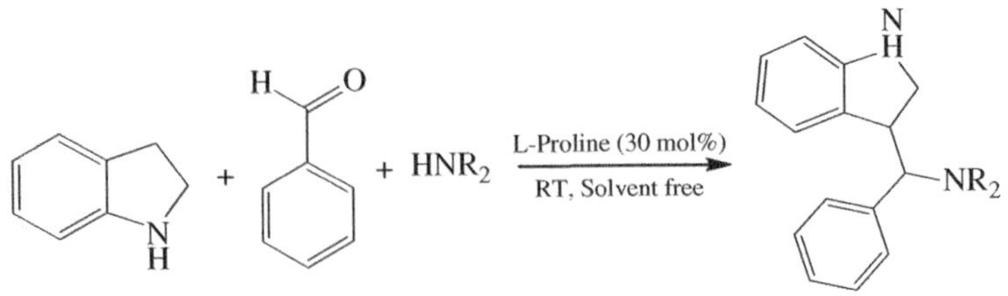

NH + H, O + HNR_2 — L-Proline (30 mol%) / RT, Solvent free → NH, NR_2

Scheme 9.7 Preparation of 3-amino-alkylated indoles.

8) Ultrasonic condensation of an aldehyde, malonitrile, and 2-mercaptoindole in the presence of boric acid and cetyltrimethylammonium bromide (CTAB) as catalysts at 80°C (Scheme 9.8) [26, 27]

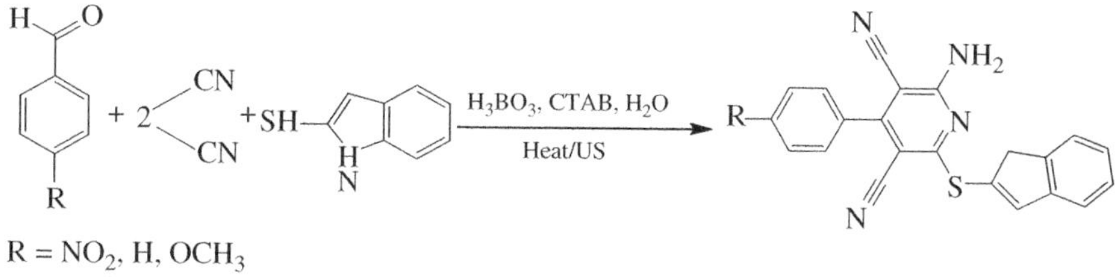

H O CN + 2 +SH— CN H N R H_3BO_3, CTAB, H_2O Heat/US R— N NH_2 N S— N

R = NO_2, H, OCH_3

Scheme 9.8 Synthetic route of Boric acid and CTAB-catalyzed indole derivatives.

9.3 A Brief Overview of Corrosion and Corrosion Inhibitors

The deterioration of metallic materials as a result of exposure to the environment still remains a subject of concern to both the academia and the industry [28]. Corrosion continues to pose a menace to individuals, industries, and the society at large. Major industries such as the metallurgical and material, chemical and petrochemical, mining, oil and gas, electronic, and transport industries spend a fortune annually to combat the effect of corrosion [29]. The loss of metallic functionality due to corrosion ends up in colossal economic losses for them. It has been reported that both industrialized and developing nations of the world are not exempted from the adverse impact of corrosion on the economy. On a global scale, the reported estimate of the loss due to metallic corrosion is around 3.4% of the world GDP [30, 31]. Besides, corrosion has environmental implications as contamination could occur due to burst pipes, buildings and bridges are also subject to collapse owing to the corrosion of metal. In a bid to control corrosion phenomena, several methodologies such as material selection, excellent equipment design, cathodic protection, use of anticorrosive coatings, and application of corrosion inhibitors have been explored by corrosion experts. Among these, corrosion inhibitors stands out as one of the best corrosion control methods in the industry because of their cost-effectiveness, in situ implementation without altering the metallic application, relative low toxicity, and good solubility. Additionally, corrosion inhibitors have shown applications in a wide range of corrosion processes [29].

Corrosion inhibitors are chemical compounds that significantly inhibit cathodic and anodic reactions at low concentrations without altering the concentration of the electrolyte [32]. Generally, these chemical inhibitors act by creating a defensive layer on the substrate surface and prevents it from the reach of corrosive ions. Certain factors are considerably important in the selection of corrosion inhibitors. According to McCafferty [33], three principal factors that influence the adsorption potential of a corrosion inhibitor are the molecular size of the inhibitor, solubility of the inhibitor in the electrolyte, and electron-donating ability of the chemical inhibitor. Besides these, other important factors that must be considered especially for manufacturing and large-scale use include the price of the inhibitor and environmental friendliness [34]. The library is filled with a plethora of studies investigating the inhibitory capacities of inorganic and heterocyclic organic compounds such as nitrites, chromates, sulfates, pyrimidines, pyridines, polymers, quinoxalines, sulfonamides, triazoles, and so on [35–40]. According to Andreatta and Fedrizzi, these corrosion inhibitors can be broadly classified based on their influence on partial electrochemical reactions, the application field, and the reaction mechanism [29].

Whereas previously, the inhibitor effectiveness of organic compounds is primal; nowadays, much emphasis is now been placed on the environmental friendliness of the organic inhibitors. Beyond the E-3 concept of efficiency, economy, and ecology proposed by Kalman [41] in 1994 in the choice of inhibitors, a new and vital consideration is the environmental friendliness of the compounds. This has mandated the development of safe, biodegradable, and nontoxic organic molecules for use as corrosion inhibitors. Green Chemistry principles [42], some of which include solvent-free synthesis, one-pot multicomponent reactions, use of safer solvents, use of bio-based catalysts, use of microwave, and ultrasonic irradiations, have now been followed in a number of reports [27, 43–45] in the preparation of indole derivatives as corrosion inhibitor formulations.

9.4 Application of Indoles as Corrosion Inhibitors

Indoles and its derivatives belong to the group of nitrogen-containing compounds like pyridines, quinolines, acridines, and pyridazines, which have proven to be efficient in repressing corrosion. Among several metal-based industries where corrosion inhibitors have found applications, the petroleum industry has been reported to be the largest consumer of corrosion inhibitors [46]. This industry is well known for operations such as acid descaling, oil well acidification, and chemical cleaning, which exposes the metal to corrosion in various degrees. Operational temperatures are also known to influence the rate of corrosion in these industries. Different concentrations of corrosive environments like sulfuric acid, phosphoric acid, nitric acid, and most commonly, hydrochloric acid have been utilized to analyze the adsorption behavior of indole and its derivatives for different metals.

Previous studies have reported that the indole nucleus has the tendency to protonate in HCl medium on the 3 C atom. The N atom possibly supplies electrons to the cyclic ring, thereby catalyzing the protonation reaction. Adsorption of this protonated indole nucleus on the metallic surface occurs by electrostatic attraction [47–49]. Weight loss (WL) technique and modern electrochemical methods such as electrochemical impedance spectroscopy (EIS) and potentiodynamic polarization (PDP) have been utilized to investigate the electrochemical behavior of metals in the presence of indoles. Spectroscopic techniques such as ultraviolet spectroscopy (UV-vis), X-ray photoelectron spectroscopy (XPS), and Fourier transform infra-red (FTIR) spectroscopy have been used in studying the interaction of organic inhibitors with the metal substrate. Investigation on the surface protection of metal by adherence of organic compounds is also carried out using microscopic techniques such as atomic force microscopy (AFM), field emission scanning electron microscopy (FESEM), and scanning electron microscopy (SEM) solution. Furthermore, theoretical methods involving density functional theory (DFT) and atomistic simulations such as Monte Carlo simulation (MCS) and molecular dynamics simulation (MDS) have been deployed as viable techniques to gain insight into the inhibition mechanism of indole compounds.

9.4.1 Indoles as Corrosion Inhibitors of Ferrous Metals

Ferrous metals are metallic materials that are iron-based. Steel, an alloy of iron, comes in various forms such as mild carbon steel, medium carbon steel, and high carbon steel [50]. It is a metal of interest that has gained wide usage in petroleum, construction, and chemical processing industries because of its many advantages such as abundance in nature, excellent mechanical strength, and prolonged existence. Besides, steel is a "green" product due to its 100% recyclability [51]. Indole derivatives have been employed to impede the corrosion of a variety of steel grades such as N80 steel, stainless steel (316L), X80 steel, gray iron, and Q235A steel. This section reviews collection of works reported in literature on corrosion inhibition of ferrous metals using indole and its derivatives.

The anticorrosive ability of indole nucleus for stainless steel (316L) degradation was investigated by PDP and EIS methods [47]. The study was conducted in a test solution of acid and alkali (0.3 M NaCl, pH = 4, 8, and 10), and the obtained results showed that at a concentration of 10^{-3} M, indole effectively impeded rusting in 0.3 M NaCl at pH = 4. However, its effectiveness in the alkaline solutions was limited, which was probably due to the strong adherence between the metallic surface and the hydroxide ions. Tussolini and coworkers also reported the inhibitive effect of indole molecule for AISI 430 stainless steel (SS) as the corrosive environment was changed from molar H_2SO_4 to molar HCl using modern electrochemical techniques such as open-circuit potential (OCP), PDP, and EIS. Surface analyses were done with the aid of optical microscopy and SEM. Their result findings revealed that the adsorption ability of indole in Cl^- solution was greater than the SO_4^{2-} electrolytic medium, which was justified on the basis of the indole action as a weak acid [48]. Other similar studies have been reported to establish the effectiveness of the indole nucleus as a corrosion inhibitor [52–59].

Khaled also employed electrochemical frequency modulation (EFM) among other electrochemical and computational methods to monitor the electrochemical behavior of pure Fe in 1M HCl by indole molecule and three related compounds [60]. In a more recent study, Lv and coworkers [61] extensively studied the effect of indole molecule on Q235 mild steel corrosion in 0.1M H_2SO_4 using WL, PDP, and EIS. The progressive increase in the concentration of indole in the corrosive environment resulted in pronounced impedance response, and the obtained depressed semicircles shown in Figure 9.2 suggest a charge transfer phenomenon. The presented phase plots show one time constant at all studied concentrations, which relates to the charge transfer process. The highest inhibition efficiency (IE) of 97.17% was obtained with a concentration of 8×10^{-3} M at 298 K. The authors concluded that the inhibitory ability exhibited by indole is traceable to the development of a shielding layer on the studied mild steel (MS) surface as demonstrated by the SEM analysis. Theoretical approaches (DFT and MCS) were also employed to clarify the mechanism of reactivity of the indole molecule.

The effects of indole-3-carboxylic acid (ICA) in 0.1M H_2SO_4 [62] and indole-5-carboxylic acid in 0.5M H_2SO_4 [63] for MS have been reported using WL and electrochemical techniques. With 4×10^{-3} M, ICA yielded its highest IE of 95.98% at 298 K, while indole-5-carboxylic offered its highest inhibition efficiency of 92.00% at 298 K. Avci reported on the surface protection of MS in 0.5M HCl by indole-3-acetic acid with the use of PDP, WL, and EIS techniques. The impacts of temperature, immersion time, and amount added on the adsorption potential of the indole compound was investigated. The study reported a maximum IE of 93% at a concentration of 1×10^{-2} M [64]. Ashhari and Sarabi purchased 2-methylindole and indole-3-carbaldehyde having $-CH_3$ and $-C=O$ functional groups, respectively, from Merck Co. and tested them against MS corrosion in molar HCl in order to study the influence of these functional groups with different induction effects on indole [65]. The choice of relatively similar molecules in terms of size and solubility was to enhance the study of the effect of the different electronic structure using several methodologies. The impedance spectra obtained showed that indole-3-carbaldehyde performed better than 2-methylindole. This was corroborated with findings from PDP and contact angle (CA) measurements. Further explanation on the observed differences in the inhibition performance was offered from the DFT results. The authors posited that the greater positive charge found on the aromatic ring of indole-3-carbaldehyde could favor the adsorption of the inhibitor on the Cl^- present on the metallic substrate better than 2-methylindole. Other obtained quantum chemical indices were employed to substantiate the better inhibitive effect of indole-3-carbaldehyde.

Some other simple derivatives of indoles such as 5-chloroindole (CI), 5-hydroxyindole (HI), 5-nitroindole (NI), and 5-aminoindole (AI) have also been investigated as potential inhibitors of MS corrosion in acidic media using different

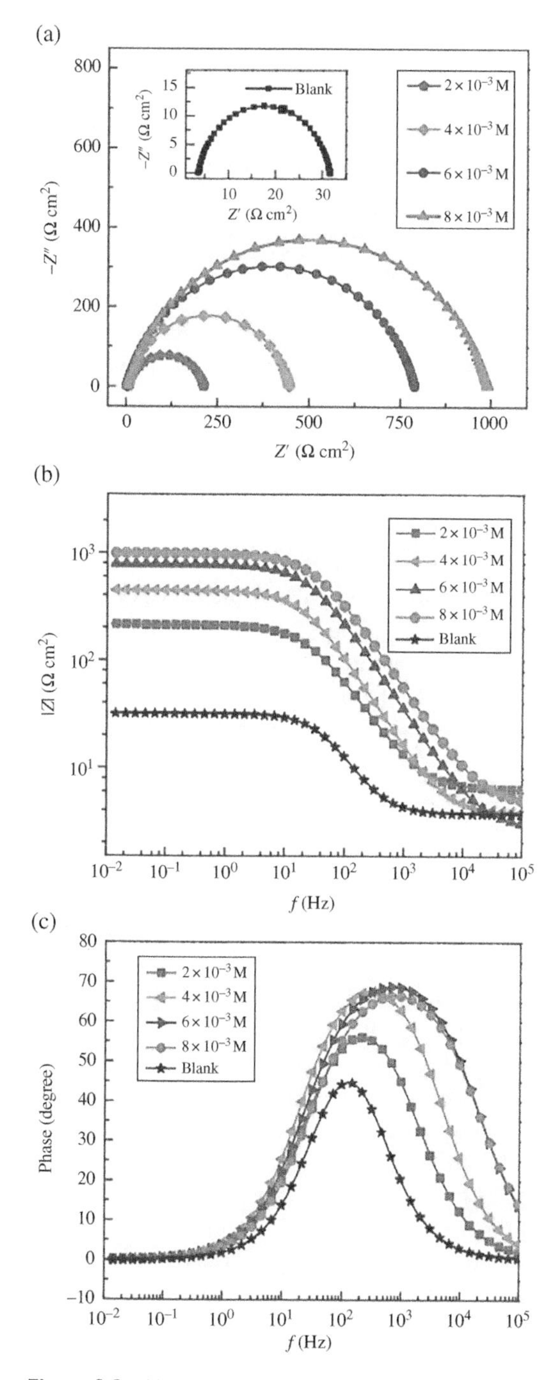

Figure 9.2 Nyquist (a), bode (b), and bode phase (c) spectra for CS obtained with and without the various amounts of indole molecule in 0.1M sulfuric acid at 298 K. *Source:* Lv et al. [61].

modern electrochemical methods [66–69]. The studies showed these indole derivatives decreased corrosion rates as the amounts of the chemical additives increased and acted as mixed-type inhibitors.

0.05M 5-aminoindole (5Ain) monomer and a 0.3 M oxalic acid were employed in the electrosynthesis of poly-5-aminoindole (P5Ain) via potentiodynamic technique. The electrosynthesized poly-5-aminoindole (P5Ain) was tested for its inhibition properties for a type-304 SS in a 3.5% NaCl solution. OCP, EIS, and PDP were used as electrochemical techniques to evaluate the inhibitive effect of the studied compounds while SEM was used to investigate the effect of rusting on the surface of the metallic substrate in the presence of P5Ain after 96 hours of immersion. The results obtained revealed that the P5Ain coating on the SS forced a barrier to inhibit corrosion on the metallic surface for a considerable period of time [70]. In yet another study, copolymer films were developed based on indole and aniline with two different monomer feed percentages to assess the behavior of SS in 3.5% NaCl [71]. The authors observed that TiO_2 pre-coated copolymer (P(In-co-An)(1:1) offered a better corrosion protection than (P(In-co-An)(1:9) evidenced by low CPE values and high charge transfer resistance for the testing time of 144 hours. Theoretical studies were used to explain that the polymer stability could be enhanced by increasing the indole monomer ratio. Similar studies involving indole-based copolymers, polymer nanocomposites, and epoxy coatings as anticorrosive agents for steel have also been reported [49, 72–79].

The presence of substituents in the aromatic ring has also been reported to alter the corrosion inhibition performance of the indole compound. Generally, electron-withdrawing substituents like $-NO_2$ have been reported to lower inhibition efficiency when contrasted to electron-donating substituents like –OH, $-CH_3$, $-OCH_3$, etc. Verma and coworkers [43] conducted an extensive study on the inhibitory actions of three 3-amino alkylated indoles (AAIs) on MS corrosion in chloride acid solution via well-documented approaches such as WL, electrochemical (PDP, EIS), surface analysis (SEM, AFM), and computational approaches (DFT and MDS). The studied compounds were designated as N-((1H-indol-3-yl)(phenyl)methyl)-N-ethylethanamine (AAI-1), 3-(phenyl(pyrrolidin-1-yl)methyl)-1H-indole (AAI-2), and 3-(phenyl(piperidin-1-yl)methyl)-1H-indole (AAI-3). The obtained results ((AAI-3 (96.95%) > AAI-2 (96.08%) > AAI-1 (94.34%)) revealed that increase in cyclic amino substituents favored higher inhibition performances than open-chain amino group. Also, due to ring size difference of the amino group, the IE of pyrrolidine-containing 3-amino alkylated indole with five rings (AAI-2) was lower than that of the piperidine-containing inhibitor having six rings (AAI-3).

Another case study is the investigation by Fouda and coworkers on the inhibitory capacities of five Schiff bases of indole nucleus via EIS, PDP, and EFM techniques. The order of IE of the studied compounds were as follows: 1(–H) > 2($-CH_3$) > 3 ($-OCH_3$) > 4(–Cl) > 5 ($-NO_2$) [80]. Three novel isatin compounds designated as 5-BEI, 5-FEI, and 5-HEI were designed and explored as chemical inhibitors of carbon steel (CS) degradation in acidic medium by Kharbach and coworkers [81]. Using gravimetric, electrochemical, surface analyses, and computational modeling techniques, the authors reported excellent inhibition performance of the compounds. The nature of substituent was found to play a role in the obtained order of IE: 5-BEI (–Br) > 5-FEI (–F)> 5-HEI (–H). The substitution of –H with –F and –Br enhanced the inhibition performance of 1-[2-(diethylamino) ethyl]-1H–indole-2,3-dione (5-HEI) as these substituents (–F and –Br) are electron donors. They resulted in increased electron cloud of the aromatic group and availed pi electrons for interaction with the CS surface. Four Mannich bases (MI-1, MI-2, MI-3, and MI-4) were prepared and applied as inhibitors of MS corrosion in 1M HCl using EIS, PDP, linear polarization resistance (LPR), and mass loss methods. At the optimum concentration of 300 ppm, authors found out that the corrosion current drastically reduced as the inhibitor was introduced into the corrosive medium. The indole derivatives were found to repress both metallic dissolution at the anode and hydrogen evolution reaction at the cathode. The authors opined that the nearly static values of anodic slope (for all studied inhibitors) and cathodic slope (for MI-1 and MI-3) suggested that the inhibitors initially adhered onto the steel substrate by blocking the active sites of the MS without altering the reaction mechanism. However, for compounds MI-2 and MI-4 with increased values of cathodic Tafel slope, the authors noted a change in the mechanism of hydrogen evolution probably due to barrier or diffusion effect [82]. Overall, the study showed agreement in the IE obtained from all the employed experimental techniques and a correlation with the DFT calculations. The sequence of IE obtained for the four studied compounds is MI-2 (–OMe)> MI-3 (–Cl)> MI-1 (–H)> MI-4 ($-NO_2$), which was elucidated in relation to the substituent present in the phenylimino group. Electron donors such as $-OCH_3$ and –Cl are believed to have increased the electron cloud on the N of azomethine moiety, thereby yielding higher IE than the electron-withdrawing group ($-NO_2$). Besides, methoxy group has a better electron-donating ability than the –Cl group, and the planarity of MI-2 (–OMe) was another possible reason for its better performance. In the same vein, three Mannich bases were prepared and studied as corrosion inhibitors of MS in HCl. The results revealed that the presence of di-n-butylamino group and a longer alkyl chain in 3-(4-chlorophenylimino)-1-[(dibutylamino)methyl]indolin-2-one (M-3) were possible factors responsible for its better inhibition performance than 3-(4-chlorophenylimino)-1-(piperidin-1-ylmethyl)indolin-2-one (M-1) and 3-(4-chlorophenylimino)-1-(morpholinomethyl)indolin-2-one (M-2) [83]. While the three inhibitors showed inhibition efficiencies greater than 90%, the presence of morpholino group in M-2 was a probable reason for its lowest performance among the three. Similar studies on the roles played by the presence of substituents have also been documented in literature by other authors [27, 84–88].

Singh and coworkers synthesized Schiff bases of isoniazid with 5-substituted indole compounds and assessed their anticorrosion activities for MS in molar HCl by computational modeling (DFT, MD), surface analyses (SEM-EDX, AFM, XPS), gravimetric method (WL), and electrochemical methods (EIS, PDP). All the compounds yielded appreciable IE of over 90% at 200 mg/l. The activation energies were found to increase ranging from 67 to 76 kJ/mol on introducing the four inhibitors to the blank HCl solution (32.8 kJ/mol). Fitting measured data into the various isotherm models proved Langmuir isotherm to be the best model as observed from the high regression coefficient values, and all the compounds were found to be mixed-type inhibitors [89]. The influence of melatonin drug on the deterioration of CS in sulfuric acid was assessed via chemical, electrochemical, and DFT procedures, and a maximum IE of 94.76% was obtained with 500 ppm at 303 K [90]. In a study conducted in 2019, Ituen and coworkers reported on the role of melatonin on the dissolution of X80 steel in 5% HCl [91]. The study reported inhibition efficiencies of 88.6 and 98.3% at 60 °C and 30 °C, respectively, with concentration of 10 mM. The adsorption of melatonin onto the metal sample obeyed the Langmuir isotherm. Furthermore, FTIR, SEM/EDS, and theoretical calculations such as DFT and MDS were employed to further study the adsorption capacity of melatonin. The study reported that adsorption occurred by the formation of a defensive layer mainly coordinated by the indole ring present in the compound. Recently, a series of 1,2,3-triazole tethered 3-hydroxy-2-oxindoles were produced and evaluated for their antibacterial, anticorrosion, and antifungal properties. The organic compounds behaved as good inhibitors of corrosion for MS in acidic medium [92].

A survey of literature also revealed some cases of synergistic studies involving indole-based compounds. While tryptophan (an indole-based amino acid) and its derivatives have been studied as effective additives of CS corrosion in HCl, NaOH, and NaCl [93–96], Mobin and coworkers conducted a synergistic study with the objective of studying the impact of surfactants addition on the inhibitory potentials of L-tryptophan [97]. Minute amounts of cetyltrimethyl ammonium bromide (cationic surfactant) and sodium dodecyl sulfate (anionic surfactant) were added to L-tryptophan in the presence of 0.1M HCl to inhibit steel corrosion. The obtained results showed an IE boost from 74.46% with 500 ppm L-tryptophan to 83 and 80.41% on addition of 1 ppm of SDS and CTAB, respectively, at 30 °C. In another study, ten isatin derivatives were prepared and characterized by mass spectrometry (MS) and nuclear magnetic resonance (NMR). They were applied as anticorrosive materials for Q235A steel in a highly corrosive environment (3M HCl) at an increased temperature of 60 °C using molecular simulation, PDP, mass loss, and molecular simulation techniques. The IE of the compounds ranged from 5.0 to 70.4% at 100 mg/l. To optimize the formulation and enhance the inhibition power of the compounds, hexamethylenetetramine (urotropine) and 1,4-dihydroxy-2-butyne were made to accompany the inhibitors, and a maximum IE of 95.5% was obtained. Synergism of indole derivatives with surfactants, halides, metal ions, and other compounds to improve the inhibitive effect of the studied compounds has also been reported by other authors [98–103]. Table 9.1 presents some other indole derivatives that have been assessed to possess anticorrosive effect on ferrous metals in different corrosive environments.

In recent years, some indole-based alkaloids extracted from plants have also been examined as green inhibitors of steel corrosion in different media. Lebrini and coworkers comprehensively explored the anticorrosion ability of norharmane (9H-pyrido[3,4-b]indole) and harmane (1-methyl-9H-pyrido[3,4-b]indole) to protect C38 steel against acidic corrosion using surface analysis, experimental, and theoretical approaches. Their results proved that the metallic surface is shielded from corrosive ions by the adsorption film created by these compounds [104, 105]. Caulerpin, a bis-indole alkaloid, was extracted from a marine alga, *Caulerpa racemosa* and characterized using NMR, UV-vis, and FTIR [34]. The inhibitory ability of the isolated indole-based alkaloid was examined using popular electrochemical methods, and the obtained EIS result was 85% at 25 ppm. Palaniappan and coworkers in 2020 carried out an elaborate study on the inhibition properties of *Catharanthus roseus* for MS in 3.5% NaCl solution. Ultrasound-assisted extraction was employed to extract the phytochemicals of the plant after which they were tested against MS corrosion via chemical (WL), electrochemical (OCP, EIS, and PDP) and surface wettability methods. Adsorption of *C. roseus* on MS was ascertained through surface analyses by attenuated total reflectance Fourier transform infra-red spectroscopy (ATR-FTIR), FESEM, AFM, XRD, and Raman mapping. They observed a drop in the corrosion potential of the system with the active vindoline of *C. roseus* due to chemisorption on the MS surface, which was also confirmed by the DFT calculations. The stem extract (IE = 96%) was adjudged to reduce corrosion rate better than the root extract (IE = 84%) on the account of its higher active area of stem phytochemicals. Phytochemical screening of *Nauclea latifolia* revealed the presence of indole alkaloids alongside other compounds. The ethanolic extracts from the barks, roots, and leaves of the plant were analyzed in sulfuric acid at 30 °C and 60 °C by WL and gasometric methods for steel corrosion [106]. Maximum inhibition efficiencies of 75.26, 94.26, and 91.58% were recorded for the roots, leaves, and barks, respectively, with 4.0 g/l. The adsorption model that best fitted the surface coverage data was found to be El-Awady adsorption model. Other indole-based alkaloids utilized as steel corrosion inhibitors have been presented in Table 9.2.

Ahmed and coworkers developed five indolium-based ionic liquids and examined them as inhibitors of MS corrosion in acid environment using chemical,

Table 9.1 Indole-based compounds as corrosion inhibitors for ferrous metals

Chemical structure/abbreviation	Metal/ medium	Study methods	Results	Ref	Chemical structure/abbreviation	Metal/ medium	Study methods	Results	Ref
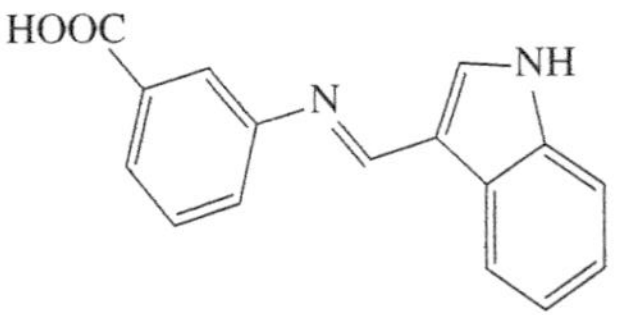 HOOC, N, NH 3-Formylindole-3-aminobenzoic acid (3FI3ABA)	Mild steel/ 1M HCl	WL, EIS, PDP	90.50% at 0.8 m M	[110]	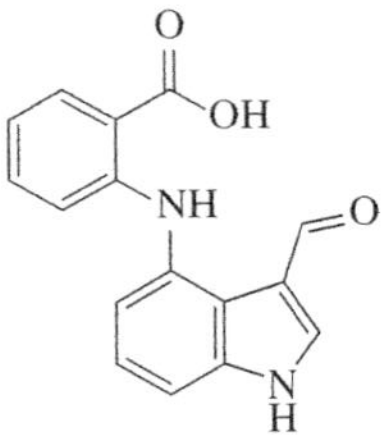 O, OH, NH, O, N H 3-formylindole-4-aminobenzoic acid (3FI4ABA)	Mild steel/1M HCl	WL, EIS, PDP	97.70% at 0.8 mM	[111]
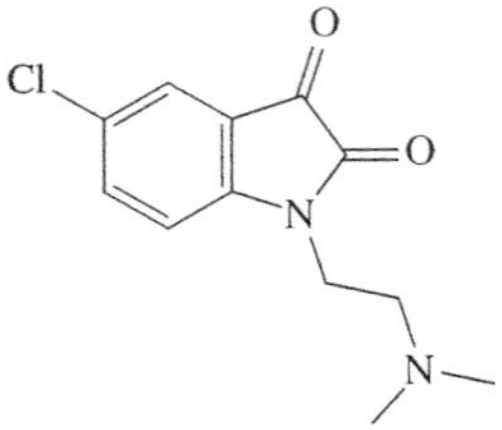 O, Cl, O, N, N 5-Chloro-1-(2-(dimethylamino) ethyl) indoline-2,3-dione (indol)	MS/1M H_3PO_4	WL, EIS, PDP, SEM	91.00% at 10^{-3} M	[112]	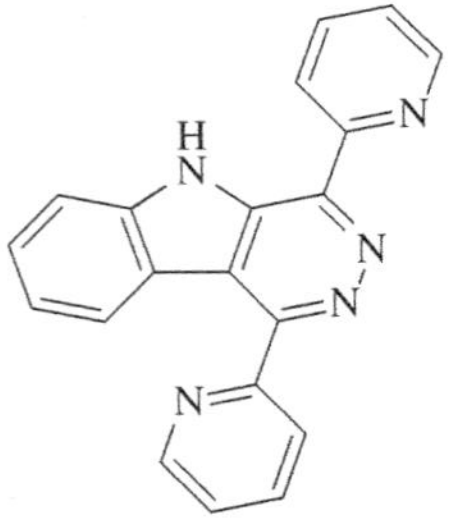 N, H N, N, N, N 1,4-bis(2-pyridyl)-5H-pyridazino[4,5-b]indole (PPI)	Mild steel/ 1M HCl	WL, EIS, PDP, XPS, DFT	90.20% at 1×10^{-4} M	[113]

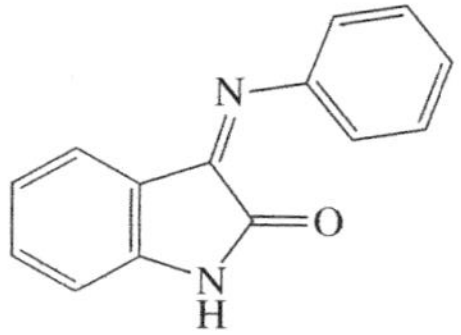 (Z)-3-(phenylimino)indolin-2-one (3-PII)	Mild steel/1 M HCl	EIS, LPR, PDP	81.91 % at 1×10^{-3} M	[114]	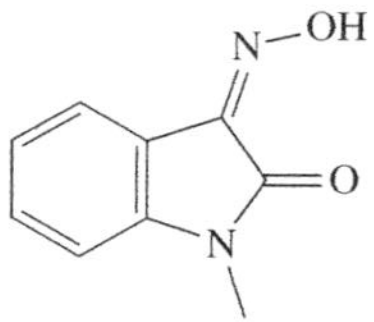 1-methyl-indolin-2-one-3-oxime	N80 steel/ 1 M HCl	WL	81.50% at 1000 mg/l	[115]
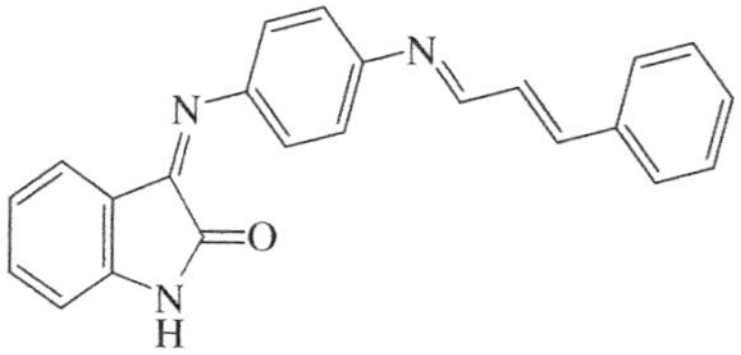 3-(4-(3-Phenylallylideneamino) phenylimino)indolin-2-one (PI)	Mild steel/ 20% H_2SO_4	WL, SEM, DFT	98.60% at 200 mg/l	[116]	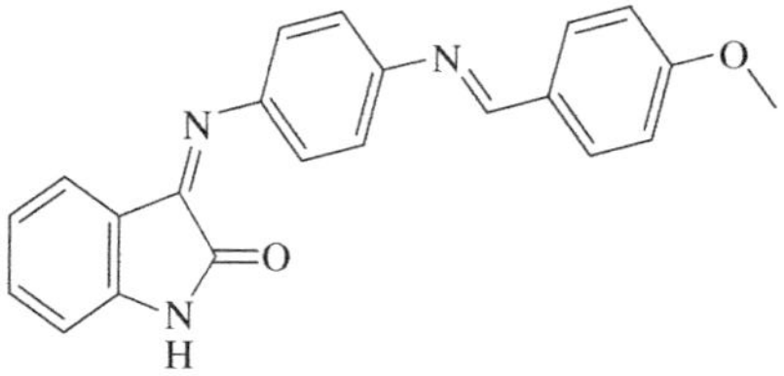 3-(4-(4-Methoxybenzylideneamino) phenylimino)indolin-2-one (MI)	Mild steel/ 20% H_2SO_4	WL, EIS, PDP, SEM, DFT	93.10% at 200 mg/l	[116]
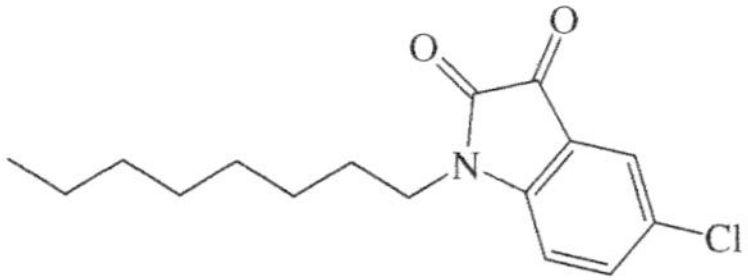 5-Chloro-1-octylindoline-2,3-dione (E1)	Mild steel/ 1 M HCl	WL, EIS,PDP, DFT, SEM	90.00% at 10^{-3} M	[117]	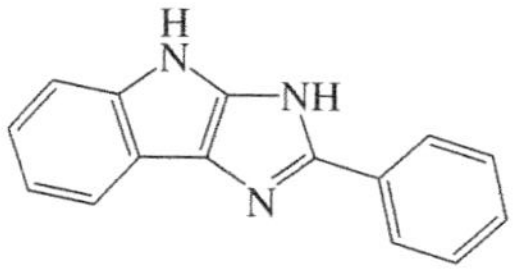 3,4-Dihydro-2-(phenyl)imidazo[4,5-b]indole (DPI)	Mild steel/0.5 M H_2SO_4	WL, EIS,PDP, DFT	67.80% at 21 ppm	[118]

(Continued)

Table 9.1 (Continued)

Chemical structure/abbreviation	Metal/medium	Study methods	Results	Ref	Chemical structure/abbreviation	Metal/medium	Study methods	Results	Ref
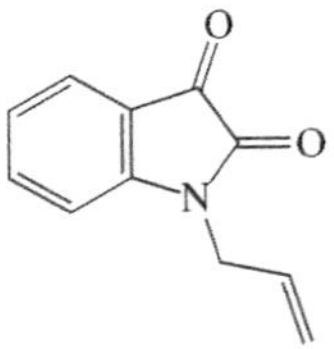 1-Allyl-1H-indole-2,3-dione (Ind1)	Mild steel/1 M HCl	WL,EIS, PDP	95.20% at 10^{-3} M	[119]	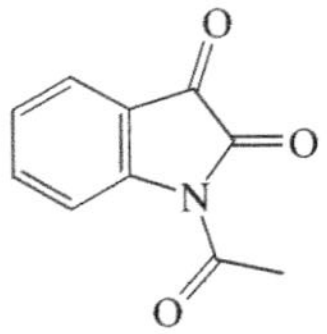 1-Acetylindoline-2,3-dione (Ind2)	Carbon steel/1M HCl	WL, EIS, PDP	93.10% at 10^{-3} M	[120]
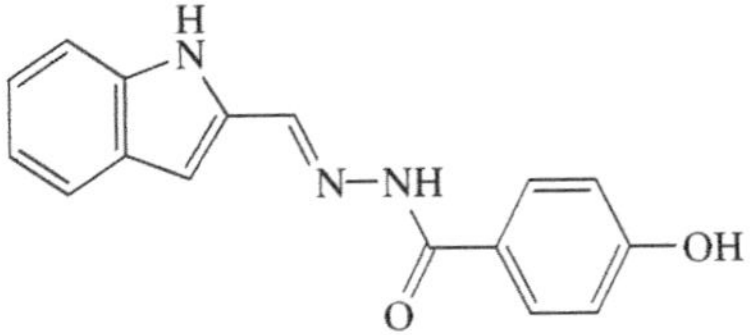 4-Hydroxy-N'-[(E)-(1H-indole-2-ylmethylidene)] benzohydrazide (HIBH)	Mild steel/1 M HCl	EIS, PDP	90.19% at 20 mg/l	[121]	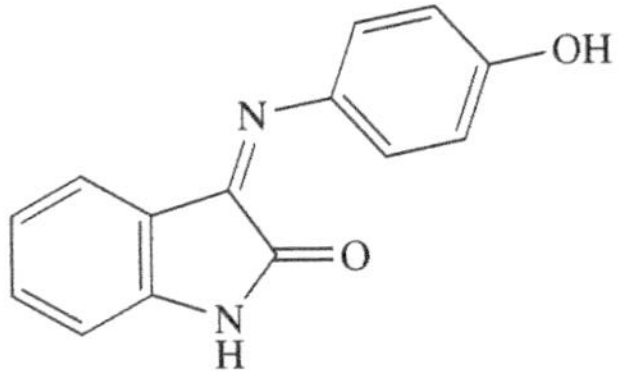 3-(4-Hydroxyphenylimino)indolin-2-one (HMTA)	Q235A steel/3M HCl	WL	77.63% at 500 mg/l	[122]
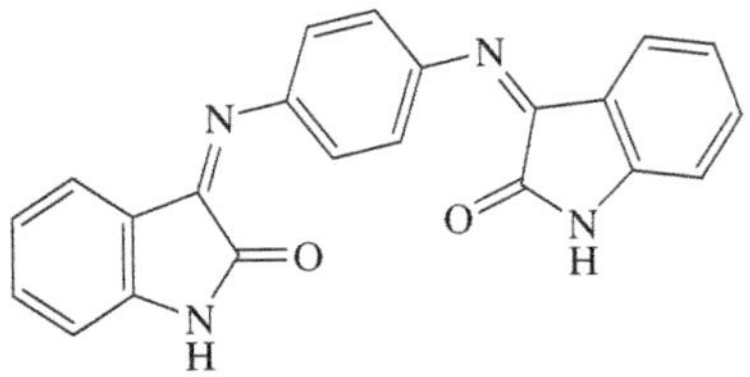 3,3'-(1,4-Phenylenebis(azan-1-yl-1-ylidene)) diindolin-2-one (PDI)	Mild steel/1M HCl	WL, EIS, PDP	87.30% at 1×10^{-3} M	[123]	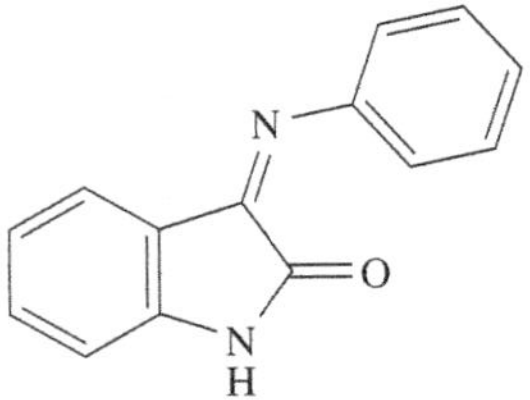 3-(Phenylimino)indolin-2-one (PII)	Mild steel/1M HCl	WL, EIS, PDP	74.80% at 1×10^{-3}M	[123]

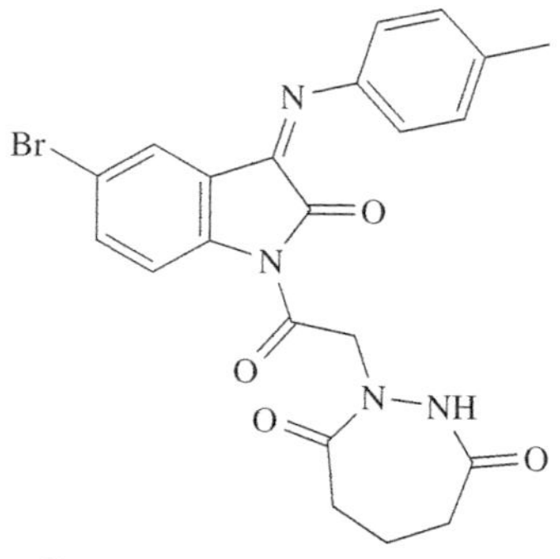 N-[(Hexahydrodiazepine-3,7-dione-1-yl) aceto]-3- (tolylimino)-5-bromo-2-oxo-indole (5BID)	Carbon steel/ 3.5% NaCl	PDP, SEM-EDX, AFM, DFT	97.80% at 20 ppm	[124]	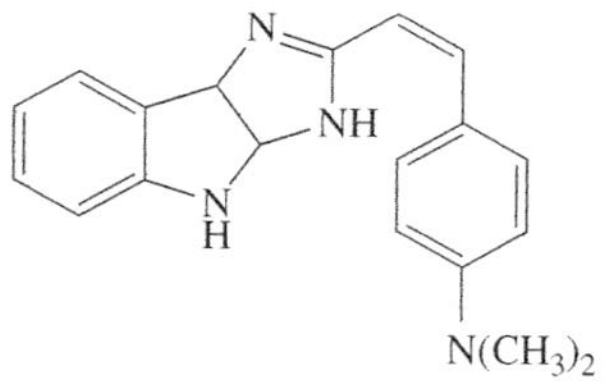 (Z)-N,N-dimethyl-4-(2-(3,3a,4,8b-tetrahydroimidazo[4,5-b]indol-2-yl)vinyl) aniline (DI)	Mild steel/1M HCl	WL, EIS, PDP, DFT, MDS	90.19% at 20 mg/l	[125]
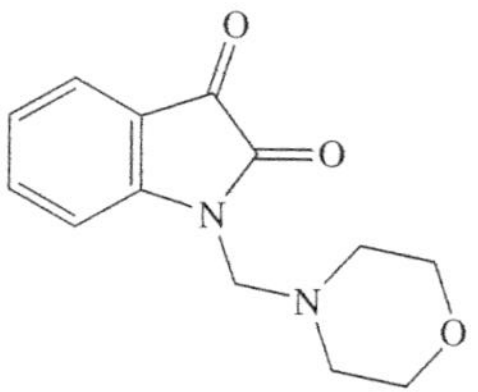 1-(Morpholinomethyl)indoline-2,3-dione	Q235A steel/1M HCl	WL	81.90% at 1000 mg/l	[126]	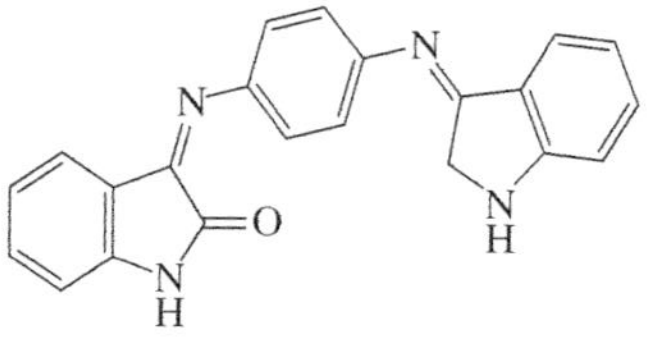 3-(4-((Z)-indolin-3-ylideneamino) phenylimino)indolin-2-one (PDBI)	Mild steel/1M HCl	WL, EIS, LPR, PDP, SEM, AFM	96.00% at 2.73 × 10^{-4} M	[127]
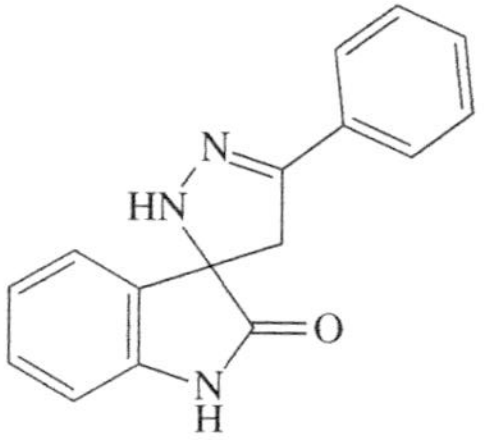 5'-Phenyl-2',4'-dihydrospiro[indole-3,3'-pyrazol]-2(1H)-one (SPAH)	Mild steel/1M HCl, 0.5M H_2SO_4	WL, EIS, PDP, SEM, DFT	40.00% (HCl) and 81.40% (H_2SO_4) at 9 ppm	[128]	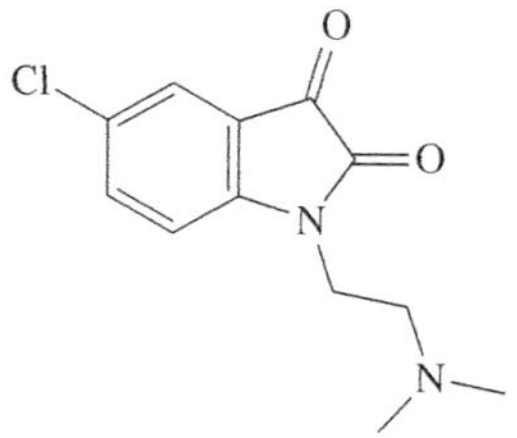 5-Chloro-1-(2-(dimethylamino) ethyl) indoline-2,3-dione	Mild steel/ 1 M HCl	WL, EIS,PDP, SEM	85.99% at 10^{-3} M	[129]

(*Continued*)

Table 9.1 (Continued)

Chemical structure/abbreviation	Metal/ medium	Study methods	Results	Ref	Chemical structure/abbreviation	Metal/ medium	Study methods	Results	Ref
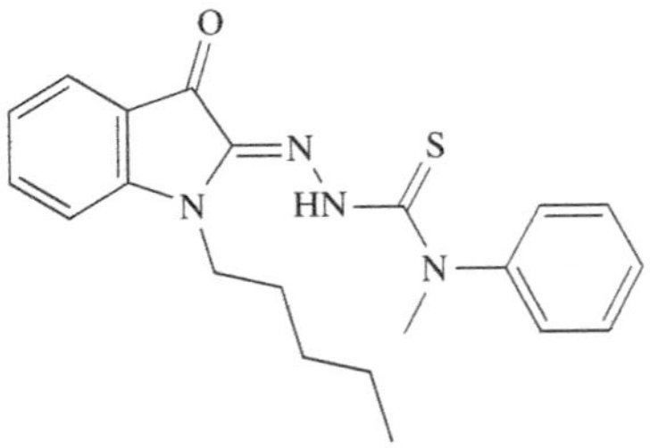 N(1)-pentylisatin-N(4)-methyl-N(4)-phenyl thiosemicarbazone(PITSc)	Mild steel/1M HCl	WL, EIS, PDP, UV, FTIR, SEM-EDX, DFT	89.54% at 100 ppm	[130]	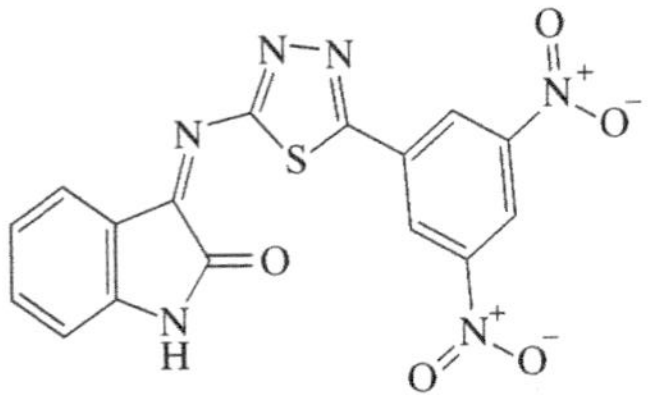 3-((5-(3,5-Dinitrophenyl)-1,3,4-thiadiazol-2-yl) imino)indolin-2-one (TDIO)	Mild steel/ 1 M HCl	WL, SEM, DFT	90.70% at 0.5 mM	[131]
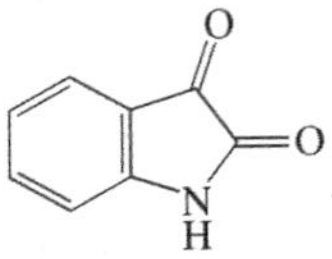 Isatin (IN)	Mild steel/0.1M HCl	WL	84.00% at 5×10^{-4} M	[132]	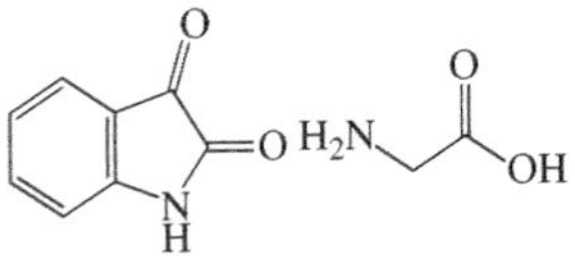 Isatin glycine (ING)	Mild steel/ 0.1 M HCl	WL	87.00% at 5×10^{-4} M	[132]
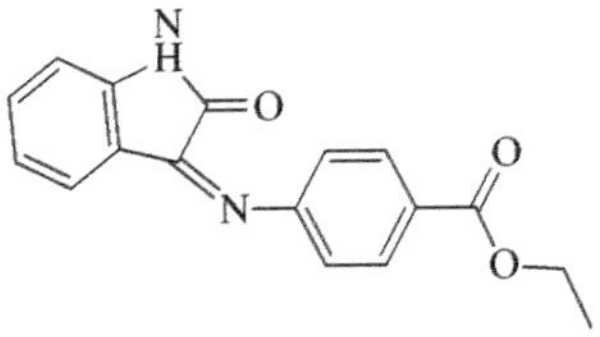 Ethyl (Z)-4-((2-oxoindolin-3-ylidene) amino)benzoate (AIB)	Mild steel/1M HCl	EIS, PDP, SEM, DFT	80.00% at 250 ppm	[133]	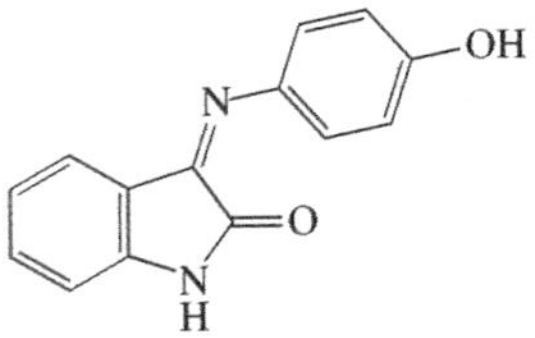 3-(4-Hydroxyphenylimino)indolin-2-one	Q235A/1 M NaCl	WL	93.90% at 1000 mg/l	[134]

Inhibitor	Metal/medium	Methods	Efficiency	Ref.	Inhibitor	Metal/medium	Methods	Efficiency	Ref.
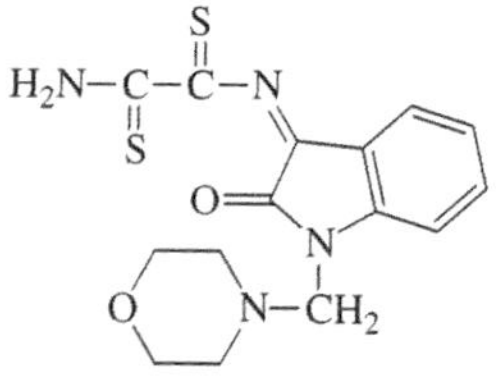 1-Morpholinomethyl-3-(1-N-dithiooxamide) iminoisatin (MMTOI)	N80 steel/15% HCl	WL, EIS, PDP, SEM	82.05% at 200 ppm	[135]	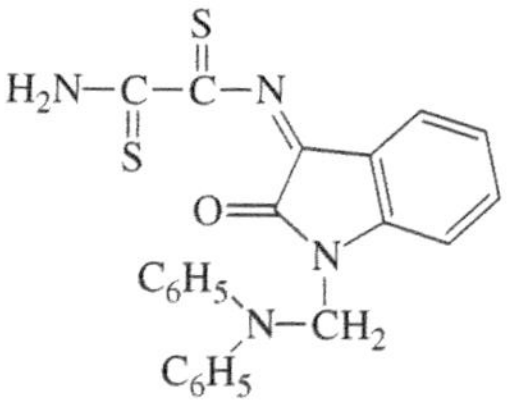 1-Diphenylaminomethyl-3-(1-N-dithiooxamide)iminoisatin (PAMTOI)	N80 steel/ 15% HCl	WL, EIS,PDP, SEM	90.35% at 200 ppm	[135]
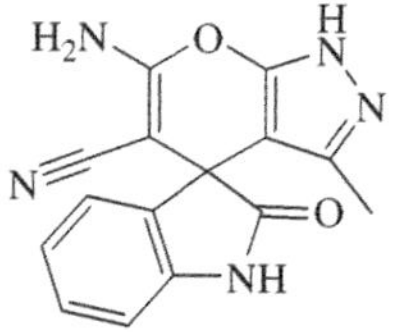 6'-Amino-3'-methyl-2-oxo-1'H-spiro[indoline-3,4'-pyrano[2,3-c] pyrazole]-5'-carbonitrile (SIPP-1)	Mild steel/1M HCl	WL, EIS, PDP, SEM, AFM, DFT, MD	95.65% at 200 ppm	[136]	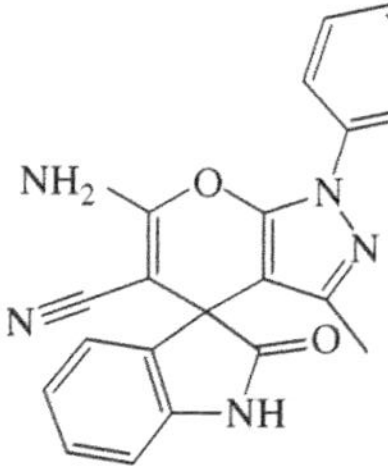 6'-Amino-3'-methyl-2-oxo-1'-phenyl-1'H-spiro[indoline-3,4'-pyrano[2,3-c]pyrazole]-5'-carbonitrile (SIPP-2)	Mild steel/1M HCl	WL, EIS,PDP, SEM, AFM, DFT, MD	96.95% at 200 ppm	[136]
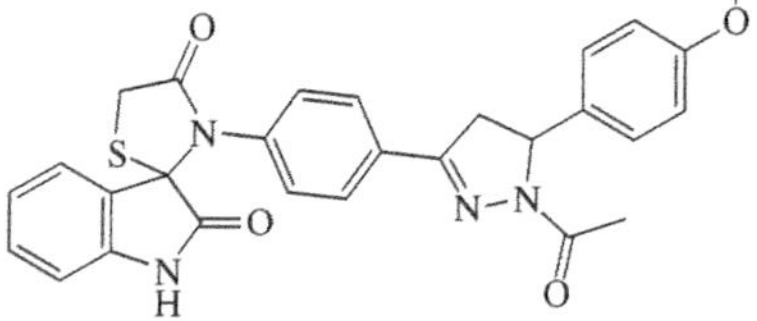 3'-(4-(1-Acetyl-5-(4-methoxyphenyl)-4,5-dihydro-1H-pyrazol-3-yl)phenyl) spiro[indoline-3,2'-thiazolidine]-2,4'-dione (MPIT)	N80 steel/15% HCl	WL, EIS,PDP, SEM-EDX, AFM, DFT	93.90% at 200 ppm	[137]	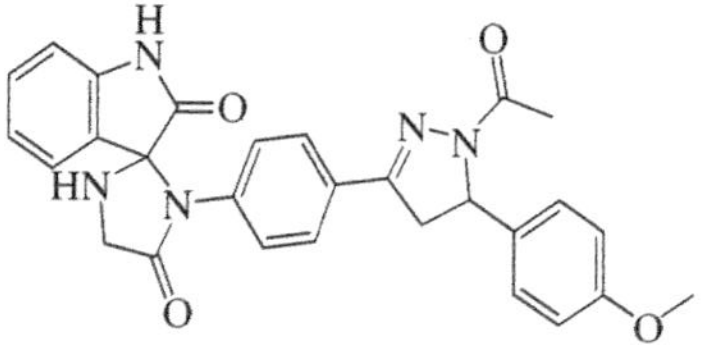 1-(4-(1-Acetyl-5-(4-methoxyphenyl)-4,5-dihydro-1H-pyrazol-3-yl)phenyl)spiro[imidazolidine-2,3'-indoline]-2',5-dione (MPII)	N80 steel/15% HCl	WL, EIS,PDP, SEM-EDX, AFM, DFT	97.20% at 200 ppm	[137]

(Continued)

Table 9.1 (Continued)

Chemical structure/abbreviation	Metal/ medium	Study methods	Results	Ref	Chemical structure/abbreviation	Metal/ medium	Study methods	Results	Ref
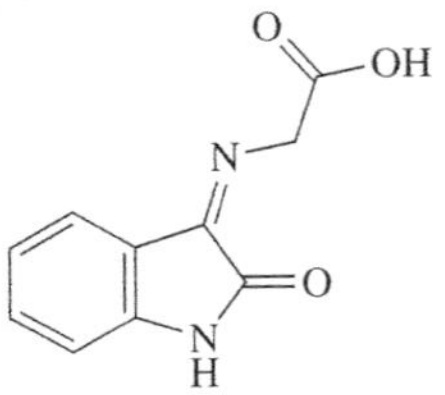 2-(2-Oxoindolin-3-ylideneamino) acetic acid (IG)	Mild steel/1M HCl	WL, Gasometric, DFT	87.07% at 5 $\times 10^{-4}$ M	[138]	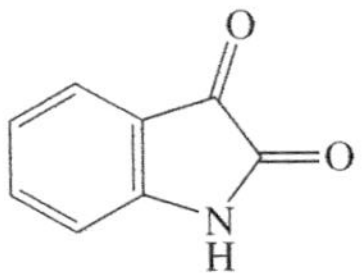 Indoline-2,3-dione (IS)	Mild steel/1M HCl	WL, Gasometric, DFT	84.04% at 5 $\times 10^{-4}$ M	[138]
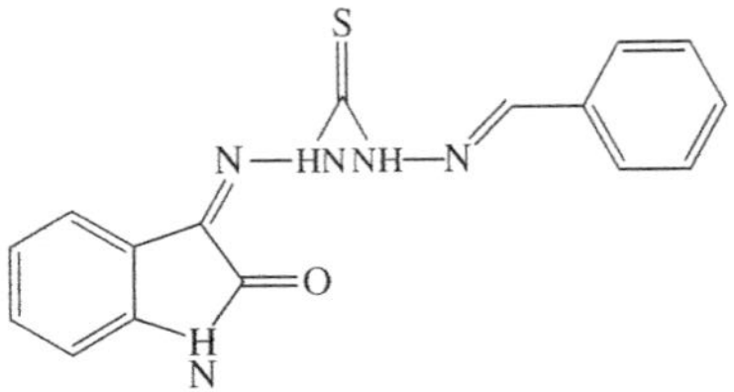 1-Benzylidene-5-(2-oxoindoline-3-ylidene) thiocarbohydrazone (TZ-1)	Mild steel/20% H_2SO_4	WL, EIS,PDP, SEM-EDX, DFT	91.50% at 300 mg/l	[45]	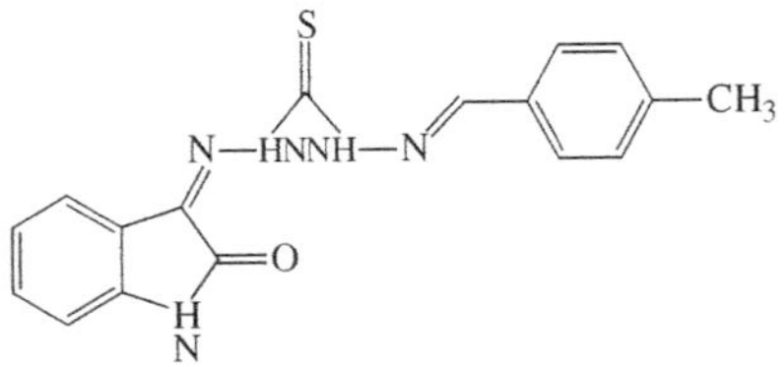 1-(4-Methylbenzylidene)-5-(2-oxoindolin-3-ylidene) thiocarbohydrazone (TZ-2)	Mild steel/20% H_2SO_4	WL, EIS, PDP, SEM-EDX, DFT	97.20% at 300 mg/l	[45]

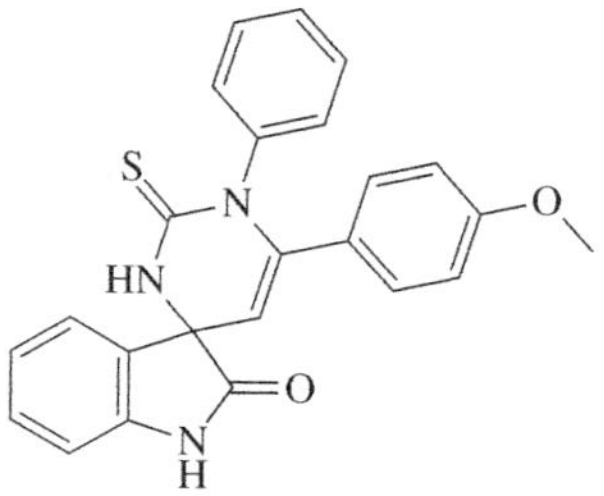 6'-(4-Methoxyphenyl)-1'-phenyl-2'-thioxo-2',3'-dihydro-1'H-spiro[indoline-3,4'-pyrimidine]-2-one (MPTS)	Mild steel/ 15%HCl	WL, EIS,PDP, FTIR, SEM- EDX, XPS, DFT	93.80% at 150 ppm	[86]	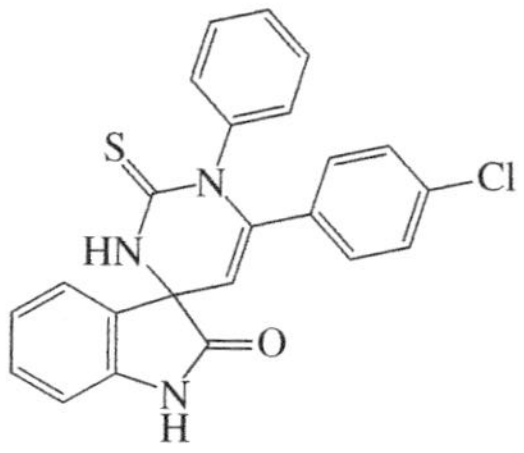 6'-(4-Chlorophenyl)-1'-phenyl-2'-thioxo-2',3'-dihydro-1'H-spiro[indoline-3,4'-pyrimidin]-2-one (CPTS)	Mild steel/15% HCl	WL, EIS, PDP, FTIR, SEM-EDX, XPS, DFT	93.20% at 150 ppm	[86]
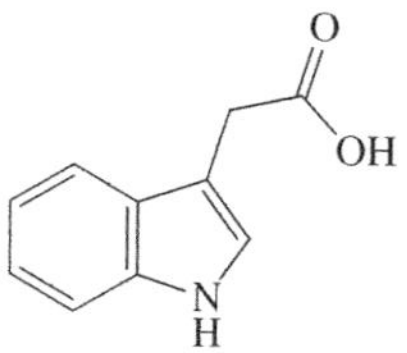 Indole-3-acetic acid	Mild steel/1M HCl	EIS, PDP, CA, DFT	97.00% at 10 mM	[139]	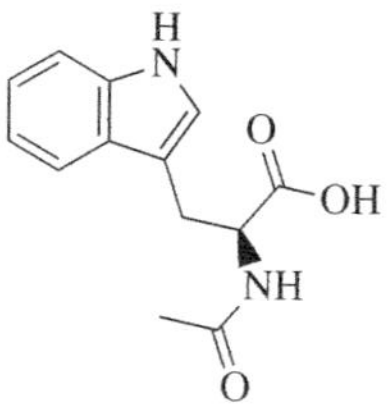 N-Acetyltryptophan	Mild steel/1M HCl	EIS, PDP, CA, DFT	80.00% at 10 mM	[139]

Table 9.2 Indole-based alkaloids as corrosion inhibitors of ferrous metal.

Chemical structure/abbreviation	Metal/medium	Study methods	Results	Nature of adsorption	Ref
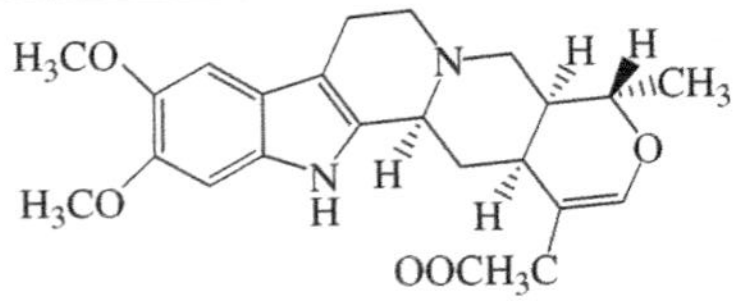 Isoreserpiline (ISR) from leaves and bark extract of *Ochrosia oppositifolia*	Mild steel/1 M HCl	EIS, PDP, SEM, FTIR, DFT	93.00% at 25 mg/l	Langmuir/mixed type	[140]
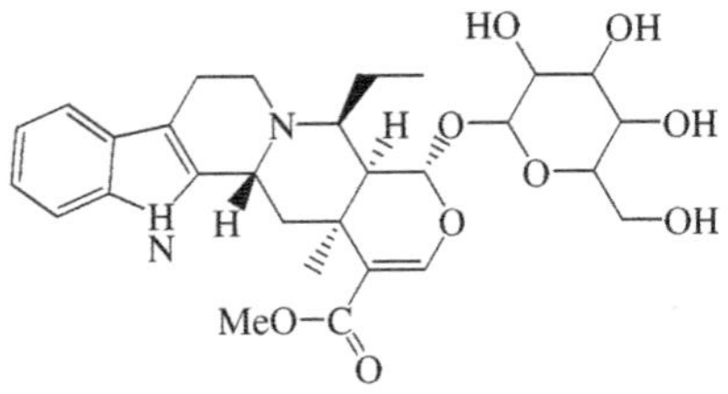 3β-isodihydrocadambine from *Neolamarckia cadamba* crude extract	Mild steel/1 M HCl	EIS, PDP, FTIR, SEM, DFT	83.00% at 5 mg/l	Langmuir/mixed type	[141]
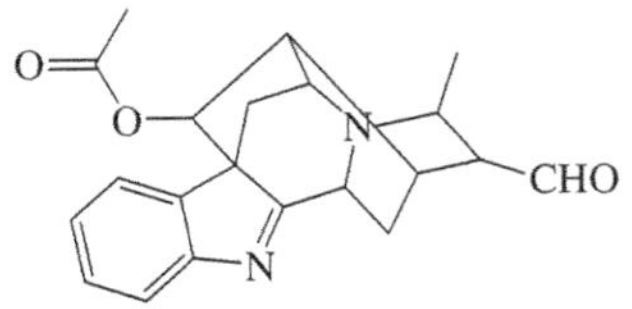 Perakine (PER)	C38 steel/1M HCl, 0.5M H_2SO_4	EIS, PDP, SEM	97.00% in 1 M HCl 93.00% in 0.5 H_2SO_4 at 200 mg/l	Langmuir/mixed type (HCl) and cathodic predominance in H_2SO_4	[98]

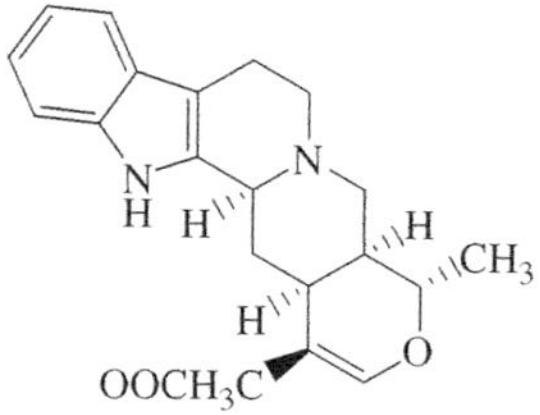 Tetrahydroalastonine (THA) from *Rauvolfia marcophylla Stapf*					
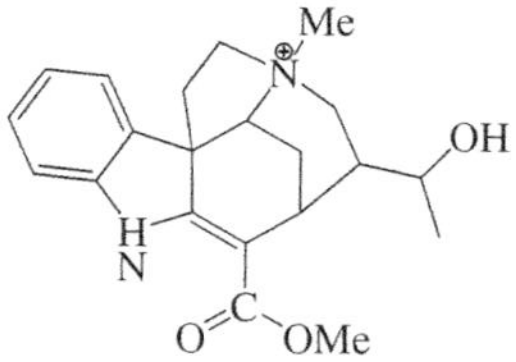 Alstogustine + 19-epialstogustine from *Alstonia angustifolia* var. *latifolia*	Mild steel/1M HCl	WL, EIS, PDP, FTIR, SEM	84.00% at 5 mg/l	Langmuir/mixed type	[142]
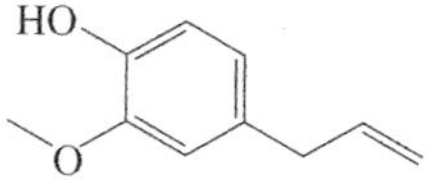 Eugenol	Mild steel/ 1M H_2SO_4	WL, DFT	94.90% at 500 mg/l	Temkin	[143]
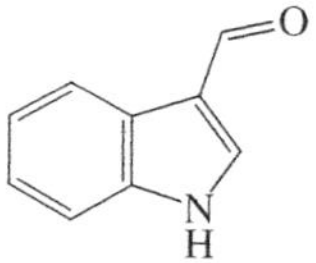 Indole-3-aldehyde from the leaves of *Ipomoea batatas*					

(Continued)

Table 9.2 (Continued)

Chemical structure/abbreviation	Metal/medium	Study methods	Results	Nature of adsorption	Ref
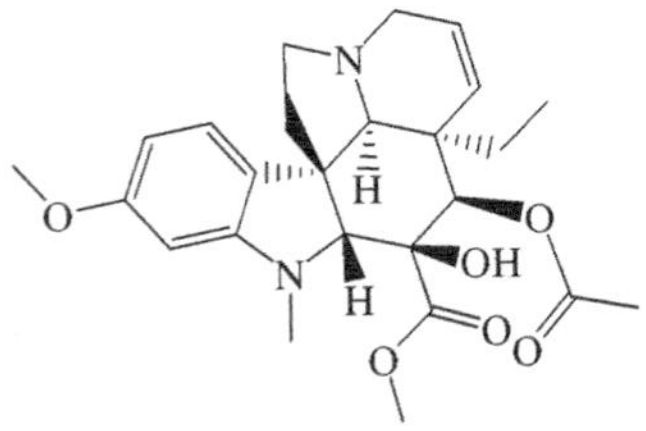 Vindoline alkaloid of *Cetharanthus roseus*	Gray iron/ 1 M HCl	WL, PDP, SEM	94.20% at 1.1×10^{-2}M	Langmuir/mixed type (predominantly anodic)	[144]
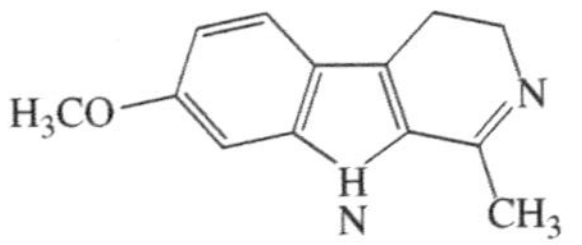 Harmaline from *Peganum harmala*	Carbon steel/ 3.5% NaCl	WL, PDP	86.62% at 8 g/l	Langmuir/Mixed type	[145]
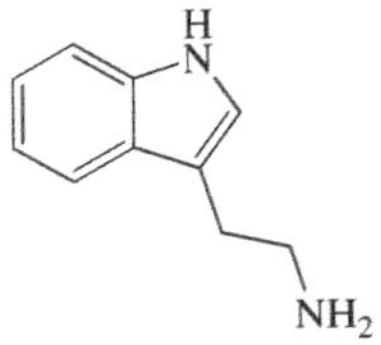 Tryptamine	Mild steel/0.5M HCl	LPR, EIS, PDP	97.00% at 500 ppm	Langmuir/mixed type	[146]

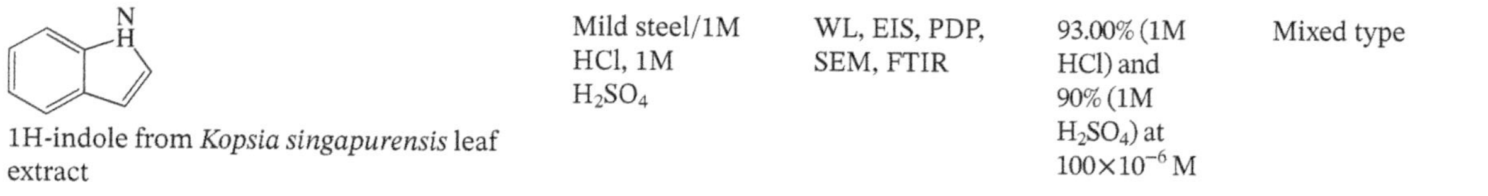

1H-indole from *Kopsia singapurensis* leaf extract	Mild steel/1M HCl, 1M H_2SO_4	WL, EIS, PDP, SEM, FTIR	93.00% (1M HCl) and 90% (1M H_2SO_4) at 100×10^{-6} M	Mixed type	[147]
Gramine	Mild steel/1M HCl	WL, EIS, PDP	98.00% at 7.5 mM	Langmuir/mixed type	[148]

electrochemical, and surface analysis methods. The compounds delivered IE within the range 56–75%. The large negative values of ΔG°_{ads} (31.4 kJ/mol) obtained from the study suggested the strong adsorption of the ionic liquids on the metal substrate, which is indicative of chemisorption and a small contribution of physisorption [107]. In addition, cyclic voltammetry (CV) at polycrystalline gold electrode was employed to further clarify the inhibition mechanism by the indolium-based ionic liquids. The ability of a surfactant 12-(2,3-dioxoind olin-1-yl)-N,N,N-trimethyldodecan-1-ammonium bromide to suppress corrosion of MS in acidic medium was studied by Abdellaoui and coworkers [108]. The report established that the synthesized surfactant exhibited IE up to 95.90% at an optimum concentration of 1 mM for MS protection after six immersions in HCl at 298 K. The study also reported the thermodynamic and kinetic parameters affecting corrosion process of MS in the studied medium. An indole-based Schiff base and its Co(II), Ni(II), Mn(II), Zn(II), and Cu(II) complexes were prepared and characterized by spectroscopic and physiochemical instruments (FTIR, UV-vis, MS, and NMR). The synthesized Schiff base was determined by single X-ray diffraction instrument. Both DFT and time-dependent DFT (TD-DFT) methods were adopted to optimize the structures of the complexes. The complexes were investigated for their abilities to suppress MS corrosion in 0.5 M sulfuric acid at 298 K via WL and electrochemical methods. The possession of octahedral geometry with two ligand molecules, large molecular size, and accessibility of adsorption sites (heteroatoms) were posited as likely reasons for the higher IE of Co(II), Mn(II), and Cu(II) complexes compared to Ni(II) and Zn(II) complexes. The obtained higher IE of the Schiff base ((E)-N'-((1H-indol-3-yl) methylene)-2-aminobenzohydrazide) than its Ni(II) and Zn(II) complexes was attributed to the availability of all the heteroatoms (O, N, S, and P) for the interaction with the empty d-orbital of the MS [109].

9.4.2 Indoles as Corrosion Inhibitors of Nonferrous Metals

Available literature revealed that while several studies have been conducted to investigate the inhibitive abilities of indole and its compounds on iron alloys, far less has been done in studying the effect of indole derivatives on nonferrous metals. Copper, aluminum, zinc, nickel, tin, brass, and other non-iron-based metallic materials fall under the category of nonferrous metals. Nonferrous metals have found wide applications in several industries. Copper, aluminum, and zinc are abundant in nature and have served numerous beneficial roles in the industry as a result of their mechanical stability, high thermal and electrical conductivities, and low weight [149–152]. Hence, researchers have channeled efforts toward combating the deterioration of these metals and their alloys [153–155]. Studies have shown that these nonferrous metals and their alloys undergo deterioration in corrosive electrolytic media despite their intrinsic corrosion-resistant properties [156, 157]. Some available works in literature that have dealt with the inhibition studies of indoles on these metals and their alloys have been discussed below.

Scendo evaluated the inhibition performance of indole molecule and 5-chloroindole for the dissolution of copper in acidic chloride medium and concluded that both molecules impeded cathodic deposition and anodic dissolution of Cu. While indole performed better than its counterpart, it was surprising to note that both compounds increased the protection efficiencies of copper via an adherent layer despite the decreasing solution pH [158]. Quartarone and coworkers also studied the inhibitive capacity of indole nucleus to impede deterioration of Cu in sulfuric acid [159].

The restraint of copper dissolution by indole 2-carboxylic acid in aerated 0.5M sulfuric acid solution was carried out by chemical and PDP methodologies. Obtained Tafel plot showed that the inhibitor impeded corrosion at both cathodic and anodic branches to the tune of 98% with a concentration of 2×10^{-3}M at 55°C. The adsorption behavior of the indole derivative was found to be both chemisorption and physisorption [160]. In a related study, indole-3-carboxylic acid and indole 5-carboxylic acid were tested and yielded IE of 92 and 95%, respectively, in a concentration of 2×10^{-3} M at 55°C [161, 162]. The ability of polyindole to prevent copper dissolution in 3.5% NaCl has also been studied and reported by Tüken and coworkers [163]. The polypyrrole/polyindole coating was found to offer better protection than polypyrrole alone due to the physical barrier effect between the substrate and the adsorbate. Ebadi and his coworkers examined the inhibitive abilities of three pyrazolylindolenine compounds on copper using OCP, linear scan voltammetry (LSV), and EIS procedures. The study reported inhibition efficiencies of 79.3, 91.4, and 94.0% for 3,3-dimethyl-2-(1-phenyl-1H-pyrazol-4-yl)-3H-indole (InPzPh), 4-(3,3-dimethyl-3H-indol-2-yl)-1H-pyrazole-1-carbothiohydrazide (InPzTH), and 4-(3,3-dimethyl-3H-indol-2-yl)-pyrazole-1-carbothioamide (InPzTAm). The study posited that the studied compounds containing S-heteroatom (an amine group) offers greater surface coverage than phenyl and hydrazine group [164]. Isatin was also reported as an excellent inhibitor of copper corrosion at a concentration range of 1.0×10^{-4} to 7.5×10^{-3}M in aerated 0.5M H_2SO_4 solutions [165].

In a recently published study, two new O and S-heterocycle-based indole derivatives were synthesized by a simple route and examined as chemical inhibitors of

Cu corrosion in acid medium using modern electrochemical procedures. The characterizations of the synthesized compounds were done using FTIR, XPS, SEM, and ellipsometric techniques. The S-heterocycle-based indole, 3,3'-(thiophen-2-ylmethylene)bis(1H-indole) (TYBI) inhibited copper corrosion better than the O-heterocycle indole, 3,3'-(furan-2-ylmethylene)bis(1H-indole) (FYBI) because of its stronger electron-donating ability. Furthermore, results from DFT and XPS show that TYBI could form more robust, hydrophobic, and anticorrosive self-assembly monolayers. Various adsorption isotherms were tested to unravel the adsorption mechanism of the indole derivatives, and the best fitted model was found to be Langmuir (R^2: FYBI = 0.9745 and TYBI = 0.9942) [166].

4-(Bis(5-bromo-1H-indol-3-yl)methyl)phenol (BMP) was newly produced and tested as inhibitor of copper corrosion using well-documented spectroscopic analyses such as FTIR, XPS, CA, and SEM-EDX and electrochemical techniques (PDP and EIS) [167]. The Tafel measurements revealed the surface coverage of copper was up to 98.34% at 10 mM. The FTIR spectra revealed that inhibitor adsorption had taken place on the Cu substrate through chelating rings in the N-heterocycle on the account of the weakening of the peaks of BMP powder (Figure 9.3). The presence of nitrogen and bromine on the XPS spectra demonstrated the adsorption of BMP on the Cu substrate. The SEM-EDX showed similar trends. The corrosion pits observed in the blank solution was more than the pits in the polished surface or inhibited surface after immersion in 0.5M sulfuric acid for 12 hours. BMP offered surface layer protection to the metal and the EDX detected new peaks of N and Br atoms as shown in Figure 9.4.

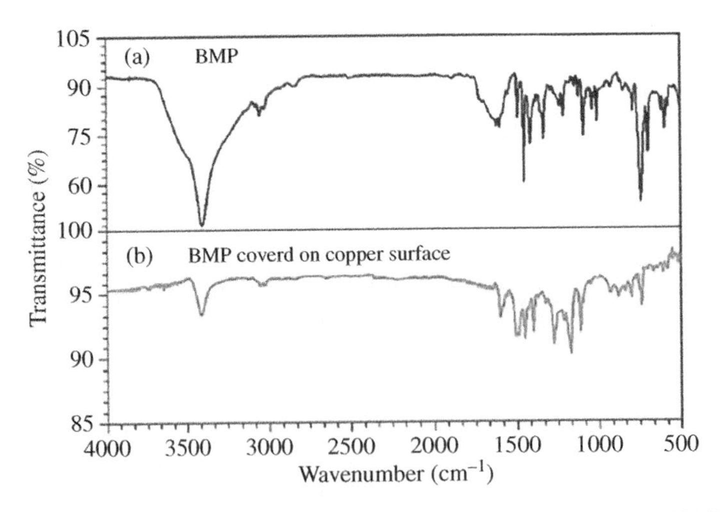

Figure 9.3 FTIR spectra of BMP powder (a) and the barrier layer of BMP adsorbed on Cu surface (b). *Source:* Feng et al. [167].

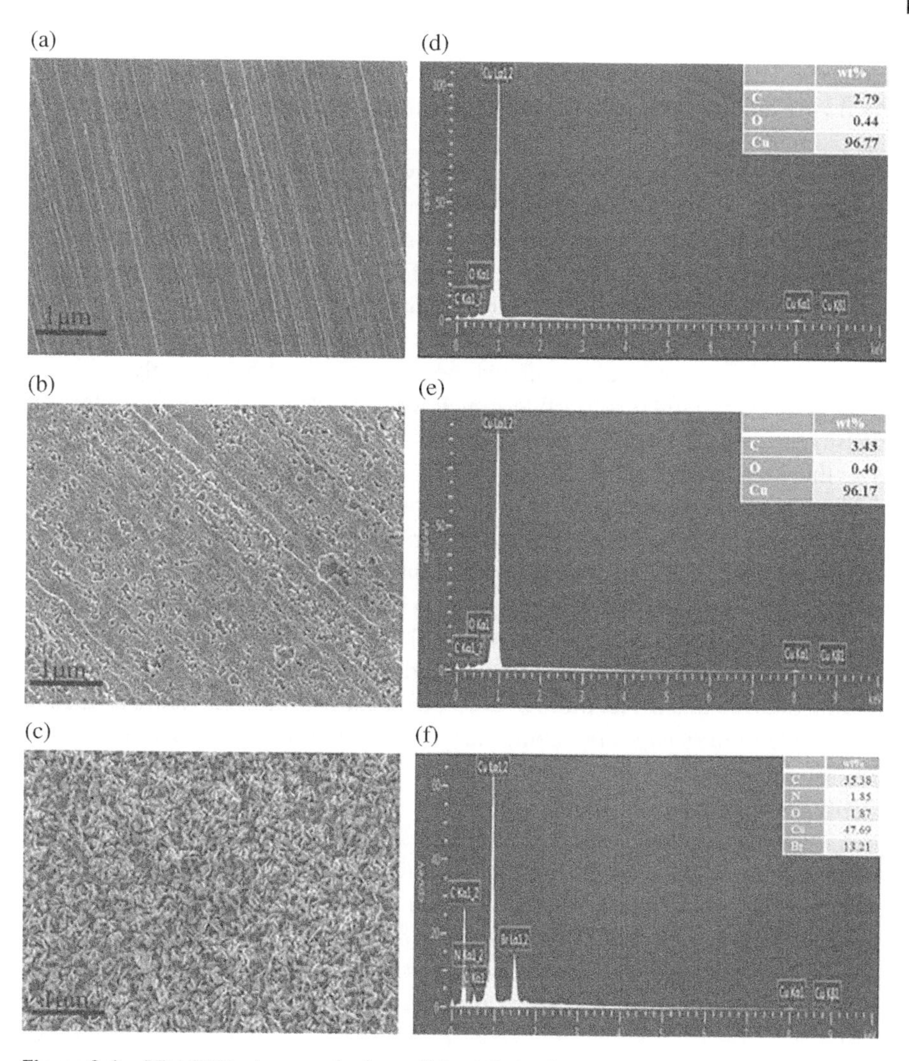

Figure 9.4 SEM-EDX micrographs for polished Cu (a, d), the polished Cu immersed in 0.5M H_2SO_4 for 12 hours (b, e), Cu modified with 10 mM BMP immersed in 0.5M H_2SO_4 for 12 hours (c, f). *Source:* Feng et al. [167].

Tryptophan has been reported as an effective inhibitor of aluminum degeneration in acidic media [168]. El-Shafei and coworkers applied indole (I), tryptamine (II), and tryptophan (III) as corrosion inhibitors of Al in neutral environments (0.1M NaCl) and found the IE to depend on the number of active centers, mode of adsorption, and molecular size as demonstrated by the increasing trend of IE: II > III > I. Furthermore, the influence of the adsorption of indole and its derivatives on ethanol oxidation in sulfuric acid at Pt electrode was investigated using CV. The order of adsorption behavior recorded from the study is as follows: I > II > III, and this was attributed to polymer film formation during cycling peculiar to indole [169].

A comprehensive study was carried out by Yuan and coworkers [170] on five synthesized isatin thiosemicarbazone compounds with diverse substituents at the 5-position of the isatin group, ISA-R (R = $-NO_2$, –H, $-CH_3$, –F and –Cl) for the inhibition of AA6060 Al alloy corrosion in HCl via electrochemical and surface analyses techniques. With the exception of ISA-CH_3, which acted as a mixed-type additive with a pronounced cathodic effect, others were cathodic-type inhibitors. It was also noted that ISA-NO_2 offered the highest protection probably due to its more positive Hammett σ_{para} value, which may have led to increased pi electron density on its isatin ring. This could have afforded a higher adsorption of ISA-NO_2 over the other studied compounds.

There have been few reports of indole alkaloids as corrosion inhibitors of aluminum. Djemoui and coworkers [171] extracted harmine and harmaline from *Peganum harmala* plant and explored its inhibitory potentials against A 6063 aluminum alloy corrosion in molar HCl using WL and PDP. Their study revealed that the additive impedes corrosion at both cathodic and anodic branches and followed Langmuir model. Besides that, at optimum concentration of 25 mg/l, the authors showed that a maximum IE of 89% could be obtained. Extracted d-lysergic acid amide from *Ipomoea involcrata* was reported to inhibit corrosion of type AA 1060 aluminum in the presence of 1M NaOH using WL methods. The adsorption ability of the extract was found to improve as the amount of the extract increased and Langmuir adsorption model was followed. The authors proposed a physisorption mechanism based on the thermodynamics data obtained [172]. An investigation into the anticorrosion activity of *Ailanthus altissima* leaf extract for Zn in 0.5M HCl using surface morphological, chemical, and electrochemical methods revealed that the presence of indoles among several other compounds was responsible for the excellent protection recorded [173].

Halides have been known to synergize with organic compounds to improve their inhibition efficacy. Nathiya and coworkers reported the synergistic influence of KI, KBr, and KCl on the inhibitive effects of two newly synthesized indole-based Schiff bases used as anticorrosive materials of Al in sulfuric acid. The synergism parameter was found to be above unity, which implies that the halide ions improved the corrosion resistance of the metal [174]. A value of synergism parameter less than unity would imply an antagonistic effect. Table 9.3 is a collection of other indole compounds utilized in repressing corrosion of nonferrous metals.

The effect of indole and 2-oxyinole on α-brass in 1M HNO_3 was assessed by WL, PDP, EIS, and EFM [175]. The inhibitors were reported to increase the IE as amount increased but decreased with rise in temperature. The Tafel plots revealed

Table 9.3 Indole-based corrosion inhibitors of nonferrous metals

Chemical structure/abbreviation	Metal/medium	Study methods	Results	Nature of adsorption	Ref
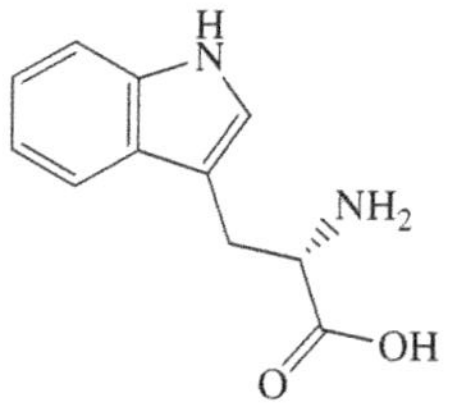 Tryptophan	Copper/ 0.5M H_2SO_4	WL, PDP	93.40% at 10^{-2} M	Mixed type	[178]
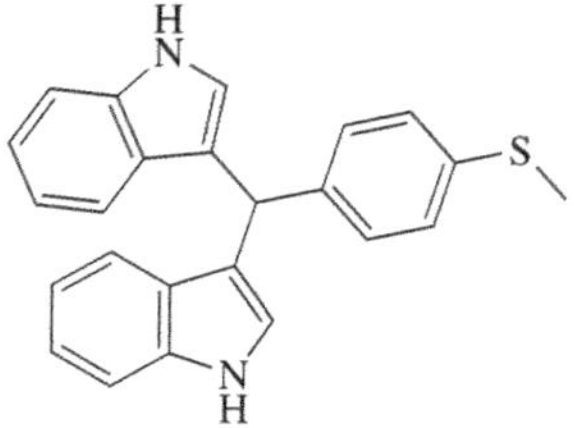 3,3-((4-(Methylthio)phenyl)methylene) bis(1H-indole) (TPBI)	Copper/0.5M H_2SO_4	EIS, PDP, FTIR, SEM-EDX, XPS, CA	99.10% at 10 mM	Langmuir/ mixed type	[152]

(*Continued*)

Table 9.3 (Continued)

Chemical structure/abbreviation	Metal/medium	Study methods	Results	Nature of adsorption	Ref
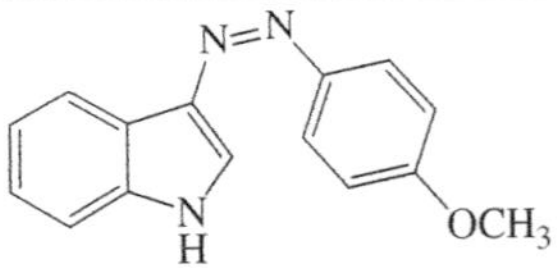 $N{=}N$, OCH_3, N H (Z)-3-((4-methoxyphenyl)diazenyl)-1H-indole (1)	Copper/2M HNO_3	WL, EIS, PDP, EFM, SEM-EDX, DFT	69.10% at 15×10^{-6}M	Temkin/mixed type	[179]
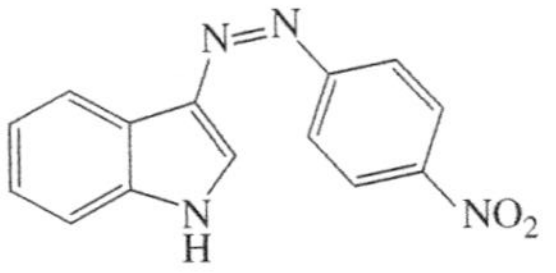 $N{=}N$, NO_2, N H (Z)-3-((4-nitrophenyl)diazenyl)-1H-indole (2)	Copper/2M HNO_3	WL, EIS, PDP, EFM, SEM-EDX, DFT	66.60% at 15×10^{-6}M	Temkin/Mixed type	[179]
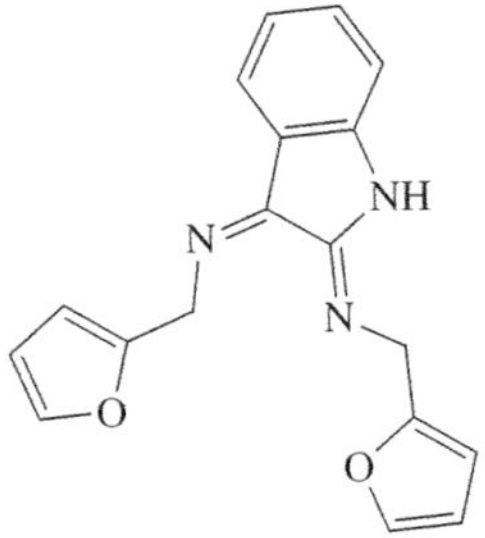 NH, N, N, O, O [(4E)-*N*-((Z)-2-((furan-2-yl)methylimino) indolin-3-ylidene)(furan-2-yl) methanamine] (SB-1)	Al/1M H_2SO_4	WL, EIS, PDP, SEM	97.34% at 500 ppm	Langmuir/ mixed type	[174]

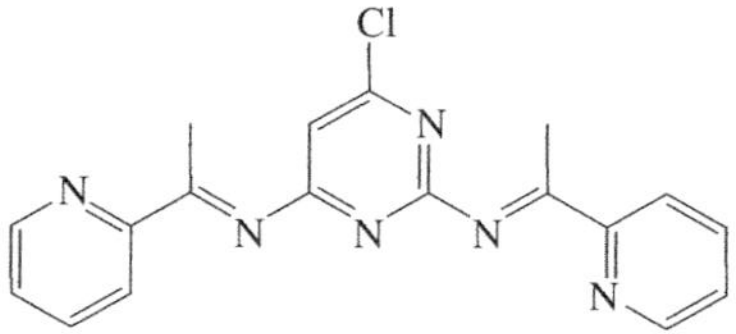 [(7Z,8Z)-6-chloro-N2,N4-bis(1-(pyridin-2-yl)ethylidene) pyrimidine-2,4-diamine] (SB-2)	Al/1M H_2SO_4	WL, EIS, PDP, SEM	94.64% at 500ppm	Langmuir/ mixed type	[174]
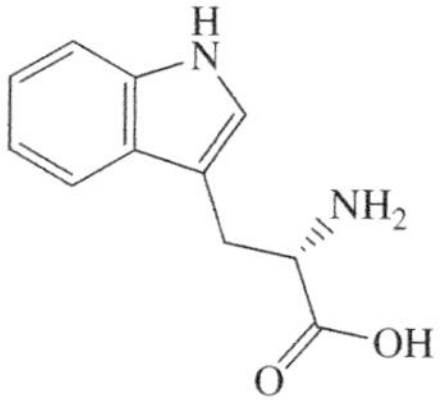 Tryptophan	Al/1M HCl, 20% $CaCl_2$, 3.5% NaCl	WL, EIS, PDP, DFT, SEM	1M HCl: 88.37% 20% $CaCl_2$:83.32% 3.5% NaCl: 80.55% at 0.008M	Langmuir/ mixed type	[180]
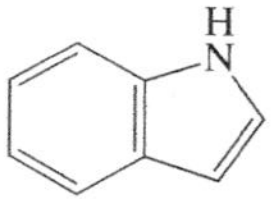 Indole	Ni/0.5M HCl	EIS, PDP	67.30% at 6×10^{-4} M	Frumkin/ Mixed type	[181]

(Continued)

Table 9.3 (Continued)

Chemical structure/abbreviation	Metal/medium	Study methods	Results	Nature of adsorption	Ref
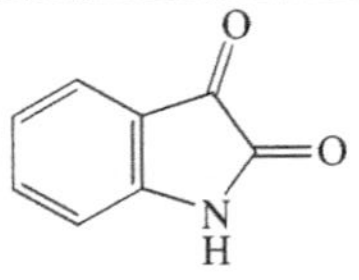 O, N, H, O Isatin	Ni/ 0.5M HCl	EIS, PDP	70.80% at 6×10^{-4} M	Frumkin/ mixed type	[181]
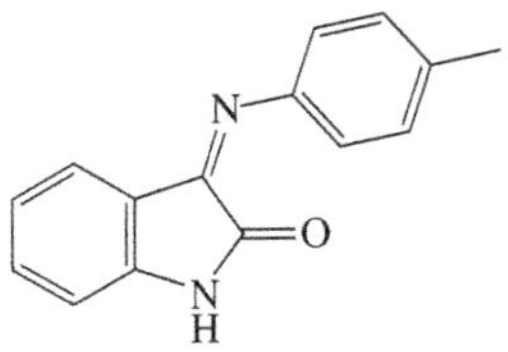 N, O, N, H 3-p-tolylimino-1,3-dihydro-indole-2-one	Ni/ 0.5M HCl	EIS, PDP	94.70% at 6×10^{-4} M	Frumkin/ mixed type	[181]

the inhibitors behaved as mixed type and gave IE of 82.5 and 97.1% for indole and 2-oxyindole in 15×10^{-5}M, respectively. In another similar study, the effect of substituted arylazo indole compounds was investigated on the deterioration of α–brass Cu-Zn, (67/33) in 2M nitric acid using electrochemical techniques [176]. The substituents were R ($-OCH_3$, $-CH_3$, –H, and $-NO_2$) and the order of decreasing IE measured was $-H > -CH_3 > -OCH_3 > -NO_2$. The inhibition abilities of the compounds were attributed to the attraction of the metal toward the pi electrons of the cyclic system and adsorption via the active centers in the arylazo indole derivatives. The influence of isatin molecule on the anticorrosion of Cu-Zn alloy in sulfuric and chloride acid solutions have also been reported [177].

9.5 Corrosion Inhibition Mechanism of Indoles

Generally, the adherence of organic molecules on metal surface can be elucidated by two major types of interaction, namely, physisorption (physical adsorption) or chemisorption (chemical adsorption). In most cases, adsorption occurs by a combination of these two processes [45, 104]. In physisorption, indoles adsorb on the substrate surface by weak electrostatic force between the charged indoles and the electrically charged substrate surface. Chemisorption, on the other hand, involves charge transfer or charge sharing between the indoles and the metal surface leading to the development of a coordinate covalent bond. The coordinate covalent bond is dependent on the electron density of the group polarizability and the functional group of the donor atoms [182]. Physisorption is often associated with low energy and is only stable at low temperature, while chemisorption has a higher adsorption energy, is irreversible, and is found to be more stable at higher temperatures [148].

The adsorption of indoles on the adsorbent can be seen as a quasi-substitution process that involves the displacement of water and other ions as shown below:

$$Org_{(sol)} + xH_2O_{(ads)} \Leftrightarrow Org_{(ads)} + xH_2O_{(sol)} \tag{9.1}$$

where x represents the size ratio of water molecules displaced by one indole adsorbate, $Org_{(sol)}$ and $Org_{(ads)}$ are the indoles in the aqueous solution and adsorbed on the metal substrate, $H_2O_{(sol)}$ and $H_2O_{(ads)}$ represent the water molecules adsorbed onto the metal substrate and that in the bulk solution. In addition, adsorption isotherms are often utilized to elucidate the adsorption process of these chemical inhibitors. Commonly used adsorption isotherms include Langmuir, Frumkin, Temkin, Freundlich, El-Awady, Bockris-Swinkels, and Dhar-Flory-Huggins [82, 148]. From literature review, it is indisputable that indoles followed Langmuir adsorption isotherm in almost all reported steel-electrolyte systems. The values of K_{ads} obtained from Langmuir isotherms indicate the

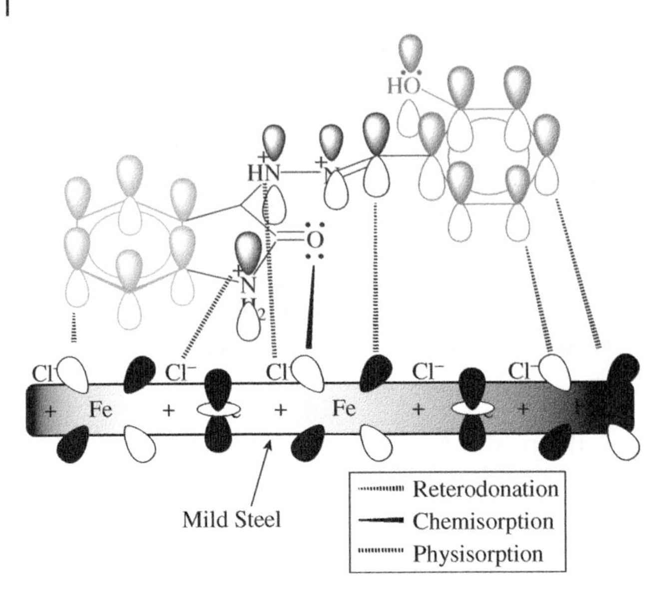

Figure 9.5 Suggested model of adsorption of bis-Schiff bases of isatin on MS in 1M HCl. *Source:* Ansari and Quraishi [44].

strength of adherence of the adsorbate on the metal substrate in the corrosive medium. Ahamad and coworkers proposed that a value of $K_{ads} \geq 100\,M^{-1}$ implies a stabler and stronger adherence of the adsorbate on the metal [82]. More so, insight is gained on the nature of interaction of the indole derivative with the adsorbent by the obtained value of ΔG^0_{ads}. Several authors have posited that pure electrostatic interaction can be said to occur if the obtained value of $\Delta G^0_{ads} < 20\,kJ/mol$ and ΔG^0_{ads} values $> 40\,kJ/mol$ account for chemisorption [89].

As shown in Figure 9.5, Ansari and Quraishi [44] proposed one or more ways by which indoles could adsorb on the metal/HCl interface, namely, (i) donor–acceptor interaction between empty d-orbital of Fe atoms and the pi electrons of aromatic ring, (ii) interaction between empty d-orbital of Fe atoms and the unshared electron pairs of heteroatoms, and (iii) electrostatic contact of protonated compounds with already adsorbed Cl^-.

9.6 Theoretical Modeling of Indole-Based Chemical Inhibitors

The limitations associated with the correlation of the structural features of organic inhibitors with obtained IE via experimental analyses brought about the use of computational modeling tools, which have proven to be powerful tools in exploring the inhibition mechanism of studied compounds at the metal/solution interface. Computational tools such as density functional theory (DFT), MCS, MDS, and quantitative structure activity relationship (QSAR) have been utilized to assess the inhibitory abilities of indoles. These tools afford corrosion experts the leverage to further understand the complex process of inhibition at the substrate/solution interface as several factors such as the influence of heteroatoms, pi bond conjugation, functional groups, alkyl side chains, and phenyl rings that play significant roles in the adsorption capacity of compounds can be investigated [183]. More so, they are crucial tools in the design of new and effective anticorrosive materials. The influence of structural features on the experimental inhibition efficiencies has been studied using several quantum chemical variables such as the lowest unoccupied molecular orbital energy (E_{LUMO}), highest occupied molecular orbital energy (E_{HOMO}), electron affinity (A), global softness (σ), global hardness (η), energy gap (ΔE), dipole moment (μ), ionization potential (I), electronegativity (χ), electrophilicity (ω), and the fraction of electrons transferred (ΔN). Furthermore, MDS and MCS are atomistic simulation tools, which have made it possible to analyze the exact orientation of indole adsorbates on metallic substrates and have become promising tools to support experimental analyses. A few publications have paid attention to the exploration of molecular modeling techniques to investigate the anticorrosive ability of indoles in different encountered media.

Two spiropyrimidinethiones were developed and tested experimentally for their inhibitive influence on MS in 15% HCl [86]. The two compounds, 6'-(4-chlorophenyl)-1'-phenyl-2'-thioxo-2',3'-dihydro-1'H-spiro[indoline-3,4'-pyrimidin]-2-one(CPTS) and 6'-(4-methoxyphenyl)-1'-phenyl-2'-thioxo-2',3'-dihydro-1'H-spiro [indoline-3,4'-pyrimidine]-2-one (MPTS) were optimized using DFT with 6-31G(d,p) basis set. The DFT-derived parameters were used to elucidate the inhibition potentials of the compounds. The authors noted that both MPTS (-4.884 eV) and CPTS (-4.998 eV) had higher values of E_{HOMO} than Fe (-5.075 eV), which implies that both compounds have the ability to donate electrons to the empty d orbital of iron. A lower E_{LUMO} value of Fe (-1.747 eV) than that of MPTS (-1.176 eV) and CPTS (-1.146 eV) also favored the MS to accept electrons. Also, the recorded lower ΔE values for both indole compounds suggest that more electron donation can occur from MPTS and CPTS to the metal. MPTS was adjudged the better inhibitor on account of its higher E_{HOMO} value and lower ΔE values. Similarly, these same authors employed DFT studies to explain the adsorption of indoline compounds, MPIT and MPII [137]. They obtained high values of μ for both inhibitors that indicated they were polar molecules that could easily donate electrons. However, the higher value of ΔN (0.94), σ (0.5066 eV^{-1}), E_{HOMO} (-5.260 eV), μ (9.819 D), and lesser values of ΔE (3.947 eV), and E_{LUMO} (-1.313 eV) recorded for MPII compared to MPIT implied that MPII has the potency to offer better corrosion protection to the N80 steel surface.

Verma and coworkers employed DFT to support their experimental analyses on some indole compounds tested as anticorrosive agents of MS in HCl [27]. Generally, E_{HOMO} is an index that measures the electron-donating capacity of inhibitors, and E_{LUMO} is a measure of the electron-accepting tendency of an organic molecule. The ΔE, which is an index of relative chemical reactivity, is a function of the difference between E_{LUMO} and E_{HOMO}. The study reported an increasing order of E_{HOMO} values as follows: TAPD-I<TAPD-II<TAPD-III, which indicates a correlation between the obtained IE and the electron donating capacity of the indole compounds. The obtained trends of E_{LUMO}, χ, ΔE, σ, and η for neutral and protonated forms of TAPDs agree with the experimental results except for μ. The optimized structures, E_{HOMO} and E_{LUMO} locations showing reactive sites of the inhibitors for neutral and protonated forms of the indoles have been presented in Figure 9.6. Other authors have also reported the use of DFT calculations in studying the action of indoles as chemical additives for combating metallic degradation in various aqueous media [43, 140, 152, 175, 184].

MDS was also employed to corroborate the experimental observations of TAPDs in HCl. MDSs are said to reach equilibrium when there is a convergence between energy change and temperature. The study also showed that the thioindole moiety adheres firmly to the Fe surface. High values of binding energies ($E_{binding}$) and interaction energies ($E_{interaction}$) for the most stable configurations generally suggest strong, stable, and spontaneous adherence of organic molecules (indoles) on the substrate. In this study, the authors reported a similar trend of $E_{binding}$ and $E_{interaction}$ for both the protonated and neutral compounds: TAPD-III>TAPD-II>TAPD-I. MCS, which attempts to estimate the lowest energy for a whole studied system, was also employed to study the indole-metal interaction and the most stable configuration (a parallel orientation) to maximize contact was presented as shown in Figure 9.7a. Alongside with the configuration was the adsorption density field of indole on Fe(110) substrate, which shows the likelihood of indole to adsorb on iron. A high value of adsorption energy (-3.10 eV) was also recorded, which indicates the inhibition effectiveness of indole [61]. Few other authors also reported the use of atomistic simulations (MDS and MCS) in investigating the inhibitive effect of indole and its derivatives [43, 60, 62, 89, 91].

QSAR, which describes a logical relationship between structural attributes and the inhibition efficiency, has often been used to correlate a composite index of DFT-based variables with measured inhibition efficiencies. Lebrini and coworkers in their report established a correlation between the studied indole molecules and their electronic structural features (E_{HOMO}, E_{LUMO}) using a linear resistance model. A correlation was obtained between the predicted and experimental values obtained [104]. Another case study is a report by Al-Fakih and coworkers, where a QSAR model was developed for 10 isatin derivatives used as inhibitors of Q235A steel corrosion in acid using partial least square (PLS) method. The study

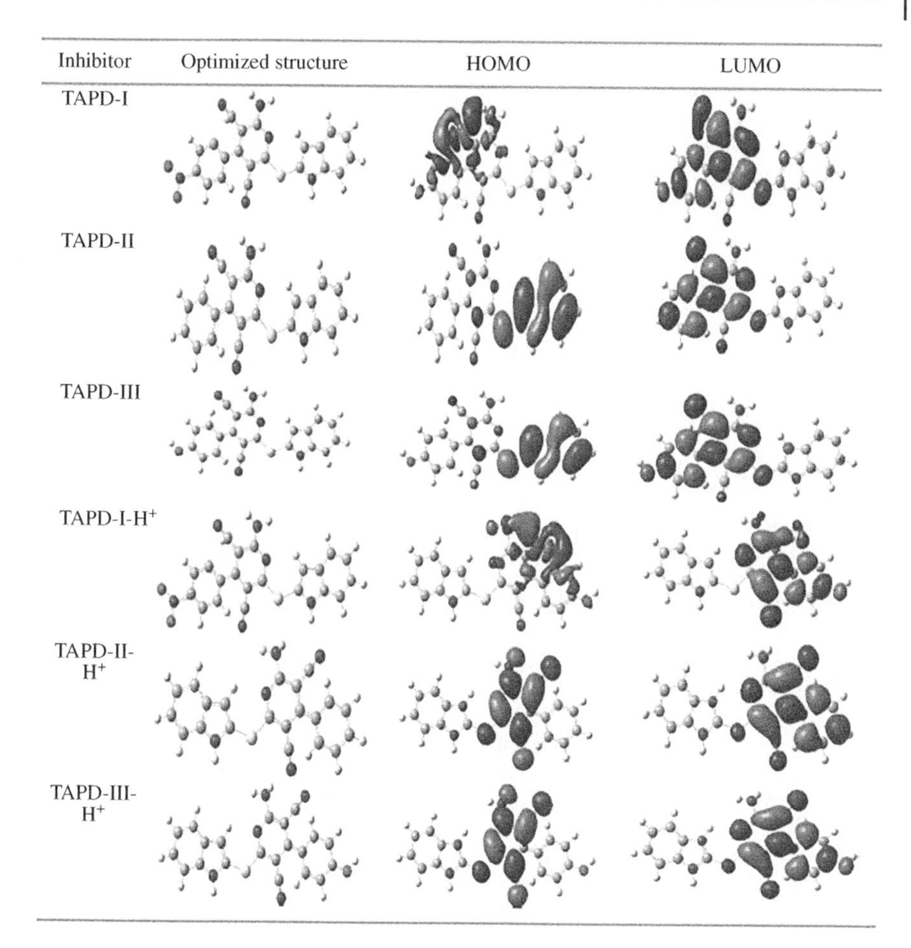

Figure 9.6 Optimized, E_{HOMO} and E_{LUMO} frontier molecular orbitals of neutral and protonated forms of TAPD-I, TAPD-II, and TAPD-III compounds. *Source:* Verma et al. [27].

reported a correlation coefficient value of 0.9676 between the selected molecular descriptors and the IE. Additionally, the authors found the results of their analysis to be consistent with reported measured data [185].

9.7 Conclusions and Outlook

Owing to the present growing demands and the forecast that boost in the economic growth of emerging economies will result in a boom for inhibitors market in the near future [186], it becomes imperative that corrosion scientists and

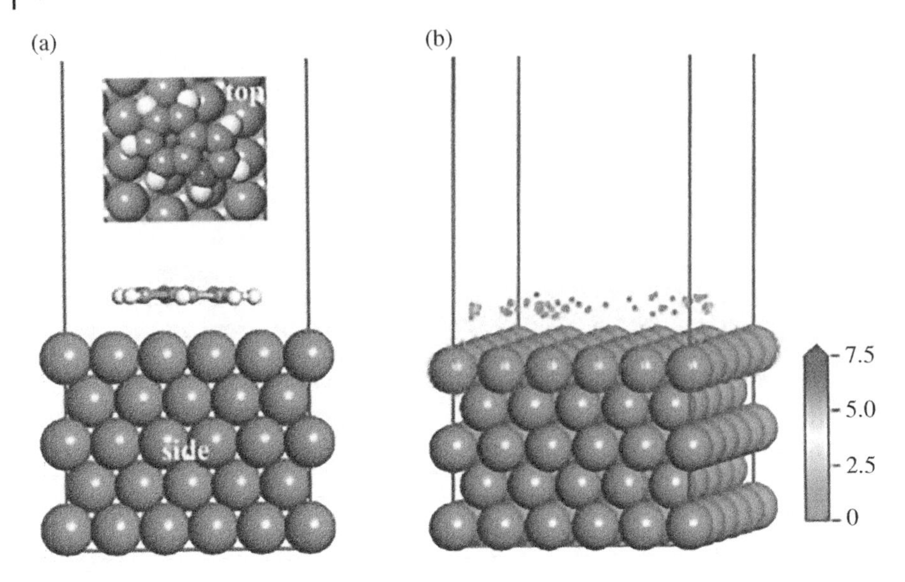

Figure 9.7 The most stable energy configuration (a) and the density field (b) for adherence of indole molecule on Fe (110) substrate. *Source:* Lv et al. [61].

engineers need to extensively explore and exploit the inhibitory capacities of environmentally friendly organic compounds. Indoles have been identified and established as cheap, safe, and effective anticorrosive agents to meet the demand. This chapter has presented a collection of works that have explored the inhibitory potentials of indoles in all media types and for all studied metal types. The indole derivatives have been investigated using testing methods such as chemical, electrochemical, surface morphological, and theoretical modeling techniques. These have clearly shown that indoles restrain metallic deterioration by adsorbing on the steel surface in order to protect against aggressive ions. In many cases, it has been reported that indoles function as mixed-type inhibitors and obey Langmuir isotherm model. There is need for more inquiries into the inhibition mechanism of indoles using computational modeling techniques such as the DFT, MCS, MDS, and QSAR as they have been scarcely reported in the inhibition behavior of indoles. Effect of substituent have been elucidated, and electron withdrawing substituents have been found to lower the inhibition efficiency, while electron releasing substituents result in improved adsorption ability of indoles on the metallic surface. The commercialization of indole compounds for large-scale industrial applications is necessary to provide safe and environmental alternatives to the conventional chemicals being used currently in the surface protection of metal.

References

1 Da Silva, J.F., Garden, S.J., and Pinto, A.C. (2001). The chemistry of isatins: a review from 1975 to 1999. *Journal of the Brazilian Chemical Society* 12 (3): 273–324.

2 Ziarani, G.M., Moradi, R., Ahmadi, T., and Lashgari, N. (2018). Recent advances in the application of indoles in multicomponent reactions. *RSC Advances 8* (22): 12069–12103.

3 Russel, J.S., Pelkey, E.T., and Yoon-Miller, S.J. (2011). Five-membered ring systems: pyrroles and benzo analogs. In: *Progress in Heterocyclic Chemistry*, vol. 22, 143–180. Elsevier.

4 Ghinea, I.O. and Dinica, R.M. (2016). Breakthroughs in indole and indolizine chemistry-new synthetic pathways, new applications. In: *Scope of Selective Heterocycles from Organic and Pharmaceutical Perspective* (ed. R. Varala), 115–142. IntechOpen.

5 Baeyer, A. (1866). Ueber die reduction aromatischer verbindungen mittelst zinkstaub. *Justus Liebigs Annalen der Chemie 140* (3): 295–296.

6 Inman, M. and Moody, C.J. (2013). Indole synthesis–something old, something new. *Chemical Science 4* (1): 29–41.

7 Singh, T.P. and Singh, O.M. (2018). Recent progress in biological activities of indole and indole alkaloids. *Mini Reviews in Medicinal Chemistry 18* (1): 9–25.

8 Sachdeva, H., Mathur, J., and Guleria, A. (2020). Indole derivatives as potential anticancer agents: a review. *Journal of the Chilean Chemical Society 65* (3): 4900–4907.

9 Tang, L., Peng, T., Wang, G. et al. (2017). Design, synthesis and preliminary biological evaluation of Novel Benzyl Sulfoxide 2-Indolinone derivatives as anticancer agents. *Molecules 22* (11): 1979.

10 Kaushik, N.K., Kaushik, N., Attri, P. et al. (2013). Biomedical importance of indoles. *Molecules 18* (6): 6620–6662.

11 Barden, T.C. (2010). Indoles: industrial, agricultural and over-the-counter uses. In: *Heterocyclic Scaffolds II: Reactions and Applications of Indoles* (ed. G.W. Gribble), 31–46. Berlin: Springer.

12 Mudila, H., Prasher, P., Kumar, M. et al. (2019). Critical analysis of polyindole and its composites in supercapacitor application. *Materials for Renewable and Sustainable Energy 8* (2): 9.

13 Costa, Â.C., Cavalcanti, S.C., Santana, A.S. et al. (2019). Insecticidal activity of indole derivatives against Plutella xylostella and selectivity to four non-target organisms. *Ecotoxicology 28* (8): 973–982.

14 Laurent, A. (1840). Recherches sur l'indigo. *Annales de Chimie et de Physique 3* (3): 393–434.

15 Grewal, A.S. (2014). Isatin derivatives with several biological activities. *International Journal of Pharmaceutical Research 6* (1): 1–7.

16 Nath, R., Pathania, S., Grover, G., and Akhtar, M.J. (2020). Isatin containing heterocycles for different biological activities: analysis of structure activity relationship. *Journal of Molecular Structure* 1222: 128900.

17 Kakkar, R. (2019). Isatin and its derivatives: a survey of recent syntheses, reactions, and applications. *MedChemComm 10* (3): 351–368.

18 Bugaenko, D.I., Karchava, A.V., and Yurovskaya, M.A. (2019). Synthesis of indoles: recent advances. *Russian Chemical Reviews 88* (2): 99.

19 Sonar, S.S., Sadaphal, S.A., Kategaonkar, A.H. et al. (2009). Alum catalyzed simple and efficient synthesis of bis (indolyl) methanes by ultrasound approach. *Bulletin of the Korean Chemical Society 30* (4): 825.

20 Yadav, J., Reddy, B.S., Singh, A.P., and Basak, A. (2007). The first one-pot oxidative Michael reaction of Baylis–Hillman adducts with indoles promoted by iodoxybenzoic acid. *Tetrahedron Letters 48* (24): 4169–4172.

21 Khan, K.M., Khan, M., Ali, M. et al. (2009). Synthesis of bis-Schiff bases of isatins and their antiglycation activity. *Bioorganic and Medicinal Chemistry 17* (22): 7795–7801.

22 Bratulescu, G. (2008). A new and efficient one-pot synthesis of indoles. *Tetrahedron Letters 49* (6): 984–986.

23 Oskooie, H.A., Heravi, M.M., and Behbahani, F.K. (2007). A facile, mild and efficient one-pot synthesis of 2-substituted indole derivatives catalyzed by Pd (PPh3) 2Cl2. *Molecules 12* (7): 1438–1446.

24 Survase, D.N., Karhale, S.S., Khedkar, V.M., and Helavi, V.B. (2019). Synthesis, characterization, and biological evaluation of indole aldehydes containing N-benzyl moiety. *Synthetic Communications 49* (24): 3486–3497.

25 Kumar, A., Gupta, M.K., and Kumar, M. (2012). L-Proline catalysed multicomponent synthesis of 3-amino alkylated indoles via a Mannich-type reaction under solvent-free conditions. *Green Chemistry 14* (2): 290–295.

26 Shinde, P.V., Sonar, S.S., Shingate, B.B., and Shingare, M.S. (2010). Boric acid catalyzed convenient synthesis of 2-amino-3, 5-dicarbonitrile-6-thio-pyridines in aqueous media. *Tetrahedron Letters 51* (9): 1309–1312.

27 Verma, C., Quraishi, M., Lgaz, H. et al. (2019). Adsorption and anticorrosion behaviour of mild steel treated with 2-((1H-indol-2-yl) thio)-6-amino-4-phenylpyridine-3, 5-dicarbonitriles in a hydrochloric acid solution: Experimental and computational studies. *Journal of Molecular Liquids 283*: 491–506.

28 Olasunkanmi, L.O., Obot, I.B., Kabanda, M.M., and Ebenso, E.E. (2015). Some quinoxalin-6-yl derivatives as corrosion inhibitors for mild steel in hydrochloric acid: experimental and theoretical studies. *The Journal of Physical Chemistry C 119* (28): 16004–16019.

29 Hughes, A.E., Mol, J.M., Zheludkevich, M.L., and Buchheit, R.G. (2016). Active protective coatings. In: *Active Protective Coatings: New-Generation Coatings for Metals, Springer Series in Materials Science*, vol. *233* (eds. A.E. Hughes, J.M. Mol,

M.L. Zheludkevich and R.G. Buchheit). Dordrecht: *Springer Science + Business Media*. ISBN: *978-94-017-7538-0*.

30 Koch, G., Varney, J., Thompson, N. et al. (2016). *International Measures of Prevention, Application, and Economics of Corrosion Technologies Study*, 1–16. Houston, TX: NACE International.

31 Verma, C., Olasunkanmi, L. O., Ebenso, E.E., and Quraishi, M. A. (2018). Substituents effect on corrosion inhibition performance of organic compounds in aggressive ionic solutions: a review. *Journal of Molecular Liquids 251*: 100–118.

32 Palanisamy, G. (2019). Corrosion inhibitors. In: *Corrosion Inhibitors* (ed. A. Singh). Intechopen.

33 McCafferty, E. (2010). *Introduction to Corrosion Science*. Springer Science + Business Media.

34 Kamal, C. and Sethuraman, M.G. (2012). Caulerpin– A bis-indole alkaloid as a green inhibitor for the corrosion of mild steel in 1 M HCl solution from the marine alga Caulerpa racemosa. *Industrial and Engineering Chemistry Research 51* (31): 10399–10407.

35 Olasunkanmi, L.O., Obot, I.B., and Ebenso, E.E. (2016). Adsorption and corrosion inhibition properties of N-{n-[1-R-5-(quinoxalin-6-yl)-4,5-dihydropyrazol-3-yl]phenyl}methanesulfonamides on mild steel in 1 M HCl: experimental and theoretical studies. *RSC Advances 6* (90): 86782–86797.

36 Verma, C., Olasunkanmi, L.O., Ebenso, E.E. et al. (2016). Adsorption behavior of glucosamine-based, pyrimidine-fused heterocycles as green corrosion inhibitors for mild steel: experimental and theoretical studies. *The Journal of Physical Chemistry C 120* (21): 11598–11611.

37 Quadri, T.W., Olasunkanmi, L.O., Fayemi, O.E. et al. (2017). Zinc oxide nanocomposites of selected polymers: synthesis, characterization, and corrosion inhibition studies on mild steel in HCl solution. *ACS Omega 2* (11): 8421–8437.

38 Frankel, G.S. and McCreery, R.L. (2001). Inhibition of Al alloy corrosion by chromates. *Interface-Electrochemical Society 10* (4): 34–39.

39 Rodič, P., Milošev, I., Lekka, M. et al. (2019). Study of the synergistic effect of cerium acetate and sodium sulphate on the corrosion inhibition of AA2024-T3. *Electrochim. Acta 308*: 337–349.

40 Murulana, L.C., Kabanda, M.M., and Ebenso, E.E. (2016). Investigation of the adsorption characteristics of some selected sulphonamide derivatives as corrosion inhibitors at mild steel/hydrochloric acid interface: experimental, quantum chemical and QSAR studies. *Journal of Molecular Liquids 215*: 763–779.

41 Kálmán, E. (1994). *"Routes to the developments of low toxicity corrosion inhibitors for use in neutral solutions, w:" "A Working Party Report on Corrosion Inhibitors"*. London: *The Institute of Materials*.

42 Anastas, P. and Eghbali, N. (2010). Green chemistry: principles and practice. *Chemical Society Reviews 39* (1): 301–312.

43 Verma, C., Quraishi, M., Ebenso, E. et al. (2016). 3-amino alkylated indoles as corrosion inhibitors for mild steel in 1M HCl: Experimental and theoretical studies. *Journal of Molecular Liquids 219*: 647–660.

44 Ansari, K. and Quraishi, M. (2014). Bis-Schiff bases of isatin as new and environmentally benign corrosion inhibitor for mild steel. *Journal of Industrial and Engineering Chemistry 20* (5): 2819–2829.

45 Ansari, K., Quraishi, M., and Singh, A. (2015). Isatin derivatives as a non-toxic corrosion inhibitor for mild steel in 20% H_2SO_4. *Corrosion Science 95*: 62–70.

46 Marinescu, M. (2019). Recent advances in the use of benzimidazoles as corrosion inhibitors. *BMC Chemistry 13* (1): 136.

47 Düdükcü, M., Yazici, B., and Erbil, M. (2004). The effect of indole on the corrosion behaviour of stainless steel. *Materials Chemistry and Physics 87* (1): 138–141.

48 Tussolini, M., Viomar, A., Gallina, A.L. et al. (2013). Electrochemical behavior of indole for AISI 430 stainless steel in changing the media from 1 mol L−1 H_2SO_4 to 1 mol L−1 HCl. *Rem: Revista Escola de Minas 66* (2): 215–220.

49 Hang, T.T.X., Truc, T.A., Olivier, M.-G. et al. (2010). Corrosion protection mechanisms of carbon steel by an epoxy resin containing indole-3 butyric acid modified clay. *Progress in Organic Coatings 69* (4): 410–416.

50 Osarolube, E., Owate, I., and Oforka, N. (2008). Corrosion behaviour of mild and high carbon steels in various acidic media. *Scientific Research and Essay 3* (6): 224–228.

51 Broadbent, C. (2016). Steel's recyclability: demonstrating the benefits of recycling steel to achieve a circular economy. *The International Journal of Life Cycle Assessment 21* (11): 1658–1665.

52 Hasan, B.O. and Sadek, S.A. (2014). The effect of temperature and hydrodynamics on carbon steel corrosion and its inhibition in oxygenated acid–salt solution. *Journal of Industrial and Engineering Chemistry 20* (1): 297–307.

53 Sanad, S. and Ismail, A. (2000). Indole and its derivatives as corrosion inhibitors for C-steel during pickling. *Journal of Materials Science and Technology* 16 (3): 291–296.

54 Popova, A. and Christov, M. (2006). Evaluation of impedance measurements on mild steel corrosion in acid media in the presence of heterocyclic compounds. *Corrosion Science 48* (10): 3208–3221.

55 Ismail, A., Sanad, S., and El-Meligi, A. (2000). Inhibiting effect of indole and some of its derivatives on corrosion of C-steel in HCl. *Journal of Materials Science and Technology 16* (4): 397–400.

56 Moretti, G., Quartarone, G., Tassan, A., and Zingales, A. (1994). Inhibition of mild steel corrosion in 1N sulphuric acid through indole. *Materials and Corrosion 45* (12): 641–647.

57 Donnelly, B., Downie, T., Grzeskowiak, R. et al. (1978). The effect of electronic delocalization in organic groups R in substituted thiocarbamoyl RCSNH2 and related compounds on inhibition efficiency. *Corrosion Science 18* (2): 109–116.

58 Brakenbury, W. and Grzeskowiak, R. (1988). Ellipsometric study of effects of organic inhibitors on mild steel in 1M hydrochloric acid. *British Corrosion Journal 23* (3): 176–180.

59 Cumper, C., Grzeskowiak, R., and Newton, P. (1982). Effect of pyrrole, indole and their dimethyl derivatives on the dissolution of magnetite. *Corrosion Science 22* (6): 551–557.

60 Khaled, K.F. (2008). Application of electrochemical frequency modulation for monitoring corrosion and corrosion inhibition of iron by some indole derivatives in molar hydrochloric acid. *Materials Chemistry and Physics 112* (1): 290–300.

61 Lv, T., Zhu, S., Guo, L., and Zhang, S. (2015). Experimental and theoretical investigation of indole as a corrosion inhibitor for mild steel in sulfuric acid solution. *Research on Chemical Intermediates 41* (10): 7073–7093.

62 Tan, J., Guo, L., and Xu, S. (2015). Investigation of indole-3-carboxylic acid as steel inhibitor in 0.1 M H_2SO_4 solution. *Journal of Industrial and Engineering Chemistry 25*: 295–303.

63 Quartarone, G., Bonaldo, L., and Tortato, C. (2006). Inhibitive action of indole-5-carboxylic acid towards corrosion of mild steel in deaerated 0.5 M sulfuric acid solutions. *Applied Surface Science 252* (23): 8251–8257.

64 Avci, G. (2008). Corrosion inhibition of indole-3-acetic acid on mild steel in 0.5 M HCl. *Colloids and Surfaces A: Physicochemical and Engineering Aspects 317* (1–3): 730–736.

65 Ashhari, S. and Sarabi, A.A. (2015). Indole-3-carbaldehyde and 2-methylindole as corrosion inhibitors of mild steel during pickling. *Pigment & Resin Technology* 44 (5): 322–329.

66 Moretti, G., Quartarone, G., Tassan, A., and Zingales, A. (1996). Some derivatives of indole as mild steel corrosion inhibitors in 0 5 M sulphuric acid. *British Corrosion Journal 31* (1): 49–54.

67 Moretti, G., Quartarone, G., Tassan, A., and Zingales, A. (1995). The inhibiting action of indole and some of its derivates on corrosion of mild steel in 1N sulphuric acid. *Materials Science Forum* 192–194: 363–378.

68 Düdükcü, M. (2011). The inhibitive effect of 5-amino-indole on the corrosion of mild steel in acidic media. *Materials and Corrosion 62* (3): 264–268.

69 Moretti, G., Quartarone, G., Tassan, A., and Zlngales, A. (1996). 5-Amino-and 5-chloro-indole as mild steel corrosion inhibitors in 1 N sulphuric acid. *Electrochimica Acta 41* (13): 1971–1980.

70 Düdükcü, M. and Avcı, G. (2016). Electrochemical synthesis and corrosion inhibition performance of poly-5-aminoindole on stainless steel. *Progress in Organic Coatings 97*: 110–114.

71 Döşlü, S.T., Mert, B.D., and Yazıcı, B. (2018). The electrochemical synthesis and corrosion behaviour of TiO2/poly (indole-co-aniline) multilayer coating: Experimental and theoretical approach. *Arabian Journal of Chemistry 11* (1): 1–13.

72 Gopi, D., Karthikeyan, P., Kavitha, L., and Surendiran, M. (2015). Development of poly (3, 4-ethylenedioxythiophene-co-indole-5-carboxylic acid) co-polymer coatings on passivated low-nickel stainless steel for enhanced corrosion resistance in the sulphuric acid medium. *Applied Surface Science 357*: 122–130.

73 Lakourj, M.M., Norouzian, R.-S., and Esfandyar, M. (2020). Conducting nanocomposites of polypyrrole-co-polyindole doped with carboxylated CNT: synthesis approach and anticorrosion/antibacterial/antioxidation property. *Materials Science and Engineering: B 261*: 114673.

74 Boomadevi Janaki, G. and Xavier, J.R. (2020). Effect of indole functionalized nano-alumina on the corrosion protection performance of epoxy coatings in marine environment. *Journal of Macromolecular Science, Part A* 57: 1–12.

75 Rahman, M.M., Islam, M.M., Khan, M.M.R. et al. (2019). IBA-modified gypsum-containing epoxy resin coating for rebar: corrosion performance and bonding characteristics. *International Journal of Plastics Technology 23* (1): 20–28.

76 Truc, T.A., Hang, T.T.X., Oanh, V.K. et al. (2008). Incorporation of an indole-3 butyric acid modified clay in epoxy resin for corrosion protection of carbon steel. *Surface and Coatings Technology 202* (20): 4945–4951.

77 Toprak Döşlü, S. and Doğru Mert, B. (2013). Polyindole top coat on TiO2 sol–gel films for corrosion protection of steel. *Corrosion Science* 66: 51–58.

78 Trinh, A.T., Nguyen, T.T., Thai, T.T. et al. (2016). Improvement of adherence and anticorrosion properties of an epoxy-polyamide coating on steel by incorporation of an indole-3 butyric acid-modified nanomagnetite. *Journal of Coatings Technology and Research 13* (3): 489–499.

79 Karthikeyan, P. and Rajavel, R. (2016). Anti-corrosion application of Poly (indole-5-carboxylic acid) coating on low nickel stainless steel in acidic medium. *Chemical Science 5* (4): 1100–1106.

80 Fouda, A., Abdallah, M., and Medhat, M. (2012). Some Schiff base compounds as inhibitors for corrosion of carbon steel in acidic media. *Protection of Metals and Physical Chemistry of Surfaces 48* (4): 477–486.

81 Kharbach, Y., Qachchachi, F., Haoudi, A. et al. (2017). Anticorrosion performance of three newly synthesized isatin derivatives on carbon steel in hydrochloric acid pickling environment: electrochemical, surface and theoretical studies. *Journal of Molecular Liquids 246*: 302–316.

82 Ahamad, I., Prasad, R., and Quraishi, M. (2010). Adsorption and inhibitive properties of some new Mannich bases of Isatin derivatives on corrosion of mild steel in acidic media. *Corrosion Science 52* (4): 1472–1481.

83 Gupta, C., Ahamad, I., Singh, A. et al. (2017). Experimental study and theoretical simulations of some indolinone based mannich bases as novel corrosion

inhibitors for mild steel in acid solutions. *International Journal of Electrochemical Science* 12: 6379–6392.

84 Verma, C., Singh, P., and Quraishi, M.A. (2016). A thermodynamical, electrochemical and surface investigation of Bis (indolyl) methanes as Green corrosion inhibitors for mild steel in 1 M hydrochloric acid solution. *Journal of the Association of Arab Universities for Basic and Applied Sciences 21* (1): 24–30.

85 Singh, A.K. and Quraishi, M. (2010). Inhibiting effects of 5-substituted isatin-based Mannich bases on the corrosion of mild steel in hydrochloric acid solution. *Journal of Applied Electrochemistry 40* (7): 1293–1306.

86 Yadav, M., Sinha, R., Kumar, S., and Sarkar, T. (2015). Corrosion inhibition effect of spiropyrimidinethiones on mild steel in 15% HCl solution: insight from electrochemical and quantum studies. *RSC Advances 5* (87): 70832–70848.

87 Singh, A.K. and Quraishi, M. (2010). Investigation of adsorption of isoniazid derivatives at mild steel/hydrochloric acid interface: electrochemical and weight loss methods. *Materials Chemistry and Physics 123* (2–3): 666–677.

88 Singh, A.K. and Quraishi, M. (2012). Study of some bidentate schiff bases of isatin as corrosion inhibitors for mild steel in hydrochloric acid solution. *International Journal of Electrochemical Science 7*: 3222–3241.

89 Singh, A.K., Pandey, A.K., Banerjee, P. et al. (2019). Eco-friendly disposal of expired anti-tuberculosis drug isoniazid and its role in the protection of metal. *Journal of Environmental Chemical Engineering 7* (2): 102971.

90 Al-Fahemi, J.H., Abdallah, M., Gad, E.A., and Jahdaly, B. (2016). Experimental and theoretical approach studies for melatonin drug as safely corrosion inhibitors for carbon steel using DFT. *Journal of Molecular Liquids 222*: 1157–1163.

91 Ituen, E., Mkpenie, V., Moses, E., and Obot, I. (2019). Electrochemical kinetics, molecular dynamics, adsorption and anticorrosion behavior of melatonin biomolecule on steel surface in acidic medium. *Bioelectrochemistry 129*: 42–53.

92 Sampath, S., Vadivelu, M., Ravindran, R. et al. (2020). Synthesis of 1, 2, 3-triazole tethered 3-hydroxy-2-oxindoles: promising corrosion inhibitors for steel in acidic medium and their Anti-Microbial Evaluation. *ChemistrySelect 5* (7): 2130–2134.

93 Fu, J.-J., Li, S.-N., Cao, L.-H. et al. (2010). L-Tryptophan as green corrosion inhibitor for low carbon steel in hydrochloric acid solution. *Journal of Materials Science 45* (4): 979–986.

94 Fu, J.-J., Li, S.-N., Wang, Y., and Cao, L.-H. (2010). Computational and electrochemical studies of some amino acid compounds as corrosion inhibitors for mild steel in hydrochloric acid solution. *Journal of Materials Science 45* (22): 6255–6265.

95 Fawzy, A., Abdallah, M., Zaafarany, I. et al. (2018). Thermodynamic, kinetic and mechanistic approach to the corrosion inhibition of carbon steel by new synthesized amino acids-based surfactants as green inhibitors in neutral and alkaline aqueous media. *Journal of Molecular Liquids 265*: 276–291.

96 Fawzy, A., Zaafarany, I., Ali, H., and Abdallah, M. (2018). New synthesized amino acids-based surfactants as efficient inhibitors for corrosion of mild steel in hydrochloric acid medium: kinetics and thermodynamic approach. *International Journal of Electrochemical Science 13* (5): 4575–4600.

97 Mobin, M., Parveen, M., and Khan, M.A. (2011). Inhibition of mild steel corrosion using L-tryptophan and synergistic surfactant additives. *Portugaliae Electrochimica Acta 29* (6): 391–403.

98 Ngouné, B., Pengou, M., Nouteza, A.M. et al. (2019). Performances of alkaloid extract from Rauvolfia macrophylla Stapf toward corrosion inhibition of C38 steel in acidic media. *ACS Omega 4* (5): 9081–9091.

99 Fawzy, A., Farghaly, T.A., Al Bahir, A.A. et al. (2021). Investigation of three synthesized propane bis-oxoindoline derivatives as inhibitors for the corrosion of mild steel in sulfuric acid solutions. *Journal of Molecular Structure 1223*: 129318.

100 AL-Mosawi, B.T.S., Sabri, M.M., and Ahmed, M.A. (2021). Synergistic effect of ZnO nanoparticles with organic compound as corrosion inhibition. *International Journal of Low-Carbon Technologies* 16 (2): 429–435.

101 Ojo, F., Adejoro, I., Akpomie, K. et al. (2018). Effect of iodide ions on the inhibitive performance of O-, M-, P-Nitroaniline on mild steel in hydrochloric acid solution. *Journal of Applied Science and Environmental Management 22* (5): 775–782.

102 Abd El-Lateef, H.M. (2020). Corrosion inhibition characteristics of a novel salycilidene isatin hydrazine sodium sulfonate on carbon steel in HCl and a synergistic nickel ions additive: a combined experimental and theoretical perspective. *Applied Surface Science 501*: 144237.

103 Guo, L., Ye, G., Obot, I.B. et al. (2017). Synergistic effect of potassium iodide with L-tryptophane on the corrosion inhibition of mild steel: a combined electrochemical and theoretical study. *International Journal of Electrochemical Science 12*: 166–177.

104 Lebrini, M., Robert, F., Vezin, H., and Roos, C. (2010). Electrochemical and quantum chemical studies of some indole derivatives as corrosion inhibitors for C38 steel in molar hydrochloric acid. *Corrosion Science 52* (10): 3367–3376.

105 Lebrini, M., Robert, F., and Roos, C. (2013). Adsorption properties and inhibition of C38 steel corrosion in hydrochloric solution by some indole derivates: temperature effect, activation energies, and thermodynamics of adsorption. *International Journal of Corrosion 2013*: 1–13.

106 Uwah, I.E., Okafor, P.C., and Ebiekpe, V.E. (2013). Inhibitive action of ethanol extracts from Nauclea latifolia on the corrosion of mild steel in H_2SO_4 solutions and their adsorption characteristics. *Arabian Journal of Chemistry 6* (3): 285–293.

107 Ahmed, S.A., Awad, M.I., Althagafi, I.I. et al. (2019). Newly synthesized indolium-based ionic liquids as unprecedented inhibitors for the corrosion of mild steel in acid medium. *Journal of Molecular Liquids 291*: 111356.

108 Arrousse, N., Taleb, M., Ghibate, R., and Senhaji, O. (2020). Study of the inhibition of corrosion of mild steel in a 1M HCl solution by a new quaternary ammonium surfactant. *Moroccan Journal of Chemistry 9* (1): 44–56.

109 Gupta, S.R., Mourya, P., Singh, M., and Singh, V.P. (2017). Structural, theoretical and corrosion inhibition studies on some transition metal complexes derived from heterocyclic system. *Journal of Molecular Structure 1137*: 240–252.

110 Paul, A., Joby Thomas, K., Raphael, V.P., and Shaju, K.S. (2012). Electrochemical and gravimetric corrosion inhibition investigations of a heterocyclic schiff base derived from 3-formylindole. *IOSR Journal of Applied Chemistry, 1* (6): 17–23.

111 Paul, A., Thomas, K.J., Raphael, V.P., and Shaju, K.S. (2012). 3-Formylindole-4-aminobenzoic acid: a potential corrosion inhibitor for mild steel and copper in hydrochloric acid media. *International Scholarly Research Notices, Corrosion* 2012: 1–9.

112 Tribak, Z., Haoudi, A., Skalli, M. et al. (2017). 5-chloro-1H-indole-2, 3-dione derivative as corrosion inhibitor for mild steel in 1M H3PO4: weight loss, electrochemical and SEM studies. *Journal of Materials and Environmental Science 8* (1): 298–309.

113 Bentiss, F., Gassama, F., Barbry, D. et al. (2006). Enhanced corrosion resistance of mild steel in molar hydrochloric acid solution by 1, 4-bis (2-pyridyl)-5H-pyridazino [4, 5-b] indole: electrochemical, theoretical and XPS studies. *Applied Surface Science 252* (8): 2684–2691.

114 Abdul Ghani, A., Bahron, H., Harun, M.K., and Kassim, K. (2012). Corrosion inhibition study of a heterocyclic Schiff base derived from Isatin. *Advanced Materials Research* 554–556: 425–429.

115 Chen, S. and Zhao, K. (2015). Synthesis, structure and anticorrosion of 1-Methyl-indolin-2-one-3-oxime. *Asian Journal of Chemistry 27* (3): 1019.

116 Ansari, K. and Quraishi, M. (2015). Experimental and quantum chemical evaluation of Schiff bases of isatin as a new and green corrosion inhibitors for mild steel in 20% H_2SO_4. *Journal of the Taiwan Institute of Chemical Engineers 54*: 145–154.

117 Tribak, Z., Rodi, Y.K., Elmsellem, H. et al. (2017). 5-chloro-1-octylindoline-2, 3-dione as a new corrosion inhibitor for mild steel in hydrochloric acid solution. *Journal of Materials and Environmental Science 8* (3): 1116–1127.

118 Sudha, D. and Nalini, D. (2013). Corrosion inhibition of indoloimidazole derivative on mild steel in H_2SO_4. *International Journal of Scientific and Engineering Research 4* (11): 1161–1165.

119 Zarrok, H., Al Mamari, K., Zarrouk, A. et al. (2012). Gravimetric and electrochemical evaluation of 1-allyl-1H-indole-2, 3-dione of carbon steel corrosion in hydrochloric acid. *International Journal of Electrochemical Science 7*: 10338–10357.

120 Al Mamari, K., Zarrok, H., Zarrouk, A. et al. (2013). Anti-corrosion properties of indole derivative for carbon steel in HCl solution. *Der Pharmacia Lettre 5* (3): 319–326.

121 Kumari, P., Shetty, P., and Rao, S.A. (2014). Corrosion inhibition effect of 4-Hydroxy-N′-[(E)-(1H-indole-2-ylmethylidene)] benzohydrazide on mild steel in hydrochloric acid solution. *International Journal of Corrosion 2014*: 1–11.

122 Hao, H.-R., Zhao, W., Zhang, J. et al. (2015). Synthesis and application of 3-(4-hydroxyphenylimino) indolin-2-one as corrosion inhibitor. *Journal of the Chemical Society of Pakistan 37* (6): 1124–1129.

123 Hidroklorik, A. (2014). Schiff bases derived from isatin as mild steel corrosion inhibitors in 1 M HCl. *Malaysian Journal of Analytical Sciences 18* (3): 507–513.

124 Kubba, R.M., Challoob, D.A.-K., and Hussen, S.M. Combined quantum mechanical and electrochemical study of new isatin derivative as corrosion inhibitor for carbon steel in 3.5% NaCl. *International Journal of Scientific Research* 6: 1656–1669.

125 Ramkumar, S. and Nalini, D. (2015). Correlation between inhibition efficiency and chemical structure of new indolo imidazoline on the corrosion of mild steel in molar HCl with DFT evidences. *Oriental Journal of Chemistry 31* (2): 1057.

126 Sun, G.-X. and Miao, Y.-Q. (2014). Structure, anticorrosion and antibacterial evaluation of 1-(morpholinomethyl) indoline-2, 3-dione. *Asian Journal of Chemistry 26* (22): 7795.

127 Singh, A.K. (2012). Inhibition of mild steel corrosion in hydrochloric acid solution by 3-(4-((Z)-indolin-3-ylideneamino) phenylimino) indolin-2-one. *Industrial and Engineering Chemistry Research 51* (8): 3215–3223.

128 Firdhouse, M.J. and Nalini, D. (2013). Corrosion inhibition of mild steel in acidic media by 5′-Phenyl-2′, 4′-dihydrospiro [indole-3, 3′-pyrazol]-2 (1H)-one. *Journal of Chemistry 2013*: 1–9.

129 Tribak, Z., Kharbach, Y., Haoudi, A. et al. (2016). Study of new 5-Chloro-Isatin derivatives as efficient organic inhibitors of corrosion in 1 M HCl medium: Electrochemical and SEM studies. *Journal of Materials and Environmental Science 7* (6): 2006–2020.

130 Muralisankar, M., Sreedharan, R., Sujith, S. et al. (2017). N (1)-pentyl isatin-N (4)-methyl-N (4)-phenyl thiosemicarbazone (PITSc) as a corrosion inhibitor on mild steel in HCl. *Journal of Alloys and Compounds 695*: 171–182.

131 Al-Azawi, K.F., Mohammed, I.M., Al-Baghdadi, S.B. et al. (2018). Experimental and quantum chemical simulations on the corrosion inhibition of mild steel by 3-((5-(3, 5-dinitrophenyl)-1, 3, 4-thiadiazol-2-yl) imino) indolin-2-one. *Results in Physics 9*: 278–283.

132 Ita, B. (2006). Inhibition of the corrosion of mild steel in hydrochloric acid by isatin and isatin glycine. *Bulletin of the Chemical Society of Ethiopia 20* (2): 253–258.

133 Al-Azawi, K.F. (2018). Corrosion inhibition of mild steel in hydrochloric acid solution by an isatin-aniline compound. *Engineering and Technology Journal 36* (2 Part (B) Engineering): 98–103.

134 Gao, L.D. (2014). Investigation of an Isatin Schiff base as corrosion inhibitor in NaCl solution. *Advanced Materials Research* 1004: 239–242.

135 Yadav, M., Sharma, U., and Yadav, P. (2013). Isatin compounds as corrosion inhibitors for N80 steel in 15% HCl. *Egyptian Journal of Petroleum 22* (3): 335–344.

136 Gupta, N.K., Haque, J., Salghi, R. et al. (2018). Spiro [indoline-3, 4′-pyrano [2, 3-c] pyrazole] derivatives as novel class of green corrosion inhibitors for mild steel in hydrochloric acid medium: theoretical and experimental approach. *Journal of Bio-and Tribo-Corrosion 4* (2): 16.

137 Yadav, M., Sarkar, T., and Purkait, T. (2015). Studies on adsorption and corrosion inhibitive properties of indoline compounds on N80 steel in hydrochloric acid. *Journal of Materials Engineering and Performance 24* (12): 4975–4984.

138 Eddy, N.O., Ita, B.I., Ibisi, N.E., and Ebenso, E.E. (2011). Experimental and quantum chemical studies on the corrosion inhibition potentials of 2-(2-oxoindolin-3-ylideneamino) acetic acid and indoline-2, 3-dione. *International Journal of Electrochemical Science 6*: 1027–1044.

139 Ashhari, S. and Sarabi, A.A. (2015). Indole-3-acetic acid and N-acetyltryptophan corrosion inhibition effects on mild steel in 1 M hydrochloric acid solution. *Surface and Interface Analysis 47* (2): 278–283.

140 Raja, P.B., Fadaeinasab, M., Qureshi, A.K. et al. (2013). Evaluation of green corrosion inhibition by alkaloid extracts of Ochrosia oppositifolia and isoreserpiline against mild steel in 1 M HCl medium. *Industrial and Engineering Chemistry Research 52* (31): 10582–10593.

141 Raja, P.B., Qureshi, A.K., Rahim, A.A. et al. (2013). Neolamarckia cadamba alkaloids as eco-friendly corrosion inhibitors for mild steel in 1 M HCl media. *Corrosion Science 69*: 292–301.

142 Raja, P.B., Qureshi, A.K., Rahim, A.A. et al. (2013). Indole alkaloids of Alstonia angustifolia var. latifolia as green inhibitor for mild steel corrosion in 1 M HCl media. *Journal of Materials Engineering and Performance 22* (4): 1072–1078.

143 Udowo, V., Uwah, I., Daniel, F. et al. (2017). Computational and experimental study of the inhibition effects of purple sweet potato leaves extract on mild steel corrosion in 1 M H_2SO_4. *Journal of Physical Chemistry Biophysics* 7: 1–6.

144 Ugi, B. Effects of nitrogen atoms in vindoline alkaloids as Fe2+ ions inhibitor in corrosion of gray iron in dilute HCl environment: potentiodynamic polarization, gravimetric analysis and SEM. *Journal of Materials and Environmental Science* 11: 1274–1285.

145 Abbas, H.A. (2015). Study the effect of peganum Harmala Seeds extracts to protect iron alloy from corrosion in salt media. *Iraqi Journal of Science 56* (3A): 1836–1843.

146 Lowmunkhong, P., Ungthararak, D., and Sutthivaiyakit, P. (2010). Tryptamine as a corrosion inhibitor of mild steel in hydrochloric acid solution. *Corrosion Science 52* (1): 30–36.

147 Raja, P., Rahim, A., Osman, H., and Awang, K. (2010). Inhibitory effect of kopsia singapurensis extract on the corrosion behavior of mild steel in acid media. *Acta Physico-Chimica Sinica 26* (8): 2171–2176.

148 Quartarone, G., Ronchin, L., Vavasori, A. et al. (2012). Inhibitive action of gramine towards corrosion of mild steel in deaerated 1.0 M hydrochloric acid solutions. *Corrosion Science 64*: 82–89.

149 Khaled, K. (2008). Guanidine derivative as a new corrosion inhibitor for copper in 3% NaCl solution. *Materials Chemistry and Physics 112* (1): 104–111.

150 Xhanari, K., Finšgar, M., Hrnčič, M.K. et al. (2017). Green corrosion inhibitors for aluminium and its alloys: a review. *RSC Advances 7* (44): 27299–27330.

151 Odnevall Wallinder, I. and Leygraf, C. (2017). A critical review on corrosion and runoff from zinc and zinc-based alloys in atmospheric environments. *Corrosion 73* (9): 1060–1077.

152 Feng, Y., Feng, L., Sun, Y., and He, J. (2020). The inhibition mechanism of a new synthesized indole derivative for copper in acidic environment via experimental and theoretical study. *Journal of Materials Research and Technology 9* (1): 584–593.

153 Pyun, S.-I. and Moon, S.-M. (2000). Corrosion mechanism of pure aluminium in aqueous alkaline solution. *Journal of Solid State Electrochemistry 4* (5): 267–272.

154 Madkour, L.H., Kaya, S., and Obot, I.B. (2018). Computational, Monte Carlo simulation and experimental studies of some arylazotriazoles (AATR) and their copper complexes in corrosion inhibition process. *Journal of Molecular Liquids 260*: 351–374.

155 Desai, M.N., Talati, J.D., and Shah, N.K. (2005). Ortho-, meta-, and para-aminophenol-N-salicylidenes as corrosion inhibitors of zinc in sulfuric acid. *Anti-Corrosion Methods and Materials 52* (2): 108–117.

156 Fouda, A., Abd El-Aal, A., and Kandil, A. (2006). The effect of some phthalimide derivatives on corrosion behavior of copper in nitric acid. *Desalination 201* (1-3): 216–223.

157 Mak, A. (2002). Corrosion of steel, aluminum and copper in electrical applications. *General Cable* 770 http://www.stabiloy.com/NR/rdonlyres/E5F38E54-48BF-43C1-9415-865B903605EE/0/CorrosioninElecApplications.pdf (accessed 18 June 2021).

158 Scendo, M., Poddebniak, D., and Malyszko, J. (2003). Indole and 5-chloroindole as inhibitors of anodic dissolution and cathodic deposition of copper in acidic chloride solutions. *Journal of Applied Electrochemistry 33* (3–4): 287–293.

159 Quartarone, G., Moretti, G., Bellomi, T. et al. (1998). Using indole to inhibit copper corrosion in aerated 0.5 M sulfuric acid. *Corrosion 54* (8): 606–618.

160 Quartarone, G., Zingales, A., Bellomi, T. et al. (2005). Corrosion inhibition of copper in aerated 0.5 M sulfuric acid by indole-2-carboxylic acid. *Corrosion 61* (11): 1041–1049.

161 Quartarone, G., Battilana, M., Bonaldo, L., and Tortato, T. (2008). Investigation of the inhibition effect of indole-3-carboxylic acid on the copper corrosion in 0.5 M H_2SO_4. *Corrosion Science 50* (12): 3467–3474.

162 Quartarone, G., Zingales, A., Bellomi, T. et al. (2000). Study of inhibition mechanism and efficiency of indole-5-carboxylic acid on corrosion of copper in aerated 0·5M H_2SO_4. *British Corrosion Journal 35* (4): 304–310.

163 Tüken, T., Yazıcı, B., and Erbil, M. (2006). The use of polyindole for prevention of copper corrosion. *Surface and Coatings Technology 200* (16-17): 4802–4809.

164 Ebadi, M., Basirun, W.J., Khaledi, H., and Ali, H.M. (2012). Corrosion inhibition properties of pyrazolylindolenine compounds on copper surface in acidic media. *Chemistry Central Journal 6* (1): 163.

165 Quartarone, G., Bellomi, T., and Zingales, A. (2003). Inhibition of copper corrosion by isatin in aerated 0.5 M H_2SO_4. *Corrosion science 45* (4): 715–733.

166 Feng, L., Ren, X., Feng, Y. et al. (2020). Self-assembly of new O-and S-heterocycle-based protective layers for copper in acid solution. *Physical Chemistry Chemical Physics 22* (8): 4592–4601.

167 Feng, Y., Feng, L., Wang, Z., and Zhang, X. (2020). Surface analysis of 4-(bis (5-bromo-1H-indol-3-yl) methyl) phenol adsorbed on copper by spectroscopic experiments. *Spectrochimica Acta Part A: Molecular and Biomolecular Spectroscopy 228*: 117752.

168 Ashassi-Sorkhabi, H., Ghasemi, Z., and Seifzadeh, D. (2005). The inhibition effect of some amino acids towards the corrosion of aluminum in 1 M HCl+ 1 M H_2SO_4 solution. *Applied Surface Science 249* (1-4): 408–418.

169 El-Shafei, A., Abd El-Maksoud, S., and Fouda, A. (2004). The role of indole and its derivatives in the pitting corrosion of Al in neutral chloride solution. *Corrosion Science 46* (3): 579–590.

170 Yuan, Q., Cheng, R., Zou, S. et al. (2020). Isatin thiosemicarbazone derivatives as inhibitors against corrosion of AA6060 aluminium alloy in acidic chloride medium: substituent effects. *Journal of Materials Research and Technology 9* (5): 11935–11947.

171 Djemoui, A., Souli, L., Djemoui, D. et al. (2017). Alkaloids Extract from Peganum harmala Plant as Corrosion Inhibitor of 6063 Aluminium Alloy in 1 M Hydrochloric Acid Medium. *Journal of Chemical and Pharmaceutical Research 9* (3): 311–318.

172 Obot, I. and Obi-Egbedi, N. (2009). Ipomoea involcrata as an ecofriendly inhibitor for aluminium in alkaline medium. *Portugaliae Electrochimica Acta 27* (4): 517–524.

173 Fouda, A., Rashwan, S., Darwish, M., and Arman, N. (2018). Corrosion inhibition of Zn in a 0.5 M HCl solution by Ailanthus altissima extract. *Portugaliae Electrochimica Acta 36* (5): 309–323.

174 Nathiya, R., Perumal, S., Moorthy, M. et al. (2020). Synthesis, characterization and inhibition performance of schiff bases for aluminium corrosion in 1M H_2SO_4 solution. *Journal of Bio-and Tribo-Corrosion 6* (1): 5.

175 Fouda, A., Shalabi, K., and Elmogazy, H. (2014). Corrosion inhibition of α-brass in HNO_3 by indole and 2-oxyindole. *Journal of Materials and Environmental Science 5*: 1691.

176 Fouda, A. and Mahfouz, H. (2009). Inhibition of corrosion of α-brass (Cu-Zn, 67/33) in HNO_3 solutions by some arylazo indole derivatives. *Journal of the Chilean Chemical Society 54* (3): 302–308.

177 Refaey, S.A., Abd El Malak, A.M., Taha, F., and Abdel-Fatah, H.T. (2008). Corrosion and inhibition of cu-zn alloys in acidic medium by using isatin. *International Journal of Electrochemical Science 3*: 167–176.

178 Moretti, G. and Guidi, F. (2002). Tryptophan as copper corrosion inhibitor in 0.5 M aerated sulfuric acid. *Corrosion Science 44* (9): 1995–2011.

179 Hassan, H.M., Al-Rashdi, A., Attia, A., and Eldesoky, A. Dry and wet lab studies for some indole derivatives as possible corrosion inhibitors for copper in 1M HNO_3. *International Journal of Scientific Engineering and Research* 6: 509–519.

180 Li, X., Xiang, B., Zuo, X.-L. et al. (2011). Inhibition of tryptophan on AA 2024 in chloride-containing solutions. *Journal of Materials Engineering and Performance 20* (2): 265–270.

181 Fouda, A., Tawfik, H., Abdallah, N., and Ahmd, A. (2013). Corrosion inhibition of nickel in HCl solution by some indole derivatives. *International Journal of Electrochemical Science 8* (3): 3390–3405.

182 Al-Amiery, A.A., Ahmed, M.H.O., Abdullah, T.A. et al. (2018). Electrochemical studies of novel corrosion inhibitor for mild steel in 1 M hydrochloric acid. *Results Physics 9*: 978–981.

183 Chauhan, D.S., Quraishi, M.A., Wan Nik, W.B., and Srivastava, V. (2020). Triazines as a potential class of corrosion inhibitors: present scenario, challenges and future perspectives. *Journal of Molecular Liquids* 321: 114747.

184 Palaniappan, N., Cole, I., Caballero-Briones, F. et al. (2020). Experimental and DFT studies on the ultrasonic energy-assisted extraction of the phytochemicals of Catharanthus roseus as green corrosion inhibitors for mild steel in NaCl medium. *RSC Advances 10* (9): 5399–5411.

185 Al-Fakih, A.M., Aziz, M., Abdallah, H.H. et al. (2016). Corrosion inhibition of Q235A steel in acid medium using isatin derivatives: A QSAR study. *Malaysian Journal of Analytical Sciences 20* (3): 484–490.

186 Montemor, M. (2016). Fostering green inhibitors for corrosion prevention. In: *Active Protective Coatings* (eds. A.E. Hughes, J.M. Mol, M.L. Zheludkevich and R.G. Buchheit), 107–137. Springer.

10

Environmentally Sustainable Corrosion Inhibitors in Oil and Gas Industry

M. A. Quraishi[1] and Dheeraj Singh Chauhan[2,3]

[1] *Interdisciplinary Research Center of Advanced Materials, King Fahd University of Petroleum and Minerals, Dhahran, Saudi Arabia*
[2] *Center of Research Excellence in Corrosion, Research Institute, King Fahd University of Petroleum and Minerals, Dhahran, Saudi Arabia*
[3] *Modern National Chemicals, Second Industrial City, Dammam, Saudi Arabia*

10.1 Introduction

The oil and gas industry reports severe corrosion issues during the upstream, midstream, and downstream processes. A major contribution of the overall corrosion damage comes from the corrosion of steel structures and during the oil-well acidizing processes [1]. The corrosion damage results in a huge economic impact due to the decrease in the life span of the metallic structures and can potentially cause the structural failure [2]. In addition, the crude oil may leak due to localized corrosion of the steel structures leading to environmental pollution and fire hazards. Despite the low corrosion resistance, the low carbon steel and other alloy steels are often employed as the preferred material for pipeline structures due to their ease of availability, good mechanical properties, and cost-effectiveness [3]. Research has shown that around 10–30% of the maintenance cost goes to the control and minimizing the corrosion damage.

Organic compounds as corrosion inhibitors are commonly used to counter the corrosion [1, 4]. Most of the aqueous organic corrosion inhibitors act by adsorption at the metal/solution interface and develop a thin protective film that can provide protection from the surrounding aggressive electrolyte. Most of the effective organic corrosion inhibitors are from the category of imidazoles, fatty acid imidazolines, triazoles, benzimidazoles, benzothiazoles, pyridines, pyrimidines,

Organic Corrosion Inhibitors: Synthesis, Characterization, Mechanism, and Applications, First Edition. Edited by Chandrabhan Verma, Chaudhery Mustansar Hussain, and Eno E. Ebenso.

etc. [5]. These inhibitors are preferred due to the presence of the heteroatoms (N, S, O), and the heterocycles, which effectively adsorb over the metal surface and provide significant corrosion inhibition efficiencies. However, a majority of these compounds are toxic and can cause harmful discharges to the aquatic environments.

In the recent years, the environmental regulations have put forward strict guidelines for the use of eco-friendly molecules as inhibitors. Already sustainable practices are being carried out in the agriculture, forestry and the energy areas. Besides, several parameters have been proposed, known as the "Green Chemistry Metrics" for the synthesis of corrosion inhibitors from the point of view of environmental sustainability [6]. Considering these regulations, a number of researches have been devoted to exploration and evaluation of environmentally safer alternatives as corrosion inhibitors [7]. Some of these include the ionic liquids [8], natural extracts [9], carbohydrate-based polymers [10], pharmaceutically active compounds, etc [11, 12]. The very recent classes include the chemically modified biopolymers [13–15] and the modified nanomaterials [16–18] as corrosion inhibitors. This chapter attempts to outline the major corrosion issues commonly encountered in the oil and gas industry, emphasizing the corrosion of steel during oil-well acidizing. Several modern chemical methods for the preparation of corrosion inhibitors are described along with the major types of environmentally sustainable corrosion inhibitors that have been applied.

10.2 Corrosion in the Oil–Gas Industry

10.2.1 An Overview of Corrosion

Corrosion can be described as a destructive attack on a material by its surrounding environment leading to reversion of the metallic material to its naturally occurring ore form. It is a naturally occurring hazard associated with the production and transportation of oil and gas [4, 19]. The annual global cost of corrosion taking place in the oil–gas industry is around US $1,372 billion [20]. The major corrosion losses occur in the production, processing, and in the pipeline systems. The various kinds of corrosion occurring in the oil–gas sector can be classified into (i) sweet corrosion, (ii) oxygen corrosion, (iii) sour corrosion, (iv) crevice corrosion, (v) galvanic corrosion, (vi) microbiologically induced corrosion, (vii) erosion corrosion, and (viii) stress corrosion cracking. In the present chapter, we have provided the detailed account of the available literature on environmentally sustainable corrosion inhibitors for corrosion occurring during the industrial process of oil-well acidizing.

10.2.2 Corrosion of Steel Structures During Acidizing Treatment

The process of acidizing is undertaken in the oil wells to stimulate the flow of oil for recovery. The acid formulation is pumped under high pressure in the pipelines via the borehole into the pores of the rock formation. The concentrated acids undergo chemical reaction with the rock formations and cause their dissolution to enlarge the flow channels in the wellbore. This process requires the use of highly concentrated mineral acids in the concentration of 5–28% [21], which can lead to an extremely corrosive environment against the tubular steel structures [22–25]. The HCl is the major acid used on account of its effectiveness in dissolving the soluble metal chlorides [26]. The most commonly used acids include HF, HCl, formic, and acetic acids. H_2SO_4, H_3PO_4, sulfamic, and chloroacetic acids are also applied for acidizing [21, 26–29]. H_2SO_4 finds little application in the oil-well acidizing, because it can form insoluble sulfate by-products with calcium and also that the acid could lead to modification of some of the oils to sludges [30, 31]. The acid solutions used for the acidizing process much be used along with the corrosion inhibitors that can bring about a mitigation of the corrosion of the well equipment. The choice of an appropriate corrosion inhibitor is dependent upon the type of the acid used.

10.2.3 Limitations of the Existing Oil and Gas Corrosion Inhibitors

One of the most effective strategies to control the menace of the corrosion is the use of inhibitors. Several corrosion inhibitors from the classes of imidazolines, amides, alkyl pyridines, quaternary amines, acetylenic alcohols, etc. have been reported [5, 32]. However, the major issue with these inhibitors is the high cost, the involvement of the multiple-step synthetic protocol, toxicity, degradation/decomposition of the corrosion inhibitor compounds to unwanted side products, leading to a lowering in the corrosion inhibitor effectiveness. Therefore, research is being focused in the recent years on the novel environmentally benign corrosion inhibitors, having low synthesis cost and effective corrosion inhibition performance.

10.3 Review of Literature on Environmentally Sustainable Corrosion Inhibitors

10.3.1 Plant Extracts

A major class of the environmentally sustainable corrosion inhibitors is the plant extracts. The plants are naturally occurring and contain a plenty of phytochemical constituents. Therefore, these inhibitors exhibit a large tendency to undergo adsorption on the metal surfaces. Although, some criticism has been made on the long-term storage stability of the plant extracts, the plant extract-based corrosion inhibitors have found a wide application on a variety of metals and alloys in diverse corrosive media. Ulaeto et al. investigated the acid extract of *Eichhornia crassipes* (water hyacinth) for corrosion of mild steel in 5M HCl medium [33]. The leaf and the root extracts were tested for inhibition performance, wherein the leaf extract showed better performance. The major active antioxidant present in the plant was studied via DFT and the molecular dynamics (MD) simulations, which revealed a flat orientation on the metal surface allowing the interaction of a large area. The group of Li et al. investigated the extracts of the *Dendrocalmus sinicus Chia* (bamboo leaves: DSCLE) for corrosion of cold-rolled steel in HCl (1–5M), and H_2SO_4 (0.5–5M) solutions [34]. Greater efficiency was observed in HCl medium compared to that observed in H_2SO_4. The efficiency increased with inhibitor dose and decreased with rise in acid concentration, and temperature. At an HCl concentration of 5M, an efficiency of <60% was noted. The same group, in a similar manner, analyzed the effect of the Ginkgo leaves extract on cold-rolled steel in HCl and H_2SO_4 at different concentrations and varying temperatures [35]. The extract showed better performance in HCl and revealed a mixed type of behavior. The efficiency decreased with rise in the HCl concentration, and about 60% inhibition was obtained at 5M concentration. The major constituents of the extract were flavonoids, ginkgo ides, and amino acids. Alcoholic extract of plant leaves (TVE-1B) were analyzed for N80 steel in 15% HCl environment along with synergism with formaldehyde [36]. A mixture of 0.8% formaldehyde with 0.2% extract produced an efficiency of 76.3% at room temperature, which lowered to 52.4% at 333K. A comparison with commercial corrosion inhibitors was also carried out, and comparable performance was observed. Crude aqueous and ethanolic extracts of date palm (seeds and leaves) were analyzed for X60 carbon steel in 15% HCl by Umoren et al. [37]. At a temperature of 60 °C, and optimum inhibitor dose of 2000 mg/l, the efficiencies were DPLAE (73.6%; date palm leaves aqueous extract) > DPLEE (62.5%; date palm leaves ethanolic extract) > DPSAE (59.9%; date palm seeds aqueous extract) > DPSEE (55.9%, date palm seeds ethanolic extract). A mixed-type performance by all the extracts was observed via PDP studies. Figure 10.1 displays some of the environmentally sustainable corrosion inhibitors developed for oil-well acidizing.

10.3.2 Environmentally Benign Heterocycles

Commonly used inhibitors in the oil–gas industry are imidazolines, amides, amines, etc. as discussed above. To circumvent the issues associated with these inhibitor types, a number of other heterocyclic molecules have been explored. The

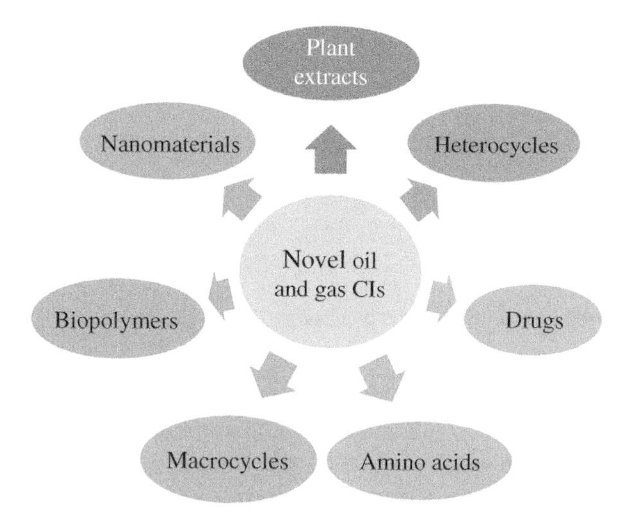

Figure 10.1 Environmentally sustainable inhibitors for oil–gas industry.

heterocyclic molecules refer to a class of organic cyclic molecules having one or more of the carbon atoms replaced by N, S, O, P, etc. atoms. The major classes of the heterocyclic corrosion inhibitors are pyridines, pyrimidines, quinolines, imidazoles, pyrazoles, triazoles, benzotriazoles, benzimidazoles, oxadiazoles, etc. [5, 38]. Due to the heteroatoms, the chemical interaction with the metallic substrate can be facilitated using the sharing of lone-pair electrons. In addition, the heterocyclic molecules consist of phenyl rings, π-bonds, and polar functional groups that can facilitate the adsorption. This section presents an overview of the environmentally benign heterocyclic compounds as corrosion inhibitors.

Singh et al. prepared two acridinone derivatives, namely, ACD-1 and ACD-2, and evaluated them for N80 steel corrosion inhibition in 15% HCl [39]. A high performance with 97.3% efficiency was observed from ACD-2 via the weight loss studies at 600 mg/l. The cathodic nature of both inhibitors was clearly observed. The MD studies indicated the chemical interaction of the metal surface and the inhibitor molecules via the radial distribution function (RDF) analysis. Two pyrazole derivatives PZ-1 and PZ-2 were evaluated for N80 steel in 15% HCl [40]. Their synthesis was carried out using environmentally benign method of ultrasonic irradiation. Mixed nature of the inhibitors was revealed with a 98.4% efficiency (PZ-1). The DFT-based computational analysis revealed a lower ΔE for PZ-1, with a strong binding and interaction ability compared to PZ-2. An environment-friendly corrosion inhibitor, namely, PCP, was evaluated for N80 steel in 15% HCl, and at 400 mg/l, an efficiency of 98.4% was obtained [41]. The inhibitor was prepared via a four-component reaction in aqueous medium at 30 °C, within

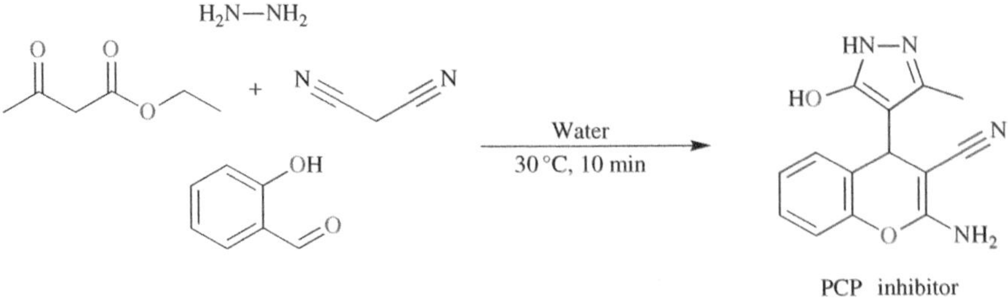

$H_2N{-}NH_2$

Water

30 °C, 10 min

PCP inhibitor

Scheme 10.1 Synthesis of chromenopyrazole derivative. *Source:* Ref. [41].

10 minutes (Scheme 10.1). The rise in the inhibitor dose produced an increment in the water contact angle from 29.31 to 112.99°, indicating an improvement in the surface hydrophobicity of the steel surface in the inhibited condition. Chromeno pyrimidine derivatives were prepared using a one-pot multicomponent reaction protocol and evaluated for N80 steel in 15% HCl medium [42]. A high performance with 96.4% efficiency was attained at a modest dose of 200 mg/l. The inhibitors revealed a mixed nature, and their adsorption was verified using AFM and XPS analyses.

The group of M. Yadav et al. have significantly contributed on environmentally benign heterocycles as corrosion inhibitors. They synthesized three benzimidazole derivatives, namely, PzMBP, MBP, and PMBP, and evaluated them on N80 steel in 15% HCl medium, using gravimetric tests, PDP and EIS methods [43]. High efficiencies of 96.3, 94.4 and 92.9% were obtained for the three studied inhibitors, at 200 mg/l, at a temperature of 303K. PDP measurements revealed a mixed-type nature of the inhibitors, and their adsorption on the metal surface obeyed the Langmuir isotherm. In a subsequent study, three more benzimidazole derivatives were analyzed for mild steel in 15% HCl by the same group [44]. The inhibitors showed the mixed nature, and their adsorption on metal surface followed the Langmuir isotherm. Quantum chemical investigations carried out using semi-empirical AM1 method supported the experimental findings.

10.3.3 Pharmaceutical Products

Drug molecules as corrosion inhibitors has gained considerable attention since the last decade [12, 45]. The major significance of the drugs arises due to the presence of heteroatoms and heterocycles in their structural framework, which provide plenty of adsorption sites. A huge plus point associated with the drugs is their appreciable aqueous solubility, because the drugs are designed to easy absorption in the body. Recently, Metformin, a medication used for type-2 diabetes, was used for steel surface in 15% HCl [46]. The performance rose with

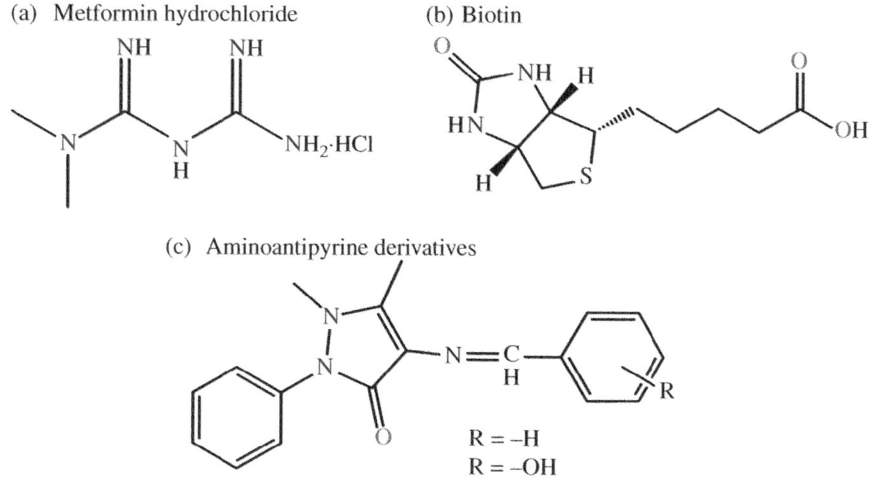

(a) Metformin hydrochloride (b) Biotin

(c) Aminoantipyrine derivatives

R = –H
R = –OH

Figure 10.2 Structures of some drugs and drug derivatives as corrosion inhibitors. *Source:* Refs. [46–48].

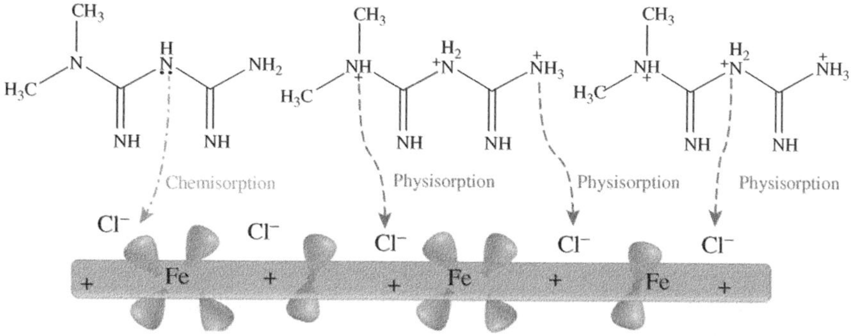

Figure 10.3 Adsorption and corrosion inhibition of Metformin drug on steel surface. *Source:* Ref. [46].

temperature elevation and the addition of KI as synergistic agent. A high efficiency of >92% was achieved at 60 °C. The structure of the drug is shown in Figure 10.2, and a plausible mechanism of adsorption is displayed in Figure 10.3. It can be shown that the drug molecule undergoes physical (via electrostatic attraction), as well as chemical adsorption (sharing lone-pair electrons) on the steel surface.

In another study, Biotin drug was evaluated as an inhibitor for mild steel in 15% HCl [47]. Weight loss tests revealed an efficiency of 95.3% at 500 mg/l. Inhibitor adsorption showed a mixed nature as revealed by the PDP study. The water contact angles measured at the inhibitor surface increased from 13.3 to 125.3° with increase in the inhibitor dosage showing a betterment in the hydrophobicity of the metal surface due to inhibitor adsorption. The measurements of scanning electrochemical microscopy (SECM) revealed a lowering in the corrosion current densities, with the inhibitor, supporting adsorption. The same group further developed a Schiff base via chemical modification of aminoantipyrine using an aromatic aldehyde and room temperature stirring [48]. The pKa analysis revealed that the major site amenable to protonation was the nitrogen atom of the imine linkage (—N=CH). The inhibitor adsorption on the metal surface was in accordance with the Langmuir isotherm, and a charge transfer control of the electrochemical process was revealed in the EIS measurements.

10.3.4 Amino Acids and Derivatives

The amino acids come into the category of biomolecules, and therefore, there is no question about their environmental sustainability. In addition, these molecules contain both the amino ($-NH_2$) and the carboxylic acid (–COOH) groups in their structural framework, which can provide them with considerable solubility. A number of amino acids and their chemically modified derivatives have been analyzed as inhibitors in acidic and neutral media [49]. Some of these are also tested for the acidizing conditions, which are described in this section.

Yadav et al. evaluated two amino acid derivatives, namely, Inh I and Inh II, using weight loss and electrochemical studies [50]. The presence of an additional phenyl group resulted in a higher efficiency of 89.36% at 200 mg/l for Inh I, compared to 83.00% for Inh II. A mixed type of nature was revealed for both the corrosion inhibitors. The same group also analyzed two benzimidazole derivatives of glycine and phenylalanine and studied in 15% HCl the obtained products for corrosion of N80 steel [51]. The potential of zero charge determined from the EIS study revealed that the steel surface assumed a positive charge in the corrosive solution, indicating the adsorption of the Cl^- ions, which facilitated the electrostatic attraction of the protonated form of amino acid derivatives. High efficiencies of 97.4, and 94.1% were obtained for the two amino acids, at a concentration of 31.71×10^{-5} M. N-acetyl cysteine (NAC)-based inhibitor formulation was studied for steels in 15% HCl in the temperature range of 30–90 °C for J55, mild, and N80 steels [52]. NAC-based inhibitor formulation produced inhibition efficiencies up to 91% at 90 °C, supporting its applicability as an effective inhibitor for oil-well acidizing. Two new inhibitors ADA and ADHP were prepared and studied for N80 steel in 15% HCl environment using experimental methods [53]. High efficiencies of 95.39% was obtained for ADHP, and 90.36% for ADA, at 25 °C at concentrations of 150 mg/l. FTIR analysis of the film deposited on the steel surface supported the inhibitor adsorption and corrosion inhibition. Two amino acid derivatives OPEM and OPEA were synthesized from glycine and cysteine and studied for mild steel corrosion in 15% HCl [54]. High efficiencies of 97.5, and 95.8% were obtained for OPEM and OPEA at 200 mg/l. OPEM and OPEA acted as mixed type of inhibitors. Subsequently, two more amino acid derivatives OYAA and OYPA were prepared from glycine and phenylalanine, and studied for mild steel in 15% HCl environment [55]. The inhibitors were evaluated using electrochemical and quantum chemical methods, and high efficiencies of 97.9% (OYPA) and 93.2% (OYAA) were obtained at a modest dose of 50 mg/l. PDP studies revealed a mixed type of nature of the two inhibitor molecules, and the inhibitor adsorption on the metallic substrate obeyed Langmuir isotherm.

10.3.5 Macrocyclic Compounds

A macrocyclic compound refers to a large ring structures such as that exists in the porphyrins, crown ethers, phthalocyanines, and so on. The large molecular structure of these molecules enables these with efficient coverage of the targeted metallic substrate [56–60]. A few of these reports on the application in the acidizing conditions are discussed herein. Three macrocyclic compounds, namely, PTAT, PTAB, and POAB (Figure 10.4), were analyzed using weight loss studies in the absence and the presence of KI at 40 °C for 3 hours in 5N HCl for mild steel [61]. At a dose of 1000 mg/l inhibitor along with 0.1% KI, the efficiencies were 80.6, 66.0, and 60.0%, respectively. The inhibitor adsorption was in agreement with the Temkin isotherm, and the Tafel investigation showed a mixed-type nature of the inhibitors. Tetramethyl-dithia-octaazacyclotetradecahexaene was evaluated for mild steel in 5N HCl [62]. At 1000 mg/l dose, the efficiency was 84.72%, whereas at 2000 mg/l dose, inhibition efficiency of 94.57% was obtained. An introduction of 0.05% of KI improved the efficiency to 90.8% at 1000 mg/l. Temkin adsorption was obvious in the inhibitor adsorption with a mixed type of behavior was revealed via the PDP measurements.

10.3.6 Chemically Modified Biopolymers

In the literature, several reports are available on starch, chitosan, pectin, Gum Arabic, cellulose, etc. as corrosion inhibitors [10]. A beneficial aspect of the higher molecular weight molecules is the large coverage of the metallic surface. However, the large molecular weight can also cause an issue with the poor solubility of the resulting polymer in the aqueous environment. A useful method in this case would be the chemical functionalization of the existing biopolymers using suitable chemical agents to improve the adsorption and inhibition behavior. Xanthan gum (XG) and its graft co-polymer with polyacrylamide was evaluated for mild steel in 15% HCl using gravimetric and electrochemical measurements [63]. A

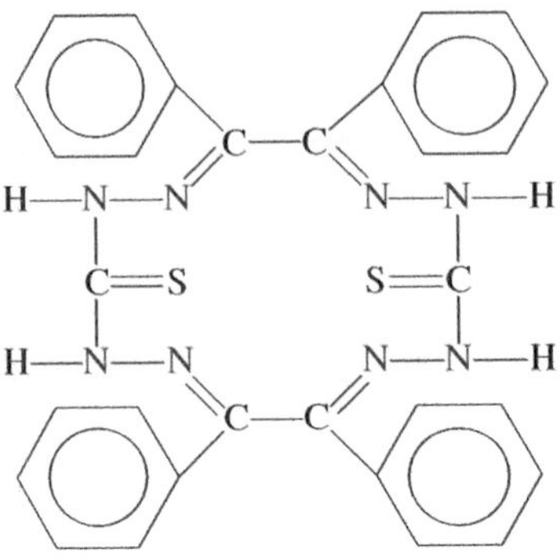

Tetraphenyl-dithia-octaaza-cyclotetradecanehexaene (PTAT)

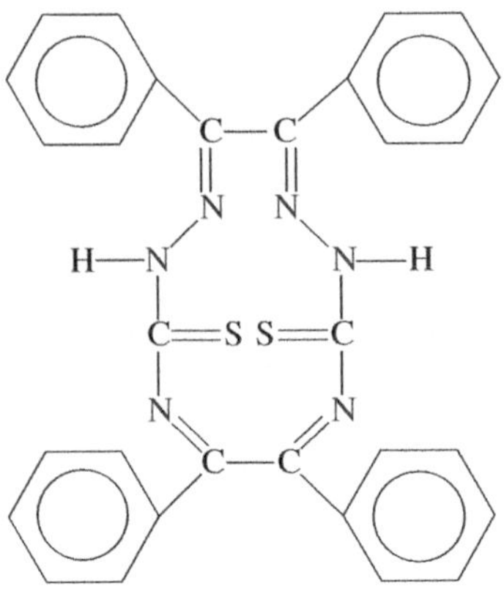

Tetraphenyl-dithia-hexaaza-cyclobidecanehexaene (PTAB)

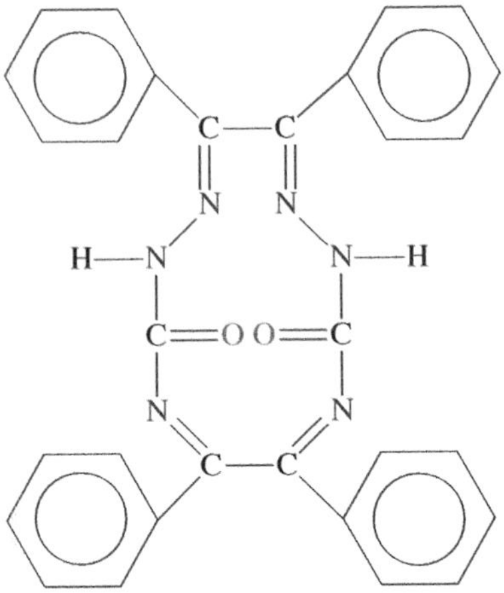

Tetraphenyl-dioxo-hexaaza-cyclobidecanehexaene (POAB)

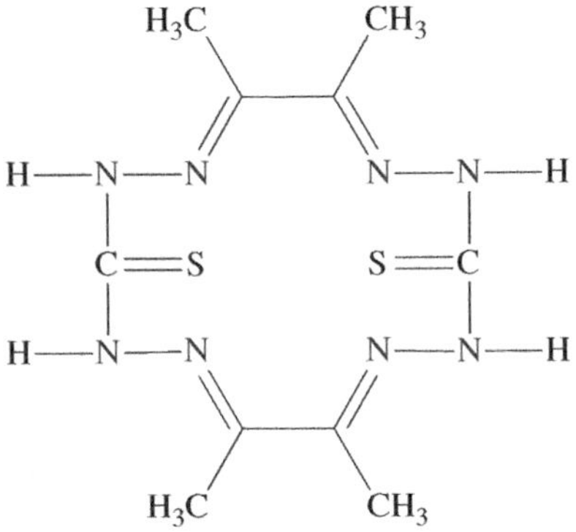

Tetramethyl-dithia-octaaza-cyclotetradecahexaene (MTAT)

Figure 10.4 Structure of some macrocyclic corrosion inhibitors.

high efficiency of 93.18% at 0.5 g/l was obtained at 298 K, which declined to 61.30% at 333 K for the copolymer. The inhibitor molecules exhibited a mixed type of nature via the PDP studies and revealed single capacitive semicircles in the EIS study. The same group of authors performed the chemical functionalization of guar gum using 3-chloro-2-hydroxypropyl trimethylammonium chloride and evaluated for mild steel in 15% HCl [64]. Corrosion testing was carried out using gravimetric, electrochemical, and surface studies. A high efficiency of 95.3% was obtained for the modified biopolymer at a dose of 0.5 g/l at 298 K, which declined to 69.39% with increase in temperature up to 333 K.

Several research reports have appeared describing the corrosion inhibition behavior of the chemically modified chitosan as described below [13, 14, 65–70]. A most commonly adopted strategy for chemical modification of chitosan is the Schiff base formation. The pyranose ring present in the chitosan contain free $-NH_2$ functional groups attached. These amine groups can be reacted with organic compounds containing free –CHO groups in a single-step procedure to introduce the imine (–N=CH) linkage in the chitosan polymer. The imine (–N=CH) linkage present in the Schiff base shows effective capability to undergo adsorption on the metal surfaces. Accordingly, several Schiff bases have been studied as efficient inhibitors for metals and alloys. The choice of Schiff bases arises because of their low toxicity, ease of synthesis, and due to the presence of the imine linkage, which can improve the film formation and corrosion inhibition [65, 71–74]. The Schiff bases of chitosan have been reported for a number of biological activities. However, it is noteworthy to mention that the use of chitosan Schiff bases as corrosion inhibitors is a recent concept. Recently, the group of Quraishi et al. have contributed on the application of the chitosan Schiff bases prepared using cinnamaldehyde [13], piperonal [14], and vanillin [70] as inhibitors for carbon steel in 15% HCl. The Schiff bases were prepared via a single-step method using microwave irradiation procedure (Scheme 10.2). The synergistic corrosion inhibition in case of vanillin was observed. The detailed analysis via the weight loss measurements, electrochemical studies, and surface analysis supported efficient adsorption of the synthesized molecules on the metal surfaces. DFT-based quantum chemical calculations, and the Monte Carlo simulations indicated that the protonation of the N atom involved in the imine (–CH=N) linkage was responsible for the physical adsorption with the metal surface.

10.3.7 Chemically Modified Nanomaterials

Nanomaterials have been applied for the acidizing corrosion inhibitor development in the past few years by the group of Umoren et al. in the form of silver nanoparticles (AgNp). The group has developed several biopolymer-AgNp nanocomposites and investigated them in 15% HCl environment. Besides, several studies on synergism in corrosion inhibition have also been reported. The major reports on the application of chitosan-AgNp in 15% HCl [75], carboxymethylcellulose-AgNp [76] in 15% HCl, and recently on dextran-AgNp [77] in 15% HCl. Another nanomaterial is the chemically modified graphene oxide (GO). Graphene shows poor aqueous solubility; although the solubility of GO is better, still the presence of a number of oxygen containing surface functional groups can be used to make it better dispersed. In this context, Quraishi et al. have

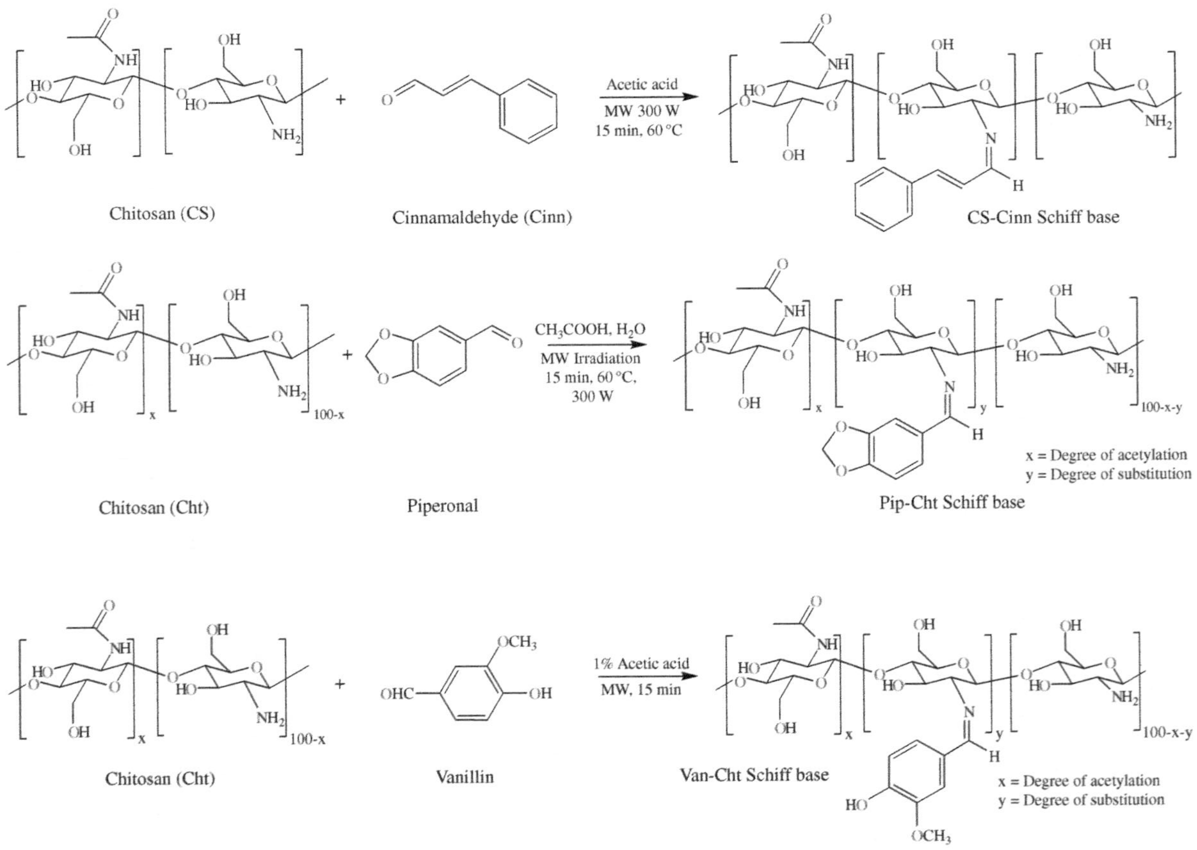

Chitosan (CS) + Cinnamaldehyde (Cinn) → (Acetic acid; MW 300 W; 15 min, 60 °C) → CS-Cinn Schiff base

Chitosan (Cht) + Piperonal → (CH_3COOH, H_2O; MW Irradiation; 15 min, 60 °C, 300 W) → Pip-Cht Schiff base

x = Degree of acetylation
y = Degree of substitution

Chitosan (Cht) + Vanillin → (1% Acetic acid; MW, 15 min) → Van-Cht Schiff base

x = Degree of acetylation
y = Degree of substitution

Scheme 10.2 Synthesis of chitosan Schiff bases for corrosion inhibition of carbon steel in 15% HCl. *Source:* Refs. [13, 14, 70].

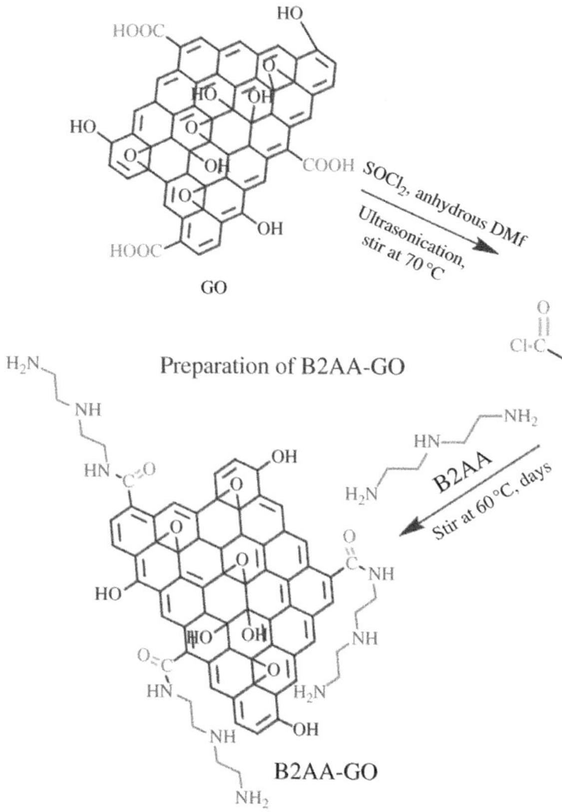

Scheme 10.3 Synthesis of bis(2-aminoethyl)amine-modified GO (B2AA-GO). *Source:* Ref. [78].

made the pioneering contribution on the application of the chemically modified GO as corrosion inhibitor. In the past couple of years, they have also analyzed bis(2-aminoethyl)amine-modified GO (Scheme 10.3) [78] and polyethyleneimine-GO (Scheme 10.4, 79] as inhibitors of carbon steel corrosion in 15% HCl. The inhibitors showed excellent efficiencies up to 65 °C, which were improved further by incorporation of KI synergistic agent to develop commercially useful corrosion inhibitor formulation.

10.4 Conclusions and Outlook

This chapter presents a collection of the environmentally benign corrosion inhibitors used in the oil and gas industries to counter the corrosion during the acidizing treatment. Different types of corrosion inhibitors developed from the classes

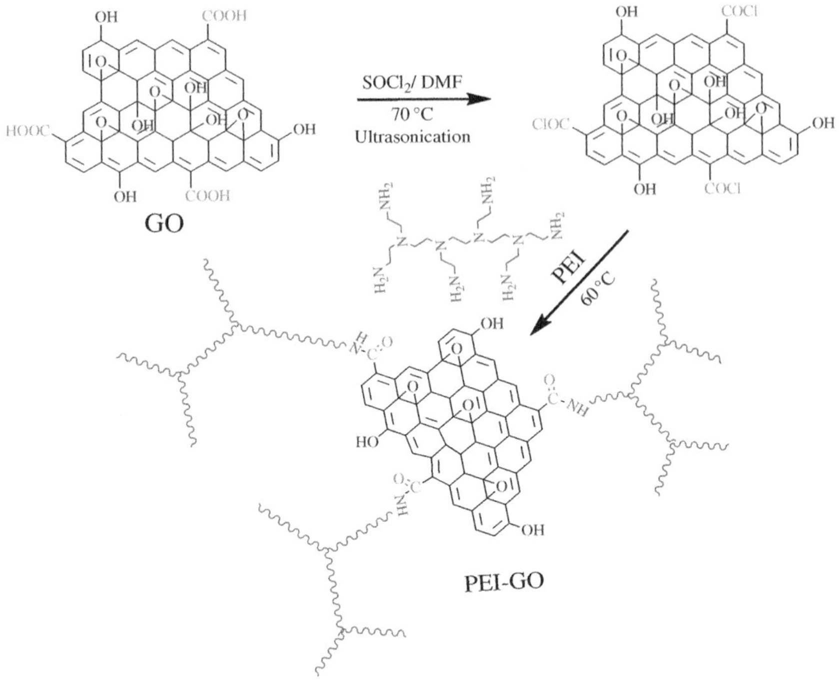

Scheme 10.4 Synthesis of polyethyleneimine-modified GO (PEI-GO). *Source:* Ref. [79].

of plant extracts, heterocyclic compounds, amino acids, drugs, biopolymers, macrocyclic compounds, and nanomaterials are presented and discussed herein. Most of the available literature in the case of environmentally benign acidizing inhibitors is reported on the steel alloys in 15% HCl. Some of the researchers have explored the effect of change in the molecular structure on the inhibition efficiency by evaluating a series of inhibitors. Several studies on the synergistic corrosion inhibition using KI are also available.

It is to be noted that considering the use of large amounts of acids, complications in preparation and handling of the concentrated acid solutions (often at high temperatures) and due to the requirement of a large amount corrosion inhibitor, the studies on acidizing are somewhat limited, especially compared to that reported on 1M HCl, for acid-pickling application. Even among these studies, there is a limited amount of literature available on the environmentally benign corrosion inhibitors. However, there is need to investigate more in this area, especially considering the fact that the corrosion of steels during oil-well acidizing is a major concern in the industry. Therefore, we hope that the collection of literature in this chapter could pave the way for more research prospects in this practically important area.

References

1 Ansari, K.R., Chauhan, D.S., Singh, A. et al. (2020). Corrosion inhibitors for acidizing process in oil and gas sectors. In: *Corrosion Inhibitors in the Oil and Gas Industry* (eds. V.S. Saji and S.A. Umoren), 153–176. Wiley-VCH Verlag GmbH & Co. KGaA.

2 Sastri, V.S. (2012). *Green Corrosion Inhibitors: Theory and Practice*. Wiley.

3 Trabanelli, G., Zucchi, F., Brunoro, G., and Rocchini, G. (1992). Corrosion inhibition of carbon and low alloy steels in hot hydrochloric acid solutions. *British Corrosion Journal* 27: 213–217.

4 Askari, M., Aliofkhazraei, M., Ghaffari, S., and Hajizadeh, A. (2018). Film former corrosion inhibitors for oil and gas pipelines-A technical review. *Journal of Natural Gas Science and Engineering* 58: 92–114.

5 Quraishi, M.A., Chauhan, D.S., and Saji, V.S. (2020). *Heterocyclic Organic Corrosion Inhibitors: Principles and Applications*. Amsterdam: Elsevier Inc.

6 Constable, D.J., Curzons, A.D., and Cunningham, V.L. (2002). Metrics to 'green'chemistry – which are the best? *Green Chemistry* 4: 521–527.

7 Haque, J., Srivastava, V., Chauhan, D.S. et al. (2020). Electrochemical and surface studies on chemically modified glucose derivatives as environmentally benign corrosion inhibitors. *Sustainable Chemistry and Pharmacy* 16: 100260.

8 Verma, C., Ebenso, E.E., and Quraishi, M.A. (2017). Ionic liquids as green and sustainable corrosion inhibitors for metals and alloys: an overview. *Journal of Molecular Liquids* 233: 403–414.

9 Verma, C., Ebenso, E.E., Bahadur, I., and Quraishi, M.A. (2018). An overview on plant extracts as environmental sustainable and green corrosion inhibitors for metals and alloys in aggressive corrosive media. *Journal of Molecular Liquids* 266: 577–590.

10 Umoren, S.A. and Eduok, U.M. (2016). Application of carbohydrate polymers as corrosion inhibitors for metal substrates in different media: a review. *Carbohydrate Polymers* 140: 314–341.

11 Chauhan, D.S., Sorour, A.A., and Quraishi, M.A. (2016). An overview of expired drugs as novel corrosion inhibitors for metals and alloys. *International Journal of Chemistry and Pharmaceutical Sciences* 4: 680–691.

12 Verma, C., Chauhan, D.S., and Quraishi, M.A. (2017). Drugs as environmentally benign corrosion inhibitors for ferrous and nonferrous materials in acid environment: an overview. *Journal of Materials and Environmental Science* 8: 4040–4051.

13 Chauhan, D.S., Mazumder, M.J., Quraishi, M.A., and Ansari, K. (2020). Chitosan-cinnamaldehyde Schiff base: a bioinspired macromolecule as corrosion inhibitor for oil and gas industry. *International Journal of Biological Macromolecules* 158: 127–138.

14 Chauhan, D.S., Mazumder, M.J., Quraishi, M.A. et al. (2020). Microwave-assisted synthesis of a new Piperonal-Chitosan Schiff base as a bio-inspired corrosion inhibitor for oil-well acidizing. *International Journal of Biological Macromolecules* 158: 231–243.

15 Chauhan, D.S., Srivastava, V., Joshi, P.G., and Quraishi, M.A. (2018). PEG cross-linked Chitosan: a biomacromolecule as corrosion inhibitor for sugar industry. *International Journal of Industrial Chemistry* 9: 363–377.

16 Baig, N., Chauhan, D.S., Saleh, T.A., and Quraishi, M.A. (2019). Diethylenetriamine functionalized graphene oxide as a novel corrosion inhibitor for mild steel in hydrochloric acid solutions. *New Journal of Chemistry* 43: 2328–2337.

17 Gupta, R.K., Malviya, M., Ansari, K.R. et al. (2019). Functionalized graphene oxide as a new generation corrosion inhibitor for industrial pickling process: DFT and experimental approach. *Materials Chemistry and Physics* 236: 121727.

18 Ansari, K., Chauhan, D.S., Quraishi M. A., and Saleh, T.A. (2020). Surfactant modified graphene oxide as novel corrosion inhibitors for mild steels in acidic media. *Inorganic Chemistry Communications* 121: 108238.

19 Devold, H. (2013). *Oil and Gas Production Handbook an Introduction to Oil and Gas Production*. Lulu. com.

20 Perez, T.E. (2013). Corrosion in the oil and gas industry: an increasing challenge for materials. *JOM* 65: 1033–1042.

21 Smith, C., Dollarhide, F., and Byth, N.J. (1978). Acid corrosion inhibitors-are we getting what we need? *Journal of Petroleum Technology* 30: 737–746.

22 Kalfayan, L. (2008). *Production Enhancement with Acid Stimulation*. Pennwell Books.

23 Finšgar, M. and Jackson, J. (2014). Application of corrosion inhibitors for steels in acidic media for the oil and gas industry: a review. *Corrosion Science* 86: 17–41.

24 Schechter, R.S. (1992). *Oil well stimulation*. Prentice Hall.

25 Schmitt, G. (1984). Application of inhibitors for acid media: report prepared for the European federation of corrosion working party on inhibitors. *British Corrosion Journal* 19: 165–176.

26 Jayaperumal, D. (2010). Effects of alcohol-based inhibitors on corrosion of mild steel in hydrochloric acid. *Materials Chemistry and Physics* 119: 478–484.

27 Barmatov, E., Geddes, J., Hughes, T. et al. (2012). Research on corrosion inhibitors for acid stimulation. In: ORROSION 2012, NACE International, 2012.

28 Huizinga, S. and Liek, W. (1994). Corrosion behavior of 13% chromium steel in acid stimulations. *Corrosion* 50: 555–566.

29 Yadav, M., Kumar, S., and Yadav, P. (2013). Corrosion inhibition of tubing steel during acidization of oil and gas wells. *Journal of Petroleum Engineering* 2013.

30 Maanonen, M. (2014) Steel pickling in challenging conditions. Materials Technology and Surface Engineering, Helsinki Metropolia University of Applied

Sciences, Thesis for Bachelor of Engineering, 1–32. https://www.theseus.fi/bitstream/handle/10024/70713/Steel_Pickling_in_Challenging_Conditions_2014_Thesis_Mika_Maanonen.pdf?sequence=1&isAllowed=y.

31 Umoren, S.A., Solomon, M.M., Obot, I.B., and Suleiman, R.K. (2019). A critical review on the recent studies on plant biomaterials as corrosion inhibitors for industrial metals. *Journal of Industrial and Engineering Chemistry* 76: 91–115.

32 Sastri, V.S. (1998). *Corrosion Inhibitors: Principles and Applications*. Wiley.

33 Ulaeto, S., Ekpe, U., Chidiebere, M., and Oguzie, E. (2012). Corrosion inhibition of mild steel in hydrochloric acid by acid extracts of Eichhornia crassipes. *International Journal of Materials and Chemistry* 2: 158–164.

34 Li, X., Deng, S., and Fu, H. (2012). Inhibition of the corrosion of steel in HCl, H_2SO_4 solutions by bamboo leaf extract. *Corrosion Science* 62: 163–175.

35 Deng, S. and Li, X. (2012). Inhibition by Ginkgo leaves extract of the corrosion of steel in HCl and H_2SO_4 solutions. *Corrosion Science* 55: 407–415.

36 Emranuzzaman, Kumar, T., Vishwanatham, S., and Udayabhanu, G. (2004). Synergistic effects of formaldehyde and alcoholic extract of plant leaves for protection of N80 steel in 15% HCl. *Corrosion Engineering, Science and Technology* 39: 327–332.

37 Umoren, S.A., Solomon, M.M., Obot, I.B., and Suleiman, R.K. (2018). Comparative studies on the corrosion inhibition efficacy of ethanolic extracts of date palm leaves and seeds on carbon steel corrosion in 15% HCl solution. *Journal of Adhesion Science and Technology* 32: 1934–1951.

38 Chauhan, D.S., Singh, P., and Quraishi, M.A. (2020). Quinoxaline derivatives as efficient corrosion inhibitors: current status, challenges and future perspectives. *Journal of Molecular Liquids* 320: 114387.

39 Singh, A., Ansari, K.R., Ituen, E. et al. (2020). A new series of synthesized compounds as corrosion mitigator for storage tanks: detailed electrochemical and theoretical investigations. *Construction and Building Materials* 259: 120421.

40 Singh, A., Ansari, K., Quraishi M. A., and Kaya, S. (2020). Theoretically and experimentally exploring the corrosion inhibition of N80 steel by pyrazol derivatives in simulated acidizing environment. *Journal of Molecular Structure* 1206: 127685.

41 Singh, A., Ansari, K.R., Chauhan, D.S. et al. (2020). Comprehensive investigation of steel corrosion inhibition at macro/micro level by ecofriendly green corrosion inhibitor in 15% HCl medium. *Journal of Colloid and Interface Science* 560: 225–236.

42 Singh, A., Ansari, K., Chauhan, D.S. et al. (2020). Anti-corrosion investigation of pyrimidine derivatives as green and sustainable corrosion inhibitor for N80 steel in highly corrosive environment: Experimental and AFM/XPS study. *Sustainable Chemistry and Pharmacy* 16: 100257.

43 Yadav, M., Kumar, S., Purkait, T. et al. (2016). Electrochemical, thermodynamic and quantum chemical studies of synthesized benzimidazole derivatives as corrosion inhibitors for N80 steel in hydrochloric acid. *Journal of Molecular Liquids* 213: 122–138.
44 Yadav, M., Behera, D., Kumar, S., and Sinha, R.R. (2013). Experimental and quantum chemical studies on the corrosion inhibition performance of benzimidazole derivatives for mild steel in HCl. *Industrial & Engineering Chemistry Research* 52: 6318–6328.
45 Gece, G. (2011). Drugs: a review of promising novel corrosion inhibitors. *Corrosion Science* 53: 3873–3898.
46 Haruna, K., Saleh, T.A., and Quraishi M. A. (2020). Expired metformin drug as green corrosion inhibitor for simulated oil/gas well acidizing environment. *Journal of Molecular Liquids* 315: 113716.
47 Xu, X., Singh, A., Sun, Z. et al. (2017). Theoretical, thermodynamic and electrochemical analysis of biotin drug as an impending corrosion inhibitor for mild steel in 15% hydrochloric acid. *Royal Society Open Science* 4: 170933.
48 Singh, A., Ansari, K., Quraishi M. A. et al. (2020). Aminoantipyrine derivatives as a novel eco-friendly corrosion inhibitors for P110 steel in simulating acidizing environment: experimental and Computational studies. *Journal of Natural Gas Science and Engineering* 83: 103547.
49 Chauhan, D.S., Quraishi M. A., Srivastava, V. et al. Virgin and chemically functionalized amino acids as green corrosion inhibitors: influence of molecular structure through experimental and in silico studies. *Journal of Molecular Structure* (2020): 129259.
50 Yadav, M. and Sharma, U. (2011). Eco-friendly corrosion inhibitors for N80 steel in hydrochloric acid. *Journal of Materials and Environmental Science* 2: 407–414.
51 Yadav, M., Sarkar, T.K., and Purkait, T. (2015). Amino acid compounds as eco-friendly corrosion inhibitor for N80 steel in HCl solution: electrochemical and theoretical approaches. *Journal of Molecular Liquids* 212: 731–738.
52 Ituen, E.B., Akaranta, O., and Umoren, S.A. (2017). N-acetyl cysteine based corrosion inhibitor formulations for steel protection in 15% HCl solution. *Journal of Molecular Liquids* 246: 112–118.
53 Yadav, M., Behera, D., and Sharma, U. (2013). Corrosion protection of N80 steel in hydrochloric acid by substituted amino acids. *Corrosion Engineering, Science and Technology* 48: 19–27.
54 Yadav, M., Kumar, S., and Gope, L. (2014). Experimental and theoretical study on amino acid derivatives as eco-friendly corrosion inhibitor on mild steel in hydrochloric acid solution. *Journal of Adhesion Science and Technology* 28: 1072–1089.
55 Yadav, M., Gope, L., and Sarkar, T.K. (2016). Synthesized amino acid compounds as eco-friendly corrosion inhibitors for mild steel in hydrochloric acid solution:

electrochemical and quantum studies. *Research on Chemical Intermediates* 42: 2641–2660.

56 Ajmal, M., Rawat, J., and Quraishi, M.A. (1998). Corrosion inhibiting properties of some polyaza macrocyclic compounds on mild steel in acid environments. *Anti-Corrosion Methods and Materials* 45: 419–425.

57 Quraishi, M.A. and Rawat, J. (2001). A review on macrocyclics as corrosion inhibitors. *Corrosion Reviews* 19: 273–299.

58 Singh, A., Lin, Y., Obot, I. et al. (2015). Corrosion mitigation of J55 steel in 3.5% NaCl solution by a macrocyclic inhibitor. *Applied Surface Science* 356: 341–347.

59 Ansari, K.R., Ramkumar, S., Chauhan, D.S. et al. (2018). Macrocyclic compounds as green corrosion inhibitors for aluminium: electrochemical, surface and quantum chemical studies. *International Journal of Corrosion and Scale Inhibition* 7: 443–459.

60 Bentiss, F., Lebrini, M., Vezin, H. et al. (2009). Enhanced corrosion resistance of carbon steel in normal sulfuric acid medium by some macrocyclic polyether compounds containing a 1, 3, 4-thiadiazole moiety: AC impedance and computational studies. *Corrosion Science* 51: 2165–2173.

61 Quraishi, M.A., Rawat, J., and Ajmal, M. (1998). Macrocyclic compounds as corrosion inhibitors. *Corrosion* 54: 996–1002.

62 Quraishi, M. A. and Rawat, J. (2000). Corrosion inhibition of mild steel in acid solutions by tetramethyl-dithia-octaazacyclotetradeca hexaene (MTAT). *Anti-Corrosion Methods and Materials* 47: 288–293.

63 Biswas, A., Pal, S., and Udayabhanu, G. (2015). Experimental and theoretical studies of xanthan gum and its graft co-polymer as corrosion inhibitor for mild steel in 15% HCl. *Applied Surface Science* 353: 173–183.

64 Biswas, A., Pal, S., and Udayabhanu, G. (2017). Effect of chemical modification of a natural polysaccharide on its inhibitory action on mild steel in 15% HCl solution. *Journal of Adhesion Science and Technology* 31: 2468–2489.

65 Haque, J., Srivastava, V., Chauhan, D.S. et al. (2018). Microwave-induced synthesis of chitosan Schiff bases and their application as novel and green corrosion inhibitors: experimental and theoretical approach. *ACS Omega* 3: 5654–5668.

66 Srivastava, V., Chauhan, D.S., Joshi, P.G. et al. (2018). PEG-functionalized chitosan: a biological macromolecule as a novel corrosion inhibitor. *ChemistrySelect* 3: 1990–1998.

67 Chauhan, D.S., Mouaden, K.E., Quraishi, M.A., and Bazzi, L. (2020). Aminotriazolethiol-functionalized chitosan as a macromolecule-based bioinspired corrosion inhibitor for surface protection of stainless steel in 3.5% NaCl. *International Journal of Biological Macromolecules* 152: 234–241.

68 Chauhan, D.S., Ansari, K.R., Sorour, A.A. et al. (2018). Thiosemicarbazide and thiocarbohydrazide functionalized chitosan as ecofriendly corrosion inhibitors

for carbon steel in hydrochloric acid solution. *International Journal of Biological Macromolecules* 107: 1747–1757.

69 Chauhan, D.S., Quraishi, M.A., Sorour, A.A. et al. (2019). Triazole-modified chitosan: a biomacromolecule as a new environmentally benign corrosion inhibitor for carbon steel in a hydrochloric acid solution. *RSC Advances* 9: 14990–15003.

70 Quraishi, M.A., Ansari, K.R., Chauhan, D.S. et al. (2020). Vanillin modified chitosan as a new bio-inspired corrosion inhibitor for carbon steel in oil-well acidizing relevant to petroleum industry. *Cellulose* 27: 6425–6443.

71 Ansari, K.R. and Quraishi, M.A. (2014). Bis-Schiff bases of isatin as new and environmentally benign corrosion inhibitor for mild steel. *Journal of Industrial and Engineering Chemistry* 20: 2819–2829.

72 Ansari, K.R., Quraishi, M.A., and Singh, A. (2014). Schiff's base of pyridyl substituted triazoles as new and effective corrosion inhibitors for mild steel in hydrochloric acid solution. *Corrosion Science* 79: 5–15.

73 Gupta, N.K., Verma, C., Quraishi, M.A., and Mukherjee, A.K. (2016). Schiff's bases derived from L-lysine and aromatic aldehydes as green corrosion inhibitors for mild steel: experimental and theoretical studies. *Journal of Molecular Liquids* 215: 47–57.

74 Singh, P. and Quraishi, M.A. (2016). Corrosion inhibition of mild steel using Novel Bis Schiff's Bases as corrosion inhibitors: electrochemical and surface measurement. *Measurement* 86: 114–124.

75 Solomon, M.M., Gerengi, H., Kaya, T., and Umoren, S.A. (2016). Performance evaluation of a chitosan/silver nanoparticles composite on St37 steel corrosion in a 15% HCl solution. *ACS Sustainable Chemistry & Engineering* 5: 809–820.

76 Solomon, M.M., Gerengi, H., and Umoren, S.A. (2017). Carboxymethyl cellulose/silver nanoparticles composite: synthesis, characterization and application as a benign corrosion inhibitor for St37 steel in 15% H2SO4 medium. *ACS Applied Materials & Interfaces* 9: 6376–6389.

77 Solomon, M.M., Umoren, S.A., Obot, I.B. et al. (2018). Exploration of dextran for application as corrosion inhibitor for steel in strong acid environment: effect of molecular weight, modification, and temperature on efficiency. *ACS Applied Materials & Interfaces* 10: 28112–28129.

78 Ansari, K.R., Chauhan, D.S., Quraishi, M.A., and Saleh, T.A. (2020). Bis (2-aminoethyl) amine-modified graphene oxide nanoemulsion for carbon steel protection in 15% HCl: effect of temperature and synergism with iodide ions. *Journal of Colloid and Interface Science* 564: 124–133.

79 Ansari, K., Chauhan, D.S., Quraishi, M.A. et al. (2020). The synergistic influence of polyethyleneimine-grafted graphene oxide and iodide for the protection of steel in acidizing conditions. *RSC Advances* 10: 17739–17751.

Part III

Organic Green Corrosion Inhibitors

11

Carbohydrates and Their Derivatives as Corrosion Inhibitors

Jiyaul Haque[1] *and M. A. Quraishi*[2]

[1] *Department of Chemistry, Indian Institute of Technology, Banaras Hindu University, Varanasi, India*
[2] *Center of Research Excellence in Corrosion, Research Institute, King Fahd University of Petroleum and Minerals, Dhahran, Saudi Arabia*

11.1 Introduction

Corrosion is a big problem worldwide, which causes enormous wastage of metallic material resulting in great impact on world economy. Several effective techniques such as cathodic protection, coating, corrosion inhibitor, design, and material selection have been used to mitigate the corrosion. In practice, corrosion inhibitor is one of the most economic and effective methods, particularly for closed systems.

Concerning the use of corrosion inhibitors in industries, toxicity is one of a major issue nowadays; because of the strict environmental rules and regulation, the discharge of toxic and environmentally harmful corrosion inhibitor has been restricted [1]. Therefore, current research is focused on developing biodegradable, cost-effective, and nontoxic corrosion inhibitors. Green chemistry serves as a source to develop environment-friendly corrosion inhibitors. To develop the green corrosion inhibitors, the researchers are focusing their attention on using naturally derived compounds including plant extracts, amino acids, and carbohydrates [2].

Carbohydrate is the most abundant and renewable material in a natural source. Carbohydrates have a different functional group in the molecular structure such as hydroxyl (alcohol), ether, ketone, aldehyde, and amine. These functional groups can interact with the metal surface and show potential inhibition performance. Besides, most of the carbohydrates are nontoxic and are widely used in medicinal

Organic Corrosion Inhibitors: Synthesis, Characterization, Mechanism, and Applications, First Edition. Edited by Chandrabhan Verma, Chaudhery Mustansar Hussain, and Eno E. Ebenso.

and industrial applications. Due to the renewable and nontoxic nature, scientists are attracted toward carbohydrate green compounds (glucose, chitosan, etc.) and studied many compounds as corrosion inhibitors for different metals and alloys [3–6]. Glucose is a simple monosaccharide, and chitosan is a polysaccharide (a polymer of glucosamine), both are the most important inhibitors due to their low-toxicity and easy availability. These carbohydrate inhibitors have some disadvantages: poor inhibition performance, poor solubility, and another important issue is the high cost (due to effective at high dose) of inhibitors. To overcome these issues, researchers have modified the structure of naturally derived simple carbohydrate inhibitors (e.g. glucose, chitosan) by the synthetic method under following green chemistry principles.

11.2 Glucose-Based Inhibitors

Glucose is the abundant carbohydrate, found in nature and is prepared by the hydrolysis of starch (starch source materials such as maize, potato, etc). Glucose-based compounds have been studied as potential corrosion inhibitors for iron alloys. Rajeswari et al. have been investigated the effect of glucose (A), gellan gum (B), and hydroxypropyl cellulose (C) inhibitors on cast iron corrosion in 1 mol/L HCl using the weight loss (WL) and electrochemical Electrochemical impedance spectroscopy (EIS) and potentiodynamic polarization (PDP) methods [7]. The results showed that polymer C Inhibition efficiency (IE 89.6%) and B (IE 80.9%) are more effective inhibitors than monomer glucose A (IE 69.5%) at 500 ppm. These inhibitors are mixed type and obeyed Langmuir adsorption isotherm.

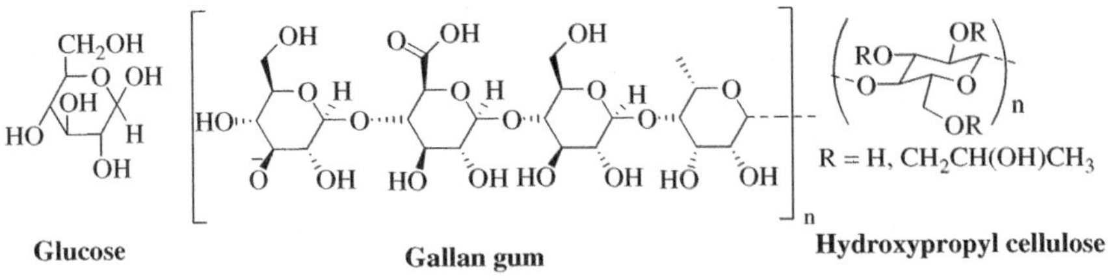

R = H, $CH_2CH(OH)CH_3$

Glucose **Gallan gum** **Hydroxypropyl cellulose**

Zhang et al. have studied a series of triazolyl glycolipid derivatives (33 compounds) as corrosion inhibitors for mild steel in hydrochloric solution [8]. The results found out that increasing the length of lipid carbon (n = 1–14), increases the IE. In pre-clicked' glucolipid derivative, the author was also added heteroatom-rich p-amino benzenesulfonamide (BSA) group and synthesized a new compound (A 31) with significantly improved inhibition performance for mild steel in HCl. The newly developed inhibitor A 31 was a mixed-type inhibitor and predominantly chemisorption on the metal surface.

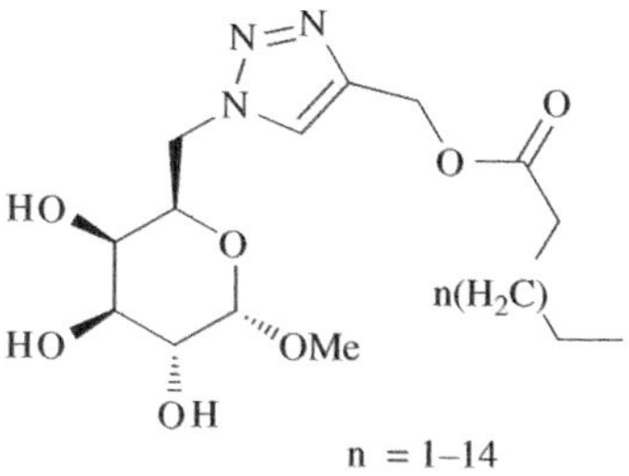

Triazolyl glycolipid derivatives

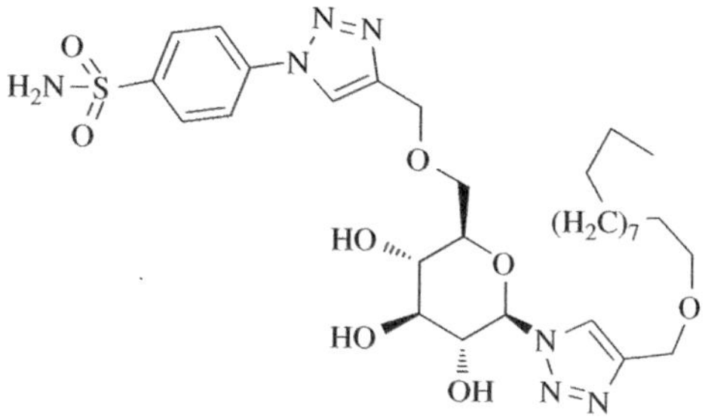

Bis-triazolyl conjugate with BSA and lipid chain

Recently, Aslam et al. have studied the sugar-based gemini surfactant as corrosion inhibitor for mild steel in 3.5% NaCl solution [9]. The results showed that Glu(12)-2-Glu(12) exhibited the maximum IE 69.1% at 2.5×10^{-3} mmol/L at 30°C. Synergistic effect also studied and results showed that by adding 10 mmol/L KI resulted in the increase of IE to 86.8%. The formulated inhibitor solution: 2.5×10^{-3} mmol/L of Glu(12)-2-Glu(12) and 10 mmol/L of KI exhibit 96.9% IE at 60°C. Both synergized and pure inhibitors act as mixed types.

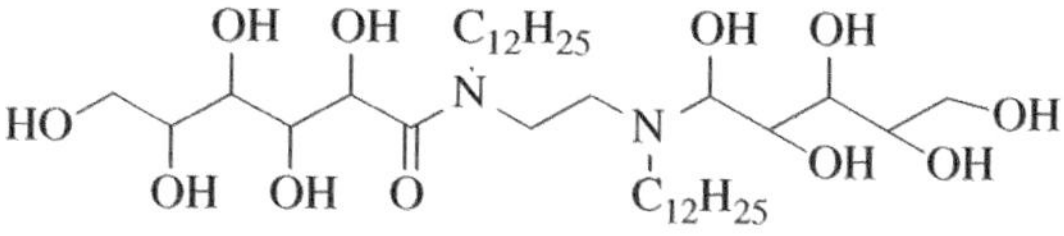

Glu(12)-Glu(12)

Verma et al. have evaluated the effect of electron withdrawing group (EWG), $-NO_3$ and electron donating group (EDG), $-CH_3$ and –OH on the on the corrosion inhibition of glucosamine-derived compounds on mild steel in hydrochloric solution [10]. The results showed that both EWG and EDG enhance the *IE*, but the EDG exhibit pronounced effect, and the order of IE is as follows chitosan Schiff base (CSB)-1 (–H) < CSB-2 ($-NO_3$) < CSB-3 ($-CH_3$) < CSB-4 (–OH). Theoretical study, Monte Carlo simulation revealed that all the four inhibitors are adsorbed flatly over the metal surface.

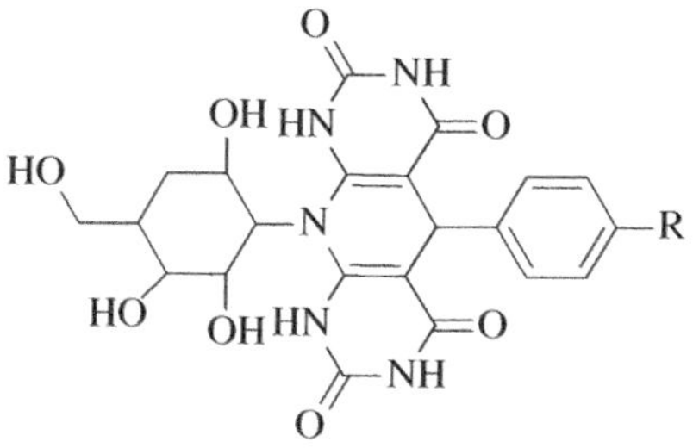

CSB-1; R=–H
CSB-2; R=–NO_3
CSB-3; R=–CH_3
CSB-4: R=–OH

More recently, Haque et al. has been investigated the three glucose derivatives as corrosion inhibitors on mild steel corrosion [11]. All three inhibitors were easily soluble in test solution. The inhibitors were synthesized using low-toxic and cheapest substrates at room temperature. The inhibition efficiency of inhibitors increases with increasing the length of carbon chain, and maximum IE 97.04% was obtained at 22.71×10^{-5} mol/L. The quantum chemical studies show that the protonated form of inhibitors adsorbed on the metal surface.

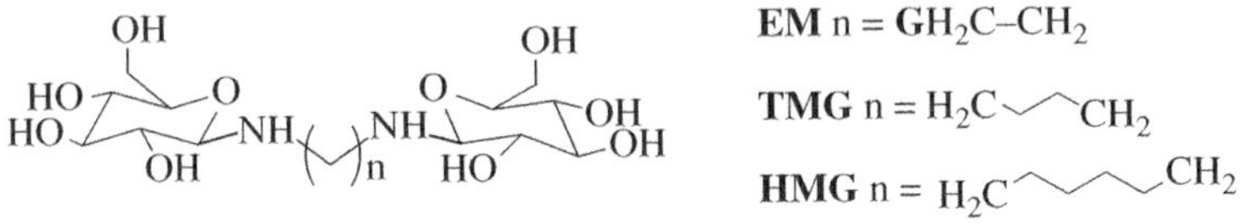

EM n = $GH_2C–CH_2$

TMG n = H_2C~~CH_2

HMG n = H_2C~~~CH_2

11.3 Chitosan-Based Inhibitors

Chitosan is derived from chitin, which is the second most abundance of natural polysaccharide. It has a number of hydroxyl groups (like other carbohydrates) and some amino groups depending upon their degree of deacetylation. In addition, chitosan has excellent properties such as biocompatibility, biodegradability, and nontoxicity [12]. Umoren et al. have investigated the inhibition effect of chitosan on mild steel corrosion in 0.1 mol/L HCl at 30–60°C using WL, electrochemical (EIS and PDP), and SEM studies [13]. The results showed that *IE* increases with inhibitor concentration and solution temperature with a maximum IE of 96% at 60°C at 800 ppm concentration, which is further decreased to 93% at higher temperature of 70°C. Electrochemical results revealed that chitosan performances as a mixed type inhibitor, and its adsorption followed the Langmuir adsorption isotherm. Further, Okoronkwo et al. studied the inhibition effect of chitosan on plane carbon steel in 1 mol/L HCl. It was found that chitosan shows the maximum IE of 93.2% at 5000 ppm at 30°C [14]. The author also tested the effect of temperature and found that the IE increases with increase in the temperature.

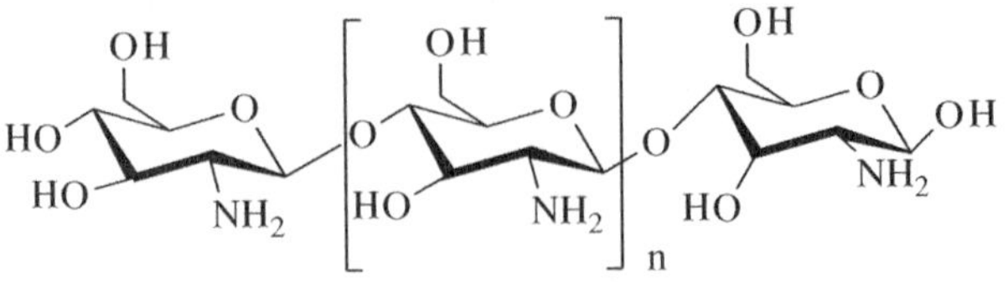

Chitosan

Many other researchers also carried out corrosion inhibition test of different chitosan derivatives on mild steel in acid solution. Cheng et al. have investigated the carboxymethyl chitosan (CMCT), Cu^{2+} and its mixture (CMCT+Cu^{+2}) as an inhibitor used for mild steel in 1 mol/L HCl [15]. The WL and electrochemical (EIS and PDP) techniques were used for inhibition tests, and their results showed that mixture of additve (CMCT + Cu^{+2}) of additive effectively controls the metal dissolution than each additive separately. The inhibition performance of a mixture of additive was found to increase with the increasing temperature from 25°C to 55°C.

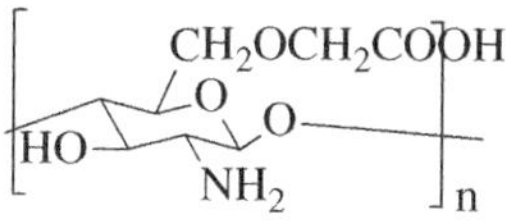

CMCT

Hussein et al. have tested the inhibition effect of two water-soluble chitosan derivatives, namely, 2-N, N-diethylbenzene ammonium chloride N-oxoethyl chitosan (compound I) and 12-ammonium chloride N-oxododecan chitosan (compound II) on carbon steel in 1 mol/L HCl [16]. The inhibition effect was tested by WL method and results showed that compound I exhibited the higher IE than the other one, may be due to it has more active groups that block the corrosive sites of metal. Compound II has a large number of carbon chains that may be oriented around the chitosan units and created steric hindrance, resulting in a lower IE. Similarly, quaternary inhibitor N-(2-hydroxy-3-trimethyl ammonium) propyl-chitosan chloride (HTACC) for mild steel in 1 mol/L HCl was studied by Sangeetha et al. [17]. Corrosion inhibition tests were evaluated by using WL and electrochemical methods. It was found that IE increases with inhibitor concentration and declines with temperature, and 98.9% IE was achieved at 500 ppm concentration at 30°C. PDP results revealed that HTACC control both anodic and cathodic reactions. EIS and SEM results exposed that the inhibitor forms a protective film on the metal surface.

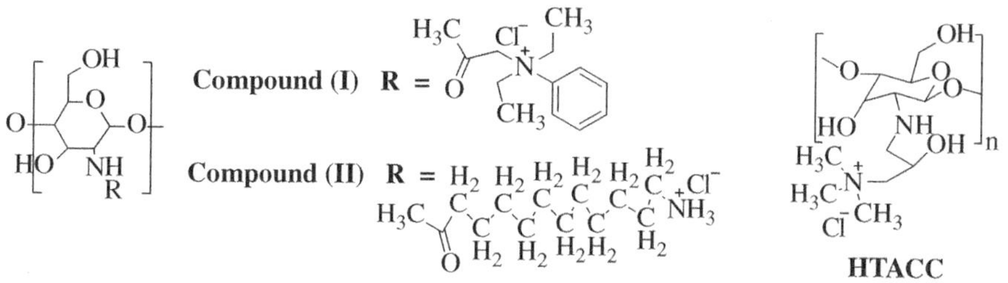

Compound (I) R =

Compound (II) R =

HTACC

Recently, Li et al. have synthesized two polyamine-grafted chitosan copolymers and evaluated as corrosion inhibitors on mild steel corrosion in 5% HCl solution, using the WL measurements, metallographic microscopy, electrochemical measurements [18]. The 0.1% mass fraction concentration of CS-MAA-EN showed the highest IE of 96% among all three: Chitosan, CS-MAA-EN, and CS-MAA-TN. The results showed that the inhibition performance of inhibitors was increased with an increase in the number of the amino groups but at a certain extent, further increase in amino group decreases the inhibition performance at higher concentrations.

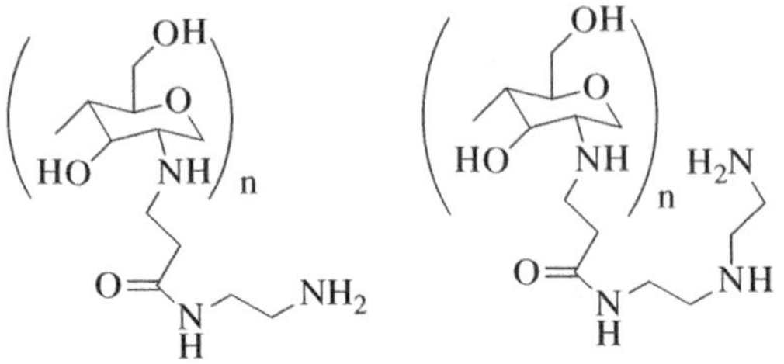

CS-MAA-TN **CS-MAA-EN**

The inhibition effect of chitosan ionic liquid: chitosan-p-toluene sulfonate salt (CSPTA) and oleic acid amidated CSPTA (CSPTA-OA) on the steel corrosion in hydrochloric acid was studied by El-Mahdy et al. [19] using the different electrochemical techniques (EIS and PDP). Electrochemical results showed that IE increases with increase in the inhibitor concentration, and 97.8% and 95.6% IE were achieved in the presence of 250 ppm of CSPTA and CSPTA-OA, respectively. Polarization results revealed that both inhibitors act as mixed type and their adsorption obeyed Langmuir isotherm.

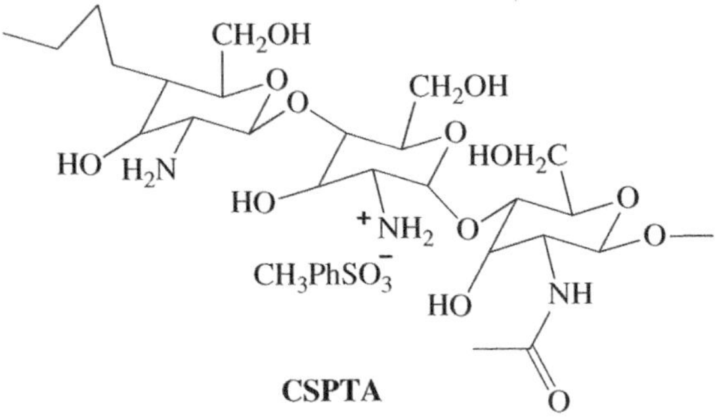

CSPTA

Liu et al. have studied inhibition effect of β-cyclodextrin-modified chitosan (β-CD-chitosan) on carbon steel in 0.5 mol/L HCl using WL, EIS, PDP, SEM, and EDX [20]. The result revealed that 96.02% IE was found at 230 ppm concentration. EIS results showed that the Nyquist curve has a single time constant. PDP results revealed that β-CD-chitosan acts as a mixed type with predominantly cathodic inhibitor. SEM and EDX results disclose the formation of inhibitor film over the metal surface.

Anticorrosion activity of two chitosan Schiff base derivatives for mild steel in 3.5% NaCl have been investigated by Ma et al., using WL and electrochemical (EIS and PDP) methods [21]. The order of IE was found to be carboxymethyl chitooligosaccharide schiff base (CM-CSB) was greater than chito-oligosaccharide schiff base (CSB). It was found that inhibition performance increases with increasing the CSB concentration up to 200 ppm and after that IE decreased with increased concentration. While in presence of CM-CSB, IE increases with concentration and maximum IE was found at 800 ppm. The author found that, in case of CSB, a higher number of inhibitor molecules reduce their solubility in test solution, which causes the desorption speed to be higher than adsorption speed. While presence of carboxymethylation in CM-CSB improves solubility and resulted in greater inhibition performance. PDP results revealed that CSB inhibits both anodic and cathodic corrosion reaction, and CM-CSB inhibits predominantly cathodic reaction. CM-CSB inhibitor obeyed the Langmuir adsorption isotherm.

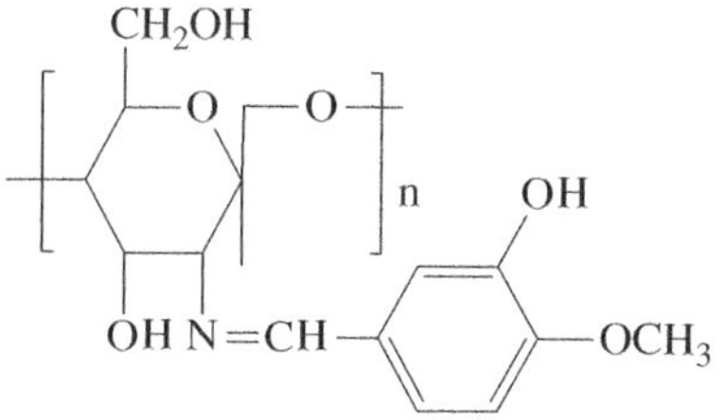

CSB 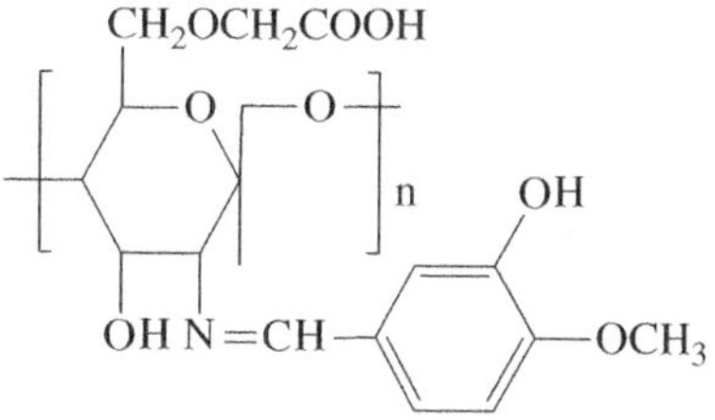 **CM-CSB**

Atta et al. synthesized two nonionic ampiphilic chitosan nanoparticles (CSLA-MPEG and CSOA-MPEG) and tested its corrosion inhibition properties on steel in 1 mol/L HCl solution using electrochemical (EIS and PDP) method [22]. Results showed that both inhibitors were effectively controlled the metal corrosion and showed maximum IE of 93.0% and 97.2% at 250 ppm in the presence of CSLA-MPEG and CSOA-MPEG, respectively. PDP results revealed that both CSLA-MPEG and CSOA-MPEG act as a mixed type with anodic predominance.

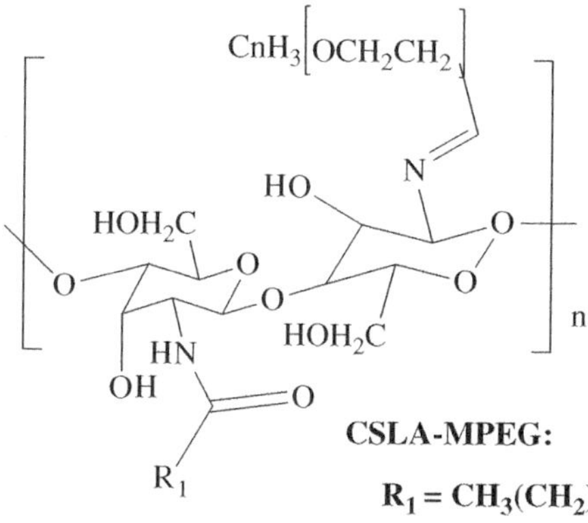

CSLA-MPEG:

$R_1 = CH_3(CH_2)_7CH = CH(CH_2)_7COOH$

COLA-NPEG:

$R_1 = CH_3(CH_2)_4\text{-}CH = CH\text{-}CH_2\text{-}CH = CH\text{-}(CH_2)_7COOH$

Suyanto et al. have investigated the two chitosan derived as corrosion inhibitors for on steel with fluidization method in various corrosive media [23]. The results showed that the inhibitors exhibited the maximum IE 80.82% (CMChi-B) and 80.62% (CMChi-UGLU) in 2% NaCl, while in 1 mol/L HCl exhibited the IE 76.55% (CMChi-B) and 68.68% (CMChi-UGLU).

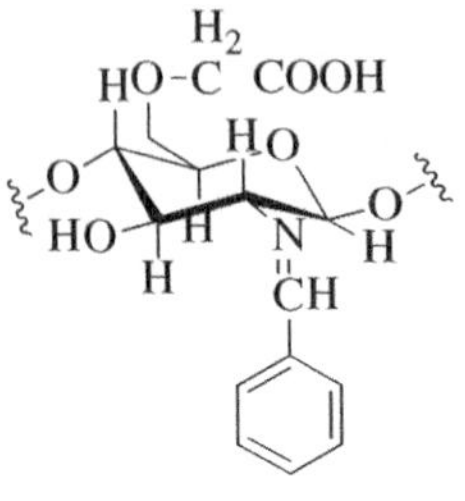

CMChi-B

Menaka et al. have examined the inhibition of chitosan thiophene carboxaldehyde Schiff base on mild steel corrosion in hydrochloric acid solution using WL and electrochemical (EIS and PDP) techniques [24]. The effects of inhibitor concentration (100–1500 ppm), exposure time (1–12 hour) and temperature (30–70°C) were investigated. The result showed that the IE increases with increase in the inhibitor concentrations, immersion times, and temperature (up to 60°C), and maximum IE of 92% was obtained at 1500 ppm at 12 hour. Polarization measurements revealed that ChTSB behaves as mixed inhibitor, and adsorption obeys the Temkin adsorption isotherm.

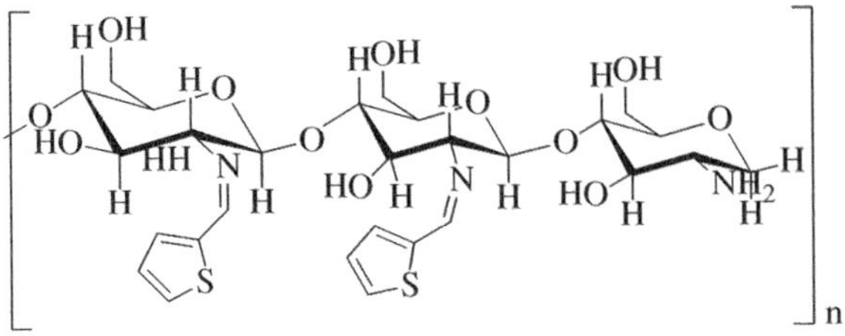

CSTSB

Recently, Haque et al. studied the modified CSBs as corrosion inhibitors [25]. These inhibitors were easily synthesized using microwave, a green technique. The inhibitors are potentially inhibited and shows 90.65% inhibition efficiency at 50 ppm (CSB-3). The inhibitors obey the Langmuir adsorption isotherm, and all three act as mixed type with cathodic predominant inhibitor. FTIR and XPS studies show that the heteroatoms of inhibitors (functional groups) interacted with metal surface, and the resulted inhibitors effectively get adsorbed on to the metal surface. The DFT and MD studies are well supported with experimental results.

Chitosan Schiff base(CSBs)

CSB-1 R = phenyl

CSB-2 R = 4-N(CH$_3$)$_2$-phenyl

CSB-3 R = 4-OH-3-OCH$_3$-phenyl

11.4 Inhibition Mechanism of Carbohydrate Inhibitor

Generally, the carbohydrate molecules inhibit corrosion through the adsorption on the metal surface. The carbohydrate molecule contains a greater number of the hydroxyl groups, ether, amine, aldehyde, and ketone. The functional group-containing lone pair of the electron can donate to the vacant metal d-orbital and get adsorbed on the surface. To enhance the solubility and inhibition performance, carbohydrates derivatives are synthesized and added a new functional group (heteroatoms) and hydrophobic carbon chain. It was found that the additional group improved the adsorption performance of inhibitors. Figure 11.1, represents the adsorption model of CSBs over the mild steel surface in HCl

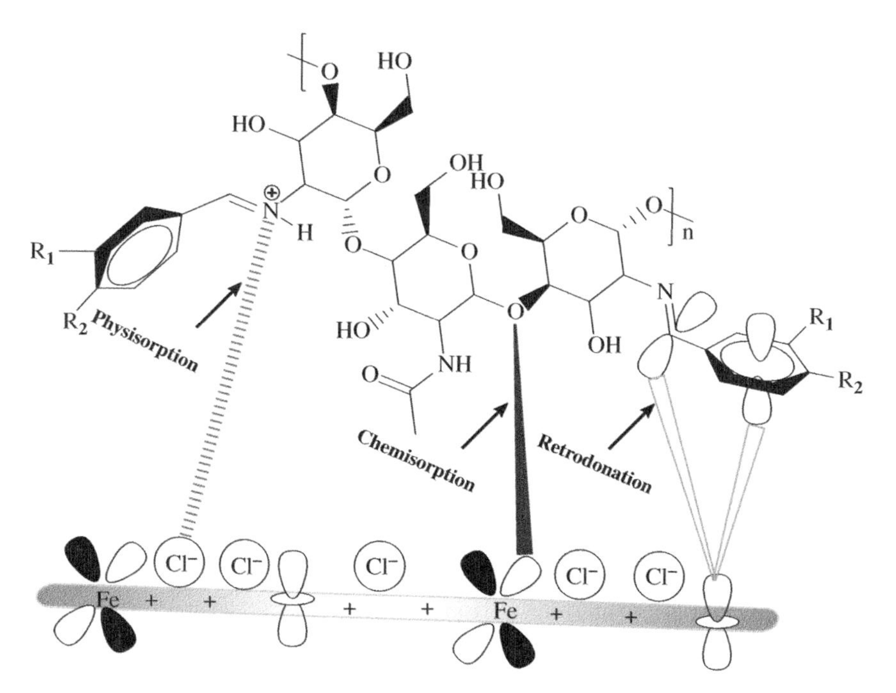

Figure 11.1 Adsorption model of chitosan Schiff base over the mild steel surface.

solution [25]. CSBs were derived from chitosan and contain a number of new functional group Schiff base. When CSBs were added in acid solution, the imine nitrogen (has highest Mulliken charge) get protonated and interacted with the negatively charged surface of mild steel (created by adsorbed chloride ion) through the physisorption mechanism. The chemisorption of inhibitor was created by the donation of the lone pair electron of hydroxyl group to the metal vacant d-orbital. Simultaneously, the antibonding orbital of π-bond can accept an electron from metal d-orbital and minimize the negative charge on metal, resulted in an increase in the chemisorption mechanism.

11.5 Conclusions

As it can be seen in this chapter, the carbohydrates, glucose and chitosan derivatives, potentially inhibit the iron alloys corrosion at lower concentration compared with its parent carbohydrates, glucose and chitosan. These inhibitors are mostly nontoxic and cheap and will be used as commercial inhibitors in place of toxic organic and inorganic inhibitors, which are currently used in many industries worldwide.

References

1 Sastri, V.S. (2012). *Green corrosion inhibitors: theory and practice*. John Wiley & Sons.
2 Sharma, S.K., Mudhoo, A., Khamis, E.K.E., and Jain, G. (2008). Green Corrosion Inhibitors: An Overview of Recent Research Green Corrosion Inhibitors: An Overview of Recent Research. *Journal of Corrosion Science and Engineering* 11: 1–14.
3 Ali-Shattle, E.E., Mami, M., and Alnaili, M. (2009). Investigation of the inhibitory effect of sucrose on corrosion of Iron (Libyan Steel) in mineral acid solutions. *Asian Journal of Chemistry* 21: 5431.
4 Solomon, M., Umoren, S., Udosoro, I., and Udoh, A. (2010). Inhibitive and adsorption behaviour of carboxymethyl cellulose on mild steel corrosion in sulphuric acid solution. *Corrosion science* 52: 1317–1325.
5 Yadav, M., Sarkar, T., and Obot, I. (2016). Carbohydrate compounds as green corrosion inhibitors: electrochemical, XPS, DFT and molecular dynamics simulation studies. *RSC advances* 6: 110053–110069.
6 Umoren, S.A. and Eduok, U.M. (2016). Application of carbohydrate polymers as corrosion inhibitors for metal substrates in different media: a review. *Carbohydrate polymers* 140: 314–341.

7 Rajeswari, V., Kesavan, D., Gopiraman, M., and Viswanathamurthi, P. (2013). Physicochemical studies of glucose, gellan gum, and hydroxypropyl cellulose—Inhibition of cast iron corrosion. *Carbohydrate polymers* 95: 288–294.

8 Zhang, H.-L., He, X.-P., Deng, Q. et al. (2012). Research on the structure–surface adsorptive activity relationships of triazolyl glycolipid derivatives for mild steel in HCl. *Carbohydrate research* 354: 32–39.

9 Aslam, R., Mobin, M., Aslam, J., and Lgaz, H. (2018). Sugar based N, N′-didodecyl-N, N′ digluconamideethylenediamine gemini surfactant as corrosion inhibitor for mild steel in 3.5% NaCl solution-effect of synergistic KI additive. *Scientific reports* 8: 3690.

10 Verma, C., Olasunkanmi, L.O., Ebenso, E.E. et al. (2016). Adsorption behavior of glucosamine-based, pyrimidine-fused heterocycles as green corrosion inhibitors for mild steel: experimental and theoretical studies. *The Journal of Physical Chemistry C* 120: 11598–11611.

11 Haque, J., Srivastava, V., Chauhan, D.S. et al. (2020). Electrochemical and surface studies on chemically modified glucose derivatives as environmentally benign corrosion inhibitors. *Sustainable Chemistry and Pharmacy* 16: 100260.

12 Hudson, S. and Smith, C. (1998). Polysaccharides: chitin and chitosan: chemistry and technology of their use as structural materials. In: *Biopolymers from renewable resources* (ed. D.L. Kaplan), 96–118. Springer.

13 Umoren, S.A., Banera, M.J., Alonso-Garcia, T. et al. (2013). Inhibition of mild steel corrosion in HCl solution using chitosan. *Cellulose* 20: 2529–2545.

14 Okoronkwo, A., Olusegun, S., and Oluwasina, O. (2015). The inhibitive action of chitosan extracted from Archachatina marginata shells on the corrosion of plain carbon steel in acid media. *Anti-Corrosion Methods and Materials* 16: 13–18.

15 Cheng, S., Chen, S., Liu, T. et al. (2007). Carboxymethylchitosan + Cu2+ mixture as an inhibitor used for mild steel in 1 M HCl. *Electrochimica acta* 52: 5932–5938.

16 Hussein, M.H., El-Hady, M.F., Shehata, H.A. et al. (2013). Preparation of some eco-friendly corrosion inhibitors having antibacterial activity from sea food waste. *Journal of surfactants and detergents* 16: 233–242.

17 Sangeetha, Y., Meenakshi, S., and SairamSundaram, C. (2015). Corrosion mitigation of N-(2-hydroxy-3-trimethyl ammonium) propyl chitosan chloride as inhibitor on mild steel. *International journal of biological macromolecules* 72: 1244–1249.

18 Li, H., Li, H., Liu, Y., and Huang, X. (2015). Synthesis of polyamine grafted chitosan copolymer and evaluation of its corrosion inhibition performance. *Journal of the Korean Chemical Society* 59: 142–147.

19 El-Mahdy, G.A., Atta, A.M., Al-Lohedan, H.A., and Ezzat, A.O. (2015). Influence of green corrosion inhibitor based on chitosan ionic liquid on the steel corrodibility in chloride solution. *International Journal of Electrochemical Science* 10: 5812–5826.

20 Lei, X., Wang, H., Feng, Y. et al. (2015). Synthesis, evaluation and thermodynamics of a 1H-benzo-imidazole phenanthroline derivative as a novel inhibitor for mild steel against acidic corrosion. *RSC advances* 5: 99084–99094.

21 Ma, F., Li, W., Tian, H. et al. (2012). Inhibition behavior of chito-oligosaccharide schiff base derivatives for mild steel in 3.5% NaCl solution. *International Journal of Electrochemical Science* 7: 10909–10922.

22 Atta, A.M., El-Mahdy, G.A., Al-Lohedan, H.A., and Ezzat, A.-R.O. (2015). Synthesis of nonionic amphiphilic chitosan nanoparticles for active corrosion protection of steel. *Journal of Molecular Liquids* 211: 315–323.

23 Suyanto, D. (2015). Application new chitosan derivatives as inhibitor corrosion on steel with coating method. *Journal Of Chemical and Pharmaceutical Research* 7: 504–516.

24 Menaka, R. and Subhashini, S. (2017). Chitosan Schiff base as effective corrosion inhibitor for mild steel in acid medium. *Polymer International* 66: 349–358.

25 Haque, J., Srivastava, V., Chauhan, D.S. et al. (2018). Microwave-induced synthesis of chitosan Schiff bases and their application as novel and green corrosion inhibitors: experimental and theoretical approach. *ACS omega* 3: 5654–5668.

12

Amino Acids and Their Derivatives as Corrosion Inhibitors

Saman Zehra and Mohammad Mobin

Corrosion Research Laboratory, Department of Applied Chemistry, Faculty of Engineering and Technology, Aligarh Muslim University, Aligarh, Uttar Pradesh, India

12.1 Introduction

The loss of materials is labeled as corrosion, as a consequence of a chemical reaction to the atmosphere [1, 2]. Corrosion has much more serious impacts than simple loss of weight of metal for the safe, stable, and effective functioning of machinery or systems and so are the results from corrosion. Multiple designs can fail and expensive replacements are required, although the amount of metal lost is relatively small [3]. The national economy is threatened by high corrosion costs, with corrosion costs of about 2–4% of the gross national product. It is therefore crucial for corrosion workers to take corrosion protection provisions in effort to reduce corrosion impact, which is about 25% of the cost that can be extricate when these arbitrations being administered [4, 5]. In addition, the avoidance of corrosion prevents metal deterioration and thus adds greatly to the protection of materials with minimal environmental impact. Health, on the other hand, is among the most serious implications of corrosion. Although protection should be uppermost in the minds of industrial workers, amid great precautions, injuries do occur. So not only is corrosion costly but also it poses threats to human life and safety. There are various methods that have been employed in order to control the corrosion, and some are outlined in Figure 12.1. Among them, one of the methods to protect the metals from corrosion is the application of the corrosion inhibitors. However, due to environmental authorities in various countries that have

Organic Corrosion Inhibitors: Synthesis, Characterization, Mechanism, and Applications,
First Edition. Edited by Chandrabhan Verma, Chaudhery Mustansar Hussain, and Eno E. Ebenso.

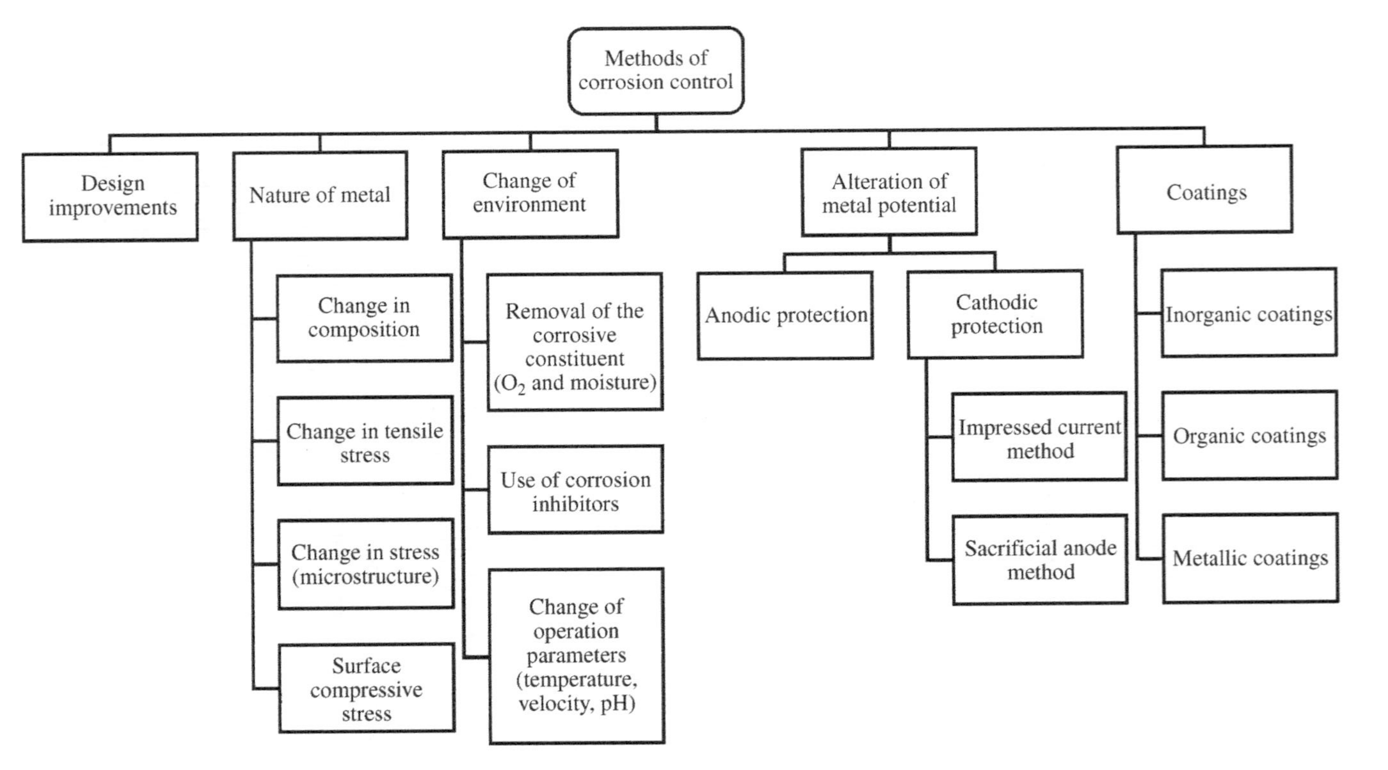

Figure 12.1 A description of the different approaches used to mitigate corrosion.

implemented stringent rules and regulations for the use and discharge of corrosion inhibitors, such as toxicity, biodegradability, and bioaccumulation requirements, the corrosion inhibitor sector is experiencing drastic changes from the point of view of environmental compatibility [6, 7]. Consequently, it has been perceived that the development of novel corrosion inhibitors with minimal to zero adverse effects is far more important and alluring. The development of the area of corrosion inhibitors has oriented over the last two decades around the use of low-cost, potent compounds with minimal to non-negative environmental effects on the substitute of environmentally harmful substances. The seek for alternative eco-friendly compounds as corrosion inhibitors has ended in the study of a number of amino acids and their derivatives as safe corrosion inhibitors for too many metals under varying severe corrosive environments.

Amino acids are one of the encouraging compounds that could be used as effective corrosion inhibitors [8–10]. Amino acids constitute a family of organic, naturally occurring, biodegradable, relatively affordable, high-purity, nontoxic organic compounds that are water soluble. There are actually more than 20 distinct amino acids that are believed to exist in nature. In the moiety of amino acids molecule structure, two ionizable functional groups with opposite chemical features are usually included: the amine group (-NH_2) with basic properties and the carboxyl group (-COOH) with acid properties [11]. Attributed to the existence of π-electrons and heteroatoms in their molecules, amino acids have exhibited an exemplary inhibition of corrosion by either adsorbing or creating an insoluble metal complex on the metal surface and resisting metal corrosion. Higher inhibition effects were demonstrated by sulfur-containing amino acids and amino acids with longer hydrocarbon chains [12]. Additional groups or groups that significantly raise the electron density at the alpha amino group often increase the reliability of amino acids.

In point of the above fact, the major focus of this chapter is to establish a working knowledge of the usefulness of amino acids and their derived compounds as corrosion inhibitors for the protection of certain metals and alloys in various applications and in various corrosive conditions, taking account of these specifics.

12.2 Corrosion Inhibitors

In general, the corrosion inhibitor is a chemical that reduces or prevents corrosion rates by producing monomolecular adsorbed films on the surface at minimum concentrations in a corrosive medium (of the metal to be protected). This, in essence, blocks the immediate (to be protected) contact of the metal with the corrosive agents in the surrounding atmosphere [13]. They were listed as synthesized or derived, depending on origin (organic or inorganic) and processes. It is therefore crucial to think for not just the efficient but also cost-effective and environmentally viable corrosion inhibitors. However, owing to their harmful effects, conventional corrosion inhibitors, such as chromates and plum, have been found to be stringent according to environmental constraints [6]. Sulfur, nitrogen, oxygen, phosphorus, and several bonds or rings are the far more powerful organic compounds comprising heteroatoms within the structures of the organic compounds. The key functional characteristics that reflect the inhibitor properties of these corrosion inhibitors are in fact the lone pairs of electrons and belatedly bound to electrons in their functional groups [14–17]. The potency of the organic inhibitor is largely based on the chemical structure, physiochemical features, and functional groups such as presence of aromatic properties, and/or mixed bonds, nature, and number of bonding atoms. Besides the above, the potency of the compounds as inhibitors of corrosion is further controlled by the factors such as molecular size and the occurrence of substituent groups and heteroatoms in the organic compounds. A further important factor is environmental acceptability when evaluating an inhibitor for the prevention of corrosion in addition to its availability and economic viability [18]. Some organic compounds used as corrosion inhibitors are therefore both harmful and destructive for humans and the atmosphere and must be substituted by nontoxic compounds that are environmentally safe and friendly. The new research trend is thus to manufacture green chemicals as inhibitors of green corrosion that are not harmful, cheaper, and more environmentally friendly. Over the next few years, organic corrosion inhibitors are expected to grow in size, and by 2024, it is expected that they will cross US $3860.0 million.

12.3 Why There Is Quest to Explore Green Corrosion Inhibitors?

In Figure 12.2, the general requirements for the most eligible corrosion inhibitors are outlined. Greenness and the cost-effectiveness are the most significant among them. The chemical products that meet the required reduced level of hazardous substance generation are typically ecologically friendly or simply green corrosion formulations. The processes involving their usage are governed by sustainable chemistry without direct or indirect negative environmental or health impacts [19]. In recent times, owing to global interest on environment safety, as well as the effect of impacting industrial activities on man's health and ecological balance, the use of toxic chemicals and operations that emit them have been minimized. On this note, while effective for the reduction of metal corrosion at lower concentrations, inorganic inhibitors and some of their dangerous organic equivalents are increasingly being replaced by environmentally safe and environment-friendly

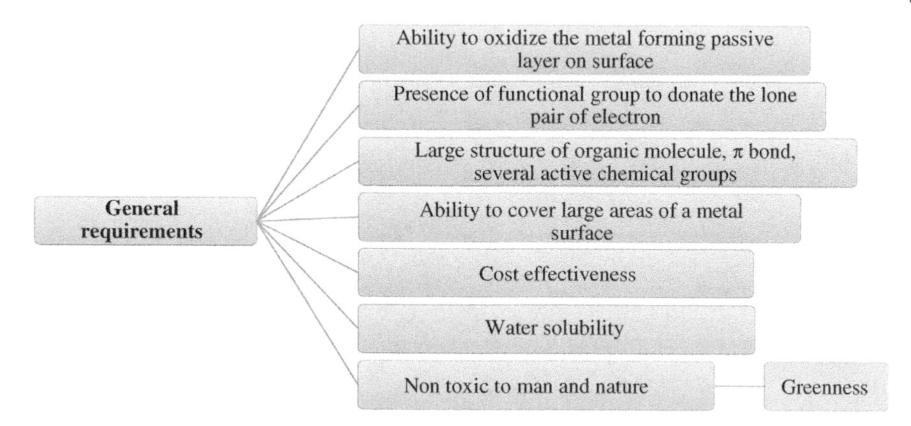

Figure 12.2 Outline of the factors that determine the performance of the corrosion inhibitors.

nonhazardous substances. In most oil field operations, corrosion preventive interventions with greener inhibitor compounds (chromate free inhibitor formulations) are required to carefully meet safety requirements and also adequately protect the intended metal substrates in their service environments with the banning of chromates [20]. Health defects with chromates vary from minor skin allergic reactions and rashes to nasal bleeding; higher doses of genetic material modification can occur with arsenates, as well as nervous breakdown and cancer. According to the US National Institute for Workplace Safety and Health, exposure amounts of arsenates and chromates were decreased to 0.002 and 0.05 $\mu g/m^3$ of air, respectively [21].

The word "green inhibitor" or "environmentally friendly inhibitor" refers to compounds that are naturally biocompatible. Naturally occurring substances of both plants and animal origin otherwise tagged "green inhibitors" are known to meet these requirements [22]. Figure 12.3 identifies some of the green compounds used as regulators of green corrosion. In the last few decades, research into green or sustainable corrosion inhibitors and environmentally friendly inhibitors has been aimed at the target of using inexpensive, effective compounds with minimal or zero negative environmental impacts. The rising publications in this area obviously reflects an interest in discovering new inhibitors in a wide spectrum of corrosive conditions to regulate the corrosion of numerous materials. Even then, there has been a dramatic spike in the number of reports on corrosion inhibitors, only 5% of the literature released in the last decade discusses green inhibitors.

As explored in previous section, due to strict rules and regulations for use and disposal of toxic chemicals, great recognition has been acquired by the advancement of environmentally benign corrosion inhibitors. Nowadays, several new

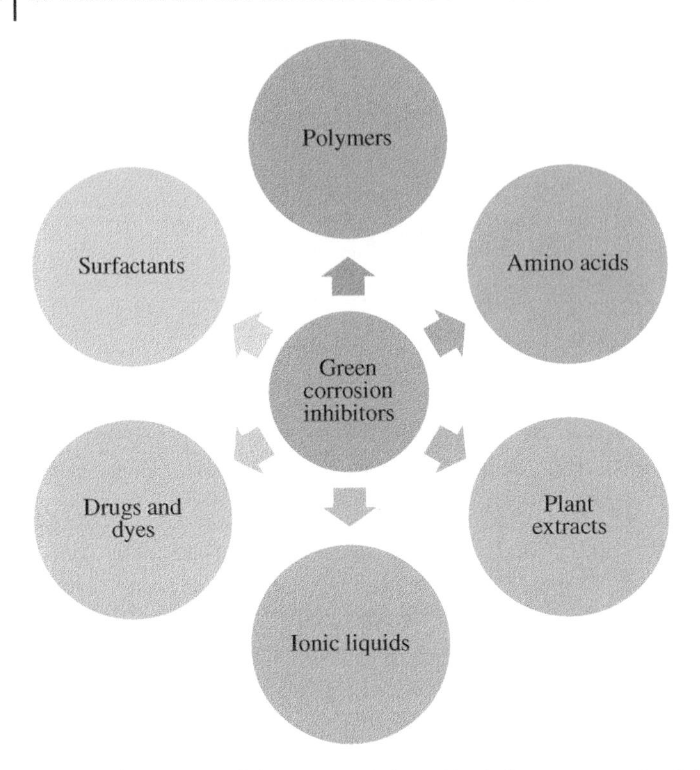

Figure 12.3 Some of the commonly investigated green corrosion inhibitors.

generations of less toxic inhibitors are being developed by the researchers working in the area of corrosion inhibition to avoid environmental toxicity and pollution. Some of the major classes explored are plant extracts, amino acids, natural polymers, drugs, green nanocomposites, etc. Due to the diversity of their structures, extracts of many common plants such as Damsissa extract [23], extract of Punica granatum peel [24], Luffa cylindrical leaf extract [25] and so on and extract of spirogyra algae [26] have been studied as economically viable potential corrosion inhibitors for many metals. In their molecular structures, plant materials contain proteins, polysaccharides, polycarboxylic acids and alkaloids, including polar functions, the main adsorption centers are atoms like nitrogen, sulfur, or oxygen atoms, as well as triple or conjugate double bonds including aromatic rings. The cost of using green inhibitors is much lower than conventional organic inhibitors, which require more processing time and additives. As amino acids being green corrosion inhibitors, certain amino acids have recently been investigated as corrosion inhibitors [26–28] due to their nontoxicity and effectiveness, polymeric compounds adsorb the metal surface more actively compared to their monomeric analogues due to various adsorption sites. At present, most researches are focused on polymers (both naturally occurring and synthetic), which are readily available, cost-effective, and ecosystem friendly. Some of the polymer inhibitors that have been examined include chitosan and their derivatives [29], xanthan gum [30], lignin [31], etc.

12.4 Amino Acids and Their Derived Compounds: A Better Alternate to the Conventional Toxic Corrosion Inhibitors

As also stated above, due to their nontoxicity and cost-effectiveness, use of such amino acids as sustainable corrosion inhibitors has drawn tremendous attention in recent times. In order to understand their behavior in different media and for the protection of the different metals, some of the basic understanding should be made, which are outlined in this section.

12.4.1 Amino Acids: A General Introduction

Amino acids are the compounds that are linked to the same carbon atom (alpha or 2-carbon) by at minimum one group of carboxyl (-COOH) and one amino ($-NH_2$) group, as can be seen in Figure 12.4, H and R-groups of different sizes, shapes, and chemical properties (side chain). In the physiological sources, there are 20 distinct amino acids, illustrated in Figure 12.5, used to make up proteins in

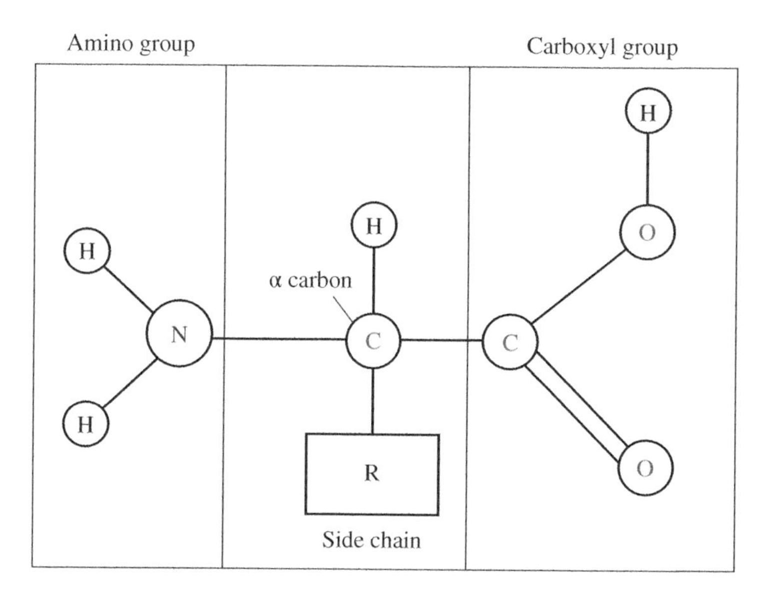

Figure 12.4 Amino acids with core asymmetrical carbon to which the amino group, the carboxyl group, the hydrogen atom, and the side chain (R group) are connected.

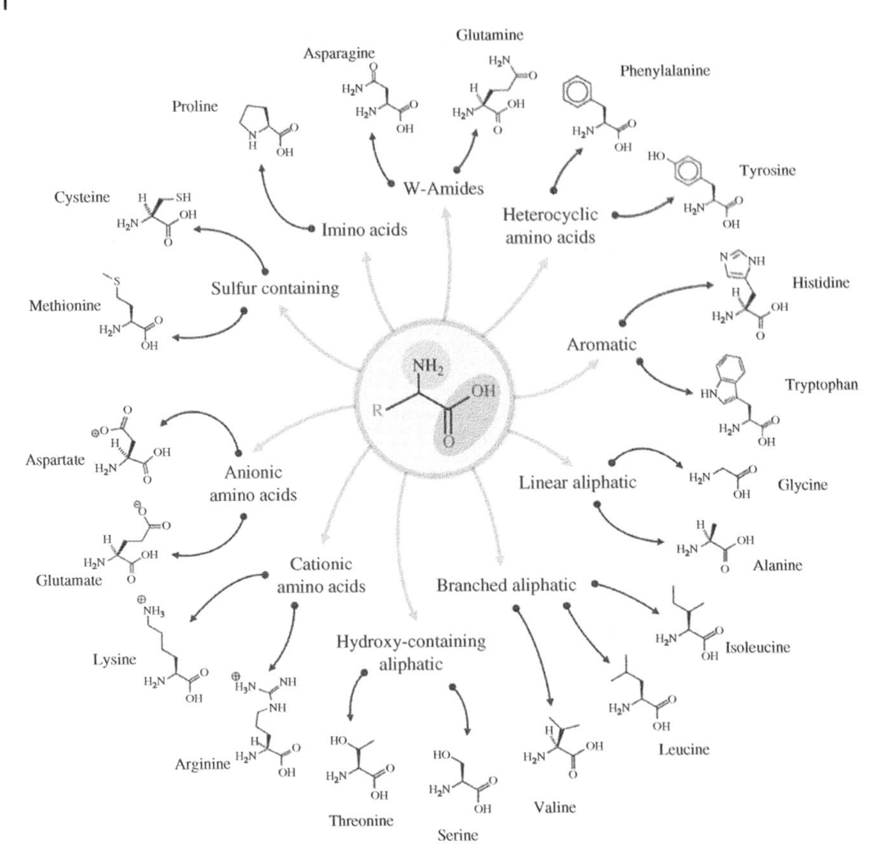

Figure 12.5 Outline of the classification of amino acids. *Source:* Republished from Ref. [28] with permission from Elsevier.

all animals, from bacteria to humans [32]. In addition to the environmental perspective, the presence of heteroatoms (e.g. S, N, and O) and subsequent conjugated π-electron structures in their molecular frameworks has attracted the attention of scientists to further study their capacity to serve as eco-friendly corrosion inhibitors. In addition, amino acids have different uses in food and feed technologies, as well as intermediates for the chemical industry (e.g. pharmaceuticals and cosmetics applications). Important amino acids have been used for human intravenous feeding, and glutamic acids and glutamates are used as taste enhancers in the food manufacturing industry. Biotechnological amino acid production is taking place on a million-ton scale, with a steady increase in demand at an annual growth rate of 5–7%. The fermentation technique yields around six million tons of L-glutamate and L-lysine per year. So many researchers and scientists worldwide have evaluated the corrosion inhibition capabilities of certain amino acids in this context. Because of their nontoxicity and biodegradability, it is specifically important to emphasize on these compounds as reliable green corrosion inhibitors. In addition, they are soluble, relatively inexpensive, and easy to manufacture in high purity in the aqueous media [33].

12.4.2 A General Mechanistic Aspect of the Applicability of Amino Acids and Their Derivatives as Corrosion Inhibitors

Based on the data available related to the literature of the interpretation of the corrosion inhibitive potential of amino acids and their derivatives, it can be concluded that these class of organic compounds exhibit excellent performance against the corrosion of different metals and alloys. This behavior can be attributable to the adsorption (directly or indirectly) of their molecules onto the surface of the substrate (which needs to be protected from corrosion). This adsorption leads to the reduction of contact between the surface of the metal and the corrosive media (nearby the metal surface), thus resulting in the formation of the protective barrier or a film, and this can be concluded on the basis of the facts available in the literature. There are various factors that directly or indirectly influence the mode of the adsorption, and some of them are (i) charge (nature and magnitude) on the surface; (ii) chemical nature of the surface; (iii) the nature of the inhibitor employed; and (iv) the protonation status of the inhibitors. The process of adsorption is carried out in two possible ways, i.e. via chemisorption and physisorption. In chemisorption mode of adsorption mechanism, the adsorption of the inhibitor is carried out by the neutral species of the molecules, while in physisorption, the adsorption is carried out by the charged metal surface electrostatically. First, the direct electrostatic interaction within the charged amino acid molecules and the opposite charged metal surface takes place. If so, the metal surface is negatively charged with respect to zero charge potential (PZC). The molecules' adsorption takes place directly onto the surface of the metal. Adsorption might occur indirectly from the already-adsorbed anions (e.g. halide ions) with respect to PZC through electrostatic interaction on the positively charged metal surface. The adsorbed anions on the surface of the metal make it negatively charged, which increases the protonated amino acid's adsorption capacity. Such behavior is observed, particularly in the acidic medium. None of the cation or anion, though, will be adsorbed on the surface if the surface charge of the metal becomes zero. In this case, the amino acid molecule is thought to be chemically adsorbed. Since the electron donors are amino acids and the metallic atoms on the surface are electron acceptors. The "donor–acceptor" interactions that transfer unshared electron pairs of heteroatoms (i.e. O, N, and S) and/or the π electrons of the amino acid aromatic ring to the empty "d" orbital of surface metal atoms are another mode.

12.4.3 Factors Influencing the Inhibition Ability of Amino Acids and Their Derivatives

In the dynamic phenomena of corrosion inhibition, there are different factors that play an important role, and the efficiency of inhibition is dependent on these factors [34]. Immersion duration, concentration of the tested inhibitors, chemical composition of the tested inhibitors and metals, metal surface, additive nature, synergistic effect, solution pH, temperature, hydrodynamic solution, aerated and de-aerated solution, etc. are some of the main variables. Immersion time can play a key role in the potential to resist corrosion. On the one side, V. Shkirskiy et al. [35] noted that cysteine had a double impact on galvanized steel corrosion, which acted both as an inhibitor of corrosion at short exposure and as an accelerator of corrosion at long immersion. In the other hand, as illustrated by the instance of alanine for nickel [36] and glutamine for mild steel [37], it has been depicted that certain amino acids' inhibition efficiency got strengthened with the duration of exposure. Such condition may be attributable to the instability/stability of the adsorbed amino acid film onto the surface of metal, and the potential to react with metallic ions in the solution is the another attributable factor.

12.5 Overview of the Applicability of Amino Acid and Their Derivatives as Corrosion Inhibitors

Amino acids are environmentally sustainable materials processed at low expense, fully soluble in aqueous media, and nontoxic with excessive purity. Their applicability as economic, green corrosion inhibitors can also be presumed by these properties. A vast array of literature has been documented as corrosion inhibitors in research incorporating their use for iron and its alloys' corrosion [38], bronze [27], zinc [39], copper [40], aluminum alloy [41], and tin [42]. Many methods are used to control the ability of amino acids and derivatives to prevent corrosion, including weight loss, electrochemical tests (such as open-circuit potential (OCP), polarization curve procedure (PDP), electrochemical impedance spectroscopy (EIS), hydrogen evolution measurements, thermometric techniques, microscopy techniques (SEM, AFM), and so on). The majority of papers written in the literature concentrating about the use of amino acids and their derivatives, either alone or with other amino acids as a corrosion inhibitor for various metals, have been compiled or discussed to understand their behavior toward the corrosion inhibition of varying metals in different media.

12.5.1 Amino Acids and Their Derivatives as Corrosion Inhibitor for the Protection of Copper in Different Corrosive Solution

Copper and its alloys are commonly used in different industries due to certain attractive characteristics, including strong corrosion resistance, high thermal and electrical conductivity, mechanical resilience, and malleability [43]. Copper and its alloys are commonly revered because of their large use in the manufacture of wire, sheets, and pipelines in the electrical industry, marine industries, power plants, heat exchangers, and cooling towers. Copper is deemed to be a noble metal, owing to the formulation of a protective passive (oxide) film or nonconductive film of corrosion products on its surface. This in turn provides adequate resistance to corrosion in the atmosphere and also in certain chemical environments [44]. Despite corrosion-resistive properties, many violent ions suffer from aqueous corrosion of copper as a strategic metal in industrial and academic fields. So copper corrosion safety is a very critical and challenging problem. For many decades, researchers have used a wide array of corrosion inhibitors to mitigate the harm that occurs in corrosive conditions.

Kumar et al. [45] have potentially evaluated amino acids such as cysteine (I), glutamic acid (II), glycine (III), and their derivative glutathione (IV) as copper corrosion inhibitors in one of their current research undertaken. By exercising the DFT approach, quantum chemical properties and adsorption of all the inhibitors (under the research) onto the surface of Cu (1 1 1), (1 0 0), and (1 1 0) are conducted out. The measured quantum chemical parameters of E_{HOMO} and Egap showed that the efficiency of inhibition varied in order: IV > I > II > III. The degree of adsorption capacity followed the above pattern as well. This finding is in line with the efficiencies of experimental inhibition reported by Zhang et al. [33] earlier. Table 13.1 indicates the outcome. The trend also persisted with the rate of adsorption potential (as already mentioned above). This observation is consistent with Zhang et al. They reported experimental inhibition efficiencies. Table 13.1 depicts the outcomes. The corrosion resistance of copper given by I, II, III, and IV solutions in a 0.5 M hydrochloric acid solution was investigated by EIS and cyclic voltammetry. The overall efficacy of cysteine inhibition was approximately 92.9% at 15 mM concentration. At a concentration of 10 mM, glutathione yielded 96.4% inhibition effectiveness.

A comparison analysis of benzotriazole, alanine, and cysteine employing weight loss, PDP, and EIS was undertaken by Zhang et al. [46] in an aerated solution (solution of 0.5 M HC). They found that cysteine > alanine > benzotriazole was the inhibition potency sequence, and they all acted as copper wire's anodic inhibitors. In one of their works, Moretti et al. [47] explored a very powerful inhibitory effect of tryptophan at short-term exposure period and at the highest corrosion concentration of pure copper in aerated sulfuric acid. The amino acid undertaken in the work (Tryptophan) was also photodegraded over time period (6 months), as illustrated by using spectrophotometric technique. But this finding did not have a sufficiently interesting effect on its inhibition efficiency. In another study, Petrović et al. [48] and Simonović et al. [49] reported an influence of the pH of 0.5M Na_2SO_4 solution on the inhibition output of cysteine in copper. Based on the result obtained, cysteine was highly physically adsorbed in neutral and alkaline solutions (pH 9) with a moderate chemical adsorption property on the copper surface, whereas it was mainly chemisorbed in acidic solution (pH 2). Furthermore, its protective effect was exemplary in alkaline and acidic as in neutral solution. Some other research results on amino acids as corrosion inhibitors for the preservation of copper and its alloys are reviewed in Table 12.1.

12.5.2 Amino Acids and Their Derivatives as Corrosion Inhibitor for the Protection of Aluminum and Its Alloys in Different Corrosive Solution

Exemplary formability, light weight (density of 2.7 g/cm^3), high electrical and thermal conductivity, and high reflectivity are evidenced by aluminum. Aluminum is quite economical and nearly double as plentiful as iron. Most aluminum alloys' high strength-to-weight ratio is also equal to that of high-strength structural steels. Moreover, owing to the existence of a resistive oxide coating, aluminum exhibits strong corrosion resistance when exposed to the atmosphere and certain aqueous conditions. In comparison, aluminum corrosion materials are colorless and nonpoisonous [65] in nature. The combination of these properties and others has made aluminum one of the most attractive materials in several applications, including as packaging for food, beverages, and pharmaceutical products, sacrificial anodes, automotive, marine, and aerospace parts. Even though oxide layer formed on the aluminum surface safeguards it in certain environments, this oxide layer is lost and the metal corrodes on exposure to some acidic and chloride-containing solutions, especially alkaline conditions.

For their (aluminum and its alloys) protection from corrosion, many approaches have been widely implemented in order to protect aluminum objects, the application of natural compounds as inhibitors of corrosion is one of the major methodology.

Alanine, leucine, methionine, valine, proline, and tryptophan were explored by Ashassi-Sorkhabi et al. [41] as corrosion inhibitors for 99.99% aluminum in a solution of HCl + H_2SO_4 (both of 1M concentration). The inhibition effectiveness increased with increasing amino acid concentration and with decreasing temperature. All amino acids acted as mixed-type inhibitors, with a significant increase in inhibition effectiveness in the case of an aromatic ring and heteroatoms, such as

Table 12.1 Depiction of scientific findings attributable to amino acids as inhibitors of corrosion for the protection of copper and its alloys.

S.No.	Metal	Medium	Inhibitors	Additives	Technique used	Outcome	References
1	Cu10Al 5Ni	Neutral 3.5% NaCl	Cysteine (I), N-acetylcysteine (II), and methionine (III)	—	OCP, PDP, EIS, and SEM	• IE (at 6.0 mM): I (96%) > II (88%) > III (77%). • Followed Langmuir adsorption isotherm • Exhibited very strong physical adsorption	[50]
2	Cu10Ni	3.5% NaCl + 20 ppm Na_2S	Glycine	S^{2-}	EFM, SSC, PDP, and SEM	• IE (at 500 ppm): 87.4% • Gly act as good inhibitor for Cu10Ni alloy • Acted as mixed-type inhibitor	[51]
3	Cu10Ni	3.5% NaCl + 16 ppm Na_2S	Valine (I), alanine (II), and glycine (III)	I^-	PP. EIS. EFM SEM-EDX, *MNDO*	• All acted as mixed-type inhibitors • IE: I > II > III • Exhibited physisorption • Between these inhibitors and I-ions, a synergistic effect was found	[52]
4	Cu	0.1 M H_2SO_4	Methionine (I), cysteine (II), Serene (III), arginine (IV), glutamine (V), and asparagine (VI)	—	PDP, EIS, DFT, and MD simulation	• IE (%) at 10^{-2} M followed: IV > V > VI > I > II > III • Arginine was a cathodic inhibitor • However, rest of the amino acids acted as mixed type corrosion inhibitors	[53]
5	Cu	3.5% NaCl	Methionine	—	PDP, EIS, and SEM	• Acted as mixed-type inhibitor • IE increases with decreasing methionine concentration • Depicted physisorption	[54]

(*Continued*)

Table 12.1 (Continued)

S.No.	Metal	Medium	Inhibitors	Additives	Technique used	Outcome	References
6	Cu	3.5% NaCl	Cysteine	—	PDP, EIS, and SEM	• Classified as anodic inhibitor • Acted as good corrosion inhibitor for Cu in the studied conditions at 10 mM • Adsorption is purely physical	[55]
7	Cu	0.5M HCl	Phenylalanine	Ce^{4+}	WL, PDP, EIS, FT-IR, and SEM-EDX	• By forming a Phe/Ce4+ complex film on the copper surface, phenylalanine with Ce^{4+}ions established a strong synergistic effect • IE (5mM Phe + 2mM Ce^{4+}): 72%	[56]
8	Cu	8M H_3PO_4	Proline (I), cysteine (II), phenylalanine (III), alanine (IV), histidine (V), and glycine (VI)	—	PDP and *PM6*	• Adsorbed via physisorption • V, II, and III illustrated maximum efficiency • Between the theoretical calculations and experimental findings, a significant correlation was observed	[57]
9	Cu	0.5, 1.0, and 1.5M of H_2SO_4	Cysteine	—	WL, PDP, and EIS	• Behaved as mixed-type inhibitor • Exhibited mixed adsorption	[58]
10	Cu	2M HNO_3	Cysteine methyl ester hydrochloride	—	WL	• Spontaneous adsorption process • The rise in temperature contributes to a decline in IE	[59]
11	Cu	0.5M HCl	Glutamic acid (I), cysteine (II), glycine (III), and glutathione (IV)	—	EIS, cyclic voltammetry (CV), *PM.*	• Adsorption is purely physisorption type • IE followed the order: IV > II > II + I + III > I > III	[60]

12	Cu	0.5 M HCl	Methionine	Cetrimonium bromide (CTAB) and cetylpyridinium bromide (CPB)	PDP, EIS, CV, and *PM3*	• Acted as cathodic inhibitor • Compared with the blended CPB/ methionine, the mixed CTAB/ methionine has a stronger synergistic effect • Exhibits strong electrostatic interaction	[61]
13	Cu	0.5M HCl	Glycine (I), threonine (II), phenylalanine (III), and glutamic acid (IV)	—	PDP	• All except III acted as cathodic inhibitors • Best inhibition effect is exhibited by IV (IE = 53.6%)	[62]
14	Cu	0.5M H_2SO_4 in O^{2-} saturated solution	Glycine (I), alanine (II), valine (III), and tyrosine (IV)		PDP, LP, EIS, and EFM	• CV, PDP, and EISActed as mixed-type inhibitors • IE dependent onto the amino acids' structure • IE: III = II << I < IV	[43]
15	Cu	0.5M HCl	Methionine	Zn^{2+}	CV, PDP, and EIS	• Limited inhibiting properties of methionine have been shown • Physisorption • Synergic effect existed between methionine and Zn^{2+} ions	[63]
16	Cu	0.5M HCl	Serine (I), threonine (II), and glutamic acid (III)	—	PDP, EIS, FTIR, and *PM3*	Exhibited as cathodic inhibitors • IE (at mM): III > II > I • Adsorption via chemisorption	[46]
17	Cu	0.6M NaCl and 1.0M HCl	Cysteine	Cu^{2+}	PDP and EIS	• Acted as cathodic inhibitor • Showed physisorption • Due to the synergistic effect, the IE increases in the presence of Cu^{2+} ions	[64]

sulfur and nitrogen on the amino acid structure. The inhibition effectiveness followed the order tryptophan > methionine > proline > leucine > valine > alanine.

In the three media, i.e. 1M HCl, 20% $CaCl_2$, and 3.5% NaCl, the inhibitive performances of the amino acid tryptophan against the corrosion of AA2024 alloy was observed [66]. The authors found that the assessed formulation avoids corrosion, and the overall inhibition efficiency obtained was 87% at 8 mM. Furthermore, it was recognized as an important part of solution, and nature of the electrochemical property of tryptophan was reported as cathodic inhibitor in HCl, while it was an anodic inhibitor in the other solutions.

Amin [67] synthesized and characterized a derivative of glycine (2-(4(dimethylamino)benzylamino) acetic acid hydrochloride in one of their corrosion inhibitor investigation exploration. In effort to get more insight into its effect on aluminum corrosion, they used electrochemical and physical techniques. The authors concluded that the amino acid derivative (under investigation) worked as mixed type of inhibitors for the suppression of the pitting of aluminum, i.e. caused by the anions of SCN^- (at pH 6.8). There was a huge improvement in the performance of the assessed inhibitor relative to glycine. However, the environmental consequences of the derivate are not well estimated.

Erosion–corrosion of AA7075 alloy experiment was executed in 3.5% NaCl solution by spinning disc electrode, and the inhibition action of glutamine was researched [68]. The findings revealed that a boost in rotation speed enhances the process of corrosion, and the performance in stagnant solution was lower but substantially increased under mid-hydrodynamic conditions.

The applicability of the glutamic acid as corrosion inhibitor toward the inhibition of the aluminum corrosion (in 0.1M HCl) was researched by the Zapata-Loria et al. [69], in which the outcomes of the polarization curves exhibited that it acted as mixed-type corrosion inhibitor. This inhibition of aluminum corrosion is owing to the chemisorption process of inhibitor molecules on the metal surface by the application of a robust chelate system, as illustrated by the surface inspection employing XPS analytical technique.

The effect of glycine, alanine, and other unrelated compounds on the electrochemical activity of aluminum-air batteries in a heavy alkaline medium was studied in a paper published by Brito et al. [70]. The electrochemical analyses showed that the measured amino acids had moderate inhibition efficiency at a concentration of 10 g/l.

Another study involving the quantum chemical analysis on the inhibition potency of six amino acids (glycine, aspartic acid, valine, alanine, phenylalanine, and glutamic acid) at pH = 5 and six pH = 8 hydroxy carboxylic acids, i.e. glucolic acid, malic acid, lactic acid, mandelic acid, benzyl acid, and citric acid, was published by Yurt et al. [71]. The analysis was carried out by employing 0.05M NaCl solution for the safety of the aluminum alloy AA7075. The results demonstrated that inhibition of pitting corrosion played a crucial role in both physisorption and chemisorption. In order to get more insights, some results are overviewed in Table 12.2.

Table 12.2 Illustration of findings of the investigations related to the amino acids as corrosion inhibitor for the protection of aluminum and its alloys.

S.No.	Metal	Medium	Inhibitor	Technique	Findings	References
1	Al	0.2–1 M HCl	Glutamic acid	WL, gasometric method, and thermometric method	• IE: 66.67% • Exhibited physical adsorption • Isotherm models studied: Langmuir, Frumkin, Flory-Huggins and kinetic and thermodynamic El-Awady	[72]
2	Al	0.1 M HCl	Glutamic acid	EIS, PDP, and XPS	• IE: 81.5% • A stable chelate was formed on to surface of aluminum • Behaved as mixed-type inhibitor • Hill-de Boer isotherm followed by the molecules • Gave chemisorption	[73]
3	Al	1M HCl + 1M H_2SO_4	Alanine (I), valine (II), leucine (III), proline (IV), methionine (V), and tryptophan (IV)	weight loss measurement, linear polarization and SEM technique	• IE: I (56.3%) < II (58.2%) < III (62.4) < IV (63.3%) < V (81.0%) < VI (90.0%) • Worked as mixed type inhibitor • Langmuir and Frumkin isotherms were followed • The SEM images affirm that aluminum corrosion in mixed acid solutions is well prevented by amino acids	[41]
4	aluminum alloy AA7075	0.05M NaCl	Glycine, aspartic acid, valine, alanine, phenylalanine, and glutamic acid	Quantum chemical calculations	• In the prevention of pitting corrosion, physical and chemical adsorption play a crucial role	[71]

12.5.3 For the Protection of Iron and Its Alloys in Different Corrosive Solution

The most desirable building materials are iron alloys, in particular steels, because of their high mechanical qualities, availability, and appropriate prices. But they possess serious threat of destruction due to high risk of corrosion, as a result of interaction with the nearby environment. Thus, half of the existing research (pertaining with the use of amino acids in various media as corrosion inhibitors) is attributable to the protection of iron and its alloys. A large number of authors working in the field of corrosion inhibitors have stated the inhibiting strength of novel amino acid derivatives against the corrosion of iron and its alloys [74–77]. The results drawn suggested that such compounds had a high propensity to serve as effective corrosion inhibitors.

In one of the work explored by Mobin et al. [12], the corrosion inhibiting efficacy of the cysteine alone and in combination of various surfactants, i.e. triton X-100 (A), sodium dodecyl sulphate (B), and cetyl pyridinium chloride (C), was explored by employing various techniques for inhibiting the corrosion of mild steel (MS) in aggressive corrosive solution (1M HCl). They reported that at the concentration of 500 ppm of cysteine, the inhibiting efficacy was found as 85.62%. At the same concentration of cysteine, the inhibiting performance was further enhanced by the addition of the surfactants A, B, and C (employed at the concentration of around 1 and 5 ppm). The inhibiting efficiency was further increased by the addition of A (95.09%), B (91.99%), and C (97.96%). The inhibiting potency of the inhibitors amended in the decreasing order: cysteine + C > cysteine + B > cysteine + A > cysteine. In some other study, Mobin et al. [78] explored the inhibitory impact of L-phenylalanine methyl ester hydrochloride on MS corrosion in corrosive solution (1M HCl solution) at varying concentrations and temperatures (30–60 °C) was also documented.

The inferred conclusions showed that the inhibitor (under the assessment) exhibited significant rating in inhibiting the corrosion of MS in HCl solution (1M concentration). It can be concluded from the outcomes of the experiments, the inhibitory effect got strengthened with concentration of the inhibitor (employed) up to 400 ppm. It also increases with the rise in temperature; this finding exhibited chemical adsorption mode. Silva et al. [79, 80] assessed the cysteine's performance for 304l stainless steel corrosion in de-aerated 1M sulfuric acid solution. The investigators indicated that the presence of cysteine at high concentrations renders the metallic surface electrochemically active by electrochemical measurements and surface analytical tools. Some more investigations related to the protection of iron and its alloys from corrosion by the applications of the amino acid and there derivatives are discussed in Table 12.3.

Table 12.3 Illustration of findings of the investigations related to the amino acids as corrosion inhibitor for the protection of Iron and its alloys.

S.No.	Metal	Medium	Inhibitors	Additives	Techniques used	Findings	References
1	Iron	1M HCl	Methionine, cysteine, cystine glycine leucine arginine serine, glutamic acid, ornithine lysine aspartic acid, alanine valine asparagine, glutamine, threonine	—	WL, EIS, PDP, and quantum chemical calculations	• The studied amino acids behaved as cathodic inhibitors • The best inhibitors are methionine, cysteine, and cystine, owing to the inclusion of sulfur atoms in their molecular structure	[81]
2	Iron	0.1M H_2SO_4	Glutamic acid derivatives *N*-Phthaloyl-L-glutamic acid (I), *N*-benzoyl-L-glutamic acid (II), and *N*-(1-oxooctadecyl)-L-glutamic acid	—	Electrochemical methods	• I (66.89%) < II (64.03%) < III (60.04%) • Development of protective films	[82]
3	Iron	9 g/l NaCl	Methionine (I) and some amino esters: methionine methyl ester (II), and methionine ethyl ester (III)	—	PDP	• IE: III (80%) > II (40%) > II (28%) at 10^{-2} M • The sulfur atom boosts the electron donor's adsorption in structure B, the electronegativity of which is reinforced by the ethyl group • Isotherm of Frumkin, assisted by the adsorption of B	[83]

(*Continued*)

Table 12.3 (Continued)

S.No.	Metal	Medium	Inhibitors	Additives	Techniques used	Findings	References
4	Iron	Citricchloride solution (pH 5)	Methionine ethyl ester	—	PDP and EIS	• Excellent inhibitor • Acted as mixed-type inhibitor • Followed Temkin adsorption isotherm • With growing temperature, IE altered slightly	[84]
5	Carbon steel	Sea water	Glutamic acid	Zn^{2+}	WL, SEM, FTIR	• 200 ppm glutamic acid with 25 ppm Zn^{2+} acted as excellent inhibitors in combination (EI = 87%) • A strong synergistic effect occurred between the glutamic acid and Zn^{2+} • Glutamic acid – Zn^{2+} worked as an anodic inhibitor	[85]
6	Carbon steel	Sea water	Glutamic acid	Zn^{2+}	WL, EIS, PDP, and AFM	• 200 ppm glutamic acid with 25 ppm Zn^{2+} (EI = 87%) • Glutamic acid – Zn^{2+} get adsorbed onto the surface of carbon steel	[86]
7	Carbon steel	0.5M H_2SO_4	Polyaspartic acid	—	WL, PDP, EIS, SEM, and FTIR	• Worked as anodic inhibitor • IE (at 10 °C) = 80% • Followed Freundlich adsorption isotherm	[87]

8	Carbon steel	1 M HCl	Alkylamides derivatives of tyrosine and glycine	—	PDP, WL	• Acted as mixed-type inhibitors • Tyrosine dodecyl amine achieved greater IE due to large steric body toward glycine • IE correlate with alkylic chain length • Followed the Temkin isotherm	[88]
9	Mild steel	1M HCl	Poly(vinyl alcoholleucine)	Br^- and I^-	WL. PDP. EIS. UV-visible	• IE relies onto the concentration, temperature, and immersion time • Acted as mixed-type inhibitor • Synergistic effect observed between the inhibitor and Br^-/I^- ions	[89]
10	Mild steel	15% HCl	Two derivatives: 2-(2- oxo-2-phenothiazin-10- yl) ethylamino)-3-mercaptopropanoic acid (I) and 2-(2-oxo-2- phenothiazin-10- yl) ethylamino) acetic acid (OPEA)	—	WL, PDP, EIS, SEM, AM1	• IE = I (97.5%) > II (95.8%) at 200 ppm • Acted as mixed-type inhibitors • Followed the Langmuir adsorption isotherm • Theoretical calculations was in line with the findings of the experiment	[90]
11	NST-44 mild steel	Cassava Fluid	Leucine (I), alanine (II), methionine (III), and glutamic acid (III)	-	WL, OM	IE increase as follow: II (50%) > I (46%) > III (45%) > IV (30%)	[91]

(Continued)

Table 12.3 (Continued)

S.No.	Metal	Medium	Inhibitors	Additives	Techniques used	Findings	References
12	NST-44 mild steel	Lime Fluid	Leucine (Leu), alanine (Ala), methionine (Met), and glutamic acid (Glu-A)	—	WL, OM	• IE: II > III > I > IV • IE increases with rise in the concentration	[92]
13	Low alloy steel	0.2M ammoniated citric acid	Tryptophan (I), tyrosine (II), and serine (III)	—	PP, EIS, EFM, and OM	• IE (at 0.06M concentration) = I (86%) > II (83%) > III (82%) • Adsorption via physisorption • As a repercussion of the displacement of inhibitor layers from the metal surface, IE tends to reduce with the enhanced stirring speed of the solution	[93]

12.6 Recent Trends and the Future Considerations

The search for corrosion inhibitors that can be classified as "green" and that can be easily found and extracted from renewable sources is undoubtedly one research area where several developments are continually evolving. Therefore, this ambiguity brings up a vast spectrum of avenues to pursue in the future. This chapter provided a detailed description of the possibilities of using amino acids and their derived compounds as an inhibitor of corrosion for the safety of various metals in a variety of media under consideration. The use of amino acids and their derivatives to safeguard from metallic corrosion, as described and discussed above, can be handled by applying appropriate several compounds to the desired environment. Recently, the studies on these topics have attracted the attention of many researchers and the number of published works increases. In the domain, there are so many recent developments that are explored below:

12.6.1 Synergistic Combination of Amino Acids with Other Compounds

Synergistic combination of corrosion inhibitors is a strategy that presents high potential in the advancement of novel corrosion protection strategies. Through this process, it is possible to combine different inhibitors, allowing them to work together in a synergistic way. This ensures higher inhibition efficiency and, eventually the use of smaller amounts of corrosion inhibitors, which might have important implications, for example, in coatings design. The synergistic effect of S-containing amino acid and surfactant self-assembly monolayers (SAMs) on corrosion inhibition of 316l stainless steel in a 0.5M NaCl solution has been investigated by Zhao et al. [94] in one of their recent studies. They found that with the increase in amino acid concentration, i.e. cysteine (I) and methionine (II) concentration, the inhibition effect basically increases with the increase in I. Further improvements in the corrosion inhibition efficiency of I and II have been documented due to the synergistic action of the surfactant SDBS. They further explained that the addition of SDBS would essentially increase the hydrophobicity of SAMs attributed to the prevalence of long hydrophobic alkyl chains within the molecules of SDBS. In some other example, the forming of hydrogen bonds between SDBS and -NH2, -COOH, as well as -SH (of CYS), head groups can contribute to the development of denser and more compact SAMs, which can previously be seen directly in SEM graphs. As a result, diffusion of the aggressive ions gets more difficult via the compact and dense SAMs. Therefore, relative to the inhibition property of SAMs of individual CYS or MET, the inhibition efficiency of SAMs was further strengthened.

Shaaban et al. [95] have been tested to prevent the degradation of mild steel in sulfuric acid solution by vanillin and other amino acids. The authors suggested researching the synergistic effect after acquiring the mild efficiency of vanillin and cysteine separately. It was shown that optimum inhibition efficiency is reached in the presence of a vanillin and cysteine mixture. An 87% inhibition efficiency was discovered in the presence of vanillin and cysteine, 0.5 mM each. Vanillin and cysteine claimed a significant synergism (synergism parameter = 1.53) for corrosion inhibition.

12.6.2 Self-Assembly Monolayers (SAMs)

The SAM is prepared by utilizing the strong interaction between the heads of the amphiphiles molecules and the surface of the solid support on the one hand, and intermolecular forces (involving van der Waals and hydrophobic forces) between backbone hydrocarbon chains in amphiphiles molecules on the other. The method of preparation of SAMs is quite simple and involves the immersion of the substrate into a solution containing the amphilphiles. This induces the self-assembly of the amphiphiles as a monolayer onto the substrate's surface, spontaneously. Among SAMs, thiols, dithiols, silanes, phosphonates, and fatty acids are perhaps the most prevalent ones. They have great potential for a wide range of application. Among the various applications of SAMs, one is the prevention of material against corrosion. SAMs are being used for this purpose as ultra-thin layers to prevent from corrosion [96]. In this framework, on certain metal surfaces, SAMs of certain amino acids and their associated compounds were formulated and their ability to regulate the corrosion rate of some metallic materials was assessed [97, 98]. The protection efficiency of the formulated SAM heavily relies on certain considerations, such as the pH and concentration of the amino acid assembly solution employed, and the contact time between the metal and the solution. In addition, as in the traditional corrosion inhibitor model, some studies [99, 100] have shown that some anions (e.g. I^-, MnO_4^-) have a synergistic effect that enhances the defense efficiency of amino acid SAM. This leads to research the optimal conditions to form the efficient monolayer that exhibits great performance. Among tested amino acids, the cysteine has exhibited the highest protection efficiency due to the high affinity of thiol group (–SH) to link with metal surface, hence to form a stable SAM on surface. These studies revealed the high ability of amino acid compounds to control the corrosion processes via this kind of treatment with low compound consumption. This understanding paves way for more research to be devoted toward the evaluation of SAMs of other amino acids as anticorrosion materials.

12.6.3 Amino Acid-Based Ionic Liquids

As it is already known, a single amino acid molecule in its structure consists of a carboxylic acid residue and an amino group. Due to this fact the amino acids can be employed as both anions/cations. Such entities are also significant because of their ability to incorporate cohesive groups. Basically, in conjunction with complex anions like bromide, chloride, and methioninate, a normal ionic liquid (IL) molecule is formed by a thick organic cation (i.e. ammonium, imidazolium, pyridinium). Easy variations in the structures of the cation and anion or the presence of the moieties added to each ion make it possible for special applications in order to the alteration of the physical properties of ILs. Despite the fact that research on ionic liquids as green inhibitors does not just go beyond 10 years, imidazole and pyridine-based ionic liquid corrosion inhibition has been documented by several researchers [101, 102]. But in spite of several advantages, these complex organic compounds are restricted to be used as corrosion inhibitors because of several reasons, among them one of the reason is their complex multistep synthesis, and another reason is they anticipated to be extremely expensive. Therefore, with simplicity in synthesis and cost considerations, there is a need to explore some more amino-acid-dependent ionic liquids exhibiting better efficiency.

12.6.4 Amino Acids as Inhibitors in Smart Functional Coatings

One of the latest coating developments is drawing researchers to build smart coating materials for corrosion mitigation [103–106]. This particular interest is attributed to the improved efficiency of these coatings in order to ensure good long-term protection against degradation of the metal surface following coating failure. The capabilities of smart hybrid coatings even include sensing and warning features, in particular to corrosion prevention and repair. The concept of such a coating is to insert corrosion inhibitors in microcapsules for protection purposes prior to their entrance into a protective layer that enables regulation of the release of the inhibitor in the event of exposure to the coating. In this sense, in order to apply this approach to galvanized steel, V. Shkirskiy et al. [35, 107] explored the idea of using cysteine in saline medium for corrosion control. As the findings show, they observed that cysteine's actions on this substance had a double effect, namely, as a high concentration of corrosion accelerator and long exposure, and as a corrosion inhibitor of low concentration and short exposure, making its practical use spurious in the smart coatings scheme. Substantial attention has been paid to hybrid coatings for corrosion suppression of metal substrates (steel, magnesium, aluminum, and their alloys) due to their ability to monitor the efficacy of cracks in protective coatings by removing active agents from micro/nanocapsules, i.e. micro/nanoparticles consisting of coatings or shells (micro/nanocontainers) and core coatings as seen in Figure 12.6. Naderi et al. [108] documented the use of L-cysteine reduced graphene oxide as a smart/control trigger nanocarrier for cerium ions to reinforce epoxy-coating anticorrosion functionality. However, to the best of

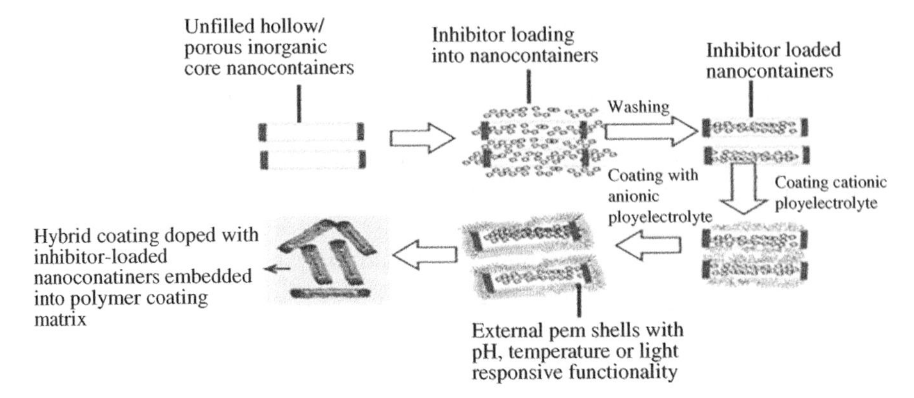

Figure 12.6 Schematic diagram for inorganic–organic core-shell nanotubes for preparation of smart functional nanocomposite coatings. Inorganic nanotubes are eventually loaded with an effective corrosion inhibitor (internal core) and afterwards coated with special multilayer polyelectrolyte receptive products (external shell). *Source:* Adapted from Ref. [109]; © 2015, Elsevier, with permission.

comprehension, use of such amino acid compounds in this mode of corrosion mitigation is not otherwise referred to in the literature. There is space in these types of coatings for more amino acids and their associated compounds to be studied as corrosion inhibitors.

12.7 Conclusion

This chapter covers the ability of amino acids and their derivatives as nontoxic inhibitors to protect various products in diverse conditions against corrosion. A strong ability to use amino acid compounds in various media as eco-friendly inhibitors against metal corrosion has been demonstrated, as is evident from the literature. In general, amino acids contain both the acid-property carboxyl group and the fundamental-property amine group, which are two functional groups of opposite chemical nature. Massive inhibition efficiency will be provided by amino acids, including additional groups such as sulfur atoms and longer hydrocarbon chains that would increase electron density on the alpha amino group. Even, there is no uniform amino acid compound common to most metal/solution systems. Therefore, the ability to prevent these compounds depends on their molecular composition, their concentration, the presence of the metallic base, the corrosive solvent, and other. While certain amino acids may act as corrosion inhibitors, they may also have the opposite effect, accelerating the rate of corrosion, depending on the operating conditions.

Acknowledgments

Authors acknowledge the financial support from Council of Scientific & Industrial Research (CSIR), New Delhi, India, through the major research project [file number: 22(0832)/20/EMR-II].

References

1 Corrosion of metals and alloys, Basic Terms and definitions (1999). European Committee for Standardization, EN ISO 8044.
2 Callister, W.D. (2007). *Materials Science and Engineering- An Introduction*, 7e. Wiley.
3 Uhlig, H.H. (1949). The cost of corrosion to the U.S. *Chem. Eng. News Arch.* 27 (39): 2764.
4 Bennet, L.H., Kruger, J., Parker, R.I. et al. (1978). *Economic Effects of Metallic Corrosion in the United States*. Washington, DC: Bureau of Standards Special Publication 511.
5 Koch, G.H., Brongers, M.P.H., Thompson, N.G. et al. (2002). Corrosion costs and preventive strategies in the United States. *Mater. Perform.* 42 (Supplement).
6 Mobin, M., Aslam, R., and Aslam, J. (2017). *Mater. Chem. Phys.* 191: 151–167.
7 Uhlig, H.H. (1965). *Corrosion and Corrosion Control: An Introduction to Corrosion Science and Engineering*. Wiley.
8 Abd-El-Nabey, B.A., Khalil, N., and Mohamed, A. (1985). *Surf. Technol.* 24: 383–389.
9 Gece, G. and Bilgiç, S. (2010). *Corros. Sci.* 52: 3435–3443.
10 Abiola, O.K., Oforka, N.C., and Ebenso, E.E. (2004). A potential corrosion inhibitor for acid corrosion of mild steel. *Bull. Electrochem.* 20: 409–413.
11 Parveen, M., Mobin, M., and Zehra, S. (2016). Evaluation of l-tyrosine mixed with sodium dodecyl sulphate or cetyl pyridinium chloride as a corrosion inhibitor for mild steel in 1 M HCl: experimental and theoretical studies. *RSC Adv.* 6: 61235–61248. https://doi.org/10.1039/C6RA10010D.
12 Mobin, M., Zehra, S., and Parveen, M. (2016). l-Cysteine as corrosion inhibitor for mild steel in 1 M HCl and synergistic effect of anionic, cationic and non-ionic surfactants. *J. Mol. Liq.* 216: 598–607. https://doi.org/10.1016/j.molliq.2016.01.087.
13 Aslam, R., Mobin, M., Zehra, S. et al. (2017). *ACS Omega* 2: 5691–5707.
14 Raja, P.B. and Mathur, G.S. (2008). Natural products as corrosion inhibitor for metals in corrosive media – A review. *Mater. Lett.* 62: 113–116.
15 Chaouiki, A., Lgaz, H., Zehra, S. et al. (2019). *J. Adhes. Sci. Technol.* 33: 921–944.
16 Lgaz, H., Zehra, S., Albayati, M.R. et al. (2019). *Int. J. Electrochem. Sci.* 14: 6667–6681.

17 Jennane, J., Touhami, M.E., Zehra, S. et al. (2019). *Mater. Chem. Phys.* 227: 200–210.
18 Bidi, H., Touhami, M.E., Baymou, Y. et al. (2020). *Chem. Engineer. Comm.* 207: 632–651.
19 Bashir, S., Lgaz, H., Chung, I.-M., and Kumar, A. (2020). *Chem. Engineer Comm.* https://doi.org/10.1080/00986445.2020.1752680.
20 Hossain, N., Chowdhury, M.A., and Kchaou, M. (2020). *J. Adhes. Sci. Technol.* https://doi.org/10.1080/01694243.2020.1816793.
21 Scheithauer, M. (2014). *Handbook of Solvents*, 2e, vol. 2. ChemTec Publishing.
22 Ashish, K.S. and Quraishi, M.A. (2010). *Corros. Sci.* 52: 152–160.
23 Abdel-Graber, A.M., Abd-El-Nabey, B.A., Sidahmed, I.M. et al. (2006). *Corros. Sci.* 62 (4): 239–299.
24 Behpour, M., Ghoreishi, S.M., Khayatkashani, M., and Soltani, N. (2012). *Mater. Chem. Phys.* 131 (3): 621–633.
25 Ogunleye, O.O., Arinkoola, A.O., Eletta, O.A. et al. (2020). *Heliyon* 6 (1): e03205.
26 Verma, D.K. and Fahmida, K. (2016). *Green Chem. Lett. Rev.* 9 (1): 52–60.
27 Hamadi, L., Mansouri, S., Oulmi, K., and Kareche, A. (2018). *Egypt. J. Pet.* 27: 1157–1165.
28 El Ibrahimi, B., Jmiai, A., Bazzi, L., and El Issami, S. (2020). *Arab. J. Chem.* 13: 740–771.
29 Zhao, Q., Guo, J., Cui, G. et al. (2020). *Colloids Surf. B* 194: 111150.
30 Biswas, A., Pal, S., and Udayabhanu, G. (2015). *Appl. Surf. Sci.* 353: 173–183. https://doi.org/10.1016/j.apsusc.2015.06.128.
31 Ren, Y., Luo, Y., Zhang, K. et al. (2008). *Corros. Sci.* 50: 3147–3153.
32 Kilberg, M.S. and Häussinger, D. (1992). *Mammalian Amino Acid Transport.* Springer Science & Business Media.
33 Zhang, D.Q., Zeng, H.J., Zhang, L. et al. (2014). *Colloids Surf. A Physicochem. Eng. Asp.* 445: 105–110.
34 Sheir, L.L., Jarman, R.A., and Burstein, G.T. (1994). *Corrosion*, 3e, vol. 2. Great Britain: Butterworth-Heinemann.
35 Shkirskiy, V., Keil, P., Hintze-Brueningb, H. et al. (2015). *Corros. Sci.* 100: 101–112.
36 Hamed, E., El-REhim, S.S.A., El-Shahat, M.F., and Shaltot, A.M. (2012). *Mater. Sci. Eng. B* 177: 441–448.
37 Singh, A. and Ebenso, E.E. (2013). *Int. J. Electrochem. Sci.* 8: 12874–12883.
38 Goni, L.K.M.O., Mazumder, M.A.J., Ali, S.A. et al. (2019). *Int. J. Miner. Metall. Mater.* 26: 467–482.
39 Nady, H. (2017). *Egypt. J. Pet.* 26: 905–913.
40 Barouni, K., Kassale, A., Albourine, A. et al. (2014). *J. Mater. Environ. Sci.* 5 (2): 456–463.
41 Ashassi-Sorkhabi, H., Ghasemi, Z., and Seifzadeh, D. (2005). *Appl. Surf. Sci.* 249: 408–418.
42 El Ibrahimi, B., Jmiai, A., El Mouaden, K. et al. (2019). *J. Mol. Struct.* 1196: 105–118.

43 Amin, M.A. and Khaled, K.F. (2010). *Corros. Sci.* 52: 1194–1204.
44 Antonijevic, M.M. and Petrovic, M.B. (2008). *Int. J. Electrochem. Sci.* 3: 1–28.
45 Kumar, D., Jain, N., Jain, V., and Rai, B. (2020). *Appl. Surf. Sci.* 514: 145905.
46 Zhang, D.Q., Cai, Q.R., Gao, L.X., and Lee, K.Y. (2008). *Corros. Sci.* 50: 3615–3621.
47 Moretti, G. and Guidi, F. (2002). *Corros. Sci.* 44: 1995–2011.
48 Petrović, M.B., Radovanović, M.B., Simonović, A.T. et al. (2012). *Int. J. Electrochem. Sci.* 7: 9043–9057.
49 Simonović, A.T., Petrović, M.B., Radovanović, M.B. et al. (2014). *Chem. Pap.* 68: 362–371.
50 El-Hafez, G.M.A. and Badawy, W.A. (2013). *Electrochim. Acta* 108: 860–866.
51 Nazeer, A.A., Nageh, K.Y., Gehan, I.A., and Elsayed, A. (2011). *Ind. Eng. Chem. Res.* 50: 8796–8802.
52 Fouda, A.S., Nazeer, A.A., and Ashour, E.A. (2011). *Mater. Prot.* 52: 21–34.
53 Zhang, D.Q., Cai, Q.R., He, X.M. et al. (2008). *Mater. Chem. Phys.* 112: 353–358.
54 Kılınççeker, G. and Demir, H. (2013). *Prot. Met. Phys. Chem.* 49: 788–797.
55 Kılınççeker, G. and Demir, H. (2013). *Anti-Corros. Methods Mater.* 60: 134–142.
56 Zhang, D.Q., Wu, H., and Gao, L.X. (2012). *Mater. Chem. Phys.* 133: 981–986.
57 Rahman, H.H.A., Moustafa, A.H.E., and Awad, M.K. (2012). *Int. J. Electrochem. Sci.* 7: 1266–1287.
58 Kuruvilla, M., John, S., and Joseph Res, A. (2012). *Chem. Int.* https://doi.org/10.1007/s11164-012-0860-y.
59 Zarrouk, A., Hammouti, B., Zarrok, H. et al. (2011). *Int. J. Electrochem. Sci.* 6: 6261–6274.
60 Zhang, D.Q., Xie, B., Gao, L.X. et al. (2011). *Thin Solid Films* 520: 356–361.
61 Zhang, D.Q., Xie, B., Gao, L.X. et al. (2011). *J. Appl. Electrochem.* 41: 491–498.
62 Makarenko, N.V., Kharchenko, U.V., and Zemnukhova, L.A. (2011). *Russ. J. Appl. Chem.* 84: 1362–1365.
63 Zhang, D.Q., Cai, Q.R., He, X.M. et al. (2009). *Mater. Chem. Phys.* 114: 612–617.
64 Ismail, K.M. (2007). *Electrochim. Acta* 52: 7811–7819.
65 Xhanari, K. and Finšgar, M. (2016). *Arab. J. Chem.* https://doi.org/10.1016/j.arabjc.2016.08.009.
66 Li, X., Xiang, B., Zuo, X.L. et al. (2011). *J. Mater. Eng. Perform.* 20: 265–270.
67 Amin, M.A. (2010). *Corros. Sci.* 52: 3243–3257.
68 Ashassi-Sorkhabi, H. and Asghari, E. (2010). *J. Appl. Electrochem.* 40: 631–637.
69 Zapata-Loria, A.D. and Pech-Canul, M.A. (2014). *Chem. Eng. Commun.* 201: 855–869.
70 Brito, P.S.D. and Sequeira, C.A.C. (2014). *J Fuel. Cell Sci. Technol.* 11: 011008.
71 Yurt, A., Bereket, G., and Ogretir, C. (2005). *J. Mol. Struct. THEOCHEM* 725: 215–221.
72 Muhammad, A.A., Uzairu, A., Iyun, J.F., and Abba, H. (2014). *IOSR J. Appl. Chem.* 7: 50–62.

73 Zapata-Lori, A.D. and Pech-Canul Chem, M.A. (2014). *Eng. Comm.* 201: 855–869.

74 Amin, M.A. and Ibrahim, M.M. (2011). *Corros. Sci.* 53: 873–885.

75 Amin, M.A., Khaled, K.F., and Fadl-Allah, S.A. (2010). *Corros. Sci.* 52: 140–151.

76 Deng, Q. et al. (2012). *Corros. Sci.* 64: 64–73.

77 Srivastava, V., Haque, J., Verma, C. et al. (2017). *J. Mol. Liq.* 244: 340–352.

78 Mobin, M., Zehra, S., and Aslam, R. (2016). *RSC Adv.* 6: 5890–5902.

79 Selva kumar, P., Karthik, B.B., and Thangavelu, C. (2013). *Res. J. Chem. Sci.* 3: 87–95.

80 Silva, A.B., Agostinho, S.M.L., Barcia, O.E. et al. (2006). *Corros. Sci.* 48: 3668–3674.

81 Aouniti, K.F. and Khaled, B.H. (2013). *Int. J. Electrochem. Sci.* 8: 5925–5943.

82 Zhang, Z., Yan, G., and Ruan, L. (2012). *Adv. Mater. Res.* 417: 964–967.

83 Bouzidi, D., Chetouani, A., Hammouti, B. et al. (2012). *Int. J. Electrochem. Sci.* 7: 2334–2348.

84 Zerfaoui, M. et al. (2002). *Rev. Metal. Paris J.* 99: 1105–1110.

85 Gowri, S., Sathiyabama, J., Rajendran, S. et al. (2013). *Chem. Sci. Trans.* 2: 275–281.

86 Gowri, S., Sathiyabama, J., and Rajendran, S. (2013). *Int. J. Chem. Tech. Res.* 5: 347–352.

87 Cui, R., Gu, N., and Li, C. (2011). *Mater. Corros.* 62: 362–369.

88 Olivares-Xometl, O., Likhanova, N.V., Dominguez-Aguilar, M.A. et al. (2008). *Mater. Chem. Phys.* 110: 344–351.

89 Rahiman, A.F.S.A. and Subhashini, S. (2013). *J. Appl. Polym. Sci.* 127: 3084–3092.

90 Yadav, M., Kumar, S., and Gope, L. (2014). *J. Adhes. Sci. Technol.* 28: 1072–1089.

91 Alagbe, M., Umoru, L.E., Afonja, A.A., and Olorunniwo, O.E. (2006). *J. Appl. Sci.* 6: 1142–1147.

92 Alagbe, M., Umoru, L.E., Afonja, A.A., and Olorunniwo Anti-Corros, O.E. (2009). *Met. Mater.* 56: 43–50.

93 Abdel-Fatah, H.T.M., Abdel-Samad, H.S., Hassan, A.A.M., and El-Sehiety, H.E.E. (2014). *Res. Chem. Intermed.* 40: 1675–1690.

94 Zhao, R., Xu, W., Yu, Q., and Niu, L. (2020). *J. Mol. Liq.* 318: 114322.

95 Awad, M.I., Saad, A.F., Shaaban, M.R. et al. (2017). *Int. J. Electrochem. Sci.* 12: 1657–1669.

96 Ariga, K. and Kunitake, T. (2006). *Supramolecular Chemistry – Fundamentals and Applications: Advanced Textbook*. Springer Berlin Heidelberg.

97 Migahed, M.A., Azzam, E.M.S., and Morsy, S.M.I. (2009). *Corros. Sci.* 51: 1636–1644.

98 Zhang, D.Q., He, X.M., Cai, Q.R. et al. (2009). *J. Appl. Electrochem.* 39: 1193–1198.

99 Zhang, D.Q., He, X.M., Cai, Q.R. et al. (2010). *Thin Solid Films* 518: 2745–2749.

100 Zhang, D.Q., Liu, P.H., Gao, L.X. et al. (2011). *Mater. Lett.* 65: 1636–1638.

101 Subasree, N. and Arockia Selvi, J. (2020). *Heliyon* 6 (2): 03498.

102 Likhanova, N.V., Domínguez-Aguilar, M.A., Olivares-Xometl, O. et al. (2010). *Corros. Sci.* 52: 2088–2097.

103 Chen, Z., Yang, W., Chen, Y. et al. (2020). *J. Colloid Interface Sci.* 579: 741–753.

104 Dou, S., Zhao, J., Zhang, W. et al. (2020). *ACS Appl. Mater. Interfaces* 12: 7302–7309.

105 Feng, W., Patel, S.H., Young, M.Y. et al. (2007). *Polym. Tech.* 26: 1–13.

106 Kendig, M., Hon, M., and Warren, L. (2013). *Prog. Org. Coat.* 47: 183–189.

107 Shkirskiy, V. (2015). *Corrosion Inhibition of Galvanized Steel by LDH – Inhibitor Hybrids: Mechanisms of Inhibitor Release and Corrosion Reaction*. Universite Pierre Et Marie Curie.

108 Javidparvar, A.A., Naderi, R., and Ramezanzadeh, B. (2020). *J. Hazard. Mater.* 389: 122135.

109 Abu-Thabit, N.Y. and Makhlouf, A.S.H. (2015). Recent advances in nanocomposite coatings for corrosion protection applications. In: *Handbook of Nanoceramic and Nanocomposite Coatings and Materials* (eds. A.S.H. Makhlouf and D. Scharnweber), 515–549. Butterworth-Heinemann.

13

Chemical Medicines as Corrosion Inhibitors

Mustafa R. Al-Hadeethi[1], *Hassane Lgaz*[2], *Abdelkarim Chaouiki*[3], *Rachid Salghi*[3], *and Han-Seung Lee*[2]

[1] *Department of Chemistry, College of Education, Kirkuk University, Kirkuk, Iraq*
[2] *Department of Architectural Engineering, Hanyang University-ERICA, Ansan, Korea*
[3] *Laboratory of Applied Chemistry and Environment, ENSA, University Ibn Zohr, Agadir, Morocco*

13.1 Introduction

Corrosion inhibitors (CIs) are chemical substances that added in small concentrations/amount to a corrosive environment to reduce or stop electrochemical corrosion reactions occurring on a metal surface [1]. They could be classified in different ways such as the original source (organic or inorganic) and preparation techniques (synthesized or extracted). Organic CIs have achieved outstanding importance in recent years attributed to their many achievements as anticorrosion agents under a broad spectrum of corrosive environments [2–8]. However, the recent rise of the "green" chemistry concept in the fields of science, technology, and engineering has restrained the application of some commercial, inorganic, and organic corrosion inhibitors [9–11]. In line with this research trend, the toxicity of commonly used anticorrosion compounds like chromate-based corrosion inhibitors has encouraged synthetic chemists to search for eco-friendly compounds that are effective and contain no toxic units and are biodegradable [12–15].

Several categories of green corrosion inhibitors have been developed in recent decades. These include plant extracts, oils, fresh and expired drugs, surfactants, amino acids, biopolymers, ionic liquids, etc. [16–24]. The preparation of "green" compounds with higher anticorrosion properties, low cost, and biodegradability has become an exciting new science and a challenging task for chemists and technologists today.

Organic Corrosion Inhibitors: Synthesis, Characterization, Mechanism, and Applications, First Edition. Edited by Chandrabhan Verma, Chaudhery Mustansar Hussain, and Eno E. Ebenso.

Drugs are one of the exciting alternatives that have attracted attention of researchers because of their eco-friendly nature and their electron-rich molecular structures, which can make them excellent corrosion inhibitors. In this chapter, attempts were made to present an overview of corrosion inhibitors based on chemical medicines, i.e. fresh and expired drugs. Their application in the mitigation of corrosion of different metal and alloys has been reviewed and discussed. Further, synthesis and application of organic compounds derived from drugs as new promising corrosion inhibitors have also been considered.

13.2 Greener Application and Techniques Toward Synthesis and Development of Corrosion Inhibitors

Several newer environment-friendly approaches have been developed for the synthesis of organic compounds for various applications. Due to the rigorous environmental regulations and increase in the awareness toward the environment, conventional synthesis and usage of the corrosion inhibitors have come under heavy scrutiny across the world. Nowadays, researchers are working on the development of new synthesis methodologies for synthesis of corrosion inhibitors with low toxicity toward the living beings and to the environment [25]. In the following sections, we represent an overview of the most important techniques:

13.2.1 Ultrasound Irradiation-Assisted Synthesis

Ultrasound irradiation techniques have gained more attention in organic synthesis strategies because of many facilities and features. The most important features of these techniques are the low reaction time and the high selectivity. In comparison with other traditional synthesis methods, ultrasound irradiation is regarded as green efficient method because it minimizes the waste and the energy requirements of the reaction. The importance and advantages of these techniques are clearly appeared when compared with traditional procedures that require long time and drastic conditions [26–28].

Sonochemistry can be defined as the application of ultrasound to enhance and encourage the chemical processes. Ultrasonic waves are generated through chains of compressions and expansion. These chains or series are induced in the structure of molecules medium of which it passes through. When the power is high, the cavitation bubbles will be formed leading to a rarefaction cycle to break and weaken the inter-attractive Van der Walls forces of the liquid molecules. The cavitation bubbles will increase and grow within few cycles to obtain an equilibrium size for the applied frequency. The ultrasound effects that are clearly shown within organic reactions are because of cavitation, which can be defined as a physical process that is creating, enlarging, and imploding vaporous and gaseous cavities in an irradiated medium. The cavitation process will generate very high temperatures and pressures within the bubbles, causing a turbulent flow of the liquid and improved mass transfer [29].

US irradiation method can be used in two essential techniques, namely, the bath and the probe. The probe technique is more effective than bath methodology because of the direct irradiation to the treated molecules or medium [30]. Hydrazones, Schiff bases, chalcones, and many other functional groups have been synthesized using US technique in good yield [30, 31]. A detailed description of ultrasound irradiation techniques can be found in [29].

13.2.2 Microwave-Assisted Synthesis

Microwave (MW)-assisted synthesis is regarded as one of the greener applications in the organic synthesis because of its ability to reduce the reaction time and to modify the chemo-selectivity, stereo-selectivity, and regio-selectivity. Moreover, it allows the solvent-free reactions, catalyst-free reactions, high yields in addition to its simplicity [32].Traditional heating methods are very slow compared to MW, and they make hot spots and surfaces resulting in decomposition of the reagents and products. But MW enters the reaction pot or reactor uniformly through the walls of the reactor resulting in direct heating to the reactor contents [33].

This technique has played since 1980s a crucial role in organic synthesis and many other scientific fields such as inorganic, solid dehydration, medicinal and combinatorial chemistry, polymers cyclo-addition, and so on. These facts are clearly confirmed by the growing number of published articles related to MW in last two decades, especially in 2003 when the reliable MW equipment were available [34, 35].

In comparison with traditional heating techniques, US- and MW-assisted synthesis techniques have several features and advantages such as homogeneous heating, lower waste production, cheap, excellent purity, higher yields. Using US and MW methods, almost all types of chemical reactions and processes could be done. The high selectivity of microwave and ultrasound irradiation reactions can be adjusted by the careful selection of the reaction conditions [36].

13.2.3 Multicomponent Reactions

Multicomponent reactions (MCRs) can be defined generally, as a reaction where three or more compounds can be mixed in one pot to provide an efficient and single final product with excellent modular approach toward molecular diversity. They became an effective technique for the drug industry because they produce faster, diverse, and more efficient methods to build a wide library of simple organic compounds from well-known intermediate precursors. Sometimes, MCRs lead to strange or unexpected, or may be unique products. So the functionalization or the modification of these products represents the key to discover the utility of these products in the field of biological applications.

The most important traditional MCRs involves Passerini, Mannich, Kabachnik-Fields, Ugi, Biginelli, Strecker, Bucherer-Bergs, and Van Leusen reactions. However, the Isocyanide MCRs represent the essential reactions for the recent MCR techniques. The principal feature of these reactions is the functionality of isocyanide that could be electrophilic and nucleophilic making the building of linear peptide structure more possible. There are two principle examples of isocyanide MCRs. The Passerini reaction and the Ugi reaction [37, 38].

Passerini reaction was the first isocyanide MCR discovered by Passerini in 1921. Since its discovery, Passerini reaction had an essential role in drug synthesis and in all combinatorial chemistry aspects. The general description of this reaction includes the reaction of three components (isocyanide, carboxylic acid, and aldehyde or ketone) to obtain α-hydroxy carboxamides directly (Scheme 13.1).

Ugi reaction is an important strategy to build peptide-like structures. The first description of this reaction was in 1959 by Ugi. It is a four-component reaction that includes carboxylic acid, aldehyde, isocyanide, and an amine to obtain α-acylomino amides (Scheme 13.2) [40].

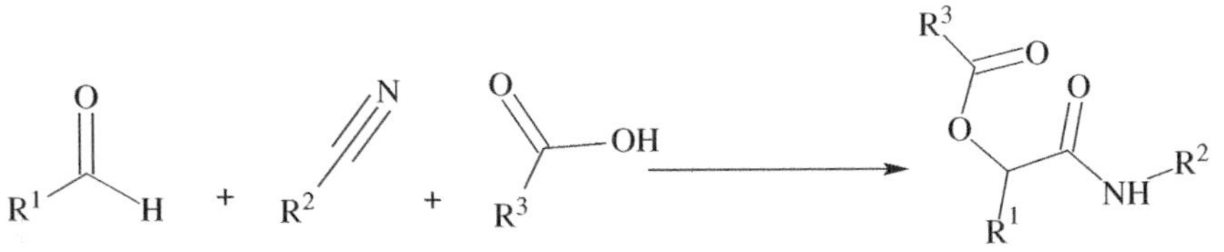

Scheme 13.1 The simple description of Passerini reaction. *Source:* Ref. [39].

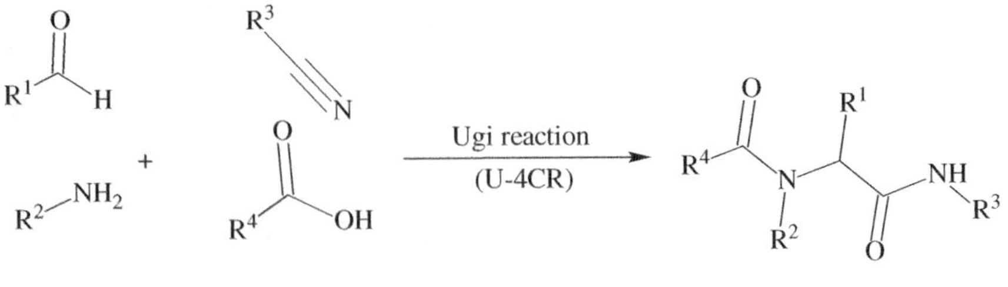

Scheme 13.2 The simple description of Ugi multicomponent reaction. *Source:* Ref. [40].

13.3 Types of Chemical Medicine-Based Corrosion Inhibitors

13.3.1 Drugs

Drugs are nontoxic, cheap, with negligible negative effects on environment, and thus they are suggested to replace the traditional toxic corrosion inhibitors. Many researchers in this field generally agree that drugs are effective corrosion inhibitors and can compete favorably with other green corrosion inhibitors owing to their efficacy and the fact that most of them synthesized from natural products. The choice of drugs used as corrosion inhibitors is mainly based on the following: (i) drug molecules contain oxygen, nitrogen, sulfur, etc. as the main active centers, (ii) drugs are reported to be environmentally friendly and important in biological reactions, and (iii) some drugs are easy to reproduce [41]. As we will demonstrate in following sections, a lot of researches have been carried out to describe the drugs as efficient corrosion inhibitors [42–46].

13.3.2 Expired Drugs

Fresh drugs are, in general, expensive, so their use as corrosion inhibitors is not appropriate due to economic reason. Most of the pharmaceutically active substances are far more costly than the organic inhibitors presently employed. Thus, using fresh drugs as corrosion inhibitors is not economically viable [47]. Therefore, it is thought worthwhile to investigate the corrosion inhibition properties of expired drugs, which are of no use. It is reported that the active constituent of a drug degrades only infinitesimally, and more than 90% of the drugs maintain stability long time after the expiration date. Physicians, however, never recommend the practice of administering a medicine past its expiration date. Hence, these products can be either discharged in wastewater where it could have negative effect on environment or valorized in other applications. The discharge of these chemicals can induce an uncontrolled change in the chemical constituents of the drug, e.g. degradation, impurity formation or change in dissolution profile, and so on. Hence, these biocompatible substances may directly get converted to potential biohazard substances [48].

It is well reported that drugs retain at least 90% of its original potency even after expiry date, but their use for the medicinal purpose is restricted due to the professional restrictions and liability concerns. Use of expired drugs as corrosion inhibitors can solve two major environmental and economic problems: limitation of environmental pollution with pharmaceutically active compounds and reduction of the disposal costs of expired drugs [49].

13.3.3 Functionalized Drugs

Functionalized drugs are organic compounds containing functional electronegative groups, π-electrons in triple or conjugated double bonds along with S, N, and O atoms and can be modified or functionalized to more efficient corrosion inhibitors [50, 51]. Many researchers have used the derivatives of many functional groups such as Schiff base, azoles, 2-benzylaminopurine, thiophene derivatives, pyrazole derivatives, pyrimidine derivatives, pyridine derivatives, and bipyrazole derivatives in acidic media as efficient corrosion inhibitors for steel through functionalization of many drugs [52]. For instance, Dapsone drug has been functionalized to a Schiff base to afford a new effective inhibitor [53]. Due to the presence of conjugated double bonds and highly delocalized π-electrons on two aromatic rings, chalcones are one of the organic classes that possess the probabilities of electron transfer reactions, so they have been used as excellent functionalized inhibitors [54].

More recently, nonsteroidal anti-inflammatory drugs (NSAIDs) have been functionalized to many hydrazones via many steps resulting in more effective corrosion inhibitors. NSAIDs are classified as an important class of therapeutic agents, and one of the mostly used drugs to treat mild to moderate pain along with other diseases. The conversion of carboxylic group into hydrazone derivatives is a useful way to control the most effective side effect of these drugs such as gastrointestinal toxicity. The resulting compounds, i.e. hydrazones, are extremely versatile class of compounds in heterocyclic chemistry because of their wide range of biological and pharmaceutical applications. In terms of their electronic properties, they are known to exhibit both hard and soft base characters, which can facilitate donor–acceptor interactions with electron-deficient metal centers. Therefore, the functionalization of drugs could be used as a useful approach for developing potent corrosion inhibitors that satisfy the ever-increasing requirements of the environmental regulation. To the best knowledge of authors, the first functionalized NSAID was mefenamic acid using conventional techniques according to the below Scheme 13.3 [51, 55, 56].

Moreover, in 2020, the same group of researchers has functionalized another NSAID, Naproxen, to many hydrazones using the same procedure (Scheme13.3) [57–60]. Principally, the application of these two functionalized drugs as corrosion inhibitors will be reviewed and discussed in below sections.

13.4 Application of Chemical Medicines in Corrosion Inhibition

13.4.1 Drugs

Recent developments in corrosion protection field point to an emerging interest in eco-friendly corrosion inhibitors; those derived from natural source and nontoxic type. While the eco-friendly concept could help in providing nontoxic

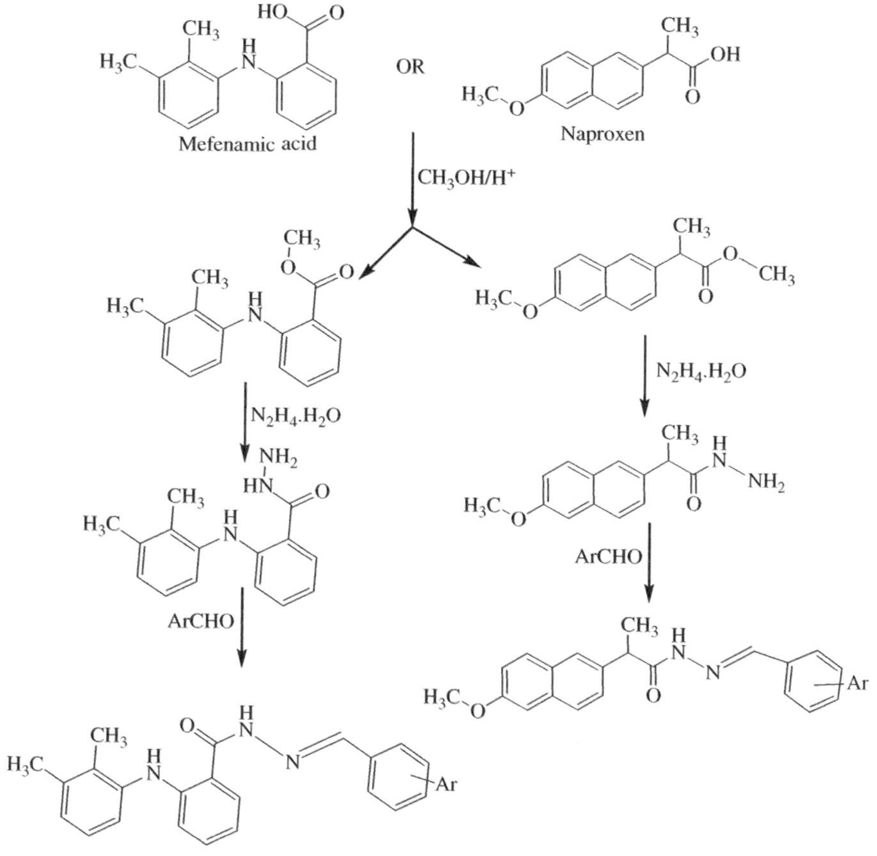

Scheme 13.3 The general pathway of functionalization of mefenamic acid and naproxen to different hydrazones.

compounds with a reduced cost, some questions may arise when dealing with their corrosion inhibition performances. Thus, the interest in drugs as corrosion inhibitors comes from two sides. Drugs are mainly from natural sources or prepared via greener techniques, nontoxic, and with no effect on the aquatic environment. Besides, drugs' molecular structures have many similarities with traditional corrosion inhibitors. Most drugs are heterocyclic and/or carboxylic compounds bearing five- and six-membered systems, polar functional groups, and heteroatoms such as oxygen, nitrogen, sulfur, and so on. These attributes are the characteristics of corrosion inhibitors with a superior performance, and this is the main reason why many researchers have investigated drugs as corrosion inhibitors.

In reviewing the literature, many works have been done on corrosion inhibition of metals and alloys using drugs. In 2007, El-Naggar [42] reported the corrosion inhibition properties of four sulfa drugs (e.g. sulfadiazine, sulfamethoxazole, sulfaguanidine, sulfamethazine) for mild steel 1.0M HCl solution using electrochemical measurements. Those drugs contain several reactive centers like heteroatoms, $-SO_2-NH-$ and $-NH_2$ groups, which can enhance their adsorption on the steel surface. The author found that all sulfa drugs reduced the corrosion rate with sulfadiazine having the highest inhibition efficiency (94%) followed by sulfamethazine (92%) and sulfamethoxazole (89%), and sulfaguanidine had the lowest performance (34%), all at 5 mM. The reason for the higher anticorrosion properties of these drugs was attributed by the author to the nature of the substituent group, the molecular size, and reactive adsorption sites. In the same context, Shukla & Quraishi [61] studied a third-generation cephalosporin antibiotic named ceftriaxone for the corrosion mitigation of mild steel in HCl solution. Results from electrochemical techniques (potentiodynamic polarization and electrochemical impedance spectroscopy) and weight loss showed that 400 ppm of cephalosporin can significantly reduce the corrosion rate with inhibition efficiency of greater than 90%. The ceftriaxone's molecule (Figure 13.1) contains several functional groups and heteroatoms such as $-NH_2$, nitrogen, sulfur, and oxygen atoms, which could facilitate the adsorption of its molecule on steel surface. The authors found that the drug's molecule adsorbed physically on the steel surface and control both cathodic and anodic corrosion reactions. The same research group investigated the corrosion inhibition of mild steel in HCl using other drugs [62–66]. Among them, they studied the anticorrosion properties of streptomycin drug [67]. Authors found that the streptomycin drug is a mixed-type inhibitor, its adsorption follows the Langmuir isotherm model, and its inhibition efficiency reaches 88.5% at 500 ppm.

Another second-generation cephalosporin antibiotic called cefuzonam has been investigated as corrosion inhibitor of mild steel in HCl [68]. Like other drugs, the molecular structure of cefuzonam is rich in reactive sites such as heteroatoms, aromatic rings, and functional groups. Interestingly, authors concluded that the presence of several sulfur and nitrogen atoms enhanced the adsorption of the drug on the steel surface. This adsorption was achieved via a combined physical and chemical interaction between drug's molecule and the steel surface. Besides, electrochemical measurements showed that the inhibitor had a mixed-type characteristic, blocking anodic and cathodic reactions, and an inhibition efficiency of 93%.

Phenytoin, which is a compound from the antiepileptic drugs, has been considered for corrosion mitigation of low carbon steel in 1M H_2SO_4 solution [69]. Its molecular structure (Figure 13.1) mimics a wide range of excellent corrosion inhibitors, and that's why authors have encouraged to study its corrosion inhibition properties. Based on the obtained results, authors proposed that the phenytoin drug adsorbed

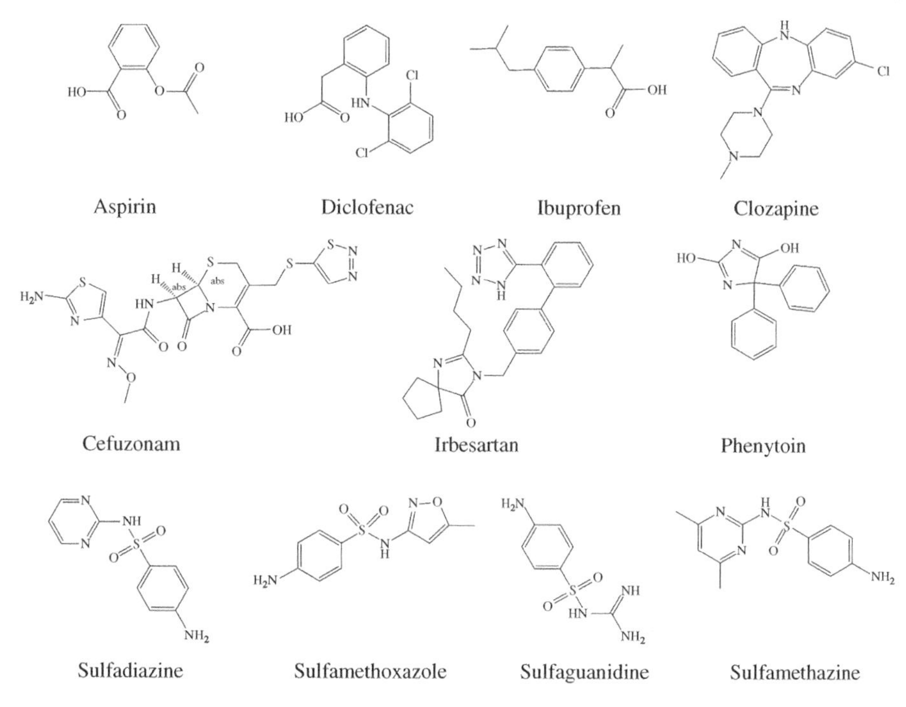

Figure 13.1 Molecular structures of some drugs used as corrosion inhibitors.

mostly through physical adsorption, following both Langmuir and Temkin adsorption isotherm models. Besides, electrochemical results confirmed a mixed-type nature with a predominant anodic effect, and an inhibition efficiency of 89% at 7 mM.

More recently, Srivastava et al. [70] showed that the addition of irbesartan, which corresponds to angiotensin II receptor antagonists, to 1M HCl and 0.5M H_2SO_4 solutions could reduce effectively the dissolution of the mild steel. It has been claimed that, in the two mediums, Irbesartan drug tended to adsorb through both physical and chemical adsorption. Maximum inhibition efficiencies of 94 and 83% were reached in HCl and H_2SO_4 at 300 mg/L, respectively, and it was understood from electrochemical measurements that the addition of the irbesartan drug increased the polarization resistance and reduced the double-layer capacitance while acting as mixed-type inhibitor in both acid solutions. Further, Patel et al. [71] introduced some nonsteroidal anti-inflammatory drugs such as ibuprofen, diclofenac, and aspirin as corrosion inhibitors of mild steel in 1.0M HCl. They have also attributed the higher anticorrosion properties of these drugs to their electron-rich molecular structure that facilitate their adsorption on the steel surface. Along with these studies, there were several other works focused on the corrosion inhibition of different steels by drugs [46, 66, 72–82].

Besides steel alloys, few studies have been conducted on the application of drugs for the corrosion protection of copper, aluminum, and zinc [15, 43, 45, 83–91]. Among these, Samide et al. [84] reported the potent inhibitory effect of metronidazole drug for copper in HCl solution. They showed that its inhibition performance increased with increase of its concentration, reaching the highest corrosion protection of copper (90%) at 1 mM. Authors found through the UV-Vis spectra that there was a formation of a metronidazole-copper complex, confirming the formation of a protective layer that prevents the corrosion attack. Authors suggested from theoretical calculations by density functional theory (DFT) that the reactive atomic centers in the drug's molecule are responsible on its higher effectiveness. Besides, they proposed that "bridged chloride ions" played a significant role in enhancing its adsorption over the copper surface.

Clozapine is an atypical antipsychotic medication; its molecular structure contains aromatic benzenoid rings along with heteroatoms N and Cl. The corrosion inhibition behavior of clozapine for copper in 1.0M nitric acid and 0.5M sulfuric acid solutions was investigated by Kumar et al. [85]. Authors suggested that the clozapine's molecule can be protonated in studied mediums due to the presence of nitrogen atom, and physical adsorption may take place at first stage. They also concluded that the presence of several reactive sites and a free energy of adsorption close to −40 kJ/mol favored the formation of covalent bonds between copper and clozapine's molecule.

The azithromycin is macrolide-type antibiotic, and it is used for treating variety of bacterial infections [92]. More recently, Tasić et al. [86] investigated azithromycin action on copper in 0.9% NaCl by applying electrochemical and surface characterization techniques. At 8 mM of the used drug, they reported maximum efficiency of 96%. On the other hand, authors suggested that the azithromycin molecule can be protonated in the studied medium (pH near 5.5), and thus physisorption may take place along with chemisorption.

Cefuroxime axetil (CA), which is a second-generation oral cephalosporin antibiotic, is used to treat a wide variety of bacterial infections. Its inhibitive and adsorption properties on aluminum in HCl solution were made by Ameh and Sani [87]. It was found that it acted through physical adsorption following a Langmuir adsorption model. A maximum inhibition efficiency of 90% was achieved at 0.5 g/l based on the results of gasometric method. Its inhibition efficiency was found to slightly decrease at a higher temperature, reaching 84% at 0.5 g/l.

The corrosion inhibition of aluminum in 1.0M HCl has also been investigated using cefixime drug at different temperatures [88]. Authors used electrochemical and DFT calculations to study the adsorption behavior of the used drug over the aluminum surface. The data gathered from these studies showed that the corrosion inhibition performance of the cefixime drug increase as the added amount to

HCl solution increase, reaching a maximum efficiency of 90% at 2 mM. It has been noticed that its efficiency decreased at higher temperatures, while it adsorbed on the aluminum surface via a physical process.

Investigations of the anticorrosion and adsorption behavior of drugs for zinc are very limited. Hebbar et al. [89] introduced tenofovir disproxil fumarate (TDF), which is an anti-HIV drug, as corrosion inhibitor of zinc in 0.1M HCl using electrochemical and weight loss techniques. At low concentration of 10 ppm, the maximum inhibition efficiency was about 57%, which significantly decreased at higher temperatures. Authors suggested that the chemical interaction between reactive sites of TDF and zinc atoms is the main route by which this drug acts against corrosion of zinc in HCl solution.

13.4.2 Expired Drugs

It has been recognized that most of the fresh drugs have excellent corrosion inhibition and adsorption properties, owing to their electron-rich molecular structures that mimic those from commercial sources. However, there are several challenges in using fresh drugs as corrosion inhibitors. Some of these challenges are related to the high cost needed for the preparation of a new drug, the time-consuming clinical validation studies, and importantly there is a critical need for these drugs as therapeutics in combating different diseases. Taking these into consideration, many researchers have directed their research efforts toward evaluating the potential of expired drugs to prevent corrosion of metals in different corrosive media.

Recently, several studies have been undertaken to investigate the application of expired drugs in corrosion inhibition. Review of the literature showed that most of those studies conducted on steel alloys in acidic medium while few have been performed for other metals. The Ambroxol drug is used for the treatment of respiratory diseases. The adsorption and corrosion inhibition behavior of the expired Ambroxol drug was studied by Geethamani and Kasthuri [93]. They reported that the addition of this expired drug to 1.0M HCl and 1.0M H_2SO_4 reduced the corrosion rate of mild steel. The maximum inhibition efficiencies of 95 and 94% were achieved after 6 and 4 hours of immersion in HCl and H_2SO_4 containing 9.0% (v/v) of the expired drug, respectively. It has been also found that the expired drug had a higher anticorrosive activity at higher temperatures, reaching 93.35 and 92.00% in HCl and H_2SO_4 solutions at 333 K. Authors explained the excellent adsorption properties of the expired drug by its richest molecular structure that favored the formation of donor–acceptor interactions along with electrostatic interactions with iron atoms. However, it should be noted that a lower inhibition efficiency was obtained from electrochemical measurements. Similar studies were also performed to investigate the suitability of an expired antimalarial drug, named Lumerax (LX) for mitigating the corrosion of mild steel in HCl [94]. Interestingly, the expired drug was found to be an excellent corrosion inhibitor, providing a higher inhibition efficiency of 98%. This anticorrosion activity is obviously higher than many traditional corrosion inhibitors. A combined physical and chemical adsorption of the expired drug on steel surface was suggested due to, on one hand, the presence of several aromatic rings rich in π-electrons along with heteroatoms and, on the other hand, the electrostatic attraction between positively charged drug molecules and negatively charged metal surface. Rajeswari et al. [95] were the first researchers who investigated the inhibitory effects of linezolid (LZ), ofloxacin (OX), levofloxacin (LV), and cefpodoxime (CF) on corrosion of cast iron in 1.0M HCl. The cefpodoxime was found to be the excellent corrosion inhibitor after the addition of 240 ppm to HCl solution, having an inhibition efficiency over 95%, while the lowest anticorrosion activity (87%) was obtained by the linezolid based on electrochemical impedance results, which were in line with weight loss and potentiodynamic polarization measurements. At this point, the molecular structure of these expired drugs played a significant role in the observed difference in their corrosion inhibition performance. The cefpodoxime having several functional groups and sulfur atoms is expected to form strong bonds with iron atoms. Authors also investigated the effect of the halide ions on the anticorrosion performance of tested expired drugs and noticed that the addition of 10 and 14 ppm of KCl and KI increased the inhibition efficiency because of the synergistic effect. Similar studies were also performed to investigate the corrosion inhibition performance of neomycin [44], metronidazole [45], atorvastatin [48], ethambutol [96], pantoprazole sodium [97], nitroimidazole antibiotics [98], and other expired drugs [99–107] for steel alloys in acidic solutions.

Of the few studies performed on the corrosion inhibition of copper using expired drugs, Fouda and Badawy tested the simvastan expired drug as potential corrosion inhibitor of copper in nitric acid solution using various electrochemical techniques supported by weight loss measurements. Based on EIS and weight loss results, maximum efficiencies of 66 and 78% were achieved at a concentration of 300 ppm. The temperature was found to have a great effect on the inhibition efficiency of the expired drug. A dramatic decrease in the inhibition efficiency values was observed at higher temperatures. Additionally, based on the thermodynamic study, authors suggested that the physisorption is the main adsorption mechanism.

More recently, in 2020, the inhibition potential of oseltamivir expired drug for aluminum in 3M HCl was apparently described for the first time by Raghavendra et al. [108]. This compound, which is a widely used antiviral drug to treat influenza A and B, showed excellent corrosion protection (98%) at only 2 mg/L. It was found that increasing its dose resulted in a significant reduction of the corrosion current density of the aluminum and, at the same time, blocking both cathodic and anodic corrosion processes. These authors proposed that the inhibition action of the expired drug is a result of physical interactions between pre-adsorbed chloride ions on aluminum surface and protonated form of the drug's molecule and chemical adsorption through electron transfer from reactive inhibitor sites and aluminum atoms. Tables 13.1 and 13.2 list some of the fresh and expired drugs used as corrosion inhibitors of different metals and alloys.

Table 13.1 Names and chemical structure of some drugs used as corrosion inhibitors.

Name and structure of drug	Metal/ alloy; corrosive medium	Techniques used	References
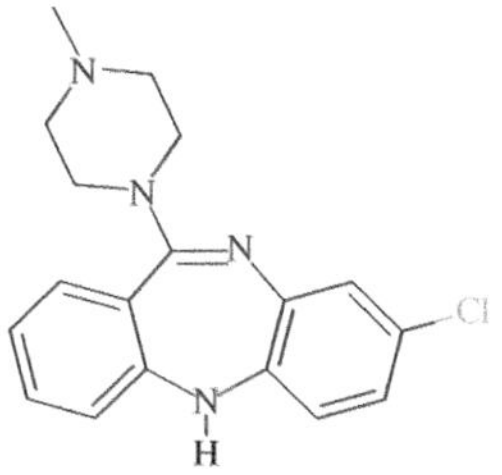 Clozapine	Copper/ 1.0M HNO_3 and 0.5M H_2SO_4	Weight loss, electrochemical impedance spectroscopy, potentiodynamic polarization, and DFT	[85]
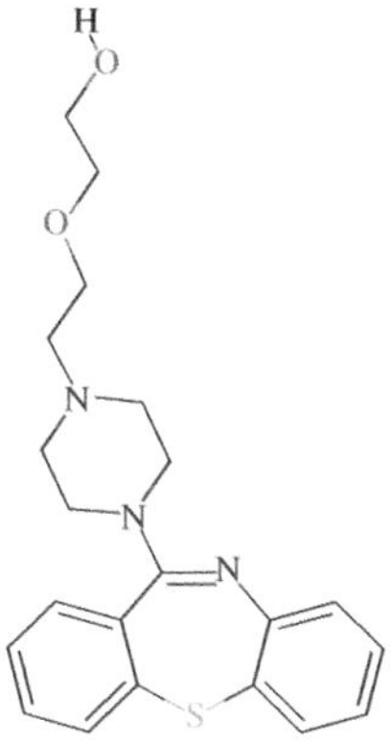 Seroquel	Zinc/ 0.1M HCl	Weight loss, electrochemical impedance spectroscopy, potentiodynamic polarization, SEM and FTIR	[90]
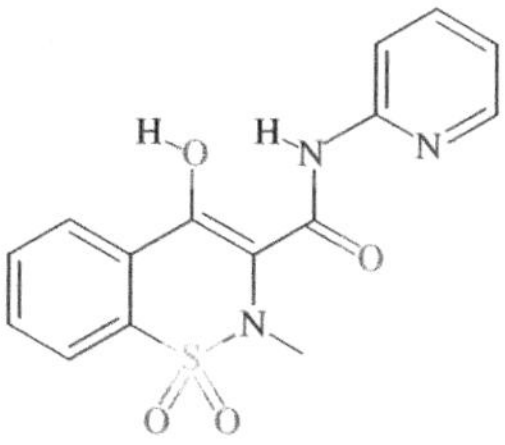 Piroxicam	Mild steel/1.0M HCl	Electrochemical impedance spectroscopy, potentiodynamic polarization, and SEM	[109]

(Continued)

Table 13.1 (Continued)

Name and structure of drug	Metal/alloy; corrosive medium	Techniques used	References
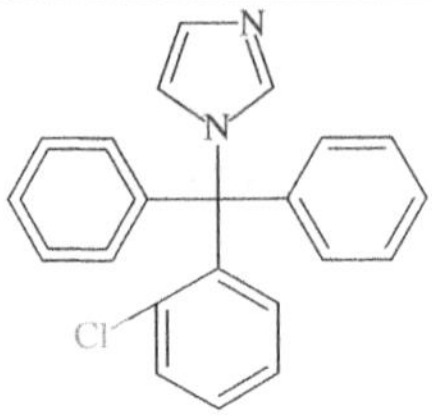 Clotrimazole	Zinc/0.1M HCl	Weight loss, electrochemical impedance spectroscopy, potentiodynamic polarization, FTIR, contact angle, SEM, and DFT	[91]
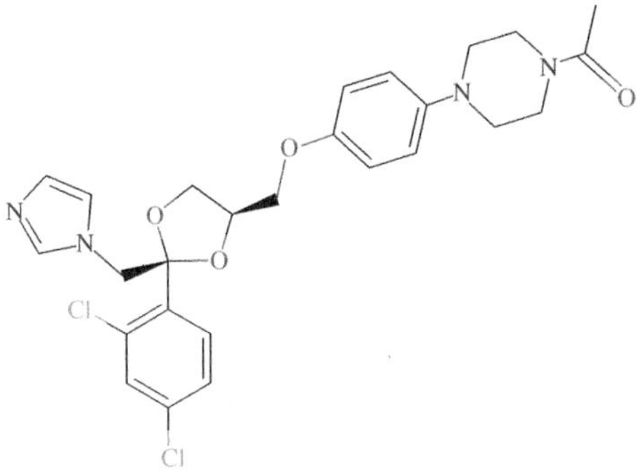 Ketoconazole	Mild steel/1.0M HCl	Weight loss, electrochemical impedance spectroscopy, potentiodynamic polarization, and AFM	[80]
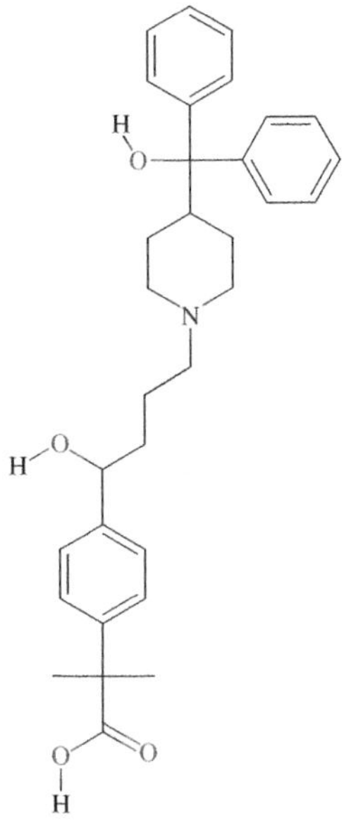 Fexofenadine	Mild steel/1.0 M HCl	Weight loss, electrochemical impedance spectroscopy, potentiodynamic polarization, FTIR, and DFT	[62]

Names and chemical structure of some expired drugs used as corrosion inhibitors.

Name and structure of drug	Metal/alloy; corrosive medium	Techniques used	Reference
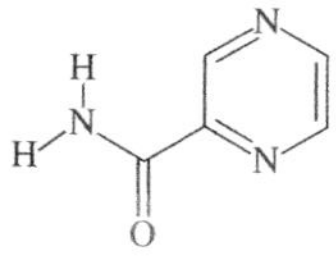 Pyrazinamide 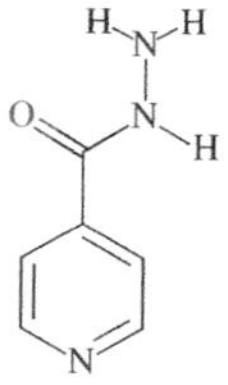 Isoniazid 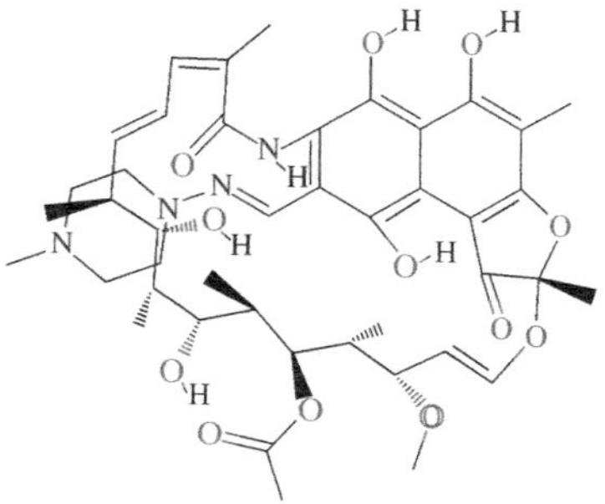 Rifampicin	Mild steel/ 0.5M HCl	Weight loss, electrochemical impedance spectroscopy, potentiodynamic polarization, SEM, molecular dynamics, and DFT	[110]

(Continued)

Table 13.2 (Continued)

Name and structure of drug	Metal/alloy; corrosive medium	Techniques used	Reference
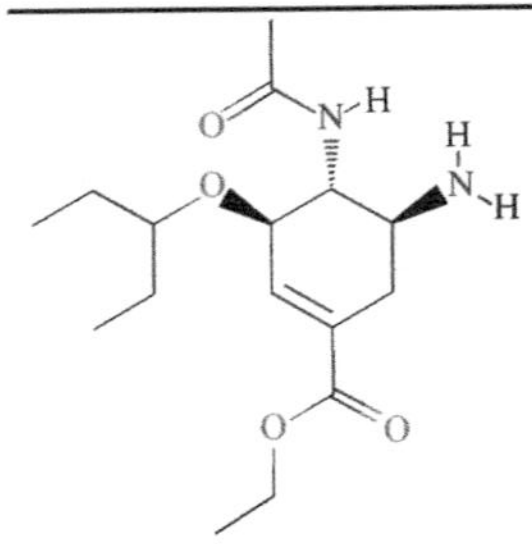 Oseltamivir	Aluminum/ 3.0 M HCl	Weight loss, atomic absorption spectroscopy, potentiodynamic polarization, AC impedance spectroscopy, DFT, and SEM/ EDX.	[108]
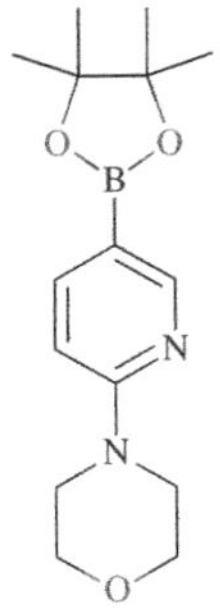 Rabeprazolesodium	Mild steel/ 1.0M HCl	Weight loss, potentiodynamic polarization, AC impedance spectroscopy, DFT, SEM, XRD, FTIR, UV	[100]

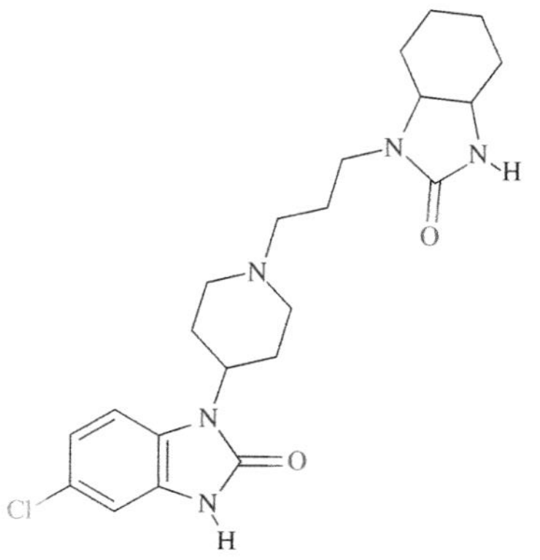

Domperidone

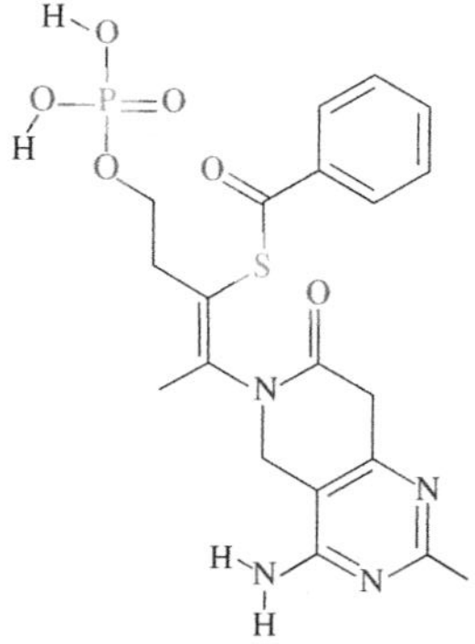

Benfotiamine

(Continued)

Table 13.2 (Continued)

Name and structure of drug	Metal/alloy; corrosive medium	Techniques used	Reference
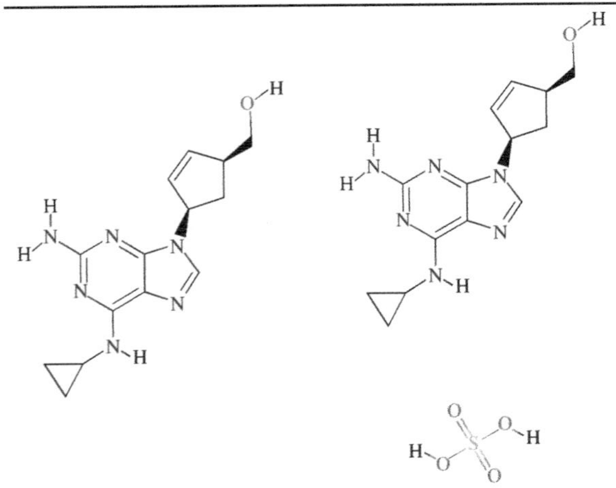 Abacavir sulfate	Mild steel/3.0M HCl	Weight loss, atomic absorption spectroscopy, potentiodynamic polarization, AC impedance spectroscopy, and SEM	[101]
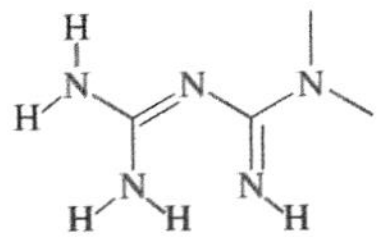 Metformin	Mild steel/15% HCl	Weight loss, potentiodynamic polarization, AC impedance spectroscopy, SEM/EDS, AFM, and FT-IR	[107]

13.4.3 Functionalized Drugs

Expired drugs are, due to their molecular structures, eco-friendly characteristics, and availability, have proven useful in mitigating corrosion of metals in different corrosive mediums. But there are obviously some issues in using these products for large-scale applications. In addition, their effectiveness is not always guaranteed. Recently, there has been a new research direction in the synthesis of corrosion inhibitors, which aims to design new potent corrosion inhibitors via the functionalization of existing drugs. This approach is very interesting in two ways. First, it guaranteed the synthesis of eco-friendly chemical compounds with higher efficacy. Second, the synthesis procedure can be easily reproduced, which solve the problem of availability of drug-based corrosion inhibitors.

The research works conducted by Lgaz et al. are one of the recent efforts in this research direction [51, 55, 56, 111–114]. In a recent report, Lgaz et al. [56] evaluated the corrosion inhibition effect of three hydrazone derivatives synthesized by the functionalization of a nonsteroidal anti-inflammatory drug (NSAID), named mefenamic acid. The three compounds, i.e. (*E*)-*N′*-benzylidene-2-((2,3-dimethylphenyl)amino)benzohydrazide (HDZ-3), (*E*)-2-((2,3-dimethylphenyl)amino)-*N′*-(4-methylbenzylidene)benzohydrazide, (HDZ-2) and (*E*)-*N′*-(4-(dimethylamino)benzylidene)-2-((2,3-dimethylphenyl)amino)benzohydrazide (HDZ-1), exhibited a higher inhibition efficiency, and their efficacy remained stable for 24 hours immersion time. Numerically, HDZ-1 showed an inhibition efficiency of 95%, followed by HDZ-2, 90%, and HDZ-3, 84%, after addition of 5 mM of inhibitors to 1.0M HCl. In addition, they proposed based on XPS results that all inhibitors acted by physicochemical adsorption on steel surface. The experimental results were supplemented by theoretical calculations using DFT and molecular dynamics simulation; both showed that these compounds contain several reactive sites and can adopt a parallel disposition over the steel surface. The same research team conducted a similar study on four hydrazone compounds derived from the NSAID mefenamic acid, in which excellent corrosion inhibition was reported for all tested compounds [51]. A full computational study was performed to investigate the electronic and adsorption properties of these hydrazone-based corrosion inhibitors [114]. From DFT calculations, authors concluded that the electron-accepting ability had a significant effect in increasing the inhibition effect. The compound with widespread reactive sites was found to have the highest inhibition efficiency. Furthermore, they studied the effect of the corrosion inhibition layer on the diffusion of corrosive particles by molecular dynamics simulation and found that the inhibition effect is controlled by the interaction of corrosion particles with inhibitor molecules.

Besides mefenamic acid, other corrosion inhibitors were synthesized via functionalization of the naproxen and tested for the mitigation of mild steel corrosion in HCl [112, 113]. In an attempt to examine the effect of these naproxen-based corrosion inhibitors on the corrosion rate of mild steel in acidic solutions, authors conducted several electrochemical and weight loss tests. A highest inhibition efficiency between 90 and 96% was achieved by the compounds bearing the methoxy groups. It has been noticed that the electron donating effect played a principal role in the efficacy of tested hydrazones. All compounds were found to act simultaneously on cathodic and anodic corrosion reactions. Authors concluded that the inhibition action was due to an effective adsorption of investigated compounds on steel surface through complex chemical and physical interactions between inhibitor's active sites and iron atoms. Other functionalized corrosion inhibitors were reported, with excellent results [50, 53, 54, 58–60].

The later class of drug-based corrosion inhibitors represented a promising approach not only to synthesis potent corrosion inhibitors but to provide compounds that meet eco-friendly criteria in both synthesis procedure and application. However, review of the literature demonstrates that many efforts are needed toward the development of this type of corrosion inhibitors.

Acknowledgments

This research was supported by basic science research program through the National Research Foundation (NRF) of Korea funded by the Ministry of Science, ICT and Future Planning (No. 2015R1A5A1037548).

References

1 Tamalmani, K. and Husin, H. (2020). Review on corrosion inhibitors for oil and gas corrosion issues. *Applied Sciences* 10: 3389.

2 Bouayed, M., Rabaa, H., Srhiri, A. et al. (1998). Experimental and theoretical study of organic corrosion inhibitors on iron in acidic medium. *Corrosion Science* 41: 501–517.

3 Quraishi, M.A. and Jamal, D. (2001). Corrosion inhibition of N-80 steel and mild steel in 15% boiling hydrochloric acid by a triazole compound – SAHMT. *Materials Chemistry and Physics* 68: 283–287.

4 Arjomandi, J., Moghanni-Bavil-Olyaei, H., Parvin, M.H. et al. (2018). Inhibition of corrosion of aluminum in alkaline solution by a novel azo-schiff base: experiment and theory. *Journal of Alloys and Compounds* 746: 185–193.

5 Badr, E.A., Bedair, M.A., and Shaban, S.M. (2018). Adsorption and performance assessment of some imine derivatives as mild steel corrosion inhibitors in 1.0 M HCl solution by chemical, electrochemical and computational methods. *Materials Chemistry and Physics* 219: 444–460.

6 Boucherit, L., Douadi, T., Chafai, N. et al. (2018). The inhibition Activity of 1,10-bis(2-formylphenyl)-1,4,7,10-tetraoxadecane (Ald) and its Schiff base (L) on the Corrosion of Carbon Steel in HCl: experimental and theoretical studies. *International Journal of Electrochemical Science* 13: 3997–4025.

7 Gouron, A., Le Mapihan, K., Camperos, S. et al. (2018). New insights in self-assembled monolayer of imidazolines on iron oxide investigated by DFT. *Applied Surface Science* 456: 437–444.

8 Haque, J., Srivastava, V., Chauhan, D.S. et al. (2018). Microwave-induced synthesis of chitosan schiff bases and their application as novel and green corrosion inhibitors: experimental and theoretical approach. *ACS Omega* 3: 5654–5668.

9 Pareek, S., Jain, D., Hussain, S. et al. (2019). A new insight into corrosion inhibition mechanism of copper in aerated 3.5 wt.% NaCl solution by eco-friendly Imidazopyrimidine Dye: experimental and theoretical approach. *Chemical Engineering Journal* 358: 725–742.

10 Ramezanzadeh, M., Bahlakeh, G., Sanaei, Z., and Ramezanzadeh, B. (2019). Corrosion inhibition of mild steel in 1 M HCl solution by ethanolic extract of eco-friendly Mangifera indica (mango) leaves: electrochemical, molecular dynamics, Monte Carlo and ab initio study. *Applied Surface Science* 463: 1058–1077.

11 Wang, W.Y., Song, Z.J., Guo, M.Z. et al. (2019). Employing ginger extract as an eco-friendly corrosion inhibitor in cementitious materials. *Construction and Building Materials* 228: 116713.

12 Lgaz, H., Salghi, R., Jodeh, S., and Hammouti, B. (2017). Effect of clozapine on inhibition of mild steel corrosion in 1.0 M HCl medium. *Journal of Molecular Liquids* 225: 271–280.

13 Dehghani, A., Bahlakeh, G., Ramezanzadeh, B., and Ramezanzadeh, M. (2020). Potential role of a novel green eco-friendly inhibitor in corrosion inhibition of mild steel in HCl solution: detailed macro/micro-scale experimental and computational explorations. *Construction and Building Materials* 245: 118464.

14 Gao, L.Z., Peng, S.N., Huang, X.M., and Gong, Z.L. (2020). A combined experimental and theoretical study of papain as a biological eco-friendly inhibitor for copper corrosion in H2SO4 medium. *Applied Surface Science* 511: 145446.

15 Li, H., Zhang, S.T., Tan, B.C. et al. (2020). Investigation of Losartan Potassium as an eco-friendly corrosion inhibitor for copper in 0.5 M H2SO4. *Journal of Molecular Liquids* 305: 112789.

16 Verma, C., Ebenso, E.E., and Quraishi, M.A. (2017). Ionic liquids as green and sustainable corrosion inhibitors for metals and alloys: an overview. *Journal of Molecular Liquids* 233: 403–414.

17 Hamadi, L., Mansouri, S., Oulmi, K., and Kareche, A. (2018). The use of amino acids as corrosion inhibitors for metals: a review. *Egyptian Journal of Petroleum* 27: 1157–1165.

18 Verma, C., Ebenso, E.E., Bahadur, I., and Quraishi, M.A. (2018). An overview on plant extracts as environmental sustainable and green corrosion inhibitors for metals and alloys in aggressive corrosive media. *Journal of Molecular Liquids* 266: 577–590.

19 Umoren, S.A., Solomon, M.M., Obot, I.B., and Suleiman, R.K. (2019). A critical review on the recent studies on plant biomaterials as corrosion inhibitors for industrial metals. *Journal of Industrial and Engineering Chemistry* 76: 91–115.

20 Verma, C., Haque, J., Quraishi, M.A., and Ebenso, E.E. (2019). Aqueous phase environmental friendly organic corrosion inhibitors derived from one step multicomponent reactions: a review. *Journal of Molecular Liquids* 275: 18–40.

21 El Ibrahimi, B., Jmiai, A., Bazzi, L., and El Issami, S. (2020). Amino acids and their derivatives as corrosion inhibitors for metals and alloys. *Arabian Journal of Chemistry* 13: 740–771.

22 Verma, C., Quraishi, M.A., and Ebenso, E.E. (2020). Quinoline and its derivatives as corrosion inhibitors: a review. *Surfaces and Interfaces* 21: 100634.

23 Wei, H., Heidarshenas, B., Zhou, L. et al. (2020). Green inhibitors for steel corrosion in acidic environment: state of art. *Materials Today Sustainability* 10: 100044.

24 Chaubey, N., Savita, A., Qurashi, D.S., and Chauhan, M.A. (2021). Quraishi, Frontiers and advances in green and sustainable inhibitors for corrosion applications: a critical review. *Journal of Molecular Liquids* 321: 114385.

25 Quraishi, M.A., Chauhan, D.S., and Saji, V.S. (2020). *Heterocyclic Organic Corrosion Inhibitors: Principles and Applications*. Elsevier.

26 Ding, L., Wang, W., and Zhang, A. (2007). Synthesis of 1, 5-dinitroaryl-1, 4-pentadien-3-ones under ultrasound irradiation. *Ultrasonics Sonochemistry* 14: 563–567.

27 Li, J.-T., Chen, G.-F., Yang, W.-Z., and Li, T.-S. (2003). Ultrasound promoted synthesis of 2-aroyl-1, 3, 5-triaryl-4-carbethoxy-4-cyanocyclohexanols. *Ultrasonics Sonochemistry* 10: 123–126.

28 Li, J.-T., Yang, W.-Z., Wang, S.-X. et al. (2002). Improved synthesis of chalcones under ultrasound irradiation. *Ultrasonics Sonochemistry* 9: 237–239.

29 Cella, R. and Stefani, H.A. (2009). Ultrasound in heterocycles chemistry. *Tetrahedron* 65: 2619–2641.

30 Cancio, N., Costantino, A.R., and Silbestri, G.F. (2019). Ultrasound-assisted syntheses of chalcones: experimental design and optimization. *Multidisciplinary Digital Publishing Institute Proceedings*, p. 13.

31 Al-Rasheed, H.H., Al Alshaikh, M., Khaled, J.M. et al. (2016). Ultrasonic irradiation: synthesis, characterization, and preliminary antimicrobial activity of novel series of 4, 6-disubstituted-1, 3, 5-triazine containing hydrazone derivatives. *Journal of Chemistry* 2016.

32 Majumder, A., Gupta, R., and Jain, A. (2013). Reviews, microwave-assisted synthesis of nitrogen-containing heterocycles. *Green Chemistry Letters and Reviews* 6: 151–182.

33 Shah, J. and Mohanraj, K. (2014). Comparison of conventional and microwave-assisted synthesis of benzotriazole derivatives. *Indian Journal of Pharmaceutical Sciences* 76: 46.

34 de la Hoz, A., Diaz-Ortiz, A., and Moreno, A. (2005). Microwaves in organic synthesis. Thermal and non-thermal microwave effects. *Chemical Society Reviews* 34: 164–178.

35 Vicente, I., Salagre, P., Cesteros, Y. et al. (2010). Microwave-assisted synthesis of saponite. *Applied Clay Science* 48: 26–31.

36 Quraishi, M.A., Chauhan, D.S., and Saji, V.S. (2020). Environmentally benign heterocyclic corrosion inhibitors. In: *Heterocyclic Organic Corrosion Inhibitors* (eds. M.A. Quraishi, D.S. Chauhan and V.S. Saji), 225–271. Elsevier.

37 Zarganes-Tzitzikas, T., Clemente, G.S., Elsinga, P.H., and Dömling, A. (2019). MCR Scaffolds Get Hotter with (18)F-Labeling. *Molecules* (Basel, Switzerland) 24.

38 Sadjadi, S., Heravi, M.M., and Nazari, N. (2016). Isocyanide-based multicomponent reactions in the synthesis of heterocycles. *RSC Advances* 6: 53203–53272.

39 Váradi, A., Palmer, T.C., Notis Dardashti, R., and Majumdar, S. (2016). Isocyanide-based multicomponent reactions for the synthesis of heterocycles. *Molecules* 21: 19.

40 Rocha, R.O., Rodrigues, M.O., and Neto, B.D.A. (2020). Review on the Ugi multicomponent reaction mechanism and the use of fluorescent derivatives as functional chromophores. *ACS Omega* 5: 972–979.

41 Aliofkhazraei, M. (2018). *Corrosion Inhibitors, Principles and Recent Applications*. BoD–Books on Demand.

42 El-Naggar, M.M. (2007). Corrosion inhibition of mild steel in acidic medium by some sulfa drugs compounds. *Corrosion Science* 49: 2226–2236.

43 Fouda, A.S., Ibrahim, H., and Atef, M. (2017). Adsorption and inhibitive properties of sildenafil (Viagra) for zinc in hydrochloric acid solution. *Results in Physics* 7: 3408–3418.

44 Samide, A., Iacobescu, G.E., Tutunaru, B. et al. (2017). Inhibitory properties of neomycin thin film formed on carbon steel in sulfuric acid solution: electrochemical and AFM investigation. *Coatings* 7.

45 Samide, A., Ilea, P., and Vladu, A.C. (2017). Metronidazole performance as corrosion inhibitor for carbon steel, 304L stainless steel and aluminum in hydrochloric acid solution. *International Journal of Electrochemical Science* 12: 5964–5983.

46 Fouda, A.S., Eissa, M., and El-Hossiany, A. (2018). Ciprofloxacin as eco-friendly corrosion inhibitor for carbon steel in hydrochloric acid solution. *International Journal of Electrochemical Science* 13: 11096–11112.

47 Vaszilcsin, N., Ordodi, V., and Borza, A. (2012). Corrosion inhibitors from expired drugs. *International Journal of Pharmaceutics* 431: 241–244.

48 Singh, P., Chauhan, D.S., Srivastava, K. et al. (2017). Expired atorvastatin drug as corrosion inhibitor for mild steel in hydrochloric acid solution. *International Journal of Industrial Chemistry* 8: 363–372.

49 Gupta, N.K., Gopal, C., Srivastava, V., and Quraishi, M.A. (2017). Application of expired drugs in corrosion inhibition of mild steel. *Ionics: International Journal of IonicsThe Science and Technology of Ionic Motion* 4: 8–12.

50 Kaddouri, M., Bouklah, M., Rekkab, S. et al. (2012). Thermodynamic, chemical and electrochemical investigations of calixarene derivatives as corrosion inhibitor for mild steel in hydrochloric acid solution. *International Journal of Electrochemical Science* 7: 9004–9023.

51 Lgaz, H., Chaouiki, A., Albayati, M.R. et al. (2019). Synthesis and evaluation of some new hydrazones as corrosion inhibitors for mild steel in acidic media. *Research on Chemical Intermediates* 45: 2269–2286.

52 Kathiresan, A. (2016). Electrochemical and quantum chemical studies of 1, 5-bis (2-nitrophenyl)-1, 4-pentadien-3-one as corrosion inhibitors for mild steel in hydrochloric acid solution. *International Journal of Engineering Science* 11: 8892–8913.

53 Singh, A., Avyaya, J., Ebenso, E.E., and Quraishi, M.A. (2013). Schiff's base derived from the pharmaceutical drug Dapsone (DS) as a new and effective corrosion inhibitor for mild steel in hydrochloric acid. *Research on Chemical Intermediates* 39: 537–551.

54 Verma, C.B., Reddy, M.J., and Quraishi, M.A. (2014). Microwave assisted eco-friendly synthesis of chalcones using 2, 4-dihydroxy acetophenone and aldehydes as corrosion inhibitors for mild steel in 1M HCl. *Analytical and Bioanalytical Electrochemistry* 6: 321–340.

55 Lgaz, H., Zehra, S., Albayati, R. et al. (2019). Corrosion inhibition of mild steel in 1.0 M HCl by two Hydrazone derivatives. *International Journal of Electrochemical Science* 14: 6667–6681.

56 Lgaz, H., Chung, I.M., Albayati, M.R. et al. (2020). Improved corrosion resistance of mild steel in acidic solution by hydrazone derivatives: an experimental and computational study. *Arabian Journal of Chemistry* 13: 2934–2954.

57 AbdelRaheem, S.K., Ali, I.H., Ebraheem, S.A. et al. (2020). Exploring the corrosion inhibition effect of two hydrazone derivatives for mild steel corrosion in 1.0 M HCl solution via electrochemical and surface characterization studies. *International Journal of Electrochemical Science* 15: 9354–9377.

58 Chafiq, M., Chaouiki, A., Albayati, M.R. et al. (2020). Unveiled understanding on corrosion inhibition mechanisms of hydrazone derivatives based on naproxen for mild steel in HCl: a joint experimental/theoretical study. *Journal of Molecular Liquids* 320: 114442.

59 Chafiq, M., Chaouiki, A., Al-Hadeethi, M.R. et al. (2020). Evaluation of the effect of two Naproxen-Based Hydrazones on the corrosion inhibition of Mild Steel in 1.0 M HCl. *International Journal of Electrochemical Science* 15: 9335–9353.

60 Chafiq, M., Chaouiki, A., Al-Hadeethi, M.R. et al. (2020). A joint experimental and theoretical investigation of the corrosion inhibition behavior and mechanism of hydrazone derivatives for mild steel in HCl solution. *Colloids and Surfaces A: Physicochemical and Engineering Aspects* 610: 125744.

61 Shukla, S.K. and Quraishi, M.A. (2009). Ceftriaxone: a novel corrosion inhibitor for mild steel in hydrochloric acid. *Journal of Applied Electrochemistry* 39: 1517–1523.

62 Ahamad, I., Prasad, R., and Quraishi, M.A. (2010). Experimental and theoretical investigations of adsorption of fexofenadine at mild steel/hydrochloric acid interface as corrosion inhibitor. *Journal of Solid State Electrochemistry* 14: 2095–2105.

63 Ahamad, I. and Quraishi, M.A. (2010). Mebendazole: new and efficient corrosion inhibitor for mild steel in acid medium. *Corrosion Science* 52: 651–656.

64 Shukla, S.K. and Quraishi, M.A. (2010). Cefalexin drug: a new and efficient corrosion inhibitor for mild steel in hydrochloric acid solution. *Materials Chemistry and Physics* 120: 142–147.

65 Shukla, S.K., Singh, A.K., and Ebenso, E.E. (2011). Pharmaceutically active compound as corrosion inhibitor for mild steel in acidic medium. *International Journal of Electrochemical Science* 6: 4276–4285.

66 Quraishi, M.A., Sudheer, and Ebenso, E.E. (2012). Ketorol: new and effective corrosion inhibitor for mild steel in hydrochloric acid solution. *International Journal of Electrochemical Science* 7: 9920–9932.

67 Shukla, S.K., Shigh, A.K., Ahamad, I., and Quraishi, M.A. (2009). Streptomycin: a commercially available drug as corrosion inhibitor for mild steel in hydrochloric acid solution. *Materials Letters* 63: 819–822.

68 Singh, A.K., Ebenso, E.E., and Quraishi, M.A. (2013). Corrosion inhibition behavior of cefuzonam at mild steel/HCl acid interface. *Research on Chemical Intermediates* 39: 3033–3042.

69 Eduok, U.M. and Khaled, M. (2015). Corrosion inhibition for low-carbon steel in 1 M H2SO4 solution by phenytoin: evaluation of the inhibition potency of another "anticorrosive drug". *Research on Chemical Intermediates* 41: 6309–6324.

70 Srivastava, M., Tiwari, P., Srivastava, S.K. et al. (2017). Electrochemical investigation of Irbesartan drug molecules as an inhibitor of mild steel corrosion in 1 M HCl and 0.5 M H2SO4 solutions. *Journal of Molecular Liquids* 236: 184–197.

71 Patel, D., Makwana, K., Shirdhonkar, M.B., and Kuperkar, K.C. (2019). A new insight into non-steroidal anti-inflammatory drugs (NSAIDs) as modulated green inhibitory agent on mild steel corrosion. *ChemistrySelect* 4: 5799–5809.

72 Naqvi, I., Saleemi, A.R., and Naveed, S. (2011). Cefixime: a drug as efficient corrosion inhibitor for mild steel in acidic media. electrochemical and thermodynamic studies. *International Journal of Electrochemical Science* 6: 146–161.

73 Samide, A., Tutunaru, B., and Negrila, C. (2011). Corrosion inhibition of carbon steel in hydrochloric acid solution using a sulfa drug. *Chemical and Biochemical Engineering Quarterly* 25: 299–308.

74 Singh, A.K. and Quraishi, M.A. (2011). Adsorption properties and inhibition of mild steel corrosion in hydrochloric acid solution by ceftobiprole. *Journal of Applied Electrochemistry* 41: 7–18.

75 Ahamad, I., Prasad, R., Ebenso, E.E., and Quraishi, M.A. (2012). Electrochemical and quantum chemical study of albendazole as corrosion inhibitor for mild steel in hydrochloric acid solution. *International Journal of Electrochemical Science* 7: 3436–3452.

76 Al-Sawaad, H.Z. (2013). Evaluation of the ceftriaxone as corrosion inhibitor for carbon steel alloy in 0.5M of hydrochloric acid. *International Journal of Electrochemical Science* 8: 3105–3120.

77 Singh, A.K., Khan, S., Singh, A. et al. (2013). Inhibitive effect of chloroquine towards corrosion of mild steel in hydrochloric acid solution. *Research on Chemical Intermediates* 39: 1191–1208.

78 Samide, A. and Tutunaru, B. (2014). Quinine sulfate: a pharmaceutical product as effective corrosion inhibitor for carbon steel in hydrochloric acid solution. *Central European Journal of Chemistry* 12: 901–908.

79 Adejoro, I.A., Ojo, F.K., and Obafemi, S.K. (2015). Corrosion inhibition potentials of ampicillin for mild steel in hydrochloric acid solution. *Journal of Taibah University for Science* 9: 196–202.

80 Nouri, P.M. and Attar, M.M. (2016). An imidazole-based antifungal drug as a corrosion inhibitor for steel in hydrochloric acid. *Chemical Engineering Communications* 203: 505–515.

81 Akpan, I.A. and Offiong, N.A.O. (2017). Inhibitory effect of amlodipine drug on corrosion of mild steel in HCl solution. *Indian Journal of Chemical Technology* 24: 107–110.

82 Al-Nami, S.Y. (2021). Corrosion inhibition of low carbon steel in 1 M HCl solution by cephalexin lionohydrate drug and synergistic iodide additives. *Biointerface Research in Applied Chemistry* 11: 8550–8563.

83 Fouda, A.S., Elewady, G.Y., Abdallah, Y.M., and Hussien, G.M. (2016). Anticorrosion Potential of salazopyrin drug for copper in nitric acid solution. *Research Journal of Pharmaceutical, Biological and Chemical Sciences* 7: 267–281.

84 Samide, A., Tutunaru, B., Dobritescu, A. et al. (2016). Electrochemical and theoretical study of metronidazole drug as inhibitor for copper corrosion in hydrochloric acid solution. *International Journal of Electrochemical Science* 11: 5520–5534.

85 Kumar, P.E., Govindaraju, M., and Sivakumar, V. (2018). Experimental and theoretical studies on corrosion inhibition performance of an environmentally friendly drug on the corrosion of copper in acid media. *Anti-Corrosion Methods and Materials* 65: 19–33.

86 Tasic, Z.Z., Mihajlovic, M.B.P., Radovanovic, M.B., and Antonijevic, M.M. (2018). Electrochemical investigations of copper corrosion inhibition by azithromycin in 0.9% NaCl. *Journal of Molecular Liquids* 265: 687–692.

87 Ameh, P.O. and Sani, U.M. (2016). Cefuroxime axetil: a commercially available drug as corrosion inhibitor for aluminum in hydrochloric acid solution. *Portugaliae Electrochimica Acta* 34: 131–141.

88 Diki, N.Y.S., Gbassi, G.K., Ouedraogo, A. et al. (2018). Aluminum corrosion inhibition by cefixime drug: experimental and DFT studies. *Journal of Electrochemical Science and Engineering* 8: 303–320.

89 Hebbar, N., Praveen, B.M., Prasanna, B.M., and Venkatesha, V.T. (2015). Inhibition effect of an Anti-HIV drug on the corrosion of Zinc in acidic medium. *Transactions of the Indian Institute of Metals* 68: 543–551.

90 Guruprasad, A.M., Sachin, H.P., Swetha, G.A., and Prasanna, B.M. (2019). Adsorption and inhibitive properties of seroquel drug for the corrosion of zinc in 0.1M hydrochloric acid solution. *International Journal of Industrial Chemistry* 10: 17–30.

91 Guruprasad, A.M., Sachin, H.P., Swetha, G.A., and Prasanna, B.M. (2020). Corrosion inhibition of zinc in 0.1 M hydrochloric acid medium with clotrimazole: experimental, theoretical and quantum studies. *Surfaces and Interfaces* 19: 100478.

92 Parnham, M.J., Haber, V.E., Giamarellos-Bourboulis, E.J. et al. (2014). Azithromycin: mechanisms of action and their relevance for clinical applications. *Pharmacology & Therapeutics* 143: 225–245.

93 Geethamani, P. and Kasthuri, P.K. (2015). Adsorption and corrosion inhibition of mild steel in acidic media by expired pharmaceutical drug. *Cogent Chemistry* 1: 1091558.

94 Dohare, P., Chauhan, D.S., Hammouti, B., and Quraishi, M.A. (2017). Experimental and DFT investigation on the corrosion inhibition behavior of expired drug Lumerax on mild steel in hydrochloric acid. *Analytical & Bioanalytical Electrochemistry* 9: 762–783.

95 Rajeswari, V., Devarayan, K., and Viswanathamurthi, P. (2017). Expired pharmaceutical compounds as potential inhibitors for cast iron corrosion in acidic medium. *Research on Chemical Intermediates* 43: 3893–3913.

96 Dahiya, S., Saini, N., Dahiya, N. et al. (2018). Corrosion inhibition activity of an expired antibacterial drug in acidic media amid elucidate DFT and MD simulations. *Portugaliae Electrochimica Acta* 36: 213–230.

97 Fouda, A.S., Ibrahim, H., Rashwaan, S. et al. (2018). Expired drug (pantoprazole sodium) as a corrosion inhibitor for high carbon steel in hydrochloric acid solution. *International Journal of Electrochemical Science* 13: 6327–6346.

98 Shi, Y.L., Bai, W., Guo, J.M. et al. (2018). Inhibition of three kinds of expired nitroimidazole antibiotics. *Anti-Corrosion Methods and Materials* 65: 398–407.

99 Motawea, M.M. (2019). Corrosion inhibition efficiency of expired nitazoxanide drug on carbon steel in hydrochloric acid solution. *International Journal of Electrochemical Science* 14: 6682–6698.

100 Palaniappan, N., Alphonsa, J., Cole, I.S. et al. (2019). Rapid investigation expiry drug green corrosion inhibitor on mild steel in NaCl medium. *Materials Science & Engineering, B: Advanced Functional Solid-State Materials* 249.

101 Raghavendra, N. (2019). Expired abacavir sulfate drug as non-toxic corrosion inhibitor for mild steel (MS) in 3 M Hydrochloric acid system. *Gazi University Journal of Science* 32: 1113–1121.

102 Raghayendra, N. (2019). The corrosion inhibition study of expired doxercalciferol drug as nontoxic inhibitor for mild steel (Ms) In 3 M HCL medium. *Heterocyclic Letters* 9: 185–190.

103 Singh, A.K., Pandey, A.K., Banerjee, P. et al. (2019). Eco-friendly disposal of expired anti-tuberculosis drug isoniazid and its role in the protection of metal. *Journal of Environmental Chemical Engineering* 7: 114423.

104 Singh, P., Chauhan, D.S., Chauhan, S.S. et al. (2019). Chemically modified expired Dapsone drug as environmentally benign corrosion inhibitor for mild steel in sulphuric acid useful for industrial pickling process. *Journal of Molecular Liquids* 286: 110903.

105 Alfakeer, M., Abdallah, M., and Fawzy, A. (2020). Corrosion inhibition effect of expired ampicillin and flucloxacillin drugs for mild steel in aqueous acidic medium. *International Journal of Electrochemical Science* 15: 3283–3297.

106 Fouda, A.S., El-Dossoki, F.I., El-Hossiany, A., and Sello, E.A. (2020). Adsorption and anticorrosion behavior of expired meloxicam on mild steel in hydrochloric acid solution. *Surface Engineering and Applied Electrochemistry* 56: 491–500.

107 Haruna, K., Saleh, T.A., and Quraishi, M.A. (2020). Expired metformin drug as green corrosion inhibitor for simulated oil/gas well acidizing environment. *Journal of Molecular Liquids* 315: 113716.

108 Raghavendra, N., Hubikar, L.V., Ganiger, P.J., and Bhinge, A.S. (2020). Prevention of aluminum corrosion in hydrochloric acid using expired oseltamivir drug as an inhibitor. *Journal of Failure Analysis and Prevention* 20: 1864–1874.

109 Addoun, A., Trari, M., and Ferroukhi, O. (2020). Corrosion control of mild steel material in HCl electrolyte by a non-steroidal anti-inflammatory drug: electrochemical and kinetic study. *Protection of Metals and Physical Chemistry of Surfaces* 56: 826–833.

110 Dahiya, S., Pahuja, P., Lgaz, H. et al. (2019). Advanced quantum chemical and electrochemical analysis of ravage drugs for corrosion inhibition of mild steel. *Journal of Adhesion Science and Technology* 33: 1066–1089.

111 Chafiq, M., Chaouiki, A., Lgaz, H. et al. (2020). Synthesis and corrosion inhibition evaluation of a new schiff base hydrazone for mild steel corrosion in HCl medium: electrochemical, DFT, and molecular dynamics simulations studies. *Journal of Adhesion Science and Technology* 34: 1283–1314.

112 Chaouiki, A., Chafiq, M., Al-Hadeethi, M.R. et al. (2020). Exploring the corrosion inhibition effect of two hydrazone derivatives for mild steel corrosion in 1.0 M HCl solution via electrochemical and surface characterization studies. *International Journal of Electrochemical Science* 15: 9354–9377.

113 Chaouiki, A., Chafiq, M., Lgaz, H. et al. (2020). Green corrosion inhibition of mild steel by hydrazone derivatives in 1.0 M HCl. *Coatings* 10: 640.

114 Lgaz, H., Salghi, R., Masroor, S. et al. (2020). Assessing corrosion inhibition characteristics of hydrazone derivatives on mild steel in HCl: insights from electronic-scale DFT and atomic-scale molecular dynamics. *Journal of Molecular Liquids* 308: 112998.

14

Ionic Liquids as Corrosion Inhibitors

Ruby Aslam[1], Mohammad Mobin[1], and Jeenat Aslam[2]

[1] *Corrosion Research Laboratory, Department of Applied Chemistry, Faculty of Engineering and Technology, Aligarh Muslim University, Aligarh, Uttar Pradesh, India*
[2] *Department of Chemistry, College of Science, Taibah University, Yanbu, Al-Madina, Saudi Arabia*

14.1 Introduction

Corrosion is a thermodynamically spontaneous process that results in the transition of metals or alloys to a more stable state through their chemical or electrochemical reaction with the surrounding constituent. The negative resultant due to the occurrence of corrosion severely affects metallic framework, renovation expenses, and public welfare. It is a common major concern that has a serious impact on natural and industrial environments. Corrosion and pollution are closely linked to hazardous activities because certain contaminants cause corrosion, and corrosion products such as rust also contaminate water bodies [1]. Both are disruptive practices that harm environmental sustainability and the durability of infr astructure services. Besides it can contribute to economic and security problems. The issue of corrosion and corrosion resistance of metals has gained a lot of interest due to the extensive use of metals in various industries; therefore, several studies on this issue have been carried out to date and are still underway. Figure 14.1 provides a graphical illustration of numerous sectors suffering from corrosion attacks. It is, therefore, necessary to establish and apply methodologies for corrosion engineering control. For industrial use of metallic materials, corrosion prevention is an important concern [2].

Organic Corrosion Inhibitors: Synthesis, Characterization, Mechanism, and Applications, First Edition. Edited by Chandrabhan Verma, Chaudhery Mustansar Hussain, and Eno E. Ebenso.

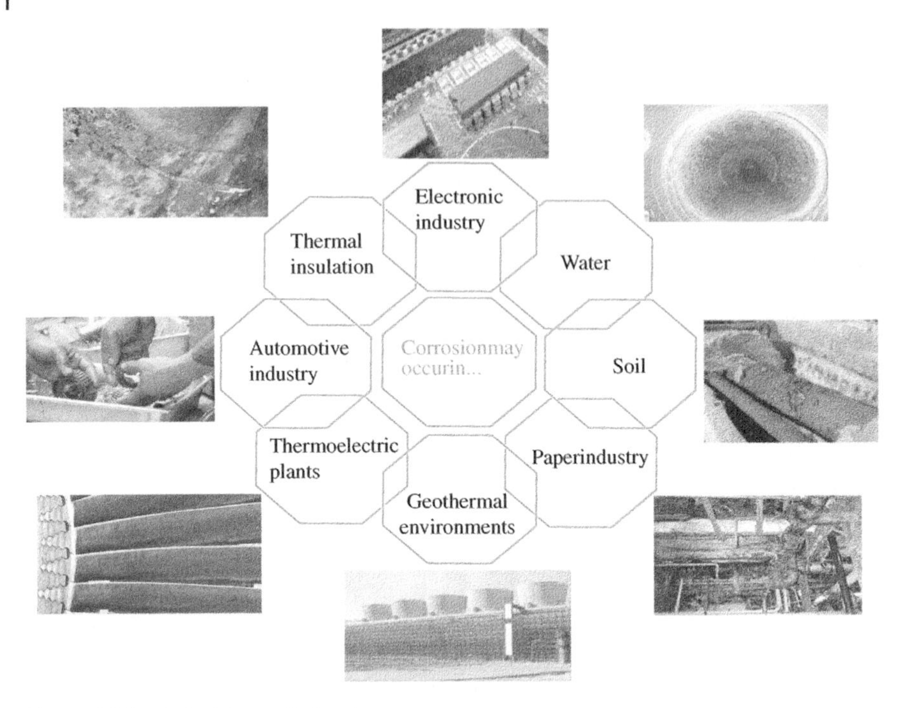

Figure 14.1 Industries suffering from corrosion.

14.2 Inhibition of Metal Corrosion

Several safety systems such as the introduction of suitable inhibitors, cathodic protection, and coatings [3] strive to reduce the rate of corrosion to a value that will allow the substance to maintain its usual or acceptable lifetime. The use of corrosion inhibitors as an efficient and economical approach is envisaged to mitigate the corrosion problem [4, 5]. Choosing suitable inhibitors for particularly in aggressive media (acidic medium, NaCl solution, etc.) is of great importance among these methods of protection because it facilitates high performance, practicality, and safety. A corrosion inhibitor is described as any chemical substance that subsequently reduces the corrosion rate when applied to an aggressive system in small quantities. Such compounds generally possess polar functional groups having O, N, S, and P atoms and also conjugated double/triple bonds/aromatic rings, which serve as adsorption centers, through which they preferentially adsorb onto the metal/solution interface [6]. The process of adsorption is contingent upon discrete factors such as the nature of metallic adsorbent, the density of electronic cloud around inhibitor molecule, the composition of the electrolytic solution, and the electrochemical potential of metal/electrolyte interface [7]. The trending anticorrosive research formulations prioritize the establishment of nontoxic, relatively cheaper, greener, and environmentally benign organic compounds with better inhibitory performance.

14.3 Ionic Liquids as Corrosion Inhibitors

Ionic fluids (ILs), from various ways of synthetic materials used as anticorrosive agents, are considered as an emerging class of minimal harmful effects. These can be best defined as low-melting salts having cations and anions with an organic moiety whose physical appearance is generally liquid at or under 100°C. These can be termed in several ways like ionic melts, liquid electrolytes, ionic fluids, fused salts, liquid salts, or ionic glasses. Out of the various applications of ILs, the best one perhaps is its anticorrosive application [8–11]. Based on their design, ILs can be split up into different groups. Figure 14.2 reflects a brief category of ILs.

ILs have been accredited as green chemicals affording surety of harmless consequences on the environment. They exhibit unique characteristics, viz., very low

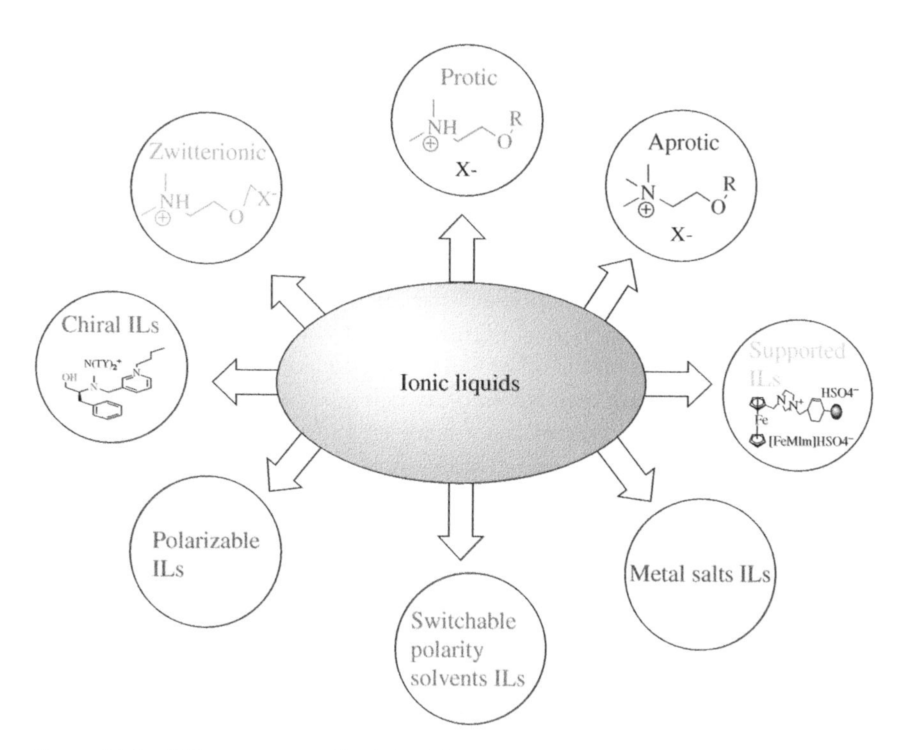

Figure 14.2 Different types of ionic liquids.

vapor pressure, low toxicity, higher thermal stability, higher ionic conductivity, nonflammability, lower melting point, etc. [12–15]. The most common ILs contain a cationic part that is of organic origin like phosphonium, ammonium, pyridinium, imidazolium, pyrrolidinium in addition to anions [16]. Figure 14.3 shows various cationic and anionic parts of ILs used in corrosion inhibition formulations.

Owing to their ionic nature, ILs are strongly soluble in an aggressive polar environment [17]. This makes them more powerful in contrast to traditional organic inhibitors. The steady rise in the number of publications covering ILs as corrosion inhibitors in recent years (Figure 14.4) (Scopus source) indicates the relevance of this topic for researchers.

In this segment, an overview of the anticorrosive application of ILs will be discussed in detail.

14.3.1 In Hydrochloric Acid Solution

HCl solution is used as a common practice in different industrial sectors such as steel pickling, descaling, and oil well acidizing [18, 19]. Such acid remedies, however, cause significant harm to metallic structures due to considerable metal losses, which are transformed into economic deficits and indirect costs associated with safety, health, and environmental issues. The inclusion of low concentrations of corrosion inhibitors is persuasive because of the aggressive nature of such corrosive media, impacting the integrity of various alloys. The application of corrosion control techniques has been a crucial issue for decades in various applications. The use of corrosion inhibitors has been an excellent approach due to their high performance and low cost [20]. During cleaning and descaling processes, concentrated HCl is used to eliminate surface impurities [21]. For the industrial acid pickling process, even more concentrated HCl solutions are used. Low concentrated HCl solutions, on the other hand, are used as electrolytes for the study of newly discovered alternative corrosion inhibitors by academics. Nonetheless, both of these solutions are extremely corrosive, creating enormous financial harm for particular industries. The use of inhibitors is the main source of protection against damage from corrosion.

Paulina et al. [22] proposed a study that tested the anticorrosive effect of two novel ILs, namely, N-trioctyl-N-methylammoniummethylsulfate and N-tetradecyl-N-trimethyl ammonium methylsulfate for APIX52 steel in 1M HCl. The authors preferred methylsulfate as an anion since it is organic and has low toxicity. At varying concentrations, temperatures, and immersion times, various electrochemical and surface analysis methods were used to assess the IE of the compounds. The study of the findings obtained was used to suggest a mechanism of corrosion inhibition carried out in acid medium by the ILs.

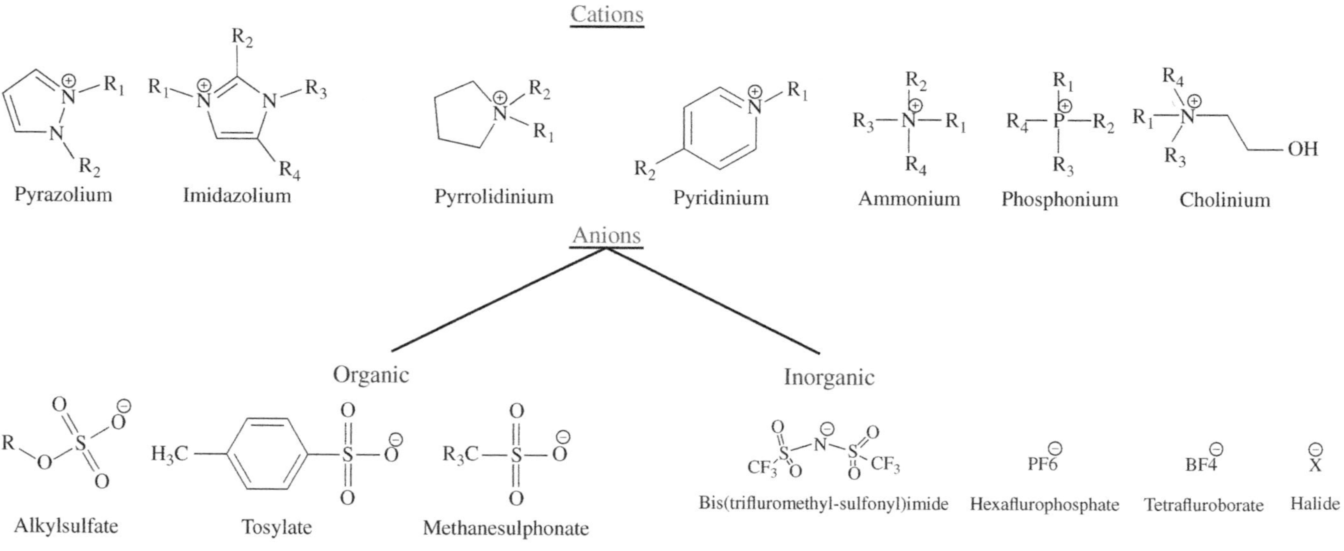

Figure 14.3 Species that are usually considered as cations and anions for designing ionic liquids.

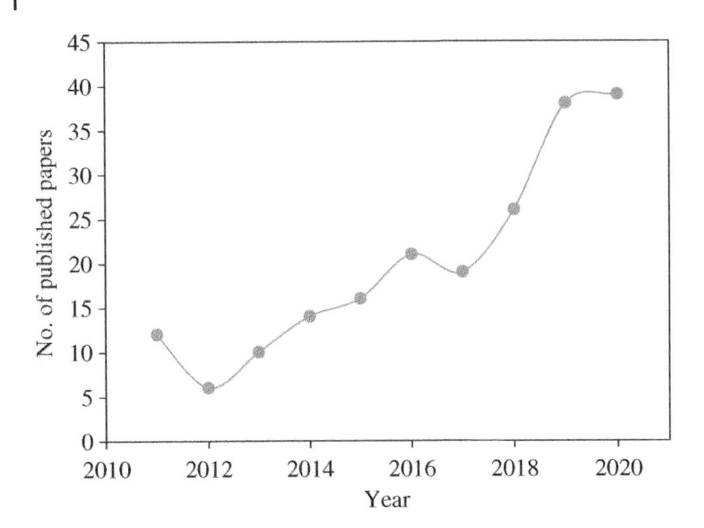

Figure 14.4 Number of papers published on the anticorrosive impact of ILs versus year of publication (until 20 October 2020).

Three IL-based gemini-cationic surfactant inhibitors N1,N1,N1,N2,N2,N2-hexadodecylhexane-1,2-diaminium bromide (G2IL), N1, N1,N1,N2,N2,N2-hexadodecylpropane-1,3-diaminium bromide (G3IL), and N1,N1,N1,N2,N2,N2-hexadodecylhexane-1,6-diaminium bromide (G6IL) were tested as corrosion inhibitors by Tawfik et al. [23]. The surface and physicochemical properties of these compounds were reported. PDP experiments suggest that the tested inhibitors were mixed-type inhibitors. The efficiency of inhibition improved with concentration changes but decreased with temperature increase and was dependent on the size of the spacer chain associated with N. The inhibition performances exhibited by three ILs followed the order: IL6 > IL3 > IL2.

Recently, the efficiency of N-methyl-2-hydroxyethylammonium oleate ([m-2HEA][Ol]) for MS was reported by Schmitzhaus et al. [24] in a 0.1 mol/l HCl. [m-2HEA][Ol] worked as a mixed-type inhibitor and achieved up to 94–97% inhibition performance. As the Raman signal confirmed, the covered, non-corroded regions were rich in [m-2HEA] [Ol].

Qiao et al. [25] reported a comparative analysis on anticorrosive applications of 1-dodecyl-2,3-dimethylimidazolium chloride ([DDMIM]Cl) and 1-benzyl-3-dode cyl-2-methylimidazol-1-ium chloride ([BDMIM]Cl) in a 15% HCl for N80 steel. Both the ILs acted as mixed-type inhibitors and [BDMIM]Cl has better inhibitive performance and lower CMC. The theoretical results suggest that the benzyl substituent and the imidazole ring are responsible for the inhibition of CS corrosion in an acidic medium.

The three ILs, namely, [(CH_2)COOHMIm][HSO_4], [(CH_2)$_2$COOHMIm][HSO_4], and [(CH_2)$_3$$SO_3$HMIm][$HSO_4$], were tested as a corrosion inhibitor to protect CS corrosion in HCl medium performing experimental and analytical methods [26]. The IEs were measured by electrochemical calculation in compliance with the order: [(CH_2)$_3$$SO_3$HMIm][$HSO_4$] > [($CH_2$)$_2$COOHMIm][$HSO_4$] > [($CH_2$) COOHMIm][$HSO_4$]. Besides, to create the structure–corrosion inhibition relationship, different theoretical parameters were determined. For the inhibition actions of the ILs, the imidazole ring, -COOH, and -SO_3H groups (to be specific, the O atoms) are responsible.

The corrosion inhibition properties of ethoxy carbonyl methyl triphenylphosphonium bromide referred to as ECMTPB for MS and aluminum (Al) in 0.5M HCl and 0.5M H_2SO_4 solutions were reported at room temperature [27]. Increasing concentration of ECMTPB from 10^{-5} M to 10^{-2} M enhanced the protective rate of both materials. ECMTPB chemisorbed on both the metal surface and obeyed Langmuir adsorption isotherm.

Hajjaji et al. [28] reported the corrosion protection performance of 1-ethyl-4-(2-(4-fluorobenzylidene)hydrazinecarbonyl)pyridine-1-ium iodide (IPyrC_2H_5) and 1-butyl-4-(2-(4-fluorobenzylidene)hydrazinecarbonyl)pyridine-1-ium iodide (IPyr-C_4H_9) against MS corrosion in 1M HCl utilizing electrochemical methods. PDP studies found that IPyr-C2H5 and IPyr-C4H9 acted as anodic type inhibitors. EIS measurements further confirmed the inhibition ability of both compounds for MS corrosion and showed IE of 88.8% for IPyr-C_2H_5 and 92.3% for IPyr-C_4H_9 at 10^{-3} M. The DFT calculations were employed to correlate the molecular structure of ILs with their IEs. MD studies revealed that IPyr-C_4H_9 exhibited higher interaction energy compared to IPyr-C_2H_5; however, both the inhibitor molecules adsorb via the entire molecular skeleton on Fe (110) surface (Figure 14.5). RDF was further used to confirm the chemical adsorption of the inhibitors on the metallic surface.

Nesane et al. [29] prepared 1-(benzyloxy)-1-oxopropan-2-aminium 4-methylbenzenesulfonate (BOPAMS) and 4-(benzyloxy)-4-oxobutan-1-aminium 4-methylbenzenesulfonate (BOBAMS) and evaluated them as harmless aluminum corrosion inhibitors. As the concentrations increased, the inhibition efficacy of studied compounds increased and decreased with the temperature rise. The mixed-type inhibitor activity of BOPAMS and BOBAMS was verified by Tafel plots (Figure 14.6).

Aslam et al. [30] synthesized amino acid ester saccharinate [AAE][Sac] (Figure 14.7): L-leucine methyl ester saccharinate, L-phenyl alanine methyl ester saccharinate, and L-alanine methyl ester saccharinate referred to as ([LeuME][Sac]), ([PheME][Sac]), and ([AlaME][Sac]) respectively. Using different experimental and theoretical measurements, they were identified as efficient green corrosion inhibitors for MS in 1M HCl solution.

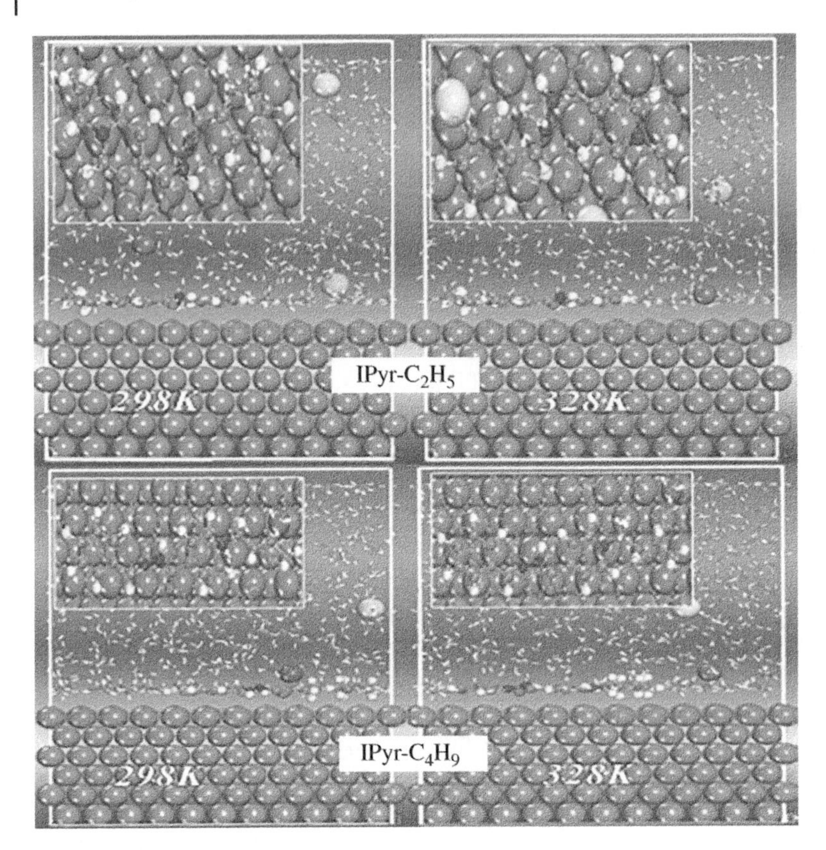

Figure 14.5 Constant adsorption configurations of IPyr-C_2H_5 and IPyr-C_4H_9 molecules on the Fe (110) surface in HCl at 25°C and 55°C. *Source:* Republished from Ref. [28] with permission from Elsevier.

All the [AAE][Sac] ILs exhibited higher IE up to 60°C but have less inhibiting properties above 60°C, i.e. 70°C and 80°C and followed the order: [PheME][Sac] > [LeuME][Sac] > [AlaME][Sac]. From Figure 14.8, it is clear that the [AAE][Sac] acted as mixed-type inhibitors. The experimental results were also adequately validated by the theoretical descriptors given by quantum chemical measurements and advanced MC simulation.

The given Table 14.1 contains some other ILs [31–49] synthesized in line for two decades, and they are efficiently exploited as corrosion inhibitors for various metals in acidic media.

14.3.2 In Sulfuric Acid Solution

H_2SO_4 is also highly corrosive, similar to HCl, especially during industrial processes, so certain procedures involving H_2SO_4 need the application of some

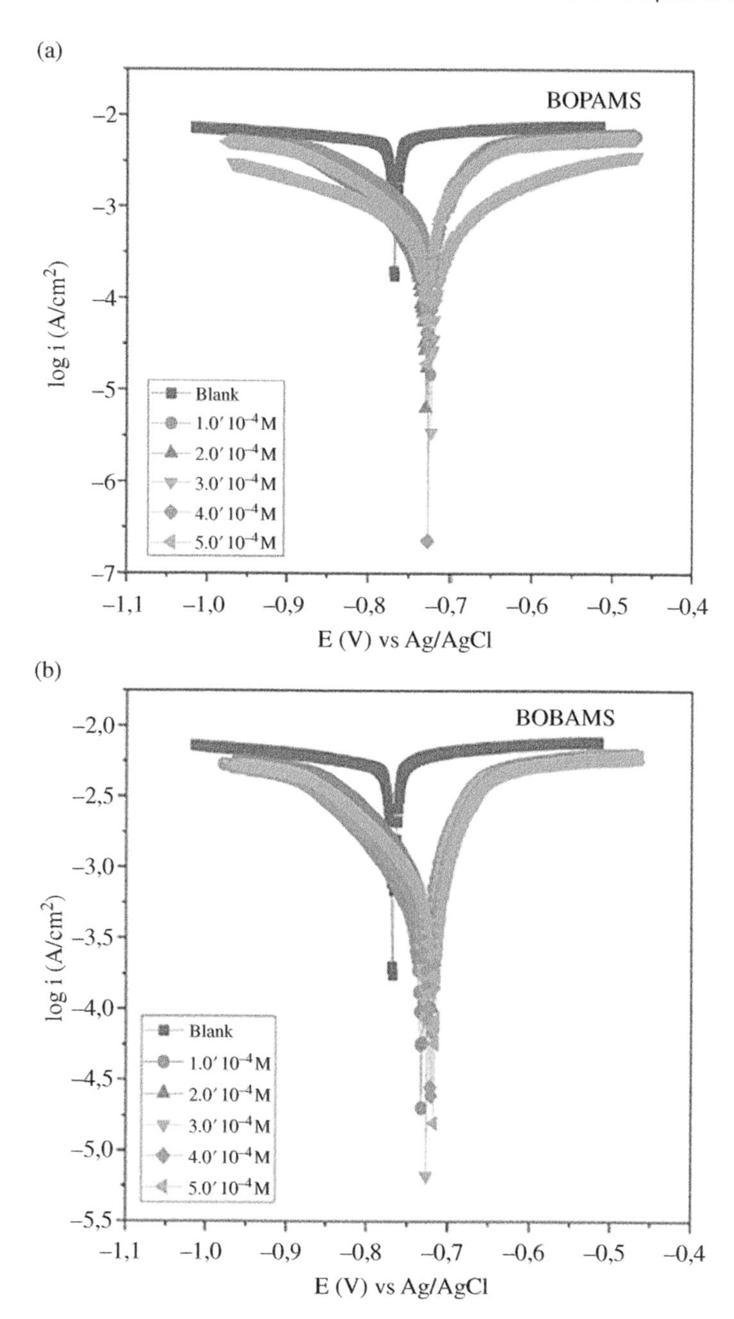

Figure 14.6 Tafel plots of Al utilizing 1M HCl for the process without and with various molarities of (a) BOPAMS and (b) BOBAMS. *Source:* Republished from Ref. [29] with permission from Elsevier.

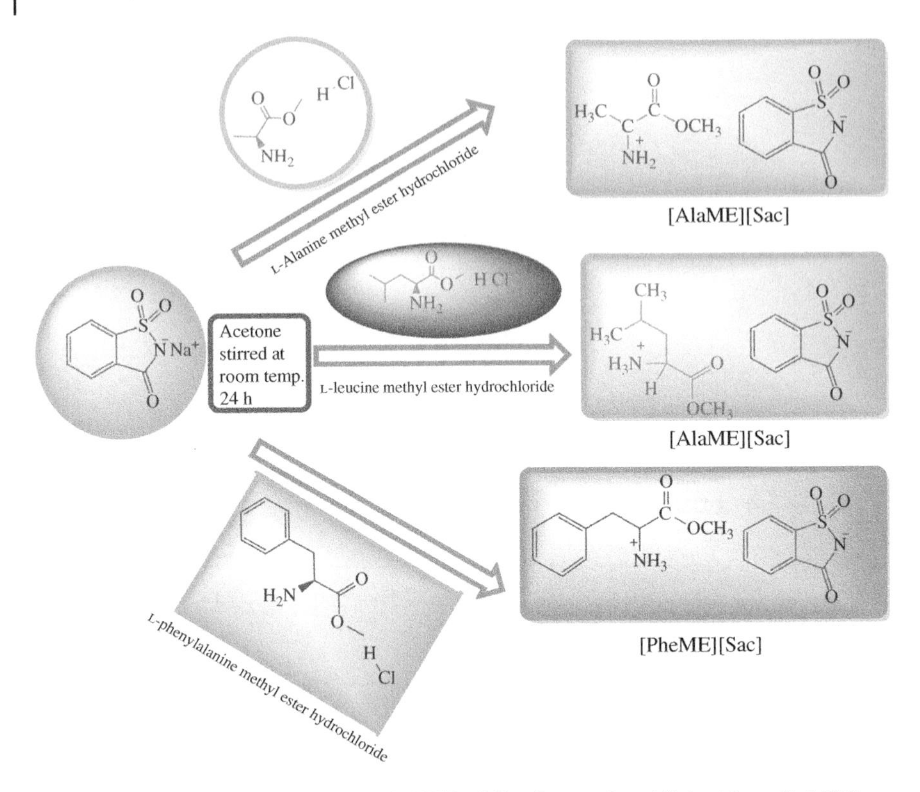

Figure 14.7 Synthesis route for the [AAE][Sac] ILs. *Source:* Republished from Ref. [30] with permission from Elsevier.

external additives called corrosion inhibitors [50]. In H_2SO_4, many types of organic and inorganic inhibitors are historically developed and used. A variety of ILs are used as corrosion inhibitors and are discussed below:

Gomez-Sanchez et al. [51] prepared 1-methyl-3-benzylimidazolium chloride (MBIC), 1-methyl-3-hexylimidazolium imidazolate (MIDI), and 1-butyl-3-benzylimidazolium acetate (BBIA) and developed them as corrosion inhibitors for AISI 1018 steel in 0.5 and 1.0 M H_2SO_4. Figure 14.9 illustrated the IE effect of the ILs as a function of temperature, finding that the compounds were more effective at 45°C in 1.0 M H_2SO_4. The assessment of the electrochemical findings showed the mixed-type action of corrosion inhibitors and the adsorption of the ILs well matched to the Temkin adsorption model. The surface characterization of the samples covered with CIs was carried out using the Mössbauer method, which helped infer that rozenite, goethite, and akaganeite/lepidocrocite were the major corrosion products. Finally, the corrosion inhibition mechanism of the ILs was proposed.

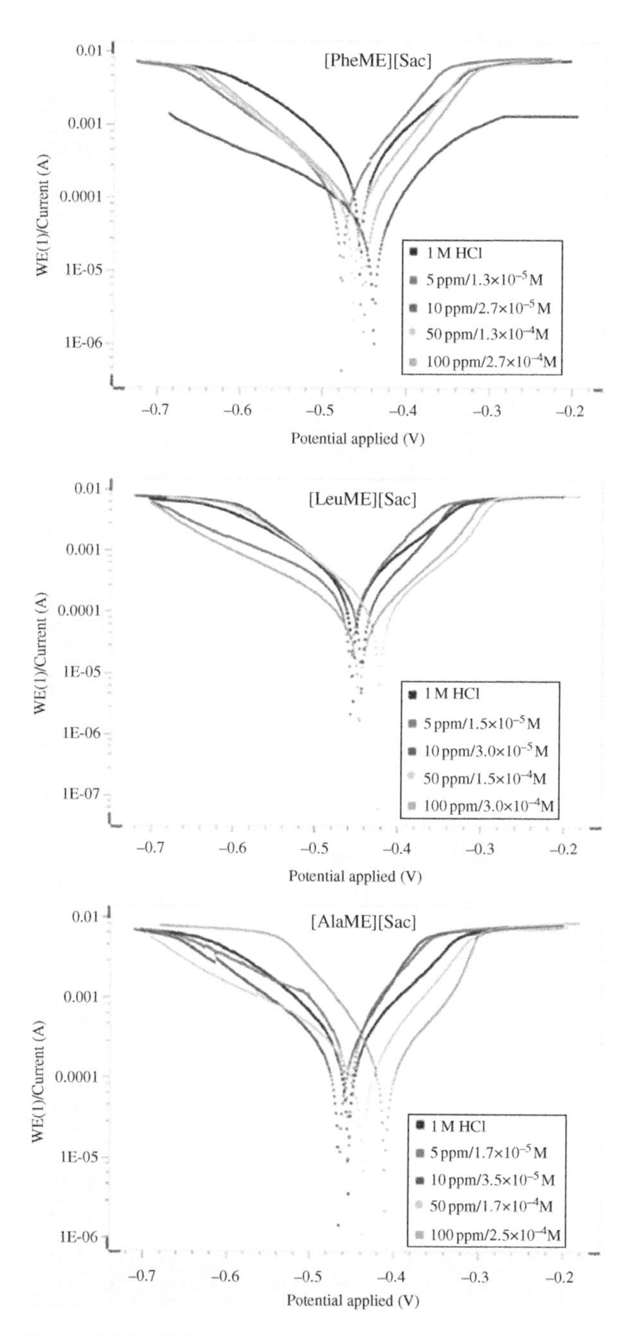

Figure 14.8 Tafel curves for mild steel in 1M HCl solution containing different concentrations of [AAE][Sac] ILs at 30°C. *Source:* Republished from Ref. [30] with permission from Elsevier.

Table 14.1 ILs used to inhibit various metal corrosion in HCl medium.

S.No.	Name	Material/ electrolyte	Inhibitor conc.	Techniques	Nature of adsorption	IE_W (%)	IE_E (%)	IE_P (%)	Ref
1.	Ethoxyloctadecyl ammonium tosylate (ODPEG-TS)	Steel/1M HCl	0.098 M	WL at 30°C, EIS, PP, SEM	Langmuir, mixed type	—	96.7	96.5	[31]
2.	Poly [3-butyl-1-vinylimidazolium bromide]	MS/1MHCl	400 ppm	EIS, PP, SEM	Frumkin, mixed type predominantly cathodic	—	92	96.4	[32]
3.	Chitosan-p-toluene sulfonate salt with oleic acid (CSPTA-OA)	Steel/HCl	250 ppm	EIS, PP	Langmuir, mixed type	—	95.6	97.8	[33]
4.	Trimethyl tetradecylammonium methyl sulfate	API-X52 steel/1M HCl	100 ppm	EIS, PP at T=25°C	Langmuir, mixed type	—	68	70	
5.	Octadecyl ammonium tosylate (ODA-TS)	Steel/1M HCl	0.341 mM	EIS, PP at T=25°C	Langmuir, mixed type	—	98.0	98.8	[34]
6.	Oleyl ammonium tosylate (OA-TS)	Steel/1M HCl	0.341 mM	EIS, PP at T=25°C	Langmuir, mixed type	—	96.6	96.3	
7.	3-((4-amino-2-Methylpyrimidin-5-yl)methyl)-5-(2-hydroxyethyl)-4-methylthiazol-3-ium chloride (AMPMHMC)	C-steel/1M HCl	40 ppm	WL, EIS, PP, DFT, XRD, SEM-EDS	Langmuir, mixed type	91.1	90.3	91.4	[35]
8.	3-(4- Chlorobenzoylmethyl)-1-methylbenzimidazoliumbromide ([BMMB]+Br−)	C-steel/1M HCl	6.84×10-4 M	EIS, PP, DFT, FTIR, SEM-EDS	Langmuir, mixed type	—	97.9 and decreased with temperature reached to 35.8 at 60 °C	84.0	[36]

9.	Task-specific ionic liquid (TSIL)	MS/1M HCl	100 ppm	WL at T = 25°C, EIS, PP, SEM, AFM, contact angle	Langmuir, mixed type predominantly anodic	—	78.9	78.7	[37]
10.	1-Hexyl-3-methylimidazolium trifluoromethane sulfonate [HMIM][TfO]	MS/1M HCl	500 ppm	WL at T = 30°C, EIS, PP, FT-IR, DFT	Langmuir, mixed type	79.88	79.88	81.16	[38]
11.	1-Hexyl-3-methylimidazolium tetrafluoroborate [HMIM][BF4],	MS/1M HCl	500 ppm	WL at T = 30°C, EIS, PP, FT-IR, DFT	Langmuir, mixed type	—	78.54	78.20	
12.	1-Hexyl-3-methylimidazolium hexafluorophosphate [HMIM][PF6]	MS/1M HCl	500 ppm	WL at T = 30°C, EIS, PP, FT-IR, DFT	Langmuir, mixed type	—	71.55	73.54	
13.	1-Hexyl-3-methylimidazolium iodide [HMIM][I]	MS/1M HCl	500 ppm	WL at T = 30°C, EIS, PP, FT-IR, DFT	Temkin, mixed type predominantly anodic	—	79.81	79.47	
14.	N-Methyl-N, N, N-trioctylammonium chloride (Aliquat 336)	MS/1M HCl	5.94μM	WL at T = 35°C, EIS, PP, FTIR, UV, DFT, AFM, MD	Langmuir, mixed type predominantly cathodic	95	95.6	93.2	[39]
15.	1-Ethyl-3-methylimidazolium chloride (EMIm Cl)	MS/2M HCl	1%	WL at T = 25°C, EIS, PP, DLS, AFM, FT-IR, DFT, QSAR	Langmuir, mixed type	—	74	73	[40]
16.	1-Butyl-3-methylimidazolium chloride (BMIm Cl	MS/2M HCl	1%	WL at T = 25°C, EIS, PP, DLS, AFM, FT-IR, DFT, QSAR	Langmuir, mixed type	—	74	75	

(Continued)

Table 14.1 (Continued)

S.No.	Name	Material/ electrolyte	Inhibitor conc.	Techniques	Nature of adsorption	IE_W (%)	IE_E (%)	IE_P (%)	Ref
17.	1-Butyl-3- methylimidazolium hexafluorophosphate (BMIm PF6)	MS/2M HCl	1%	WL at T = 25°C, EIS, PP, DLS, AFM, FT-IR, DFT, QSAR	Langmuir, mixed type	—	79	84	
18.	1-Butyl-3-methylimidazolium tetrafluoroborate (BMIm BF4)	MS/2M HCl	1%	WL at T = 25°C, EIS, PP, DLS, AFM, FT-IR, DFT, QSAR	Langmuir, mixed type	—	77	84	
19.	1-Butyl-3-methylimidazolium bromide (BMIm Br)	MS/2M HCl	1%	WL at T = 25°C, EIS, PP, DLS, AFM, FT-IR, DFT, QSAR	Langmuir, mixed type	—	74	75	
20.	1-Hexyl- 3-methylimidazolium chloride (HMIm Cl)	MS/2M HCl	1%	WL at T = 25°C, EIS, PP, DLS, AFM, FTIR, DFT, QSAR	Langmuir, mixed type	—	88	84	
21.	N-ethyl-N,N,N-trihexylammoniumadipate	X60 steel/1M H2SO4	100 ppm	WL at T = 25°C, EIS, PP, diffuse reflectance infrared Fourier transform spectroscopy (DRIFTS), DFT, XPS	Langmuir, mixed type	68		73	[41]
22.	N-ethyl-N,N,N-trioctylammonium ethyl sulfate					78		83	
23.	N-methyl-2-hydroxyethylammonium oleate ([m-2HEA][Ol])	MS/0.1M HCl	2.5 mM	WL, PP, contact angle, Raman	Mixed type	—		93.6	[42]
24.	1,1′ (1,4 Phenylenebis(methylene)) bis(3 (carboxymethyl) 1H imidazole-3-ium) chloride	CS/0.5M HCl	0.1 mM	EIS, PP, SEM-EDS, UV-vis, AFM, XPS, DFT	Langmuir, mixed type	—	94.8	88.5	[43]

25.	Benzyl tributylammonium tetrachloroaluminate [BTBA]+[AlCl4]	CS/2M HCl	400 ppm	EIS, EN, SEM-EDS, UV-vis, DFT, MC simulation	Langmuir, mixed type	—	97	—	[44]
26.	1-Butyl-1H-imidazol-1-ium 3-carboxy-2-((2-carboxylatoethyl) thio)propanoate ([Bhim]CETSA)	A3 steel/1M HCl	10 mM	WL at T = 25–45°C, IR, XPS, SEM, contact angle, DFT	Langmuir, mixed type	90.11 and decreases with temperature reaches to 79.19 at 45°C	—	—	[45]
27.	2-Benzyl-1- butyl-3-(3-(triethoxysilyl)propyl)-1H-benzo[d]imidazolium chloride (BTOSPB)	CS/1M HCl	3 mM	EIS, WL at T = 30–60°C, PP XPS, contact angle, SEM-EDS	Langmuir, mixed type	99.04 and increases slightly with temperature reaches to 99.77 at 60°C	99.55	—	[46]
28.	1-Butyl-3-methylimidazolium chloride	MS/2M HCl	2500 ppm	PP	Langmuir	—	82.29	86.7	[47]
29.	1-Butyl-3-methylimidazolium chloride	Aluminum/2M HCl	2500 ppm	PP	Langmuir	—	88.43	90.0	
30.	1-Butyl-3-methylimidazolium chloride	SS/2M HCl	2500 ppm	PP	Langmuir	—	94.38	99.3	
31.	1-Ethyl-3-methyl-imidazolium hexafluorophosphate (EMIMPF6),	MS/1M HCl	5×10^{-3}M	PP at T-25°C, EIS, SEM, FTIR, DFT	Langmuir, mixed type	—	82.29	82.95	[48]
32.	1-Methyl-3-pentyl-imidazolium hexafluorophosphate (PenMIMPF6),	MS/1M HCl	5×10^{-3}M	PP at T-25°C, EIS, SEM, FTIR, DFT	Langmuir, mixed type	—	88.43	87.60	

(Continued)

Table 14.1 (Continued)

S.No.	Name	Material/ electrolyte	Inhibitor conc.	Techniques	Nature of adsorption	IE_W (%)	IE_E (%)	IE_P (%)	Ref
33.	1-Methyl-3-octylimidazolium hexafluorophosphate (OctMIMPF6)	MS/1M HCl	5×10^{-3}M	PP at T-25°C, EIS, SEM, FTIR, DFT	Langmuir, mixed type	—	94.38	94.62	
34.	1-Octylpyridinium hexafluorophosphate(OctpyPF6)	MS/1M HCl	5×10^{-3}M	PP at T-25°C, EIS, SEM, FTIR, DFT	Langmuir, mixed type	—	92.05	92.23	
35.	1-Methyl-1-octyl-pyrrolidinium hexafluorophosphate (OctMpyrPF6)	MS/1M HCl	5×10^{-3}M	PP at T-25°C, EIS, SEM, FTIR, DFT	Langmuir, mixed type	—	93.57	93.94	
36.	1-Methyl-1-octyl-pyrrolidinium thiocyanate (OctMpyrSCN)	MS/1M HCl	5×10^{-3}M	PP at T-25°C, EIS, SEM, FTIR, DFT	Langmuir, mixed type	—	95.31	95.74	
37.	Methyl-1-octyl-pyrrolidinium dicyanamide (OctMpyrN	MS/1M HCl	5×10^{-3}M	PP at T-25°C, EIS, SEM, FTIR, DFT	Langmuir, mixed type	—	96.07	96.21	
38.	1-(Benzyloxy)-1-oxopropan-2-aminium 4-methylbenzenesulfonate (BOPAMS)	Aluminum/1M HCl	5×10^{-4}M	PP, WL at T- 30–60°C, EIS, FT-IR	Langmuir, mixed type	93.52 and decreased with temperature reached to 63.86 at 60°C	96.43	97.59	[49]
39.	4-(Benzyloxy)-4-oxobutan-1-aminium 4-methylbenzenesulfonate	Aluminum/1M HCl	5×10^{-4}M	PP, WL at T-30–60°C, EIS, FT-IR	Langmuir, mixed type	91.94 and decreased with temperature reached to 68.75 at 60°C	97.84	96.95	

η_W(%)- percent inhibition efficiency calculated by weight loss method; η_E(%)- percent inhibition efficiency calculated by electrochemical impedance measurement; η_P(%)- percent inhibition efficiency calculated by potentiodynamic polarization measurement.

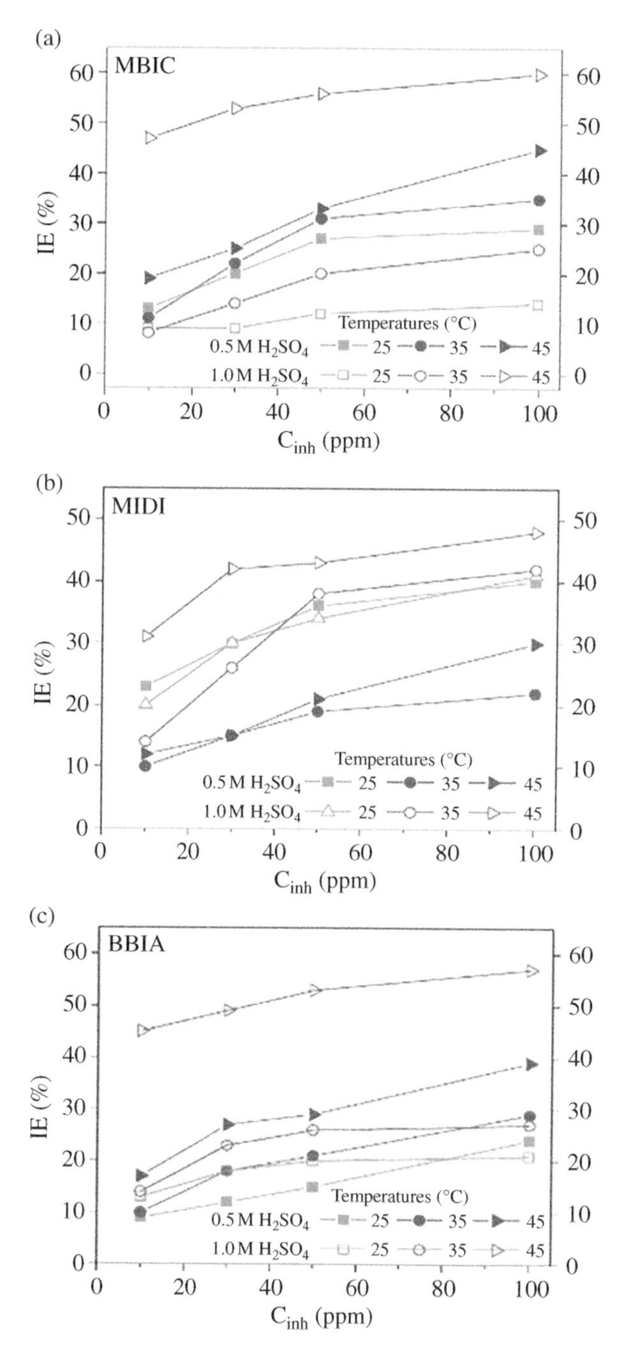

Figure 14.9 Inhibition efficiency of the ILs using AISI 1018 steel in 0.5 and 1.0 M H_2SO_4: (a) MBIC, (b) MIDI and (c) BBIA. *Source:* Republished from Ref. [51] Licensed under CC BY-4.0.

Ahmed et al. [52] synthesized (Figure 14.10) five new environmentally friendly indolium-based ILs: 5-methoxy-1,2,3,3-tetramethyl-3H-indolium iodide (IBIL-I), 1-(2-carboxyethyl)-2,3,3-trimethyl-3H-indolium iodide (IBIL-II), 2,3,3-trimethyl-

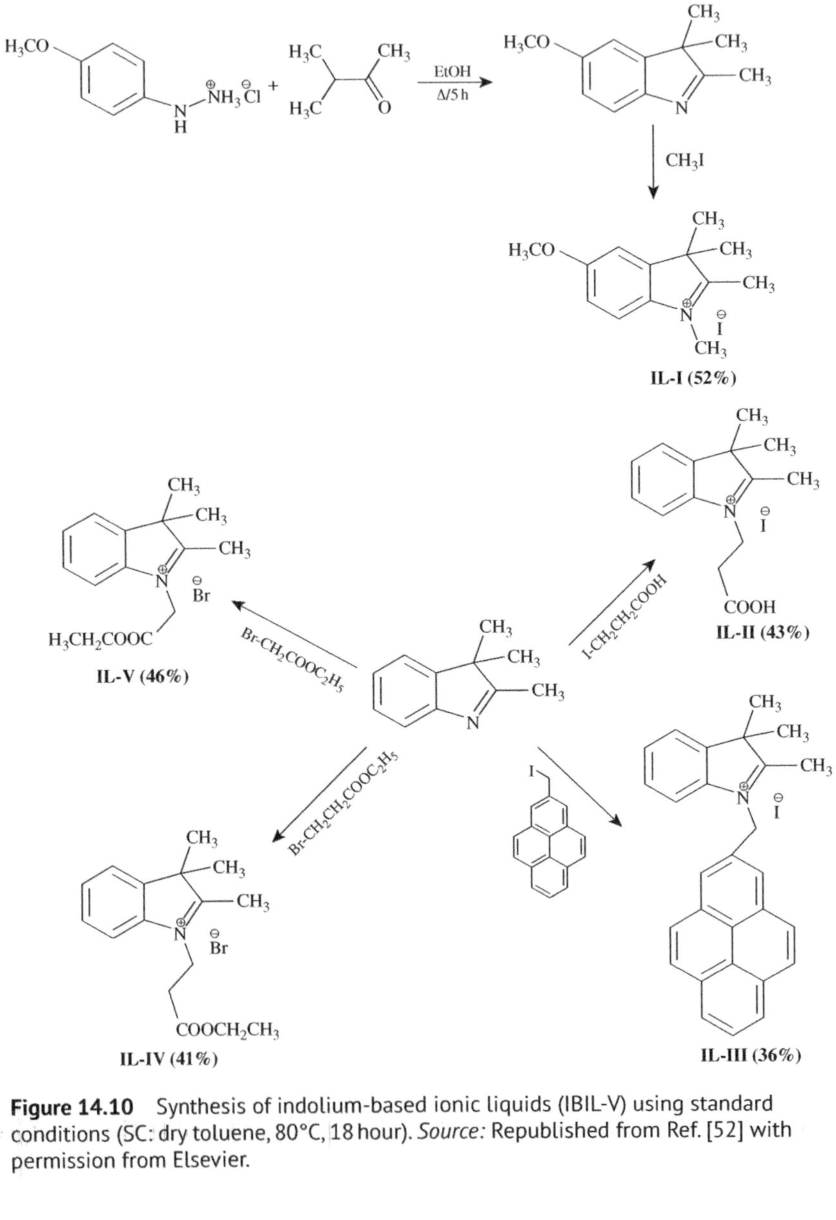

Figure 14.10 Synthesis of indolium-based ionic liquids (IBIL-V) using standard conditions (SC: dry toluene, 80°C, 18 hour). *Source:* Republished from Ref. [52] with permission from Elsevier.

1-(pyren-2-yl methyl)-3H-indolium iodide (IBIL-III), 1-(3-ethoxy-3-oxopropyl)-2, 3,3-trimethyl-3H-indolium bromide (IBILIV), and 1-(2-ethoxy-2-oxoethyl)-2,3,3-t rimethyl-3H-indolium bromide (IBIL-V), and established them as MS corrosion inhibitor in 0.5 M H_2SO_4.The inhibition performance depends on the molecular structure of the studied IBILs; IBILs consisting of bromide (as an anionic head) are efficient than those IBILs consisting of iodide. Electrochemical measurements find that all inhibitors primarily acted as a mixed type inhibitor, and IBIL-V exhibited the greatest effectiveness.

Arellanes-Lozada et al. [53] studied the inhibition of aluminum alloy AA6061 corrosion in sulfuric acid using three ILs, namely, poly(1-vinyl-3-dodecyl-imidazolium) (PImC12), poly(1-vinyl-3-octylimidazolium) (PImC8), and poly(1-vinyl-3-butylimidazolium) (PImC4) hexafluorophosphate. The following order demonstrated the effectiveness of these compounds: PImC12 > PImC8 > PImC4. ICP-OES found that dissolved Al^{3+} in solution decreased after PILs addition, supporting their inhibitory activity. In another study, Arellanes-Lozada et al. [54] studied anticorrosive properties of the imidazolium-derived cations and iodide anion-based IL, namely, (1-butyl-2,3-dimethylimidazolium iodide $[DBIM]+I^-$ and 1-propyl-2,3-dimethylimidazolium iodide $[DPIM]^+I^-$) for API 5L X52 steel in 1M H_2SO_4. The IE adopted the order [DBIM]+I- > [DPIM]+I- at stationary condition, but the IE was reversed under the laminar flow method. Likewise, the analyses of XPS, SEM, and AFM confirmed the IEs obtained.

The anticorrosive impact of 1,2- dimethyl-3-decylimidazolium iodide IL was tested in H_2SO_4 and HCl aqueous media by Olivares-Xometl et al. [55]. With efficiencies of up to 95% at 100 ppm, the IL demonstrated excellent corrosion inhibition outcomes in H_2SO_4. In another study, Olivares-Xometl et al. [56] synthesized a novel class of four new-free-halide ILs and tested their anticorrosive application for API 5L X60 steel in 1M sulfuric acid. The findings indicated that the IE exhibited by the four ILs is a function of their molecular structure and concentration. The IE obtained ranged from 51 to 89%. Corrosion inhibition behavior of the IL (EMIM) (SCN) was studied on API 5L X52 steel in 0.5M H_2SO_4 medium [57].The inhibition efficacy of the inhibitor was reported to be concentration and temperature-dependent. To explain the relationship between the experimental results and electronic properties, a stochastic simulation study was used.

Goyal et al. [58] presented two quaternary phosphonium salts, namely, (1-napthy lmethyl)-triphenylphosphonium chloride (1-NpMe-TPC) and (4-methoxybenzyl)-triphenylphosphonium bromide (4-MeOBz-TPB) as corrosion inhibitors for MS in 0.5M H_2SO_4. 10^{-3} M of 1-NpMe-TPC and 4-MeOBz-TPB exhibited the greatest IE of 99.72% and 97.76%, respectively, at 298 K. Moreover, the adsorption capability of both the inhibitors was explained by various thermodynamic and semi-empirical AM1 and MD parameters.

Ma et al. [59] tested the corrosion behavior of three ILs 1-(4-sulfobutyl)-3-meth ylimidazolium hydrogen sulfate ([BsMIM][HSO4]) and 1-(4-sulfobutyl)-3-methyl imidazolium tetrafluoroborate ([BsMIM][BF4]) in 1.0 M H_2SO_4 toward 304 stainless steel. With the dosage of the inhibitor, the IE rises and drastically declines above 60°C and followed the order: [BsMIM][BF4] < [BsMIM][HSO4]. Nessim et al. [60] prepared 3,3'-(1,4-phenylenebis(methylene))bis(1-alkyl-1H-imidazol-3-ium)bromide referred as IL1, IL2, and IL3, characterized, and tested the 304 SS corrosion inhibitor against 0.5 M H_2SO_4 solution. The gemini ILs showed corrosion inhibition values of 90.7%, 97.3%, and 82.6% for IL1, IL2, and IL3, respectively. Zaky et al. [61] tested the anticorrosive efficiency of three dicationic imidazolium abbreviated as IL1, IL2, and IL3 for SS corrosion in an acidic environment. The prepared ILs showed 91.5% (IL1), 98.4% (IL2), and 83.3% (IL3) efficiencies at 100 ppm.

14.3.3 In NaCl Solution

Through electrochemical characterization, Vega et al. [62] tested 2-hydroxyethylammonium oleate (2HEAOl) as an aluminum corrosion inhibitor in the neutral 0.5 mol/l NaCl medium. 5×10^{-4} mol/l of inhibitor was reported to be a suitable concentration to promote corrosion inhibition until 72 hour. A novel IL, namely, 1-butyl-1H-imidazol-1-ium-3-carboxy-2-((2-carboxylatoethyl) thio)propanoate, referred to as ([Bhim]CETSA) was reported to be a green and proficient corrosion inhibitor for A3 steel in 1M HCl solution [63], and IE reached up to 90.11% at 10 mM at 298 K.

Cao et al., [64] tested three imidazolium-type ILs (Figure 14.11), to inhibit the corrosion of CS in a 0.3 M NaCl saturated $Ca(OH)_2$ solution. Imidazolium IL acted as a mixed type of inhibitor and followed Langmuir adsorption isotherm.

Chong et al. [65] prepared and developed 2-methylimidazolinium 4-hydroxycinnamate (2-MeHImn 4-OHCin) as a highly effective MS inhibitor against neutral chloride conditions. The efficiency of the inhibition of the individual

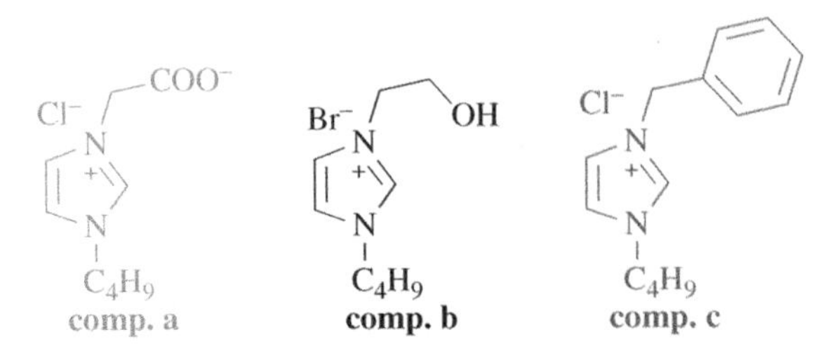

Figure 14.11 Imidazolium ionic liquids. *Source:* Republished from Ref. [64] Licensed under CC BY-4.0.

components of this salt, namely, 2-methylimidazolinium bromide (2-MeHImn Br) and 4-hydroxycinnamate sodium (Na 4-OHCin) salts, was also explored. Abbas et al. [66] studied the IE of 1-butylpyrrolidinium chloride [BPm1,1] Cl in 3.5% NaCl medium to inhibit CS corrosion. The protection performance increased by an increase of [BPm1,1] Cl concentration and solution temperature. The reported IL can be regarded as an efficacious biocide for bacterial strains. The anticorrosion impact of 1-butyl-1-methylpyrrolidiniumtrifluoromethylsulfonate ([Py1,4]TfO) was reported in 3.5% NaCl for MS corrosion [67]. The findings indicate that the IL used can efficiently inhibit the growth of planktonic and sessile bacteria, and the efficiency of inhibition depends on the concentration of IL. At 100 ppm, [Py1,4]TfO effectively inhibited MS corrosion with more than 80% performance. The effect of [Py1,4]TfO on corrosion inhibition is due to the adsorptive contact of the IL species with MS surface and the development of a protective barrier toward corrosion attack. Huang et al. [68] investigated the viability of film formation on AA5083 aluminum alloy by electrochemical treatments in the presence of trihexyl(tetradecyl)phosphoniumdiphenylphosphate ([P6,6,6,14][dpp]) IL. Lower rates of corrosion were seen on all samples treated at room temperature in IL. The surface characterization revealed a nonuniform porous film with a film thickness varying from 37 nm to 155 nm on an IL-treated sample at 50°C. Velrani et al. [69] tested 1-butyl-3-methylimidazolium chloride (1-BMIC) as anticorrosive material for MS corrosion in 3.5% NaCl solution at 35–55°C. 1-BMIC adsorption on the surface of MS obeys Freundlich isothermal adsorption. 1-BMIC functions as a mixed-type inhibitor with anodic dominance. SEM and FT-IR techniques affirm the protective film formation against the chloride attack.

Two different ILs, namely, 1-ethyl-3-methylimidazolium chloride ([EMIm]Cl) and 1-butyl-1-methylpyrrolidinium chloride ([Py1,4]Cl), were tested to prevent the corrosion of cast iron in the Arabian Gulf seawater [70]. In comparison with [Py1,4]Cl, [EMIm]Cl demonstrates a higher IE at a short immersion time which increases on increasing immersion time, which may be due to the formation of a more stable corrosion barrier on the cast iron base.

Su et al. [71] reported benzyl triphenylphosphoniumbis(trifluoromethylsulfonyl)amide ([BPP][NTf2]) IL as a corrosion inhibitor of AZ31B Mg alloy in 0.05 wt% NaCl solution, and the best IE (91.4%) was observed at room temperature. The 1-decyl-3-methylimidazolium chloride (DMICL) was investigated as an MS inhibitor in a CO_2-saturated NaCl solution [72]. The higher IE values (99%) was obtained at pH 3.8.

14.4 Conclusion and Future Trends

ILs are recognized as one of the most potent categories of corrosion inhibitors that are highly soluble, economic, and eco-friendly. According to the findings, different ILs are used for the protection of metals from corrosion attacks. Nevertheless, there are few aggressive environments under which ILs are used, primarily HCl, H_2SO_4, NaCl solution. The majority of the ILs identified were mixed-type inhibitors. IL-inhibited medium showed a decrease in corrosion current density values and an improvement in charge transfer resistance values. Various surface analytical, i.e. SEM, EDX, AFM, and spectroscopic, i.e. XPS, FT-IR, and UV-vis, studies confirmed that ILs inhibit corrosion by adsorbing on the metal surface and thereby forming a protective layer over the metal surface. In most cases, the adsorption of ILs on the metal surface followed the Langmuir adsorption isotherm model. Only a few of them showed Temkin, El-Awady, and Flory–Huggins adsorption model. The literature analysis revealed that most of the ILs examined were based on imidazolium, so the use of other IL groups, including cholonium, pyrrolidinium, phosphonium, tetra-ammonium, and pyridinium, should be further studied. In various earlier studies, the corrosion inhibition effects of ILs on iron alloys were the key consideration because of their relatively low price, so their application for other metals and alloys, mainly for copper, aluminum, brass, and magnesium, should therefore be extended. Besides, future work may focus on the development of new corrosion inhibitors relying on designer ILs, along with the use of the synergistic effect of multicomponent corrosion inhibitors. As aqueous phase corrosion inhibitors, many ILs are being used, so utilization of ILs as anticorrosive coating products is strongly recommended.

Acknowledgment

Ruby Aslam acknowledges the Council of Scientific & Industrial Research, New Delhi, India, for providing financial aid under the Research Associate fellowship (09/112(0616)2K19 EMR-I).

Abbreviations

C	steel-Carbon steel
MS	Mild steel
MD	Molecular dynamics
RDF	Radial distribution feature
NaCl	Sodium chloride
HCl	Hydrochloric acid
IE	Inhibition efficiency
OCP	Open circuit potential
PDP	Potentiodynamic polarization

EIS Electrochemical impedance spectroscopy
DFT Density functional theory
MD Molecular dynamics

References

1 Valdez, B., Schorr, M., Zlatev, R. et al. (19). Corrosion control in industry. In: *Environmental and industrial corrosion – practical and theoretical aspects* (eds. B. Valdez and M. Schorr), 54–2012. Rijeka, Croatia: Intech.

2 Koch, G.H., Brongers, M.P.H., Thompson, N.G. et al. (2002). Corrosion Cost and Preventive Strategies in the United State. National Technical Information Service Report No. FHWA-RD-01-156, R315-0. https://ntrl.ntis.gov/NTRL/dashboard/searchResults/titleDetail/PB2002106409.xhtml.

3 NACE (2000). *International Basic Corrosion Course Handbook Houston.* TX: NACE.

4 Olivares-Xometl, O., Likhanova, N.V., Domínguez-Aguilar, M.A. et al. (2006). Surface analysis of inhibitor films formed by imidazolines and amides on mild steel in an acidic environment. *Appl. Surf. Sci.* 252: 2139–2152.

5 Zuriaga-Monroy, C., Oviedo-Roa, R., Montiel-Sánchez, L.E. et al. (2016). Theoretical study of the aliphatic-chain length's electronic effect on the corrosion inhibition activity of methylimidazole-based ionic liquids. *Ind. Eng. Chem. Res.* 55: 3506–3516.

6 Garverick, L. (2011). *Corrosion in Petrochemical Industry*. United States of America: ASM International.

7 Sankarapapavinasam, S., Pushpanaden, F., and Ahmed, M.F. (1991). Piperidine, piperidones and tetrahydrothiopyrones as inhibitors for the corrosion of copper in H2SO4. *Corros. Sci.* 32: 193–203.

8 Martínez-Palou, R. and Sánche, P.F. (2011). *Perspectives of ionic liquids applications for clean oilfield technologies*. Rijeka, Croatia: INTECH Open Access Publisher.

9 Zhang, Q. and Hua, Y. (2009). Corrosion inhibition of mild steel by alkylimidazolium ionic liquids in hydrochloric acid. *Electrochim Acta* 54: 1881–1887.

10 Yuan, J. and Antonietti, M. (2011). Poly(ionic liquid)s: Polymers expanding classical property profiles. *Polymer* 52: 1469–1482.

11 Plechkova, N.V. and Seddon, K.R. (2008). Applications of ionic liquids in the chemical industry. *Chem. Soc. Rev.* 37: 123–150.

12 Wasserscheid, P. and Welton, T. (2002). *Ionic Liquids in Synthesis.* Wiley-VCH Verlag GmbH & Co. KGaA ISBNs: 3-527-30515-7.

13 Hart, W.E.S., Harper, J.B., and Aldous, L. (2015). The effect of changing the components of an ionic liquid upon the solubility of lignin. *Green Chem.* 17: 214–218.

14 Cevascoa, G. and Chiappe, C. (2014). Are ionic liquids a proper solution to current environmental challenges? *Green Chem.* 16: 2375–2385.
15 Gu, Y. and Jerome, F. (2013). Bio-based solvents: an emerging generation of fluids for the design of eco-efficient processes in catalysis and organic chemistry. *Chem. Soc. Rev.* 42: 9550–9570.
16 Verma, C., Ebenso, E.E., and Quraishi, M.A. (2017). Ionic liquids as green and sustainable corrosion inhibitors for metals and alloys: an overview. *J. Mol. Liq.* 233: 403–414.
17 Lopes, J.N.C., Gomes, M.F.C., and Padua, A.A.H. (2006). Nonpolar, polar, and associating solutes in ionic liquids. *The Journal of Physical Chemistry B.* 110: 16816–16818.
18 Eid, A.M., Shaaban, S., and Shalabi, K. (2020). Tetrazole-based organoselenium bi-functionalized corrosion inhibitors during oil well acidizing: experimental, computational studies, and SRB bioassay. *J. Mol. Liq.* 298: 111980.
19 Arellanes-Lozadaa, P., Díaz-Jiméneza, V., Hernández-Cocoletzia, H. et al. (2020). Corrosion inhibition properties of iodide ionic liquids for API 5L X52 steel in acid medium. *Corros. Sci.* 175: 108888.
20 Cao, C. (1996). On electrochemical techniques for interface inhibitor research. *Corros. Sci.* 38: 2073–2082.
21 Verma, C., Quraishi, M.A., and Ebenso, E.E. (2020). Corrosive electrolytes. *Int. J. Corros. Scale Inhib.* 9: 1261–1276.
22 Arellanes-Lozadaa, P., Olivares-Xometl, O., Likhanova, N.V. et al. (2018). Adsorption and performance of ammonium-based ionic liquids as corrosion inhibitors of steel. *J. Mol. Liq.* 265: 151–163.
23 Tawfik, S.M. (2016). Ionic liquids based gemini cationic surfactants as corrosion inhibitors for carbon steel in hydrochloric acid solution. *J. Mol. Liq.* 216: 624–635.
24 Schmitzhaus, T.E., Vega, M.R.O., Schroeder, R. et al. (2020). N-methyl-2-hydroxyethylammonium oleate ionic liquid performance as corrosion inhibitor for mild steel in hydrochloric acid medium. *Mater. Corros.* 71: 1–18.
25 Qiao, K. and Zeng, Y. (2020). Comparative study on two imidazolium-based ionic liquid surfactants as corrosion inhibitors for N80 steel in 15% hydrochloric acid solution. *Mater. Corros.* 71: 1–14.
26 Cao, S., Liu, D., Ding, H. et al. (2020). Towards understanding corrosion inhibition of sulfonate /carboxylate functionalized ionic liquids: an experimental and theoretical study. *J. Colloid Interface Sci.* 579: 315–329.
27 Goyal, M., Vashist, H., Kumar, S. et al. (2018). Acid corrosion inhibition of ferrous and non-ferrous metal by nature friendly Ethoxy carbonylmethyl triphenylphosphonium Bromide (ECMTPB): Experimental and MD simulation evaluation. *J. Mol. Liq.* 315: 113705.

28 Hajjaji, F., Ech-chihbi, E., Rezki, N. et al. (2020). Electrochemical and theoretical insights on the adsorption and corrosion inhibition of novel pyridinium derived ionic liquids for mild steel in 1M HCl. *J. Mol. Liq.* 314: 113737.

29 Nesane, T., Mnyakeni-Moleele, S.S., and Murulana, L.C. (2020). Exploration of synthesized quaternary ammonium ionic liquids as unharmful anti-corrosives for aluminium utilizing hydrochloric acid medium. *Heliyon* 6: e04113.

30 Aslam, R., Mobin, M., Obot, I.B., and Alamri, A.H. (2020). Ionic liquids derived from α-amino acid ester salts as potent green corrosion inhibitors for mild steel in 1M HCl. *J. Mol. Liq.* 318: 113982.

31 Atta, A.M., El-Mahdy, G.A., Allohedan, H.A., and Abdullah, M.M.S. (2016). Adsorption characteristics and corrosion inhibition efficiency of ethoxylatedoctadecylamine ionic liquid in aqueous acid solution. *Int. J. Electrochem. Sci.* 11: 882–898.

32 Ardakani, E.K., Kowsari, E., and Ehsani, A. (2019). Imidazolium-derived polymeric ionic liquid as a green inhibitor for corrosion inhibition of mild steel in 1.0 M HCl: Experimental and computational study. *Colloids Surf. A.* 586: 124195.

33 El-Mahdy, G.A., Atta, A.M., Al-Lohedan, H.A., and Ezzat, A.O. (2015). Influence of green corrosion inhibitor based on chitosan ionic liquid on the steel corrodibility in chloride solution. *Int. J. Electrochem. Sci.* 10: 5812–5826.

34 Atta, A.M., El-Mahdy, G.A., Al-Lohedan, H.A., and Ezzat, A.R.O. (2015). A new green ionic liquid-based corrosion inhibitor for steel in acidic environments. *Molecules* 20: 11131–11153.

35 Farag, A.A., Migahed, M.A., and Badr, E.A. (2019). Thiazole Ionic liquid as corrosion inhibitor of steel in 1M HClsolution: gravimetrical, electrochemical, and theoretical studies. *Journal of Bio- and Tribo-Corrosion* 5: 53.

36 Kannan, P., Karthikeyan, J., Murugan, P. et al. (2016). Corrosion inhibition effect of novel methyl benzimidazolium ionic liquid for carbon steel in HCl medium. *J. Mol. Liq.* 221: 368–380.

37 Kowsari, E., Payami, M., Amini, R. et al. (2014). Task-specific ionic liquid as a new green inhibitor of mild steel corrosion. *Appl. Surf. Sci.* 289: 478–486.

38 Mashuga, M.E., Olasunkanmi, L.O., Adekunle, A.S. et al. (2015). Adsorption, thermodynamic and quantum chemical studies of 1-hexyl-3-methylimidazolium based ionic liquids as corrosion inhibitors for mild steel in HCl. *Materials* 8: 3607–3632.

39 Quraishi, M.A., Haque, J., Srivastava, V. et al. (2017). N-Methyl-N, N, N-trioctylammonium chloride as novel and green corrosion 2 inhibitor for mild steel in acid chloride medium: Electrochemical, DFT and 3 MD studies. *New J. Chem.* 41: 13647–13662.

40 Yousefi, A., Javadian, S., Dalir, F.N. et al. (2015). Imidazolium-based ionic liquids as modulators of corrosion inhibition of SDS on mild steel in hydrochloric acid solutions: experimental and theoretical studies. *RSC Adv.* 5: 11697–11713.

41 Likhanova, N.V., Arellanes-Lozada, P., Olivares-Xometl, O. et al. (2019). Effect of organic anions on ionic liquids as corrosion inhibitors of steel in sulfuric acid solution. *J. Mol. Liq.* 279: 267–278.

42 Schmitzhaus, T.E., Vega, M.R.O., Schroeder, R. et al. (2020). N-methyl-2-hydroxyethylammonium oleate ionic liquid performance as corrosion inhibitor for mild steel in hydrochloric acid medium. *Mater. Corros.* 71: 1885–1902.

43 Cao, S., Liu, D., Ding, H. et al. (2019). Corrosion inhibition effects of a novel ionic liquid with and without potassiumiodide for carbon steel in 0.5MHCl solution: An experimental study and theoretical calculation. *J. Mol. Liq.* 275: 729–740.

44 Kannan, P., Varghese, A., Palanisamy, K., and Abousalem, A.S. (2019). Evaluating prolonged corrosion inhibition performance of benzyltributylammoniumtetrachloroaluminate ionic liquid using electrochemical analysis and Monte Carlo simulation. *J. Mol. Liq.* 297: 111855.

45 Gao, H., Xie, N., Wang, H. et al. (2020). Evaluation of corrosion inhibition performance of a novel ionic liquid based on synergism between cation and anion. *New J. Chem.* 44: 7802–7810.

46 Zafari, S., Sarabi, A.A., and Movassagh, B. (2020). A novel green corrosion inhibitor based on task-specific benzimidazolium ionic liquid for carbon steel in HCl. *Corros. Eng. Sci. Technol.* 55: 589–601.

47 Husin, H., Solo, B.B., Ibrahim, I.M. et al. (2018). Weight loss effect and potentiodynamic polarization response of 1-butyl-3-methylimidazolium chloride ionic liquid in highly acidic medium. *JESTC* 13: 1005–1015.

48 Al-Rashed, O.A. and Nazeer, A.A. (2019). Ionic liquids with superior protection for mild steel in acidic media: effects of anion, cation, and alkyl chain length. *J. Mol. Liq.* 288: 111015.

49 Azeez, F.A., Al-Rashed, O.A., and Nazeer, A.A. (2018). Controlling of mild-steel corrosion in acidic solution using environmentally friendly ionic liquid inhibitors: Effect of alkyl chain. *J. Molliq. Liq.* 265: 654–663.

50 Finsgar, M. and Jackson, J. (2014). Application of corrosion inhibitors for steels in acidic media for the oil and gas industry: a review. *Corros. Sci.* 86: 17–41.

51 Gómez-Sánchez, G., Likhanova, N.V., Arellanes-Lozada, P. et al. (2019). Electrochemical, surface and 1018-steel corrosion product characterization in sulfuric acid with new imidazole-derived inhibitors. *Int. J. Electrochem. Sci.* 14: 9255–9272.

52 Ahmed, S.A., Awad, M.I., Althagafi, I.I. et al. (2019). Newly synthesized indolium-based ionic liquids as unprecedented inhibitors for the corrosion of mild steel in acid medium. *J. Mol. Liq.* 291: 111356.

53 Arellanes-Lozada, P., Olivares-Xometl, O., Guzmán-Lucero, D. et al. (2014). The inhibition of aluminum corrosion in sulfuric acid by poly(1-vinyl-3-alkyl-imidazolium Hexafluorophosphate). *Materials* 7: 5711–5734.

54 Arellanes-Lozada, P., Díaz-Jiménez, V., Hernández-Cocoletzi, H. et al. (2020). Corrosion inhibition properties of iodide ionic liquids for API 5L X52 steel in acid medium. *Corros. Sci.* 175: 108888.

55 Olivares-Xometl, O., López-Aguilar, C., Herrastí-González, P. et al. (2014). Adsorption and corrosion inhibition performance by three new ionic liquids on API 5L X52 steel surface in acid media. *Ind. Eng. Chem. Res.* 53: 9534–9543.

56 Olivares-Xometl, O., Álvarez-Álvarez, E., Likhanova, N.V. et al. (2017). Synthesis and corrosion inhibition mechanism of ammonium based ionic liquids on API 5L X60 steel in sulfuric acid solution. *J. Adhes. Sci. Tech.* 32: 1092–1113.

57 Luna, M.C., Le Manh, T., Sierra, R.C. et al. (2019). Study of corrosion behavior of API 5L X52 steel in sulfuric acid in the presence of ionic liquid 1-ethyl 3-methylimidazolium thiocyanate as corrosion inhibitor. *J. Mol. Liq.* 289: 111106.

58 Goyal, M., Kumar, S., Verma, C. et al. (2019). Interfacial adsorption behavior of quaternary phosphoniumbased ionic liquids on metal-electrolyte:Electrochemical, surface characterization and computational approaches. *J. Mol. Liq.* 298: 111995.

59 Ma, Y., Han, F., Li, Z., and Xia, C. (2016). Acidic-functionalized ionic liquid as corrosion inhibitor for 304 stainless steel in aqueous sulfuric acid. *ACS Sustainable Chem. Eng.* 4: 5046–5052.

60 Nessim, M.I., Zaky, M.T., and Deyab, M.A. (2018). Three new gemini ionic liquids: Synthesis, characterizations and anticorrosion applications. *J. Mol. Liq.* 266: 703–710.

61 Zaky, M.T., Nessim, M.I., and Deyab, M.A. (2019). Synthesis of new ionic liquids based on dicationic imidazolium and their anti-corrosion performances. *J. Mol. Liq.* 290: 111230.

62 Vega, M.R.O., Mattedi, S., Schroeder, R.M., and Malfatti, C.F. (2020). 2-hydroxyethilammonium oleateprotic ionic liquid as corrosion inhibitor for aluminum in neutral medium. *Mater. Corros.*: 1–14. https://doi.org/10.1002/maco.202011847.

63 Gao, H., Xie, N., Wang, H. et al. (2020). Evaluation of corrosion inhibition performance of a novel ionic liquid based on synergism between cation and anion. *New J. Chem.* 44: 7802–7810.

64 Cai, J., Liu, J., Mu, S. et al. (2020). Corrosion inhibition effect of three imidazolium ionic liquids on carbon steel in chloride contaminated environment. *Int. J. Electrochem. Sci.* 15: 1287–1301.

65 Chong, A.L., Mardel, J.I., MacFarlane, D.R. et al. (2016). Synergistic corrosion inhibition of mild steel in aqueous chloride solutions by an imidazolinium carboxylate salt. *ACS Sustainable Chem. Eng.* 4: 1746–1755.

66 Abbas, M.A., Zakaria, K., El-Shamy, A.M., and Abedin, S.Z. (2019). Utilization of 1-butylpyrrolidinium chloride ionic liquid as an eco-friendly corrosion inhibitor and biocide for oilfield equipment: combined weight loss, electrochemical and sem studies. *Z. Phys. Chem.* 235: 377–406.
67 El-Shamy, A.M., Zakaria, K., Abbas, M.A., and El-Abedin, S.Z. (2015). Anti-bacterial and anti-corrosion effects of the ionic liquid 1-butyl-1-methylpyrrolidinium trifluoromethylsulfonate. *J. Mol. Liq.* 211: 363–369.
68 Huang, P., Somers, A., Howlett, P.C., and Forsyth, M. (2016). Film formation in trihexyl(tetradecyl)phosphoniumdiphenylphosphate ([P6, 6, 6, 14][dpp]) ionic liquid on AA5083 aluminium alloy. *Surf. Coat. Technol.* 303: 385–396.
69 Velrani, S., Jeyaprabha, B., and Prakash, P. (2014). Inhibition of mild steel corrosion in 3.5% NaCl medium using 1-butyl-3-methylimidazolium chloride. *Int. J. Innovative Sci., Eng. Technol.* 1: 57–69.
70 Sherif, E.-S., Abdo, H.S., and Abedin, S.Z. (2015). Corrosion inhibition of cast iron in arabian gulf seawater by two different ionic liquids. *Materials* 8: 3883–3895.
71 Su, H., Liu, Y., Gao, X. et al. (2019). Corrosion inhibition of magnesium alloy in NaCl solution by ionic liquid: Synthesis, electrochemical and theoretical studies. *J. Alloys Compd.* 791: 681–689.
72 Yang, D., Zhang, M., Zheng, J., and Castaneda, H. (2015). Corrosion inhibition of mild steel by an imidazolium ionic liquid compound: the effect of pH and surface pre-corrosion. *RSC Adv.* 5: 95160–95170.

15

Oleochemicals as Corrosion Inhibitors

F. A. Ansari[1], Sudheer[2], Dheeraj Singh Chauhan[3,4], and M. A. Quraishi[5]

[1] *Department of Applied Sciences, Faculty of Engineering, Jahangirabad Institute of Technology, Barabanki, India*
[2] *Department of Chemistry, Faculty of Engineering and Technology, SRM-Institute of Science and Technology, Ghaziabad, India*
[3] *Center of Research Excellence in Corrosion, Research Institute, King Fahd University of Petroleum and Minerals, Dhahran, Saudi Arabia*
[4] *Modern National Chemicals, Second Industrial City, Dammam, Saudi Arabia*
[5] *Interdisciplinary Research Center for Advanced Materials, King Fahd University of Petroleum and Minerals, Dhahran, Saudi Arabia*

15.1 Introduction

The utilization of metal/alloys likely steel, zinc, brass, etc., for various applications is needed for a variety of industrial processes. These applications involve the cleaning of corrosion product with acids, pickling in neutral-alkaline and acidic medium, oil recovery, etc. wherein the metal surfaces come in direct contact with strong mineral acids [1, 2]. In addition, the use of heat exchangers and onshore/offshore oil recovery processes make the metal surfaces to come in direct contact with strongly saline media [3, 4]. The strong acids/alkali/saline environments cause considerable corrosion damage to the metallic structures causing huge economic losses. The application of organic compounds as additives to the corrosive environments is a potential way to combat corrosion.

The majority of organic compounds work by adsorbing on the metal top through their π-bonds, aliphatic chains, phenyl rings, etc. Heterocyclic compounds containing atoms like nitrogen, sulfur, oxygen, etc. enables the adsorption of the organic corrosion inhibitors on the metallic surface due to the lone pair electrons. Most organic complexes having such heteroatoms are likely to be azoles (di, tri, imida), thymine-based amines (pyirimidine, pyridine), etc. are synthetic organic compounds [5]. However, most of these organic compounds are extremely costly,

Organic Corrosion Inhibitors: Synthesis, Characterization, Mechanism, and Applications,
First Edition. Edited by Chandrabhan Verma, Chaudhery Mustansar Hussain, and Eno E. Ebenso.

poisonous, and from the point of view of environmental protection, their functional applicability becomes doubtful.

On this basis, the advantage of biomass, which is biodegradable, has low toxicity, easily available, is environmentally safe, and is a potent method to counter the drawbacks of the conventionally used corrosion inhibitors. This chapter attempts to outline a brief background on oleochemicals, their significance as corrosion inhibitors followed by an overview of literature on the applicability of oleochemicals as corrosion inhibitors.

15.2 Corrosion

15.2.1 Definition and Economic Impact

Corrosion can be described as a natural process leading to the deterioration of the metallic materials upon exposure to aggressive environments. These environments range from the atmospheric factors such as air, moisture, soil, and so on to highly corrosive conditions such as the mineral acids, alkali, concentrated saline solutions, high temperatures encountered in different industries [5]. Corrosion not only deteriorates monuments of historical importance, bridges, highways, electrical appliances, etc. but it also damages the environment and accelerates air pollution. The high concentration of corrosive contaminants in the air activates atmospheric corrosion. The major pollutant, which accelerates corrosion, includes sulfur dioxide, carbon dioxide, humidity, etc. Additional pollutants that result in high rates of corrosion include waste-generated hydrogen sulfide from geothermal activity, the by-product of anaerobic oxidation of living organic matter, traffic, and combustion process releases nitrogen dioxide, hydrochloric acid, acetic acid, etc., which are released in the environment as industrial waste. Such conditions can lead to worsen the process of metal loss to a great extent and cause potential structural failure and the loss of human life. The annual global loss resulted due to corrosion has been estimated to go in excess of US$ 2.5 trillion, which is about to be 3–4% GDP of the globe [6]. In practice, there are a range of corrosion-prevention techniques, including corrosion-resistant alloys, the use of anticorrosion coatings, and the most efficient application of corrosion inhibitors.

15.2.2 Corrosion Inhibitors

The implementation of organic complexes as corrosion-retarding agents is best effective means to mitigate the aqueous corrosion loss to metal surfaces. As defined "A corrosion inhibitor is a chemical substance, which, when introduced in trace amounts to a given corrosive environment, can bring down the corrosion rate, without considerably altering the concentration of the medium" [5, 7].

A number of organic compounds such as pyridines, triazoles, pyrimidines, benzimidazoles, and so on have been applied as corrosion inhibitors [5]. Different approaches have historically controlled corrosion reduction, which includes "cathodic protection, process control, reduction of the content of metal impurity, and application of surface treatment techniques, as well as introduction of appropriate alloys." The best preventive way to combat corrosion in different corrosive environment is the application of inhibitors. As in particular, inhibitors slower down deterioration of metals/alloys and thus save billions of dollars on industrial vessels, machinery, or surfaces due to metallic corrosion. The inhibition mechanism includes the adsorption of organic compounds at the active sites of the metallic alloy surface [8, 9] and reduce deposition of oxide layer.

15.3 Significance of Green Corrosion Inhibitors

Conventionally, the inorganic/organic compounds are in practice as inhibitors that are in practice are toxic, carcinogenic, and cause a number of potential human health hazards when discharged to the environment. Therefore, in recent decades, innumerable guidelines were proposed to minimize the discharge of environmentally detrimental chemicals to the soil and aquatic life [10, 11]. In addition, a set of toxicity tests have been proposed to analyze the organic corrosion inhibitors prior to use. To add to this, a variety of green chemical reactions or routes have also been proposed to analyze for the preparation of organic compounds and also to verify that safer practices have been carried out during the preparation [12]. Considering the above issue, efforts are being devoted to developing novel and cost-effective materials, eco-friendly, and biodegradable to satisfactorily fulfill requirement of effective corrosion inhibitors. Therefore, the corrosion-inhibiting compounds are from the classes of plant origin [13], amino acids [14], ionic liquids [15], biological polymers [16], drugs [17], expired pharmaceuticals [18], etc., which are in demand due to their environmentally benign nature. Herein, we have presented an overview on the oleochemical-based corrosion inhibitors.

15.4 Overview of Oleochemicals

15.4.1 Environmental Sustainability of Oleochemicals

Oleochemicals are fats and oils, which are derived from natural resources, i.e. animal and plants. "The basic units of oleochemicals are fatty acids, from which various derivatives fatty acid esters, alcohol, amine derivatives are formed by chemical and enzymatic changes. Oleochemicals have a varied applications ranging from pharmaceuticals, food industry, lubricants, paints, coatings and corrosion inhibitors." Natural oils are plant-based products and is a source for widely available soybean, palm kernel, rapeseed, and sunflower oil. All fats and oils have a distinct characteristic in their chemical composition. They are trimesters of fatty acids and glycerol. Their chemical structure makes them different and unique, can undergo many chemical changes, and thus has a broad spectrum of industrial utility.

Oleochemicals are on the ascent as practical substitutes, as nonrenewable assets become exhausted and natural guideline of usage of natural resources become stricter. The development of the oleochemical-based products showcase a promising future driven by various market variables:

- Accessibility of crude materials
- Popularity from customers
- Development of the green synthetic compounds

The interest for oleochemical-based products will increase due to increment in usage of applications daily utility-based personal care essential products, cleansers, food, and beverages.

15.4.2 Production/Recovery of Oleochemicals

Oils and fats are significant building materials for the development of oleochemicals; it includes unsaturated fats, methyl-esters, greasy alcohols and amines, and glycerol as a result. Unsaturated vegetable oils represent almost 80% of the worldwide oil and fat production; in 2019, it was accounted about 200 million tons [19]. Among oil crops, soyabeans are the most significant, trailed by seed oils, palm, and sunflower oils. Industries utilize near about 15–17 million gallon tons of vegetable oils for the creation of soaps, oils, varnishes/paints, makeup, and different items. On a fundamental level, hydrolysis/lypolysis of saturated or unsaturated fatty acids can result in any oil or fat (by reaction in presence of water at high temperature or pressure or by enzymatic reaction). It is estimated that, depending on geographical development, just about eight or so fats/oils contribute to the mass production of fatty acids.

A specific crude product is picked by the manufacturer to bring in the suitable chain length profile for the desired synthesis compound with minimal by-products and ease in parting process. The production is based on availability and cost-effective procedure; thus, the prevalence of indigenous oils utilized inside the different topographical locales. However, it is scarce that all consumer requirements should be met adapted from single source. For instance, tall oil contains critical amounts of C18 = 1, C18 = 2 and higher series of unsaturates, which likewise has some sulfur content that poison hydrogenation catalysis procedure and limits the process of tall oil fatty acid. Table 15.1 shows the composition of fatty acids present in different oils.

Table 15.1 Composition of fatty acids in various oils.

	Carbon chain length								C_{18}		
Oils	Caprylic 8.0	Capric 10.0	Lauric 12.0	Myristic 14.0	Palmitic 16.0	Staearic 18.0	20.0	22.0	Oleic 18 = 1	Linoleic 18 = 2	Linolenic 18 = 3
Coconut Oil	8	7	48	17	9	3			6	2	
Palm kernel oil	4	5	50	17	7	2			12	2	
Palm oil				2	42	2			45	10	
Corn Oil				<1	11	2			27	58	1
Cotton seed oil				1	24	2			18	54	1
Groundnut oil				<1	14	3			40	36	1
Olive oil				<1	10	2			78	7	1
Rape seed oil				2.5	40	4			26	20	
Sunflower oil				<1	6	5	2	3	60	61	
Soyabean oil				<1	11	2		1	23	51	6

Source: Ref. [20].

The majority of the oils are of Lauric family, which exhibits principally 12/14 carbon chain acids (C12 is known as lauric acid) and made from coconut/palm kernel oils. While, long chain (C16-C18) saturated (stearines), and long chain unsaturated (oleines) are hydrophobes required, all of which may be found in natural oils and fats. Unsaturates with only one double bond, which are fundamentally oleic (C18 = 1), however and erucic (C22 = 1) acid are produced from sunflower and soyabean oil. Polyunsaturation possess more than one, two, or more number of double bonds, e.g. linoleic (C18 = 2) and linolenic (C18 = 3). The development of basic oleochemical complexes like unsaturated fats, unsaturated fat methyl esters (FAME), alcohol-based, amine-based, and glycerols are produced by various chemical reactions and enzymatic changes. Several by-product and intermediate compounds also generated from fundamental oleochemical complexes incorporate alcohol-based ethoxylates, sulfates, ether sulfates, quaternary ammonium salts, monoacylglycerols, diacylglycerols, some triacylglycerides, carbohydrate-based esters, and other oleochemical compounds.

Manufacturing of oleochemicals involve two major processes, which includes hydrolysis and transesterification:

i) **Hydrolysis**

The dissociation (or hydrolysis) of the fatty substances forms unsaturated fats and glycerin:

$RCO_2CH_2–CHO_2CR–CH_2O_2CR + 3\ H_2O \rightarrow 3\ RCOOH + HOCH_2–CHOH–CH_2OH$

Triglyceride water fatty acid glycerol

2) **Transesterification**

In the process of transesterification, fats and oil combine with alcohols (R'OH) rather than with H_2O as in hydrolysis. Glycerol is being obtained along with the unsaturated fat esters. Most normally, the response involves the utilization of methyl alcohol (MeOH) to give unsaturated fatty methyl esters:

$RCO_2CH_2–CHO_2CR–CH_2O_2CR + 3MeOH \rightarrow 3\ RCO_2Me + HOCH_2–CHOH–CH_2OH$

Triglyceride Methanol Fatty acid Glycerol
Methyl ester

3) **Hydrogenation**

The unsaturated fat or greasy esters delivered by these techniques might be changed. For instance, hydrogenation changes over unsaturated fats into immersed unsaturated fats. The acids or triglycerides can likewise converts to fatty -ols. For certain applications, unsaturated fats are changed over to fatty nitriles. Hydrogenation of these nitriles gives amines of fatty acids, which shows a vivid range of utilizations [21]. A schematic of the production of fatty acid derivatives is shown in Figure 15.1.

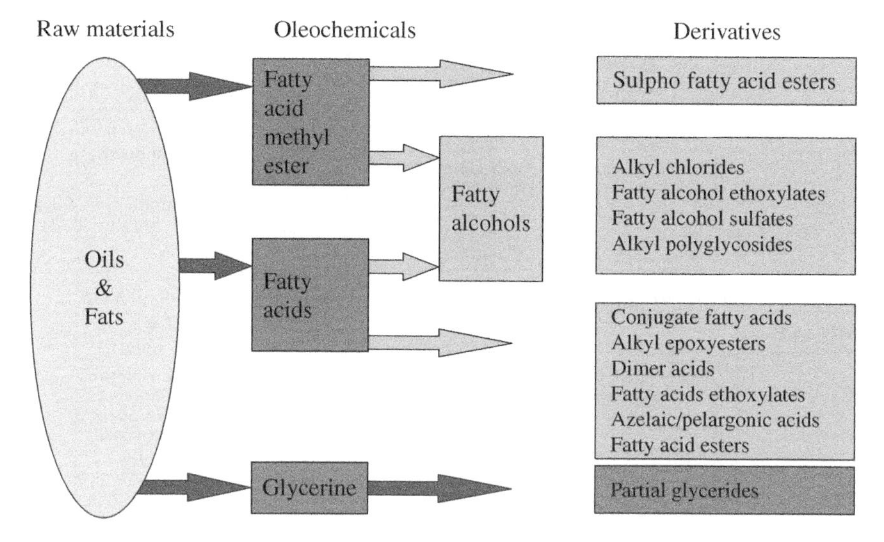

Figure 15.1 Production scheme of fatty acid and its derivatives.

15.5 Literatures on the Utilization of Oleochemicals as Corrosion Protection

Heterocyclic derivatives of various fatty acids, i.e. amines, alcohol, azoles, ethoxylates containing nitrogen, sulfur, oxygen heteroatoms, proved to be efficient corrosion inhibitors. Hence provided with more active centers in their structures, they have better ability for the process of adsorption and helps ineffective adsorption on the metal/alloy surface. A number of fatty acids and their derivatives exhibit corrosion-inhibiting property, which proves to be eco-friendly, economical, and of very low toxicity. Various researches have proved that bioproducts can be a good alternative for synthesized eco-friendly corrosion inhibitors. The efficiency of inhibition depends upon the medium (acidic or basic), temperature conditions, and time of immersion. Different types of corrosion inhibitor molecules developed from fatty acids are given in Figure 15.2 [20].

The group of Quraishi et al. has significant contribution toward synthesis and application of oleochemical-based corrosion inhibitor chemistries. They have reported oleochemicals in acidic media for acid pickling, acidizing, and as the volatile corrosion inhibitors (VCI). Recently, Tripathy et al. examined the corrosion protection impact of palmitic-acid-based imidazole to be applied during the acid-cleansing process of the multistage flash (MSF) desalination plants [22]. The inhibitor was prepared following a one-step microwave synthesis protocol within 7 min (Scheme 15.1), and high efficiency of 90.10% was afforded at 10^{-3}M

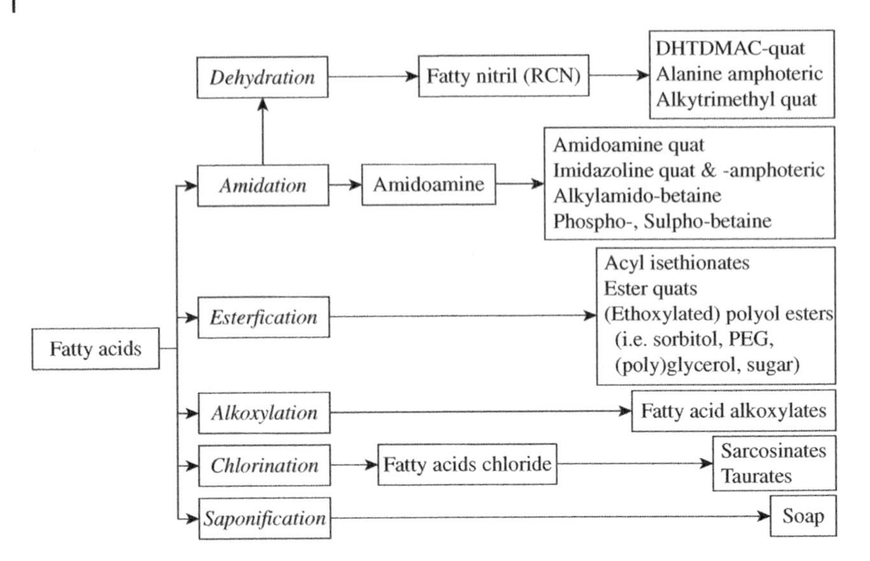

Figure 15.2 Various corrosion inhibitors derived from fatty acids. *Source:* Ref. [20].

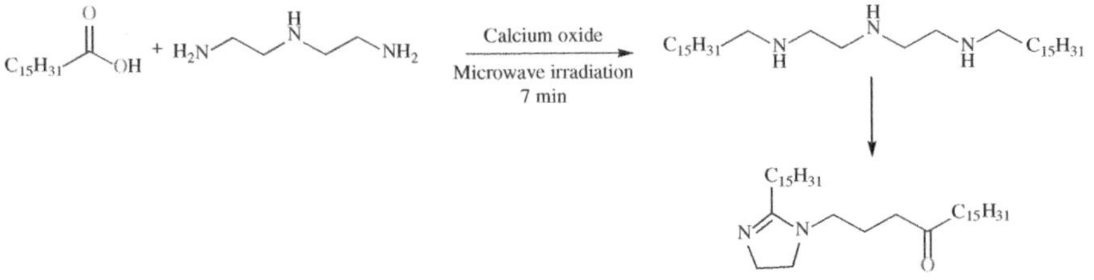

Scheme 15.1 Synthetic route of palmitic acid imidazole (PI). *Source:* Ref. [22].

concentration in 1 mol/l H_2SO_4. The efficiency increased further to 98.10% with the synergistic influence of 6×10^{-3}M KI. In the presence of an inhibitor, the water contact angle rose from 57.7° to 103.2°, which shows an enhancement in the hydrophobicity of the metal surface. Molecular dynamics explains that the adsorption of inhibitor complex on the metallic layer is in a linear fashion.

The process of acidization in oil well carried out in the industry makes use of concentrated mineral acid formulations, which causes considerable loss of the metal surface. This above process requires the inhibited mineral acids, which can allow protection to the underlying metal surface. Myristic acid imidazoline was prepared from tetradecanoic acid and diethylenetriamine as efficient compound for corrosion resistance of low carbon steel in 15% HCl environment using weight loss and electrochemical studies [23]. EIS study reflects a two time-constant process for the inhibitor adsorption and revealed an efficiency of 94.99% at a dose of

300 mg/l. Thermodynamic parameters showed the presence of a mixed mode of physiochemical adsorptions, schematically shown in Figure 15.3. FTIR and XPS examination of the chemisorbed inhibitor film indicated the adsorption and reduction in the chloride content of the adsorbed corrosion inhibitor.

Another imidazoline derivative, namely, QSI was synthesized, characterized, and evaluated for low carbon steel in 15% HCl medium under hydrodynamic and static circumstances [24]. At an inhibitor dose of 400 mg/l, an efficiency of $<50\%$ was afforded, which rose to 90% with synergistic influence of KI. The prevalent adsorption process of QSI on the metal/alloy surface was chemical adsorption. A major observation was that the carbon chain adheres to the pendant group had more profound effect on performance compared to the length of the tail group. The increase in the length of the carbon chain connected to the pendant active functional group resulted in an improvement in the efficiency. Rafiquee et al. investigated four imidazoline-based oleochemicals PDI, UDI, HDI, and NI, as environmental friendly compounds for steel in 0.5M H_2SO_4 [25]. The synthesis was accomplished in a two-step process from the respective fatty acids with $SOCl_2$,

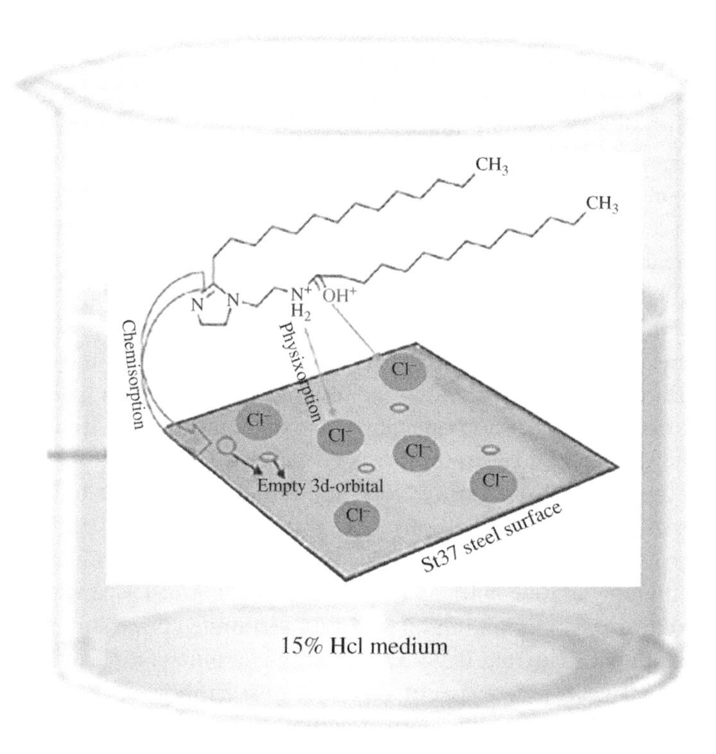

Figure 15.3 Schematic of the adsorption of the myristic acid imidazoline on steel surface. *Source:* Ref. [23].

which was reacted further with ethylenediamine to afford the corresponding imidazoline inhibitor. A high efficacy of 97.43% was afforded for the inhibitor UDI at a dose of 500 mg/l. The adsorption of inhibitor followed the adsorption isotherm of Langmuir, and the PDP studies revealed a mixed type of inhibitor.

Rice bran is a good source of vegetable oil. Its 10% is utilized as oil, and the rest is used as cattle feed. The rice bran oil is rich in oil composition with 47% monounsaturated, 33% polyunsaturated, and 10–23% fatty acid, and apart from it protein, tocopherol source are also present. E. Reyes-Dorantes et al. synthesized an environmentally being fatty amide type of corrosion inhibitor from crude rice bran oil [26]. Corrosion inhibitor of fatty amide was used as potential class inhibiting compounds for API-X 70 steel. The corrosive environment was saturated solution of 3.5% CO_2 in 3.5% NaCl. The Tafel plot and electrochemical impedance studies show inhibitors are anodic in nature and also reflects inhibition efficiency of up to 95%.

Some novel sulfonated fatty acid diethanolamide were prepared by combination of corn oil, and ethanolamine for about 8 hours at 170°–180°C, with general formula, i.e. [$R-CH-(OSO_3M)-CON-(CH_2-CH_2-OH)_2$ Figure 1(a) where M-Na, K, NH_4, – $NH-CH_2-CH_2-OH$, and $-N-(CH_2-CH_2-OH)_2$ [27]. These inhibitors were tested for mild steel in NaCl environment saturated with CO_2. These novel inhibitors showed excellent performance, which was revealed by polarization studies. They behaved as mixed corrosion inhibitors by retarding both cathodic and anodic reactions. Abbasov et al. examined the inhibitor performance by weight loss and linear potentiodynamic polarization techniques [28, 29]. The maximum efficacy was attained at a concentration close to its critical micellar concentration (CMC). Furthermore, the study shows that inhibiting efficiency increases with gradual increment in inhibitor molar concentration.

The action of inhibition is depends on two significant variables which the molecule possesses: the hydrophilic polar part that is rich in electron and fasten to the metal surface through coordination bond and the hydrophobic part that makes a barrier on metal surface and safeguard from outer alkaline environment. Furthermore, the double bond, carbon-hydrogen chain length, and functional group increased the corrosion-resistance efficiency of the corrosion inhibitors. In continuation series of environmentally friendly corrosion inhibitors, sunflower oil sulfated amine derivatives, i.e. sulfated fatty acid, -ethylamine complex; sulfated fatty acid, dimethylamine; sulfated fatty acid, diethylamine (Scheme 15.2), are proved to be effective corrosion inhibitors for CO_2-saturated brine solution.

A comparative study of sulfated fatty acid – diethanolamine complexes (DC), which are originated from four different vegetable oils: sunflower, cottonseed, corn, and palm oil, was carried out as corrosion inhibitors [30]. The best inhibition efficacy is about 99.95% at 100 ppm of the complex obtained from corn oil. The adsorption behavior followed Langmuir isotherm, which was further

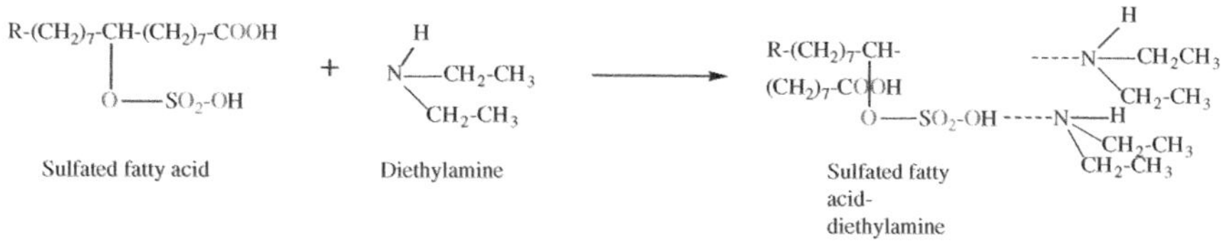

R-$(CH_2)_7$-CH-$(CH_2)_7$-COOH + H-N-CH_2-CH_3 (-CH_2-CH_3) → Sulfated fatty acid-diethylamine

Sulfated fatty acid — Diethylamine — Sulfated fatty acid-diethylamine

Scheme 15.2 Synthesis of sulfated fatty acid – diethylamine. *Source:* Modified from Ref. [28].

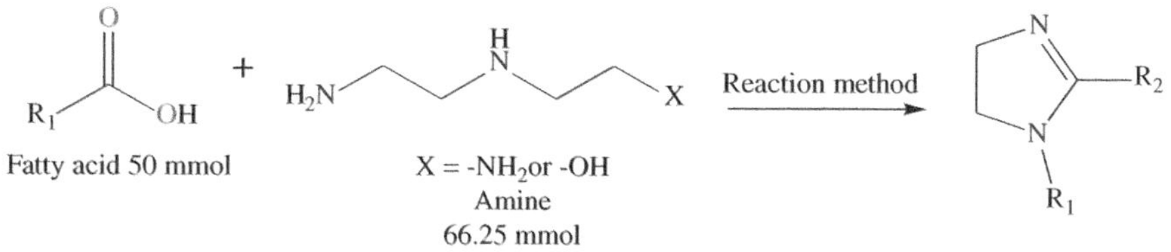

Fatty acid 50 mmol + Amine ($X = -NH_2$ or $-OH$) 66.25 mmol —Reaction method→

Scheme 15.3 Synthesis of imidazole by fatty acid, ethanolamine, and ethanamine. *Source:* Ref. [32].

evaluated by X-ray fluorescence microscopy by the presence of adsorptive film on the metallic surface. In a subsequent study, the authors developed a series of five surfactants from palm oil and evaluated for carbon steel corrosion behavior in NaCl solution infused with CO_2 at 50°C [31]. The increasing chain length played an important role in lowering the CMC of the same and led to increase in the protection of metal surface. The inhibitors followed spontaneous adsorption fits as Langmuir adsorption isotherm.

Wahyuningrum et al. developed some novel imidazoline-based corrosion inhibitors using and tested for mild steel corrosion in 1% NaCl [32]. The fatty acids were refluxed with diethylenetriamine and aminoethylethanolamine from 100°C until 230°C for approximately 13 hours (Scheme 15.3). The synthesis process was carried out using oleic and stearic acids using diethylenetriamine, and aminoethylethanolamine as reactant by both straight and microwave-assisted synthesis methods. The potentiodynamic polarization studies revealed that heptadec-8-enyl and hydroxyethyl substituents at C(2) and N(1) position of imidazoline ring proved to be the best inhibitors toward corrosion retardation of mild steel. However, moderate efficiencies of up to 55% were afforded. Daniyan et al. studied corrosion resistant of ductile iron and mild steel through palm oil in freshwater and 1M NaOH media [33]. The palm oil provided reasonable protection in both freshwater and 1M NaOH medium, thereby supporting its utility as an eco-friendly corrosion inhibitor. Zulkafi et al. analyzed the organic acid from palm kernel oil (PKO) as an effective inhibitor for SAE carbon steel 1045 in 1M HCl medium [34]. At a dose of 4 g/l, an efficiency of 90.10% was afforded for PKO, suggesting its utility as a green corrosion inhibitor. A relevant study by the same group, PKO was used for carbon steel SAE 1045 in 1M NaOH [35]. An efficiency of 96.67% was afforded using weight loss and polarization studies at 8 v/v%.

1, 3-Docosenoic acid was reacted with aliphatic/aromatic amines primary and secondary to produce amide derivatives, and the products were tested for mild steel in 1M HCl solution [36]; 96.8% of inhibition efficiency was obtained at an elevated temperature of 60°C. Thermodynamic parameters indicated physical and chemical mixed-mode adsorption, and the quantum chemical calculations showcased electron-donating nature of the studied inhibitors. Adewuyi et al. synthesized a fatty acid amide (KSFA) from the seed oil of *Khaya senegalensis* via esterification, transesterification, hydroxylation, and amidation reactions [37]. KSFA was examined for aluminum corrosion in 0.5M HCl using weight loss measurements. KFSA acted by physical adsorption mechanism, via the involvement of the hydroxyl and amide functional groups. It was proposed that due to the protonation of the nitrogen atom, in the acidic medium, KFSA could exist in the cationic form, which could adhere on the Al surface via electrostatic interaction with the Cl^- ions. Oleic acid hydrazide (OAH) was analyzed as an eco-friendly corrosion inhibitor for API X70 steel surface in an oil-field produce water under flow conditions [38]. At a dose of 0.30 g/l, the maximum efficiency of 87.7% was obtained. The metal surface is covered, and inhibitor molecules adhere at the active sites of the electrode increases with the subsequent increase in inhibitors molar concentration and gradually more corrosive sites are blocked, as shown in Figure 15.4a–c. Thus, from the lower to a higher molar concentration, the inhibition efficiency improved.

The bottom panel of Figure 15.4 shows the effect of the hydrodynamic conditions of the turbulent flow on inhibitor adsorption and corrosion inhibition. The current of the fluid ensures a better corrosion inhibition performance. However, hydrodynamic flow is promoted by the mass-transfer process of the ferrous ions (Fe^{2+}) to the concentrated solution, which can delay the formation of the metal-inhibitor complex. In addition, few factors like shear stress, which is parallel to side wall, and also turbulence effects on desorption of the inhibitor film adversely affect the inhibition efficiency.

Palm oil and its derivatives are a potential class of inhibitors. Mohammad Makrus et al. (2019) synthesized palm-oil-based ethanolamine derivatives for carbon steel protection in 0.5 M HCl [39]. The synthetic procedure involves the reaction of fatty acid and ethanolamine (Scheme 15.4), which results in the formation of 2-aminoethyl fatty esters and N-(2-hydroxyethyl) fatty amides. Around 80% efficiency was obtained at a dose 80 mg/l. The hydrolysate of the crude oil provided the major constituents as oleic and palmitic acids. A mixed inhibition behavior also fits Langmuir isotherm. A comparison of few of the oleo-chemical-based corrosion inhibitors is given in Table 15.2.

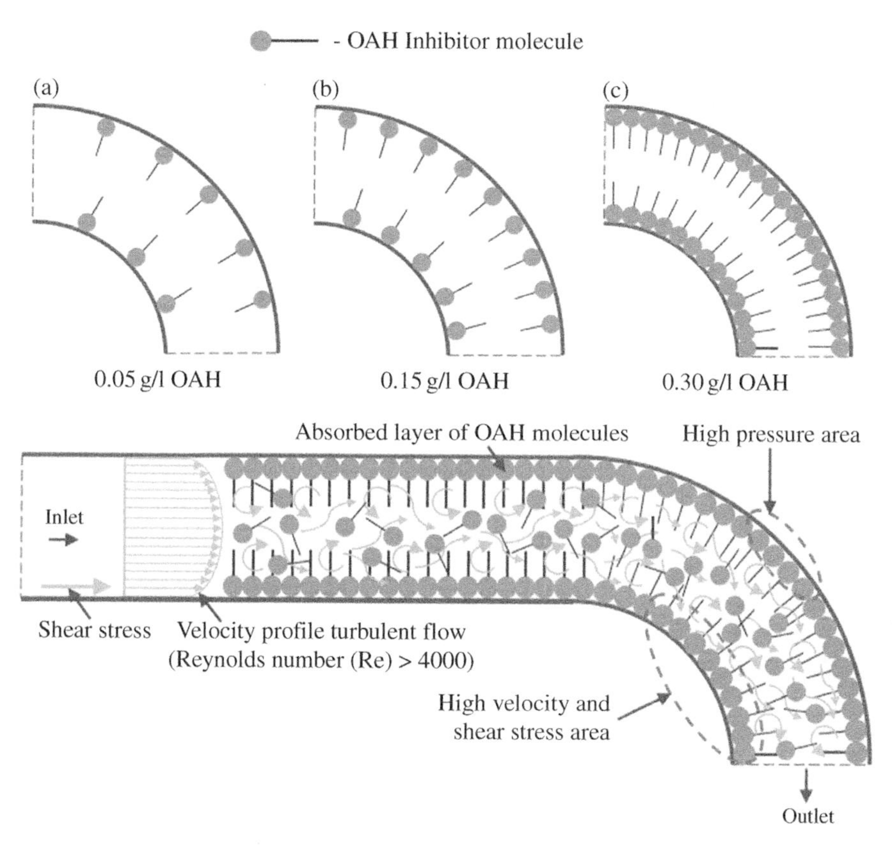

Figure 15.4 Schematic of adsorption of oleic acid hydrazide (OAH) on the elbow test section at various concentrations: (a) 0.05 g/l, (b) 0.15 g/l, and (c) 0.30 g/l. Bottom panel displays a schematic of the interaction of OAH inhibitor molecules at the internal wall of the pipe and elbow under turbulent flow condition.

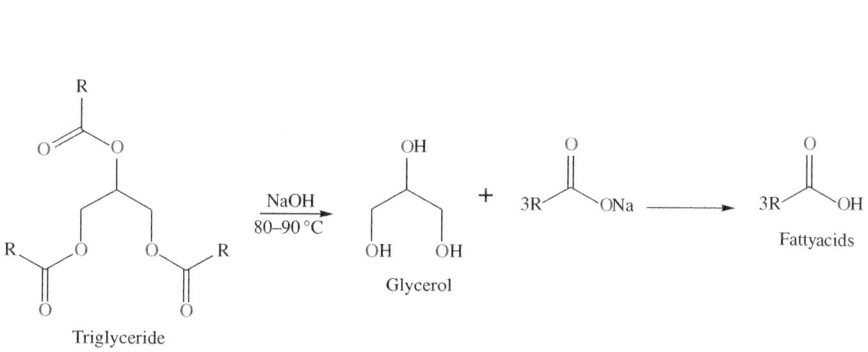

Triglyceride $\xrightarrow[80\text{–}90\,°C]{NaOH}$ Glycerol + 3RCOONa ⟶ 3RCOOH (Fattyacids)

Scheme 15.4 Hydrolysis reaction of crude Palm oil. *Source:* Ref. [39].

Table 15.2 A comparison of literature on the performance of oleochemicals as corrosion inhibitors

Inhibitor	Metal/medium	I.E. (%)/Conc.	Binding and adsorption/ inhibitor type	Reference
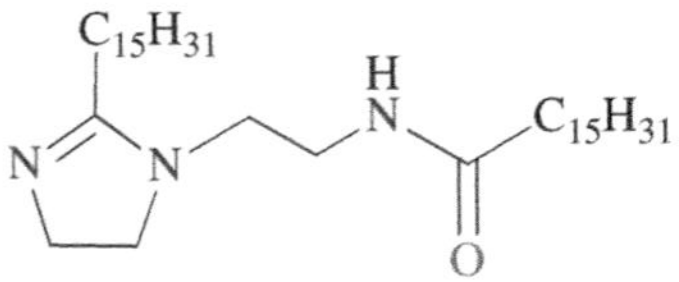 $C_{15}H_{31}$ H N N N $C_{15}H_{31}$ O Palmitic acid imidazole (PI)	Mild steel/H_2SO_4.	$90/1 \times 10^{-3}$mol/l	Langmuir adsorption isotherm/mixed type	[22, 23]
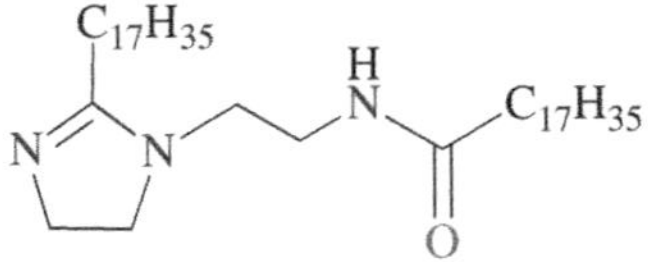 $C_{17}H_{35}$ H N N N $C_{17}H_{35}$ O	Mild steel/15 HCl	89.83 %/400 mg/ l+1 mM KI	--/Mixed type	[24]
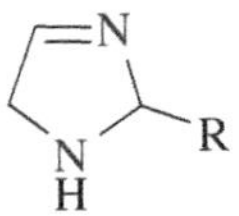 N N H R R= $-(CH_2)_8-CH_3$ $-(CH_2)_{10}-CH_3$ $-(CH_2)_{14}-CH_3$ $-(CH_2)_{16}-CH_3$ 2-Undecyl-1, 3-imidazoline (UDI) Imidazoline derivatives	Mild steel/0.5 H_2SO_4	96.2/500 ppm	Langmuir adsorption isotherm/mixed type	[25]

Fatty amide of Crude rice bran oil (CRBO)	API X70 steel/CO_2 saturated solution (3.5% NaCl)	95/100 ppm	--/Anodic type	[26]
$R-(CH_2)_7-CH(-O-SO_2OH\cdots NH(C_2H_5)_2)-(CH_2)_7-COOH\cdots NH(CH_2CH_3)_2$ Sulfated fatty acid-ethylamine complex	Mild steel/CO_2 saturated 1% NaCl solution	99.72%/150 ppm	Langmuir adsorption isotherm/ --	[28]
$R-(CH_3)_7-CH_2(-O-SO_2-N(CH_2-CH_2-OH)_2)-(CH_2)_7-CO-N(CH_2-CH-OH)_2$	CO_2-saturated 1% NaCl solution	99.9%/100 ppm	Langmuir adsorption isotherm/mixed type	[29]
$R-(CH_2)_8-CH(-O-SO_3^{-}\overset{+}{N}H_2-(CH_2-CH_2-OH)_2)-(CH_2)_7-COO^{-}\overset{+}{N}H_2(CH_2CH_2-OH)_2$ SFADC, Sulfated fatty acid	Mild steel / CO_2-saturated oilfield water	96.22 %/457 ppm	Langmuir adsorption isotherm/--	[30]
$R-(CH_2)_8-CH(-O-SO_3^{-}\overset{+}{N}H_3(CH_2CH_2OH)_2)-(CH_2)_7-COO^{-}\overset{+}{N}H_2-(CH_2-CH_2-OH)_2$ sulfated fatty acid diethanolamine complexes	Mild steel/1% NaCl	99.95%/100 ppm	Langmuir adsorption isotherm/--	[31]

(Continued)

Table 15.2 (Continued)

Inhibitor	Metal/medium	I.E. (%)/Conc.	Binding and adsorption/ inhibitor type	Reference
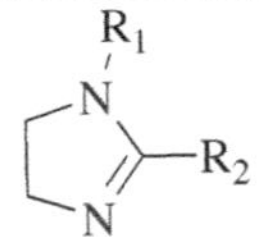 R_1, R_2 (imidazoline structure) R_1: Heptadec-8-enyl, heptadecyl R_2: Ethylamine, hydroxyethyl 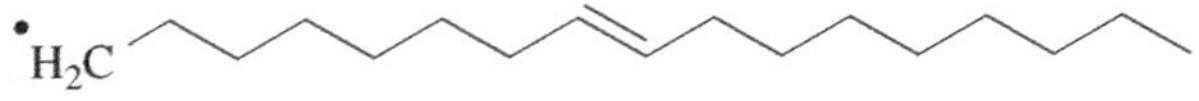 $^{\bullet}H_2C$ Heptadec-8-enyl Caution: A radical appears to be present 1b. ((Z)-2-(2-(heptadec-8-enyl)-4,5-dihydroimidazol-1-yl) ethanamine 2b. ((Z)-2-(2-(heptadec-8-enyl)-4,5-dihydroimidazol-1-yl)ethanol) 3b. (2-(2-heptadecyl-4,5-dihydroimidazol-1-yl)ethanamine)	Carbon steel (39.59%)/ 1% NaCl	1b 32.18%, 2b 39.59% and 3b 12.73%/ 8 ppm	Langmuir isotherm Adsorptions/--	[32]
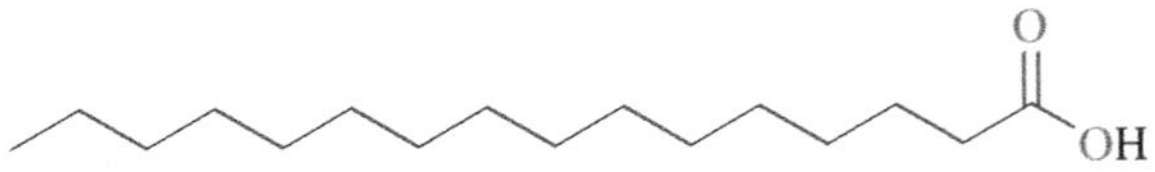 O, OH (structure) Palm oil	Ductile iron and mild steel/1M NaOH	--	--/ --	[33]
Palm kernel oil	carbon steel SAE 1045/1 M NaOH	96.67 %/8 v/v %	--/Mixed type	[34]

13-Docosenoic acid amide 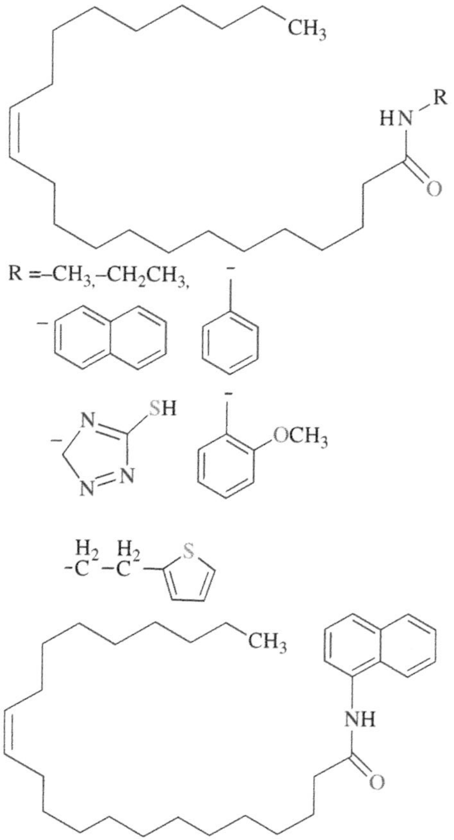 CH_3 HN R O R =$-CH_3$, $-CH_2CH_3$, SH N N N OCH_3 H_2 H_2 -C -C- S CH_3 NH O	Mild steel/1.0 HCl	96.8/ 500 ppm	Langmuir isotherm Adsorptions/--	[36]

(Continued)

Table 15.2 (Continued)

Inhibitor	Metal/medium	I.E. (%)/Conc.	Binding and adsorption/ inhibitor type	Reference
Khaya senegalensis fatty hydroxylamide (KSFA)	Al/0.5 HCl	90.43/0.1 × 10^{-5}g/l	--	[37]
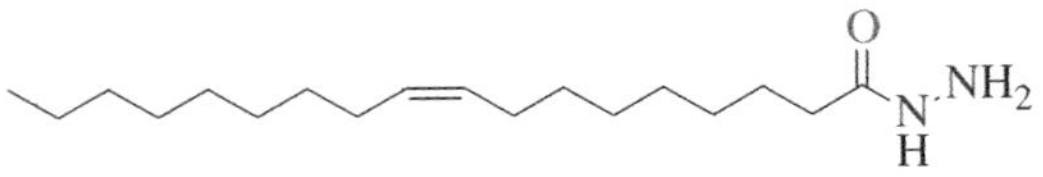 NH_2 N H O Oleic acid hydrazide (OAH)	API X70 steel/ oilfield water	87.7/0.30 g/l	Absorption by metal–inhibitor complexes/---	[38]
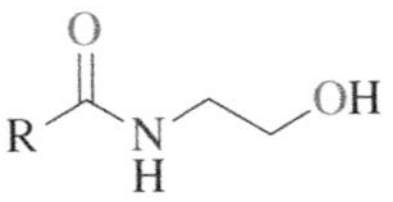 O R N H OH N-(2-hydroxyethyl) fatty amides 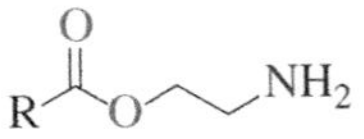 O R O NH_2 2-Aminoethyl fatty ester	Mild steel/0.5 M HCl	80/80 ppm	Langmuir adsorption isotherm/ mixed type	[39]

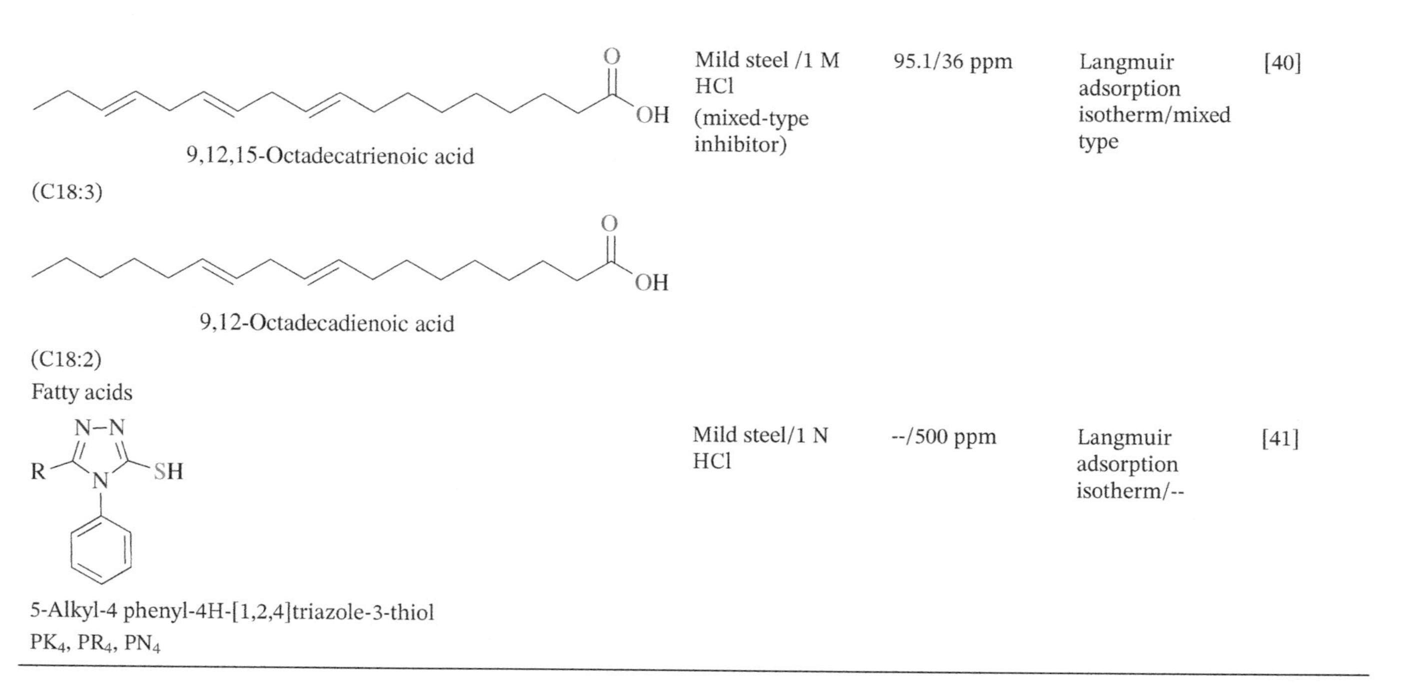 	Mild steel /1 M HCl (mixed-type inhibitor)	95.1/36 ppm	Langmuir adsorption isotherm/mixed type	[40]
 5-Alkyl-4 phenyl-4H-[1,2,4]triazole-3-thiol PK_4, PR_4, PN_4	Mild steel/1 N HCl	--/500 ppm	Langmuir adsorption isotherm/--	[41]

Microalgae developed from fatty triglyceride *Scenedesmus sp.* were studied for mild-steel corrosion in 1M HCl solution [40]. At a low dose of 36 mg/l, high efficiency of 95.1% was afforded. AFM investigations revealed a lowering in the metal surface abrasion from 138.51 to 45.09 nm, joined by the visual recording, which showed a decrease in the hydrogen evolution in the existence of *Scenedesmus sp.* Toliwal and Jadav investigated the phenyl thiosemicarbazides derived from nonconventional oils (from neem, rice bran, and karanja) and studied for mild steel corrosion in HCl medium [41]. In all the three oils, oleic acid was found to be the major component followed by palmitic and linoleic acids. All the inhibitors showed the mixed inhibition behavior and adhere on the steel surface as per the Langmuir isotherm. In a subsequent study by the same group, a new synthetic Schiff base was derived from the above nonconventional oils and evaluated for mild steel in HCl medium [42]. Weight loss tests revealed a high efficiency >92% at a dose of 500 mg/l. A predominantly physical adsorption mechanism was proposed to explain the inhibitor adsorption on the metallic surface. It has been outlined that in early 1960, J. E. O. Mayne et al. studied sodium, calcium, and lead salts of azelaic, capric, caproic, caprylic, suberic, and sebacic acids as corrosion inhibitors for iron at a pH range of 4.0–6.0 [43]. The study revealed that the inhibitive property increased with increase in carbon chain length.

Quraishi and Jamal have investigated oleochemicals as volatile corrosion inhibitors (VCIs) (Scheme 15.5) [44]. The VCIs are a category of corrosion inhibitors that are used for the protection of ferrous/nonferrous metals from deterioration and oxidation where the other surface treatments are not practical. The group developed some organic and environment-friendly inhibitors via the condensation of 2-Dec-9-enyl-2-imidazoline with various acids such as maleic acid, orthophosphoric acid, nitrobenzoic acid, phthalic acid, and cinnamic acid (Scheme 15.5). The inhibitors were evaluated for mild steel, brass, and copper metal surfaces via weight loss measurements. Eschke test methods, NaCl inoculation test, and sulfur dioxide (SO_2) tests were conducted to analyze the corrosion retardation behavior. All the tested VCIs showed good inhibition performance for the tested metal surfaces. The cinnamate salt provided the best results among all the analyzed molecules.

Sudheer et al. have investigated oleic hydrazide benzoate (OHB) and oleic hydrazide salicylate (OHS) oleochemicals as VCIs (Scheme 15.6). Atmospheric corrosion monitor cell (ACMC) was used to study the electrochemical properties of the inhibitors. The results of polarization study show that OHB is anodic type, and OHS is a mixed-type inhibitor. OHB exhibited higher inhibition efficiency of 89.31% from potentiodynamic polarization under vapor-phase condition [45].

Some novel acidizing inhibitors were developed from hydrazides and thiosemicarbazides of long-chain fatty acids [46–49]. Most of the inhibitors as acetylenic alcohols, etc., which were widely used in industries, were highly toxic. The

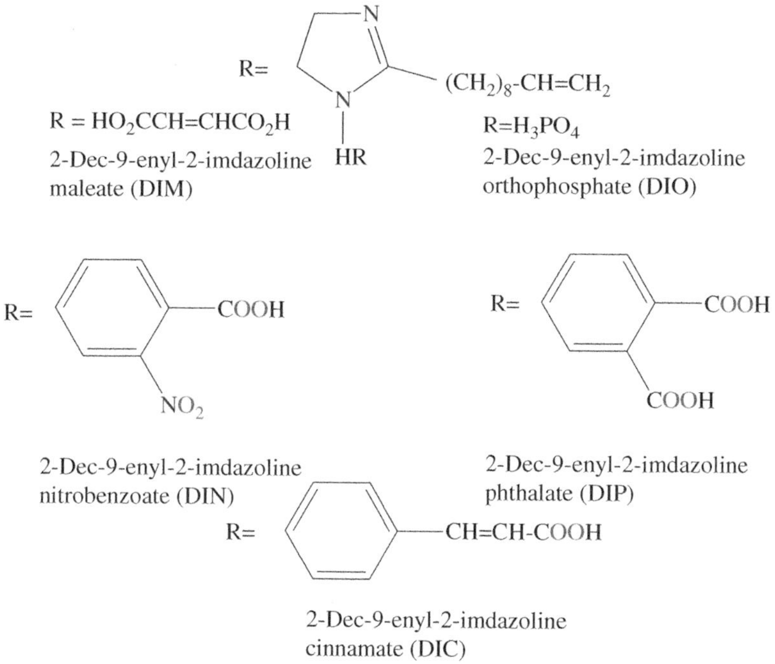

R=

$(CH_2)_8$-CH=CH_2

N

N

HR

R = HO_2CCH=$CHCO_2H$

2-Dec-9-enyl-2-imdazoline maleate (DIM)

R=H_3PO_4

2-Dec-9-enyl-2-imdazoline orthophosphate (DIO)

R= COOH NO_2

2-Dec-9-enyl-2-imdazoline nitrobenzoate (DIN)

R= COOH COOH

2-Dec-9-enyl-2-imdazoline phthalate (DIP)

R= CH=CH-COOH

2-Dec-9-enyl-2-imdazoline cinnamate (DIC)

Scheme 15.5 Structure of some oleochemical-based volatile corrosion inhibitors (VCIs). *Source:* Ref. [44].

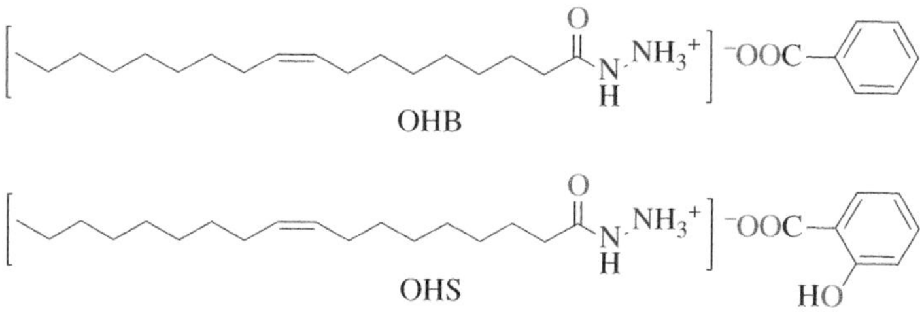

O N H NH_3^+ ^-OOC

OHB

O N H NH_3^+ ^-OOC HO

OHS

Scheme 15.6 Structure of oleic hydrazide benzoate and oleic hydrazide salicylate (OHS)-based volatile corrosion inhibitors (VCIs). *Source:* Modified from Ref. [45].

tested corrosion inhibitors were eco-friendly, had less toxic, and cheap compared to the acetylenic alcohols. The PDP studies were undertaken on the mild steel, and N80 steel at room temperature inhibitors showed mixed behavior. The inhibitor adsorption on the metal surface obeyed the Temkin isotherm. Hydrazides and thiosemicarbazides of fatty acids having 11, 12, and 18 carbon atoms were prepared and evaluated for mild steel in 15% HCl in hot-boiling

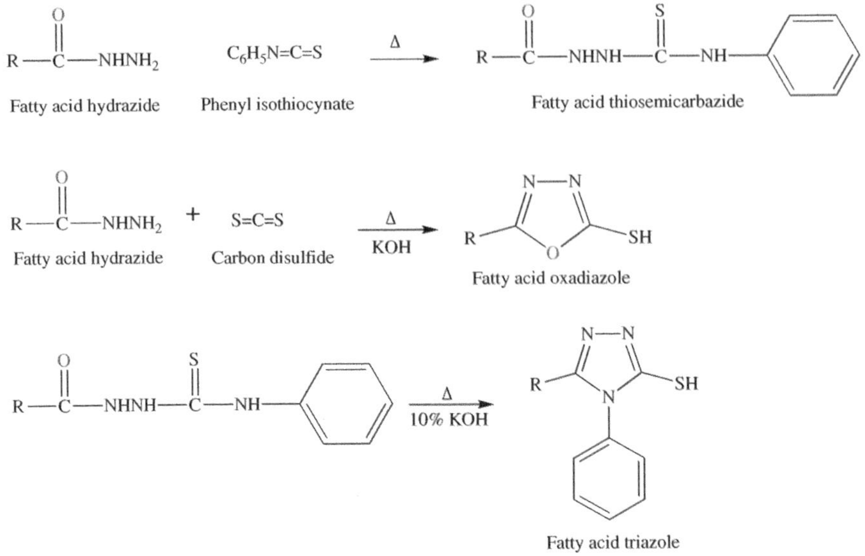

R—C(=O)—NHNH$_2$ (Fatty acid hydrazide) + $C_6H_5N{=}C{=}S$ (Phenyl isothiocynate) —Δ→ R—C(=O)—NHNH—C(=S)—NH—C_6H_5 (Fatty acid thiosemicarbazide)

R—C(=O)—NHNH$_2$ (Fatty acid hydrazide) + S=C=S (Carbon disulfide) —Δ, KOH→ Fatty acid oxadiazole

R—C(=O)—NHNH—C(=S)—NH—C_6H_5 —Δ, 10% KOH→ Fatty acid triazole

Scheme 15.7 Synthesis of hydrazides, thiosemicarbazides, oxadiazoles, triazoles. *Source:* Ref. [50–52].

conditions [46]. 1-Decene-4-phenyl-thiosemicarbazide provided the better performance with 96.0% inhibition efficiency at 5000 mg/l concentration as determined by the weight loss tests. The same group evaluated some fatty acid oxadiazoles, wherein 2-decane-5-mercapto-1-oxa-3,4-diazole (DMOD) showed the best performance with 92.50% efficiency at a dose of 500 mg/l [48]. These oxadiazoles were evaluated for N80 steel and mild steel in 15% HCl, and 2-undecane-5-mercapto-1-oxa-3,4-diazole (UMOD) provided the best results with 98.94% efficiency [49].

Quraishi et al. synthesized hydrazides, thiosemicarbazides, oxadiazoles, triazoles (Scheme 15.7) from oleic, lauric, undecenoic acids [50–52]. They did an extensive study on their corrosion inhibition behavior on mild steel in formic and acetic acid corrosive medium. Gravimetric and electrochemical methods were employed to perform corrosion inhibition studies. Among the studied types of the fatty acids, namely, undecenoic, oleic, and lauric acids, the undecenoic-acid-containing inhibitor provided the best inhibition behavior, which could be attributed to the efficient coverage of the metal surface from the undecenoic acid compared to the other fatty acids. Among the investigated cyclic oleochemicals, the fatty acid thiosemicarbazides provided the better performance. The presence of an additional –C = S group along with a benzene ring was attributed to the superior corrosion protection compared to the hydrazides.

Among the group of heterocyclic oleochemicals, the fatty acid triazoles provided the better performance compared to the oxadiazole derivatives. This was attributable to the existence of an additional benzene ring and three N atoms in comparison to the oxadiazoles, which possessed two heteroatoms (N and O) only.

15.6 Conclusions and Outlook

The oleochemicals have, recently, emerged as a potential class of corrosion inhibitors. The major benefit of the oleochemicals is their natural origin, which forms the basis of their environmentally benign nature. Bearing the structures of long chain fatty acids, the oleochemical compounds possess quite large molecular weight, which provides an efficient coverage of the metal surface. In addition, the presence of an abundance of N and O containing functionalities provide efficient adsorption centers to interact with the target metal surface. Therefore, among the classes of green corrosion inhibitors derived from natural extracts, biological polymers, drugs, amino acids, ionic liquids, etc., the oleochemicals have also formed a strong grounding. Some of the drawbacks associated with the oleochemicals could possibly be the requirement of long synthesis steps for the isolation and purifications of these molecules from the source animal/plant fat/oil. In general, for synthesis of corrosion inhibitors, considering the economic and time constraints, the multicomponent reaction schemes are preferred involving single-step reactions in most cases. Therefore, this side of the oleochemicals needs to be taken into consideration while developing a commercially useful corrosion inhibitor formulation using the oleochemicals.

The present chapter first provides an introduction to the oleochemicals, followed by their environmental sustainability and their application in corrosion inhibition. A brief review of the literature is presented in the application of oleochemicals as corrosion inhibitors. It has been observed that a number of research articles have come across in recent years describing the applicability of the oleochemical-based corrosion inhibitors. A wide variety of ferrous and nonferrous metals and alloys have been explored as substrates for studying the oleochemicals. Different corrosive media ranging from acid pickling to acidizing, from saline to sweet and alkaline environments, have been covered. Along with the studies covered by the inhibitor molecules alone, a number of synergistic corrosion inhibition studies have also been undertaken involving the oleochemicals. These molecules have been documented to provide efficient adsorption and inhibition performance at elevated temperatures and at prolonged immersion times.

References

1 Finšgar, M. and Jackson, J. (2014). Application of corrosion inhibitors for steels in acidic media for the oil and gas industry: a review. *Corrosion Science* 86: 17–41.

2 Xhanari, K. and Finšgar, M. (2019). Organic corrosion inhibitors for aluminum and its alloys in chloride and alkaline solutions: a review. *Arabian Journal of Chemistry* 12: 4646–4663.

3 Antonijević, M.M., Milić, S.M., and Petrović, M.B. (2009). Films formed on copper surface in chloride media in the presence of azoles. *Corrosion Science* 51: 1228–1237.

4 Chauhan, D.S., Quraishi, M.A., Carrière, C. et al. (2019). Electrochemical, ToF-SIMS and computational studies of 4-amino-5-methyl-4H-1, 2, 4-triazole-3-thiol as a novel corrosion inhibitor for copper in 3.5% NaCl. *Journal of Molecular Liquids* 289: 111113.

5 Quraishi, M.A., Chauhan, D.S., and Saji, V.S. (2020). *Heterocyclic Organic Corrosion Inhibitors: Principles and Applications*. Amsterdam: Elsevier Inc.

6 Koch, G., Varney, J., Thompson, N. et al. (2016). International measures of prevention, application, and economics of corrosion technologies study. *NACE International* 216: 2–3.

7 Sastri, V.S. (1998). *Corrosion inhibitors: Principles and Applications*. John Wiley & Sons.

8 Chauhan, D.S., Ansari, K.R., Sorour, A.A. et al. (2018). Thiosemicarbazide and thiocarbohydrazide functionalized chitosan as ecofriendly corrosion inhibitors for carbon steel in hydrochloric acid solution. *International Journal of Biological Macromolecules* 107: 1747–1757.

9 Chauhan, D.S., Mazumder, M.J., Quraishi, M.A., and Ansari, K. (2020). Chitosan-cinnamaldehyde Schiff base: a bioinspired macromolecule as corrosion inhibitor for oil and gas industry. *International Journal of Biological Macromolecules* 158: 127–138.

10 OSPAR Commission (2005). Protocols on methods for the testing of chemicals used in the offshore oil industry, 1–25.

11 Occupational Safety and Health Administration (2013). US, Globally harmonized system of classification and labelling of chemicals (GHS), Economic Commission for Europe, 1–568.

12 Constable, D.J., Curzons, A.D., and Cunningham, V.L. (2002). Metrics to 'green'chemistry—which are the best? *Green Chemistry* 4: 521–527.

13 Verma, C., Ebenso, E.E., Bahadur, I., and Quraishi, M.A. (2018). An overview on plant extracts as environmental sustainable and green corrosion inhibitors for metals and alloys in aggressive corrosive media. *Journal of Molecular Liquids* 266: 577–590.

14 Kaya, S., Tüzün, B., Kaya, C., and Obot, I.B. (2016). Determination of corrosion inhibition effects of amino acids: quantum chemical and molecular dynamic

simulation study. *Journal of the Taiwan Institute of Chemical Engineers* 58: 528–535.

15 Verma, C., Ebenso, E.E., and Quraishi, M.A. (2017). Ionic liquids as green and sustainable corrosion inhibitors for metals and alloys: an overview. *Journal of Molecular Liquids* 233: 403–414.

16 Umoren, S.A. and Eduok, U.M. (2016). Application of carbohydrate polymers as corrosion inhibitors for metal substrates in different media: a review. *Carbohydrate polymers* 140: 314–341.

17 Gece, G. (2011). Drugs: A review of promising novel corrosion inhibitors. *Corrosion Science* 53: 3873–3898.

18 Chauhan, D.S., Sorour, A.A., and Quraishi, M.A. (2016). An overview of expired drugs as novel corrosion inhibitors for metals and alloys. *International Journal of Chemistry and Pharmaceutical Sciences* 4: 680–691.

19 Shahbandeh, M. (2019). Vegetable oils: production worldwide 2012/13-2019/20, by type. https://www.statista.com/statistics/263933/production-of-vegetable-oils-worldwide-since-2000/

20 Pel, A. (2001). Fatty acids: a versatile and sustainable source of raw materials for the surfactants industry. *Oléagineux, Corps Gras, Lipides* 8: 145–151.

21 Roose, P., Eller, K., Henkes, E. et al. (2000). Amines, aliphatic. In: *Ullmann's Encyclopedia of Industrial Chemistry*, 1–55. Wiley-VCH Verlag GmbH & Co.

22 Tripathy, D.B., Murmu, M., Banerjee, P., and Quraishi, M.A. (2019). Palmitic acid based environmentally benign corrosion inhibiting formulation useful during acid cleansing process in MSF desalination plants. *Desalination* 472: 114128.

23 Solomon, M.M., Umoren, S.A., Quraishi, M.A., and Salman, M. (2019). Myristic acid based imidazoline derivative as effective corrosion inhibitor for steel in 15% HCl medium. *Journal of Colloid and Interface Science* 551: 47–60.

24 Solomon, M.M., Umoren, S.A., Quraishi, M.A. et al. (2020). Effect of akyl chain length, flow, and temperature on the corrosion inhibition of carbon steel in a simulated acidizing environment by an imidazoline-based inhibitor. *Journal of Petroleum Science and Engineering* 187: 106801.

25 Rafiquee, M., Khan, S., Saxena, N., and Quraishi, M. (2009). Investigation of some oleochemicals as green inhibitors on mild steel corrosion in sulfuric acid. *Journal of Applied Electrochemistry* 39: 1409–1417.

26 Reyes-Dorantes, E., Zuñiga-Díaz, J., Quinto-Hernandez, A. et al. (2017). Fatty amides from crude rice bran oil as green corrosion inhibitors. *Journal of Chemistry* 2017: 1–14.

27 Ismayilov, I.T., Abd El-Lateef, H.M., Abbasov, V.M. et al. (2013). Anticorrosion ability of some surfactants based on cornoil and monoethanolamine. *American Journal of Physical Chemistry* 1: 79–86.

28 Abbasov, V.M., Abd El-Lateef, H.M., Aliyeva, L.I. et al. (2013). Efficient complex surfactants from the type of fatty acids as corrosion inhibitors for mild steel C1018 in CO_2-environments. *Journal of the Korean Chemical Society* 57: 25–34.

29 Abbasov, V., Abd El-Lateef, H.M., Aliyeva, L. et al. (2013). A study of the corrosion inhibition of mild steel C1018 in CO_2-saturated brine using some novel surfactants based on corn oil. *Egyptian Journal of Petroleum* 22: 451–470.

30 Abd El-Lateef, H.M., Ismayilov, I., Abbasov, V. et al. (2013). Green surfactants from the type of fatty acids as effective corrosion inhibitors for mild steel in CO_2-saturated NaCl solution. *American Journal of Physical Chemistry* 2: 16–23.

31 Abd El-Lateef, H.M., Abbasov, V., Aliyeva, L. et al. (2013). Inhibition of carbon steel corrosion in CO_2-saturated brine using some newly surfactants based on palm oil: experimental and theoretical investigations. *Materials Chemistry and Physics* 142: 502–512.

32 Wahyuningrum, D., Achmad, S., Syah, Y.M. et al. (2008). The synthesis of imidazoline derivative compounds as corrosion inhibitor towards carbon steel in 1% NaCl solution. *Journal of Mathematical and Fundamental Sciences* 40: 33–48.

33 Daniyan, A., Ogundare, O., AttahDaniel, B., and Babatope, B. (2011). Effect of palm oil as corrosion inhibitor on ductile iron and mild steel. *The Pacific Journal of Science and Technology* 12: 45–53.

34 Zulkafli, M.Y., Othman, N.K., Lazim, A.M., and Jalar, A. (2013). Effect of carboxylic acid from palm kernel oil for corrosion prevention. *International Journal of Basic & Applied Sciences* 13: 29–32.

35 Zulkafli, M., Othman, N., Lazim, A., and Jalar, A. (2013). Inhibitive effects of palm kernel oil on carbon steel corrosion by alkaline solution. In: *AIP Conference Proceedings* (ed. A. Waldron), 42–47. American Institute of Physics.

36 Elsharif, A.M., Abubshait, S.A., Abdulazeez, I., and Abubshait, H.A. (2020). Synthesis of a new class of corrosion inhibitors derived from natural fatty acid: 13-Docosenoic acid amide derivatives for oil and gas industry. *Arabian Journal of Chemistry* 13: 5363–5376.

37 Adewuyi, A. and Oderinde, R.A. (2018). Synthesis of hydroxylated fatty amide from underutilized seed oil of Khaya senegalensis: a potential green inhibitor of corrosion in aluminum. *Journal of Analytical Science and Technology* 9: 1–13.

38 Ajmal, T.S., Arya, S.B., Thippeswamy, L.R. et al. (2020). Influence of green inhibitor on flow-accelerated corrosion of API X70 line pipe steel in synthetic oilfield water. *Corrosion Engineering, Science and Technology* 55: 487–496.

39 Ali, M.M., Irawadi, T.T., Darmawan, N. et al. (2019). Reaction Products of Crude Palm Oil-based Fatty Acids and Monoethanolamine as Corrosion Inhibitors of Carbon Steel. *Makara Journal of Science* 23: 155–161.

40 Khanra, A., Srivastava, M., Rai, M.P., and Prakash, R. (2018). Application of unsaturated fatty acid molecules derived from microalgae toward mild steel corrosion inhibition in HCl solution: a novel approach for metal–inhibitor association. *ACS Omega* 3: 12369–12382.

41 Toliwal, S. and Jadav, K. (2009). Inhibition of corrosion of mild steel by phenyl thiosemicarbazides of nontraditional oils. *Journal of Scientific and Industrial Research* 68: 235–241.

42 Toliwal, S., Jadav, K., and Pavagadhi, T. (2010). Corrosion inhibition study of a new synthetic Schiff base derived from nontraditional oils on mild steel in 1N HCl solution. *Journal of Scientific and Industrial Research* 69: 43–47.

43 Mayne, J. and Ramshaw, E. (1960). Inhibitors of the corrosion of iron. II. Efficiency of the sodium, calcium and lead salts of long chain fatty acids. *Journal of Applied Chemistry* 10: 419–422.

44 Quraishi, M. and Jamal, D. (2004). Synthesis and evaluation of some organic vapour phase corrosion inhibitors. *Indian Journal of Chemical Technology* 11: 459–464.

45 Sudheer, Quraishi, M.A., Ebenso, E.E., and Natesan, M. (2012). Inhibition of atmospheric corrosion of mild steel by new green inhibitors under vapour phase condition. *International Journal of Electrochemical Science* 7: 7463–7475.

46 Quaraishi, M., Jamal, D., and Tariq Saeed, M. (2000). Fatty acid derivatives as corrosion inhibitors for mild steel and oil-well tubular steel in 15% boiling hydrochloric acid. *Journal of the American Oil Chemists' Society* 77: 265–268.

47 Ajmal, M., Jamal, D., and Quraishi, M. (2000). Fatty acid oxadiazoles as acid corrosion inhibitors for mild steel. *Anti-Corrosion Methods and Materials* 47: 77–82.

48 Rafiquee, M., Saxena, N., Khan, S., and Quraishi, M. (2007). Some fatty acid oxadiazoles for corrosion inhibition of mild steel in HCl. *Indian Journal of Chemical Technology* 14: 576–583.

49 Quraishi, M.A. and Jamal, D. (2001). Corrosion inhibition by fatty acid oxadiazoles for oil well steel (N-80) and mild steel. *Materials Chemistry and Physics* 71: 202–205.

50 Quraishi, M. and Ansari, F.A. (2006). Fatty acid oxadiazoles as corrosion inhibitors for mild steel in formic acid. *Journal of Applied Electrochemistry* 36: 309–314.

51 Quraishi, M. and Ansari, F. (2003). Corrosion inhibition by fatty acid triazoles for mild steel in formic acid. *Journal of Applied Electrochemistry* 33: 233–238.

52 Quraishi, M.A. and Jamal, D. (2000). Fatty acid triazoles: Novel corrosion inhibitors for oil well steel (N-80) and mild steel. *Journal of the American Oil Chemists' Society* 77: 1107–1111.

Part IV

Organic Compounds-Based Nanomaterials as Corrosion Inhibitors

16

Carbon Nanotubes as Corrosion Inhibitors

Yeestdev Dewangan[1], Amit Kumar Dewangan[1], Shobha[2], and Dakeshwar Kumar Verma[1]

[1] *Department of Chemistry, Government Digvijay Autonomous Postgraduate College, Rajnandgaon, Chhattisgarh, India*
[2] *Department of Physics, Banasthali Vidyapith, Vanasthali, Rajasthan, India*

16.1 Introduction

Corrosion is a process occurring in the metal surface in which there is loss of metal surface and due to this, various properties of metal such as mechanical property, conductivity, durability, and others are affected [1]. Various types of processes in which acid pickling and descaling are predominant are used to remove the rust rusted over the metal. Mineral acids such as HCl, H_2SO_4, and HNO_3 and base electrolyte such as NaCl and sea water, etc. are used for acid pickling [2–4]. During the descaling process, after rust removal, corrosive ions attack the metal surface, causing direct loss of metal. Some materials are used to prevent this loss, which is deposited in the metal surface and protects it from corrosive attack. These are called corrosion inhibitors [5]. The best corrosion inhibitor is the one in which heteroatoms (S, O, N, P) and pi electron are present. Because these inhibitors share these electrons with the vacant "d" orbital present in the metal surface by these electron-rich centers and deposited on the metal surface to form a protective layer, which protects the metal surface from further corrosion [6, 7]. Hydroxamic acids, thiourea derivatives [8, 9], imidazole derivatives [10], amino acid derivatives [11], and porphyrine derivatives [12] are used as corrosion inhibitors by some of the previous researchers. Since the last decade, the use of carbon nanotubes (CNTs)-based materials and composites have been reported by some scientists as major corrosion inhibitors. The main reason for the attention of

Organic Corrosion Inhibitors: Synthesis, Characterization, Mechanism, and Applications,
First Edition. Edited by Chandrabhan Verma, Chaudhery Mustansar Hussain, and Eno E. Ebenso.

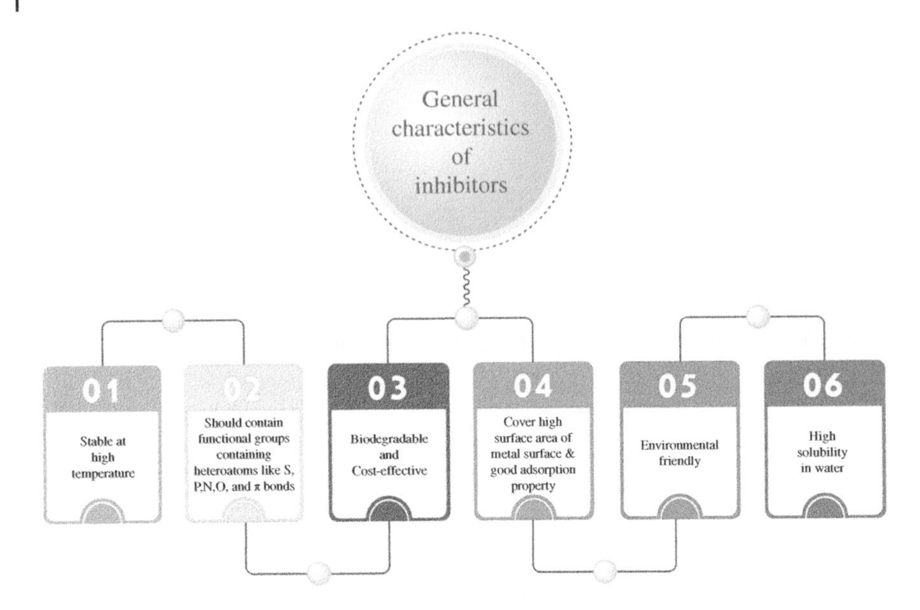

Figure 16.1 General characteristics of inhibitors.

researchers toward CNTs is its advanced physical properties such as good mechanical property, excellent corrosion protecting property, high temperature resistivity, wide surface area, and high solubility [13]. CNTs are used for various fields such as sensors, polymer-based composites, catalyst, removal of heavy metal ions, and corrosion protection methods [14–16]. CNTs and polymer-based nanomaterial composites are currently being widely used as anticorrosion protection methods, behind the enhanced mechanical property and stability of CNTs and polymer composites. Experimental, electrochemical, surface analysis, and theoretical calculation approaches are mainly used by researchers to determine the anticorrosion property of CNTs. Figure 16.1 illustrates the key characteristics of inhibitor molecules.

16.2 Characteristics, Preparation, and Applications of CNTs

Usually CNTs are stable at high thermal and mechanical strength due to their tensile elastic nature. CNTs exhibited high electrical and thermal conductivity at moderate temperature. CNTs exhibited high flexibility; hence, it can be bent at any angle without change, as well as CNTs are also good electron field emitter [17]. Research suggest the high expansion and thermal conductivity of CNTs even at <20 K. The main reason is exotic strength, unique electrical property, as similar to metals, or semiconductors. In graphene, carbon atom is present in the planner honey comb lattice, in which each carbon atom is connected to three carbon atom of neighboring chemical bond [18] (Figure 16.2).

Previously laser ablation or arc-discharge-based high temperature preparation methods were mainly used for CNTs preparation. But currently low-temperature-based chemical deposition technique is used for preparation of CNTs. Generally, most methods require supporting gases and vacuum, mainly because of the diameter, length, and purity of nanotubes. Preparation of CNTs mainly consists of chemical vapor deposition (CVD), thermal synthesis process, plasma-enhanced CVD (PE-CVD), plasma ablation method and plasma-based synthesis method, or arc discharge evaporation method [19–21]. Due to the unique properties of CNTs, they are mainly used for field emission, thermal conductivity, electrical conductivity, catalyst support, air and water filtration, fiber and rubber, thermal materials, conductor properties, corrosion protection [22–24], etc. (Figure 16.3).

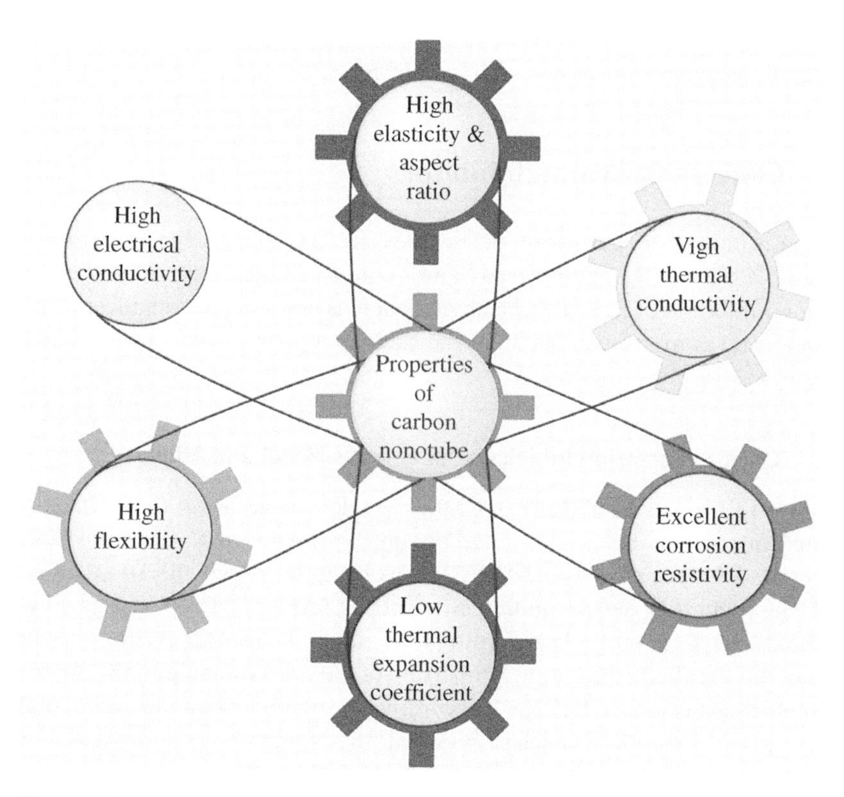

Figure 16.2 General properties of carbon nanotubes (CNTs).

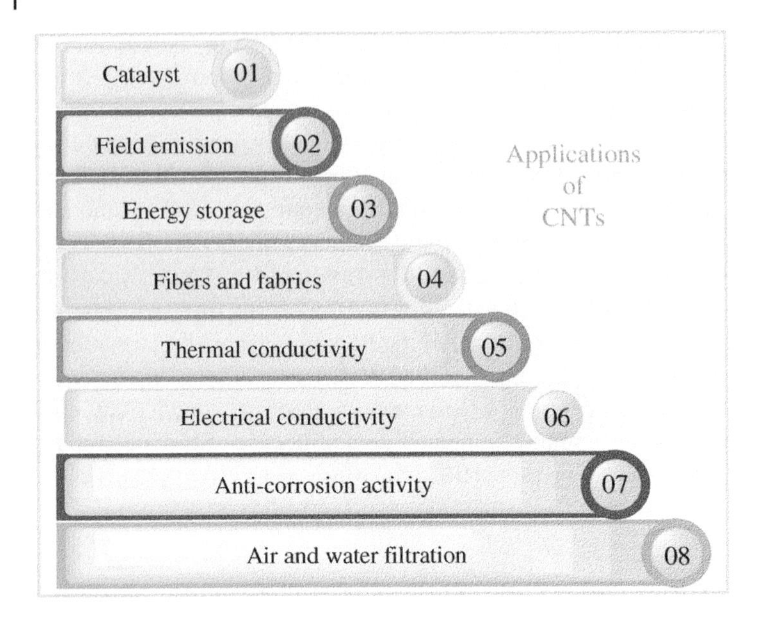

Figure 16.3 Important applications of carbon nanotubes (CNTs).

16.3 CNTs as Corrosion Inhibitors

Due to the unique properties of CNTs, they are used as excellent corrosion inhibitors. CNT-based nanomaterials and composites are commonly used as corrosion inhibitors of ferrous-based metal such as carbon steel, stainless steel, mild steel, and nonferrous-based material such as copper, aluminum, zinc, and their alloys.

16.3.1 CNTs as Corrosion Inhibitors for Ferrous Metal and Alloys

According to Cubites et al. (2016), CNTs zinc-rich epoxy resin (CNT-70ZRP) exhibited mixed-type behavior for carbon steel corrosion in NaCl media [25]. Fareg et al. (2017) applied PANI-MWCNTs as inhibitor in which PDP analysis reveals the highest 94.88% inhibition efficiency at PANI-MWCNTs-coated alkyl with the 0.05 P/B ration [26]. According to Ionita and coworkers (2011), result revealed the excellent inhibition properties of PPY-CNTs composites toward carbon steel corrosion in 35% NaCl according to theoretical and experimental analysis [27]. Also PDP analysis revealed the lowest E_{corr}(inh) value for PANI-25wt% CNT composites and highest inhibition efficiency at the above ration [28]. Similarly electrochemical analysis exhibited that PANI-3/f-CNT substrate showing 92%RS (Ω cm^2) and %η 91% for mild steel in 3.5% NaCl [29]. Parsannakumar et al. (2020) proved that the gravimetric analysis suggested the corrosion rate of 0.00480 (mm/year) for Ni-MWCNT mild steel in 3.5% chloride [30]. According to the electrochemical technique, both PPY and PPY-MWCNT composites protect the steel surface efficiently in 3% NaCl as per the investigation of Ganash (2013) [31]. Wet and coworkers research suggest that electrochemical impedance spectroscopy (EA) analysis revealed the 97.70% optimum inhibition efficiency of PU-MWCNTs for stainless steel in 3.0% NaCl electrolyte [32]. Studied technique reveals the excellent inhibition property of both POPDA/MWCNT and PoPDA/FMCNT nanocomposites for 316 L SS in 3.5 NaCl aggressive medium according to Zare et al. (2015) [33]. Experimental and surface analysis technique exhibited that the hybrid HA/F-MWCNT composite substrate decreases the metal dissolution rate and protect the 316 LSS metal surface [34]. Hierarchically structured C-pani nanobrushes prepared and applied as potential corrosion inhibitors for steel in 3.5wt% NaCl from the G. Qiu and team (2017) [35]. Study of Curtz wiler et al. (2017) reveals the excellent anticorrosion coating property of epoxy amine matrix-MWCNTs for the steel surface at concentrated chloride ion containing electrolytic medium [36]. The investigation of Sivray et al. (2019) deals with the spray pyrolysis, technique-mediated synthesized Ag-HA/f-MWCNTs' absorb efficiency on the 316 L SS metal surface at corrosive medium [37]. Table 16.1 illustrated the Inhibitors, metal/electrolyte, methods applied and outcomes of CNTs as corrosion inhibitors for ferrous metal and alloys.

16.3.2 CNTs as Corrosion Inhibitors for Nonferrous Metal and Alloys

TiO_2-coated BTESPT/MWCNTs composite material showing excellent coating property on AA2024, aluminum alloy, metal surface and exhibited 99.58% inhibition efficiency according to EIS measurement suggested by Y. Zhang and coworkers (2016) [38]. Large area synthesis of GFNACTL/ CNT achieved successfully for the anticorrosion application toward Cu alloy, namely, Cu_{11}, Sn_{11} in sea water electrolyte [39]. P3ABA@MWCNTs hybrid nanocomposites characterized through various techniques such as FTIR, SEM, EIS, PDP, and TGA toward Cu corrosion in NaCl electrolytic solution [40]. According to Palaniappan et al. (2019), experimental and computational studies suggested that 4,5 diphenyl imidazole functionalized CNTs acted as potential corrosion inhibitor for Ni-Alloy in 1M H2SO4 medium [41]. Experimental analysis shows that Cds/MWCNTs QDs show 99.19% inhibition efficiency at its optimum concentration according to K. Kandasamy and team (2020) [42]. Table 16.2 illustrated the inhibitors, metal/electrolyte, methods applied, and outcomes of CNTs as corrosion inhibitors for nonferrous metal and alloys.

Table 16.1 Inhibitors, metal/electrolyte, methods applied, and outcomes of CNTs as corrosion inhibitors for ferrous metal and alloys.

S.No.	Inhibitors	Metal/electrolyte	Methods applied	Outcomes	References
1.	CNTs zinc-rich epoxy resin (CNT-ZRPs)	Carbon steel/ NaCl	LEIS, SEM, EPS, XPS, OCP, EIS	CNTs zinc-rich epoxy resin (CNT-70ZRP) exhibited mixed-type behavior for carbon steel corrosion in NaCl media	[25]
2.	PANI/MWCNTs/ alkyl	Carbon steel/1M HCl	EIS, HRTEM, TGA, Raman, PDP	PDP analysis reveals the highest 94.88% inhibition efficiency at PANI-MWCNTs-coated alkyl with the 0.05 P/B ration	[26]
3.	CNT-PABS and CNT-CA PPY	Carbon steel (OL 48–80)/3.5% NaCl	Molecular mechanics, MD, EIS, SEM	The result revealed the excellent inhibition properties of PPY-CNTs composites toward carbon steel corrosion in 35% NaCl according to theoretical and experimental analysis	[27]
4.	PANI-CNT	Mild steel/1M HCl	FTIR, XRD, SEM, Raman spectra, OCP, PDP	PDP analysis revealed the lowest E_{corr}(inh) value for PANI-25wt% CNT composites and highest inhibition efficiency at the above ration	[28]
5.	PANI-f-CNT	Mild steel/3.5% NaCl	AIR-IR, Fe-SEM, EIS, Raman spectra, PDP	Electrochemical analysis exhibited that PANI-3/f-CNT substrate showing 92%RS (Ω cm^2) and %η 91% for mild steel in 3.5% NaCl	[29]
6.	Ni-MWCNT	Mild steel/3.5% NaCl	XRD, EIS, Fe-SEM, PDP, WL	The gravimetric analysis suggested the corrosion rate of 0.00480 (mm/year)for Ni-MWCNT mild steel in 3.5% chloride	[30]
7.	PPY & PPY-MWCNT	304 stainless steel/3% NaCl	Cyclic voltametry, FTIR, SEM, EIS	According to the electrochemical technique, both PPY and PPY-MWCNT composites protect the steel surface efficiently in 3% NaCl	[31]

8.	PU-MWCNT	Stainless steel/3.0% NaCl	SEM, TGA, FTIR, OCP, PDP	Electrochemical impedance spectroscopy (EA) analysis revealed the 97.70% optimum inhibition efficiency of PU-MWCNTs for stainless steel in 3.0% NaCl electrolyte	[32]
9.	PoPDA-MWCNTs	316 L stainless steel/3.5% NaCl	FTIR, SEM, XRD, PDP, EIS, OCP	Studied technique reveals the excellent inhibition property of both POPDA@MWCNT and PoPDA@FMCNT nanocomposites for 316 L SS in 3.5 NaCl aggressive medium	[33]
10.	HA/F-MWCNT	316L stainless steel/SBF	XRD, FTIR, Fe-SEM, PDP, EIS	Experimental and surface analysis technique exhibited that the hybrid HA/F-MWCNT composite substrate decreases the metal dissolution rate and protect the 316 LSS metal surface	[34]
11.	PANA-MWCNTs nanobrushes	Steel/3.5 wt% NaCl	TEM, FTIR, XRD, UV-Vis, CV, EIS, PDP, XPS	Hierarchically structured C-pani nanobrushes, prepared and applied as potential corrosion inhibitors for steel in 3.5 wt% NaCl	[35]
12.	Epoxy amine matrix-MWCNTs	Steel/5% NaCl	SEM, WL	Study reveals the excellent anticorrosion coating property of epoxy amine matrix-MWCNTs for the steel surface at concentrated chloride ion containing electrolytic medium	[36]
13.	Ag-substituted hydroxyapatite f-MWCNTs (Ag-HA/fMWCNTs)	316 L stainless steel	XRD, FTIR, SEM, EDS, AFM, PPDP	The present investigation deals with the spray pyrolysis, technique-mediated synthesized Ag-HA/f-MWCNTs absorb efficiency on the 316 L SS metal surface at corrosive medium	[37]

Table 16.2 Inhibitors, metal/electrolyte, methods applied, and outcomes of CNTs as corrosion inhibitors for nonferrous metal and alloys.

S.No.	Inhibitors	Metal/ electrolyte	Methods applied	Outcomes	References
14.	MWCNTs/BTESPT/ TiO_2	Aa2024 aluminum alkyl/ sea water	EIS, PDP, FTIR, SEM	TiO_2-coated BTESPT/MWCNTs composite material showing excellent coating property on AA2024, aluminum alloy, metal surface, and exhibited 99.58% inhibition efficiency as per EIS measurement	[38]
16.	GFNACTL/CNT	Copper alloys/ sea water	EIS, PDP, SEM, UV-Vis, XRD	Large area synthesis of GFNACTL/CNT achieved successfully for the anticorrosion application toward Cu alloy, namely, Cu_{11}, Sn_{11} in sea water electrolyte	[39]
17	Poly (3-aminobenzoic acid @MWCNTs)	Copper/3.5% NaCl	FTIR, XRD, TGA, OCP, SEM	P3ABA@MWCNTs hybrid nanocomposites characterized through various techniques such as FTIR, SEM, EIS, PDP, and TGA toward Cu corrosion in NaCl electrolytic solution	[40]
18.	4,5 diphenyl imidazole-F-CNTs	Ni-alloys/1M H_2SO_4	FESEM, PDP, EIS,XRD	Experimental and computational studies suggested that 4,5 diphenyl imidazole functionalized CNTs acted as potential corrosion inhibitor for Ni-Alloy in 1M H_2SO_4 medium	[41]
19.	Cds/MWCNTs/QDs	Zinc	XRD, FTIR, HRTEM, TGA, XRF, EIS	Experimental analysis showing that Cds/ MWCNTs QDs showing 99.19% inhibition efficiency at its optimum concentration	[42]

16.4 Conclusion

Present literature deals with the anticorrosion and adhesive property of CNTs and its related material for metal and alloys corrosion. Some unique physical properties such as high thermal stability, high elasticity, good electrical conductivity, good stability in most of the solvents, and mechanical properties make CNTs as prominent material for anticorrosion. CNTs early adsorb on to the metal surface by interacting with metal cation to form strong chemical bonds. CNTs incorporated with polymers are also exhibited efficient inhibition property due to the presence of functional groups and extended Pi bonds present in it. CNT polymer composites strongly adsorb to the metal surface and protect it by forming a stable protective layer on it. These protective layers are strong enough to resist the aggressive attack of various anions to the metal surface. Additionally, these CNT-based corrosion inhibitors are cost-effective, nontoxic nature from synthesis to use and most importantly environmental-friendly. Experimental and theoretical calculation is applied extensively in order to prove the inhibition nature of CNTs.

Conflict of Interest

Authors declared no conflict of interest

Acknowledgment

Authors greatly acknowledge the Principal Govt. Digvijay Autonomous PG College for providing instrumental and lab facilities.

Abbreviations

CNTs	Carbon nanotubes
WL	Weight loss
LEIS	Local electrochemical impedance spectroscopy
SEM	Scanning electron microscopy
EDS	Electron dispersion X-ray spectroscopy
XPS	X-ray photoelectron spectroscopy
OCP	Open circuit potential
EIS	Electrochemical impedance spectroscopy
PANI	Poly aniline
MWCNTs	Multiwalled carbon nanotubes

HRTEM	High resolution transmission electron microscopy
TGA	Thermogravimetric analysis
PDP	Potentiodynamic polarization
CNT-PABS	Poly pyrrole/Poly aminobenzene sulfonic acid functionalized single walled carbon nanotubes
PPY/CNT-CA	PPY/carboxylic acid functionalized single walled carbon nanotube

References

1. Sikine, M. et al. (2017). Experimental, Monte Carlo simulation and quantum chemical analysis of 1, 5-di (prop-2-ynyl)-benzodiazepine-2, 4-dione as new corrosion inhibitor for mild steel in 1 M hydrochloric acid solution. *Journal of Materials and Environmental Science* 8: 116–133.
2. Verma, D. et al. (2018). Inhibition performance of Glycine max, Cuscuta reflexa and Spirogyra extracts for mild steel dissolution in acidic medium: density functional theory and experimental studies. *Results in Physics* 10: 665–674.
3. Errahmany, N. et al. (2020). Experimental, DFT calculations and MC simulations concept of novel quinazolinone derivatives as corrosion inhibitor for mild steel in 1.0 M HCl medium. *Journal of Molecular Liquids* 312: 113413.
4. Verma, D.K. et al. (2019). Gravimetric, electrochemical surface and density functional theory study of acetohydroxamic and benzohydroxamic acids as corrosion inhibitors for copper in 1 M HCl. *Results in Physics* 13: 102194.
5. Saha, S.K. et al. (2016). Novel Schiff-base molecules as efficient corrosion inhibitors for mild steel surface in 1 M HCl medium: experimental and theoretical approach. *Physical Chemistry Chemical Physics* 18 (27): 17898–17911.
6. Verma, D.K. and Khan, F. (2016). Green approach to corrosion inhibition of mild steel in hydrochloric acid medium using extract of spirogyra algae. *Green Chemistry Letters and Reviews* 9 (1): 52–60.
7. Chauhan, D.S. et al. (2019). Triazole-modified chitosan: a biomacromolecule as a new environmentally benign corrosion inhibitor for carbon steel in a hydrochloric acid solution. *RSC Advances* 9 (26): 14990–15003.
8. Verma, D.K. et al. (2020). Experimental and computational studies on hydroxamic acids as environmental friendly chelating corrosion inhibitors for mild steel in aqueous acidic medium. *Journal of Molecular Liquids* 314: 113651.
9. Guo, L. et al. (2017). Toward understanding the anticorrosive mechanism of some thiourea derivatives for carbon steel corrosion: a combined DFT and molecular dynamics investigation. *Journal of Colloid and Interface Science* 506: 478–485.
10. Singh, A. et al. (2017). Electrochemical, surface and quantum chemical studies of novel imidazole derivatives as corrosion inhibitors for J55 steel in sweet corrosive environment. *Journal of Alloys and Compounds* 712: 121–133.

11 Kabanda, M.M., Obot, I.B., and Ebenso, E.E. (2013). Computational study of some amino acid derivatives as potential corrosion inhibitors for different metal surfaces and in different media. *International Journal of Electrochemical Science* 8 (2013): 10839–10850.

12 Singh, A. et al. (2015). Porphyrins as corrosion inhibitors for N80 Steel in 3.5% NaCl solution: electrochemical, quantum chemical, QSAR and Monte Carlo simulations studies. *Molecules* 20 (8): 15122–15146.

13 Peng, Y.G. et al. (2014). Preparation of poly (m-phenylenediamine)/ZnO composites and their photocatalytic activities for degradation of CI acid red 249 under UV and visible light irradiations. *Environmental Progress & Sustainable Energy* 33 (1): 123–130.

14 Zare, E.N., Lakouraj, M.M., and Ramezani, A. (2016). Efficient sorption of Pb (II) from an aqueous solution using a poly (aniline-co-3-aminobenzoic acid)-based magnetic core–shell nanocomposite. *New Journal of Chemistry* 40 (3): 2521–2529.

15 Baghayeri, M., Zare, E.N., and Lakouraj, M.M. (2015). Monitoring of hydrogen peroxide using a glassy carbon electrode modified with hemoglobin and a polypyrrole-based nanocomposite. *Microchimica Acta* 182 (3-4): 771–779.

16 Olad, A., Rashidzadeh, A., and Amini, M. (2013). Preparation of polypyrrole nanocomposites with organophilic and hydrophilic montmorillonite and investigation of their corrosion protection on iron. *Advances in Polymer Technology* 32 (2).

17 Hoenlein, W. et al. (2003). Carbon nanotubes for microelectronics: status and future prospects. *Materials Science and Engineering: C* 23 (6-8): 663–669.

18 Chen, M. et al. (2007). Effect of purification treatment on adsorption characteristics of carbon nanotubes. *Diamond and Related Materials* 16 (4-7): 1110–1115.

19 Dresselhaus, G., Dresselhaus, M.S., and Saito, R. (1998). *Physical Properties of Carbon Nanotubes*. World scientific.

20 Salvetat, J.-P. et al. (1999). Mechanical properties of carbon nanotubes. *Applied Physics A* 69 (3): 255–260.

21 Lu, J.P. (1997). Elastic properties of carbon nanotubes and nanoropes. *Physical Review Letters* 79 (7): 1297.

22 Yakobson, B.I. and Avouris, P. (2001). Mechanical properties of carbon nanotubes. In: *Carbon Nanotubes* (eds. M.S. Dresselhaus, G. Dresselhaus and P. Avouris), 287–327. Springer.

23 Ruoff, R.S. and Lorents, D.C. (1995). Mechanical and thermal properties of carbon nanotubes. *Carbon* 33 (7): 925–930.

24 Mintmire, J. and White, C. (1995). Electronic and structural properties of carbon nanotubes. *Carbon* 33 (7): 893–902.

25 Cubides, Y. and Castaneda, H. (2016). Corrosion protection mechanisms of carbon nanotube and zinc-rich epoxy primers on carbon steel in simulated concrete pore solutions in the presence of chloride ions. *Corrosion Science* 109: 145–161.

26 Farag, A.A. et al. (2017). Influence of polyaniline/multiwalled carbon nanotube composites on alkyd coatings against the corrosion of carbon steel alloy. *Corrosion Reviews* 35 (2): 85–94.

27 Ioniţă, M. and Prună, A. (2011). Polypyrrole/carbon nanotube composites: molecular modeling and experimental investigation as anti-corrosive coating. *Progress in Organic Coatings* 72 (4): 647–652.

28 Rajyalakshmi, T. et al. (2020). Enhanced charge transport and corrosion protection properties of polyaniline–carbon nanotube composite coatings on mild steel. *Journal of Electronic Materials* 49 (1): 341–352.

29 Kumar, A.M. and Gasem, Z.M. (2015). In situ electrochemical synthesis of polyaniline/f-MWCNT nanocomposite coatings on mild steel for corrosion protection in 3.5% NaCl solution. *Progress in Organic Coatings* 78: 387–394.

30 Prasannakumar, R. et al. (2020). Electrochemical and hydrodynamic flow characterization of corrosion protection persistence of nickel/multiwalled carbon nanotubes composite coating. *Applied Surface Science* 507: 145073.

31 Ganash, A. (2014). Electrochemical synthesis and corrosion behaviour of polypyrrole and polypyrrole/carbon nanotube nanocomposite films. *Journal of Composite Materials* 48 (18): 2215–2225.

32 Wei, H. et al. (2013). Anticorrosive conductive polyurethane multiwalled carbon nanotube nanocomposites. *Journal of Materials Chemistry A* 1 (36): 10805–10813.

33 Zare, E.N. et al. (2015). Emulsion polymerization for the fabrication of poly (o-phenylenediamine)@ multi-walled carbon nanotubes nanocomposites: characterization and their application in the corrosion protection of 316L SS. *RSC Advances* 5 (84): 68788–68795.

34 Sivaraj, D. and Vijayalakshmi, K. (2019). Novel synthesis of bioactive hydroxyapatite/f-multiwalled carbon nanotube composite coating on 316L SS implant for substantial corrosion resistance and antibacterial activity. *Journal of Alloys and Compounds* 777: 1340–1346.

35 Qiu, G., Zhu, A., and Zhang, C. (2017). Hierarchically structured carbon nanotube–polyaniline nanobrushes for corrosion protection over a wide pH range. *RSC Advances* 7 (56): 35330–35339.

36 Curtzwiler, G.W. et al. (2017). Measurable and influential parameters that influence corrosion performance differences between multiwall carbon nanotube coating material combinations and model parent material combinations derived from epoxy-amine matrix materials. *ACS Applied Materials & Interfaces* 9 (7): 6356–6368.

37 Sivaraj, D. and Vijayalakshmi, K. (2019). Enhanced antibacterial and corrosion resistance properties of Ag substituted hydroxyapatite/functionalized multiwall carbon nanotube nanocomposite coating on 316L stainless steel for biomedical application. *Ultrasonics Sonochemistry* 59: 104730.

38 Zhang, Y. et al. (2016). TiO2 coated multi-wall carbon nanotube as a corrosion inhibitor for improving the corrosion resistance of BTESPT coatings. *Materials Chemistry and Physics* 179: 80–91.

39 Jeong, N. et al. (2017). One-pot large-area synthesis of graphitic filamentous nanocarbon-aligned carbon thin layer/carbon nanotube forest hybrid thin films and their corrosion behaviors in simulated seawater condition. *Chemical Engineering Journal* 314: 69–79.

40 Zare, E.N., Lakouraj, M.M., and Moosavi, E. (2016). Poly (3-aminobenzoic acid)@ MWCNTs hybrid conducting nanocomposite: preparation, characterization, and application as a coating for copper corrosion protection. *Composite Interfaces* 23 (7): 571–583.

41 Palaniappan, N. et al. (2019). Experimental and DFT studies of carbon nanotubes covalently functionalized with an imidazole derivative for electrochemical stability and green corrosion inhibition as a barrier layer on the nickel alloy surface in a sulphuric acidic medium. *RSC Advances* 9 (66): 38677–38686.

42 Kandasamy, K. et al. (2020). Ultrasound-assisted microwave synthesis of CdS/ MWCNTs QDs: a material for photocatalytic and corrosion inhibition activity. *Materials Today: Proceedings* 26: 3588–3594.

17

Graphene and Graphene Oxides Layers Application as Corrosion Inhibitors in Protective Coatings

Renhui Zhang[1], Lei Guo[2], Zhongyi He[1], and Xue Yang[1]

[1] *School of Materials Science and Engineering, East China JiaoTong University, Nanchang, People's Republic of China*
[2] *School of Material and Chemical Engineering, Tongren University, Tongren, People's Republic of China*

17.1 Introduction

Since its discovery more than 10 years ago, graphene has showed the broad applications owing to excellent physical and chemical performance [1]. Moreover, graphene exhibits inherent unique optical, electrical, and thermal performance due to its linear-band microstructure [2–4]. On the other hand, graphene is also considered as the barrier coatings for a wide field owing to its superior properties, such as antioxidation/corrosion and molecule diffusion [5–8]. Thus, graphene is considered as the potential corrosion inhibitor in epoxy on the surface of copper, silver, and other metal alloys [9–12]. In particular, graphene-based composite materials with a multilayer structure exhibit superior anticorrosion properties, assigning to the better barrier effects for corrosive media [13–16].

Although graphene is considered as an ultrathin and light-weight corrosion barrier, its high conductivity causes the formation of the galvanic cell and aggravates the electrochemical reactions, and as a result, graphene fails to provide long-life protection of the metal substrates [17, 18]. However, graphene oxides as an insulator exhibits better corrosion-barrier properties than graphene in a long term. Mahmoudi et al. report the important role of graphene oxides layers on corrosive barriers to the substrates [19]. Li et al. report that graphene oxides as a corrosion inhibitor in epoxy layers exhibit superior corrosion resistance in 3.5 wt.% sodium chloride [9]. Dehghani et al. declare that combination of graphene oxides and beta-cyclodextrin-zinc acetylacetonate, a smart anticorrosion nanocarrier is

Organic Corrosion Inhibitors: Synthesis, Characterization, Mechanism, and Applications,
First Edition. Edited by Chandrabhan Verma, Chaudhery Mustansar Hussain, and Eno E. Ebenso.

produced [20]. Graphene oxides possessing superior anticorrosion performance mainly correspond to the oxygen groups on its surface reducing conductivity. Besides, graphene doping with N could effectively reduce its conductivity and exhibit excellent corrosion resistance as reported by Ren et al. [21]. Based on this work, Jiang et al. fabricate the composite layer including N-doped graphene quantum dots (GQDs)/PMTMS and investigate the anticorrosion performance-coated AZ91D magnesium alloy, and they find that the composite layer greatly enhances the anticorrosion performance [22].

However, the protective efficiency of single graphene or graphene oxides for metal substrates is still insufficient. Theoretically, if we combine their advantages, the N-doped graphene and graphene oxides can effectively enhance the anticorrosion performance of metals in corrosion environments. In this chapter, we will review and discuss the latest progresses in the development of N-doped graphene and graphene oxides as the corrosion inhibitors in organic coatings in corrosive media, meanwhile the challenges and prospects of graphene are discussed and highlighted.

17.2 Preparation of Graphene and Graphene Oxides

Graphene and graphene oxides are important 2D materials. Variable methods are proposed to fabricate two materials, such as chemical vapor deposition, liquid-phase exfoliation, and mechanical exfoliation.

17.2.1 Graphene

Graphene, just one-atom-thick sheet of carbon, is a potential application in many fields, especially, it is of great significance in corrosive environment. The C–C bond length of 0.142 nm is ordered in a closely hexagonal flawless honeycomb lattice structure. The pore-diameter is deduced to 0.064 nm with a theoretical geometric pore, if the van der Waals radius of C is 0.11 nm as shown in Figure 17.1.

Since graphene was first prepared by Novoselov et al. [2], the mechanical exfoliation method has become one of the often used method to make similar 2D materials from pristine bulk crystals. Graphene is regarded as "mother" of all graphitic carbon allotropes like graphite, carbon nanotubes, and fullerenes (Figure 17.2), which can be derived as reported by Novoselov and Soldano [25, 26]. Although this approach provides high-quality product for 2D materials, exfoliation yield is quite low.

Chemical vapor deposition (CVD) has been confirmed to be the most practical ways to prepare crystalline two-dimensional materials [28], which not only allows

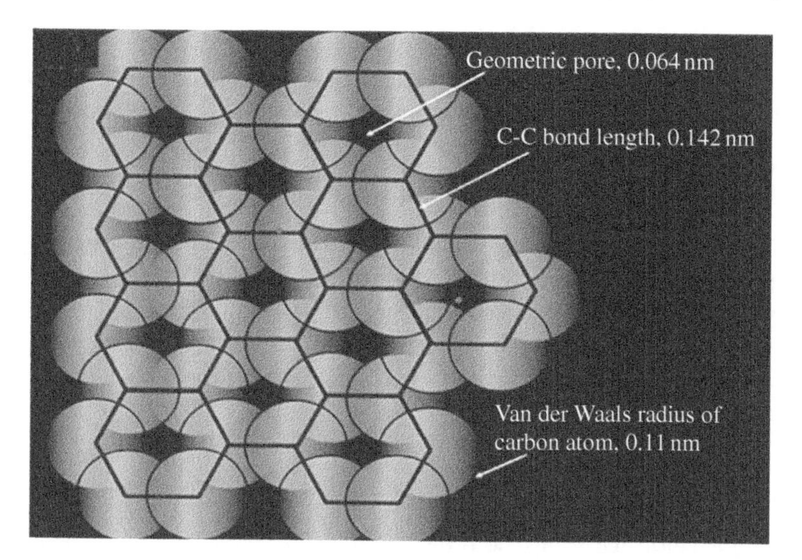

Figure 17.1 Theoretical parameters of graphene. *Source:* Refs. [23, 24].

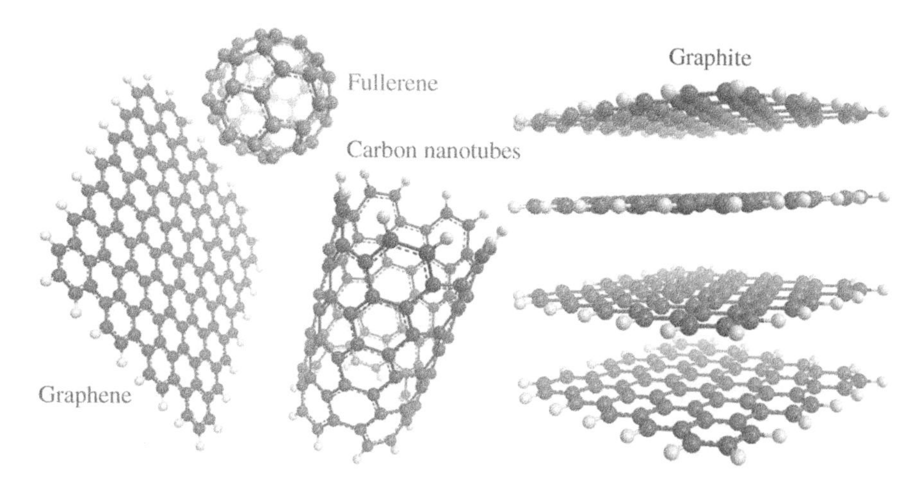

Figure 17.2 Several common carbon materials. *Source:* Ref. [27].

graphene layers depositing on large surface [29], but also possibly controlling the deposition layers. However, the procedure is too complicated to achieve ideal results. Besides, the cost of this method is quite high and the yield is quite low. Compared to chemical vapor deposition, Liquid-phase exfoliation is simple and cost-effective method. Manna et al. [30] produce the graphene and molybdenum disulfide nanosheets in the mixed solvent of water-NMP (N-methylpyrrolidinone) by liquid-phase sonication exfoliation, and these nanosheets exhibited low defects and high stability for up to 18 months. This method is treated as a useful way to fabricate the graphene layers due to its low cost.

17.2.2 N-doped Graphene and Its Composites

Figures 17.3 and 17.4 represent the production of N-doped graphene quantum dots. N-doped graphene films are fabricated on Cu foils using atomic layer deposition system [21]. The fabricated graphene layer number is controlled as 2–3 layers. N-doped graphene quantum dots is fabricated using hydrothermal method, and its composite coating mixed with PMTMS is electrodeposited on the Mg alloy surface [22].

17.2.3 Graphene Oxides

Generally, GO is fabricated using the Brodie, Staudenmaier, or Hummers methods [31]. These methods include graphite's oxidation. Brodie and Staudenmaier involve using $KClO_3$and HNO_3 to oxidize graphite powders, and Hummers method experiences the treatment of graphite with $KMnO_4$ and H_2SO_4. The GO is easily exfoliated in liquid media [32, 33]. Besides the hydroxyl, epoxy, carbonyl, and carboxyl groups are exposed on GO, which can form covalent and non-covalent bonds to the 2D materials, thus improving their physical and chemical performance [34–36]. The oxygen groups lead to the poor electrical conductivity [37]. The poor electrical conductivity is conducive to enhancing the anticorrosion performance. Fabrication process of graphene oxide is illustrated in Figure 17.5.

17.3 Protective Film and Coating Applications of Graphene

The function of graphene-based material has two forms in protective coating, one as a protective film, and the other as nanofillers for the metallic substrate.

The recent investigations indicated graphene films as protection layers, which could significantly reduce the corrosion rate of metals, due to their exceptional barrier to reactive media [38]. These barrier performances of graphene are possibly attributed to its microstructure as plotted in Figure 17.6a. Surfaces of sp^2 carbon allotropes form dense and delocalized electron cloud of the π-conjugated carbon networks, and the rings pose a repelling field to reactive targets and therefore provides a physical separation between the metal and corrosive media [23].

Although graphene films are probed to exhibit excellent barrier properties in theory, the defectiveness of the graphene greatly reduces the anticorrosion ability for metals and easy to be damaged by corrosive media. The defects are the main

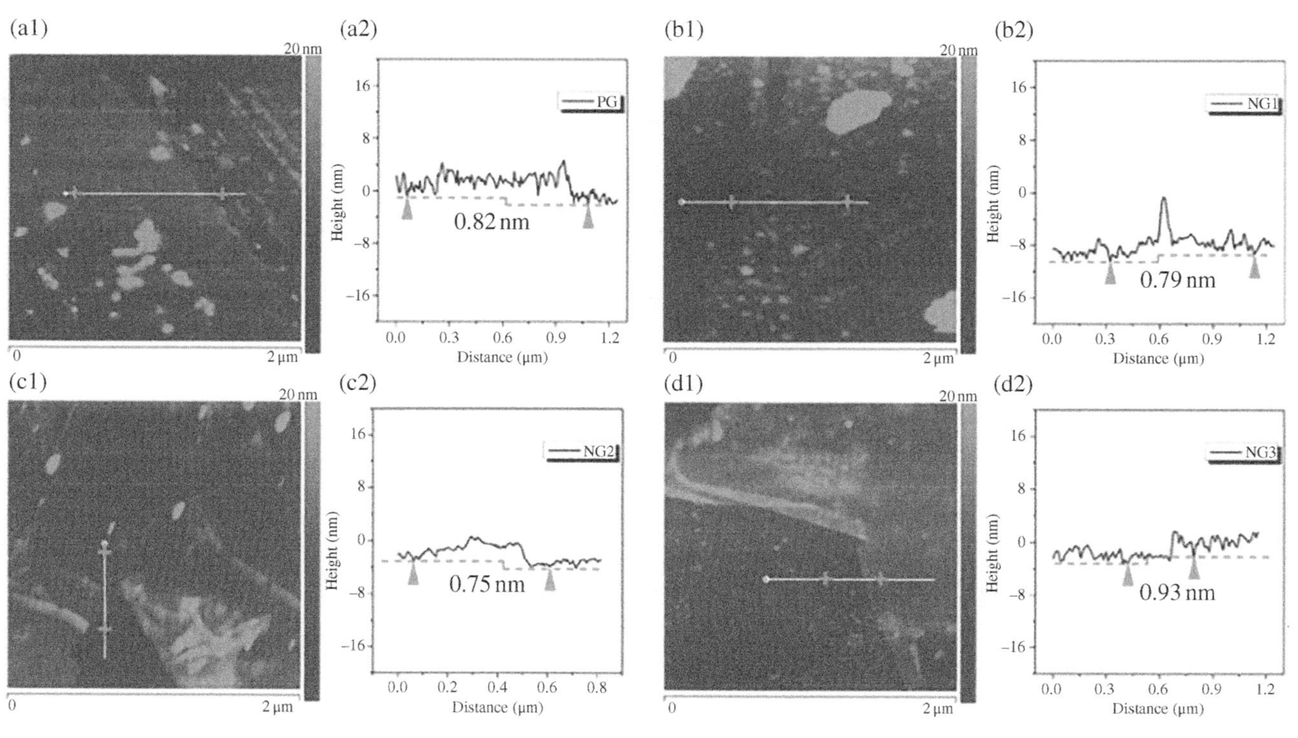

Figure 17.3 The fabricated N-doped graphene. *Source:* Ref. [21].

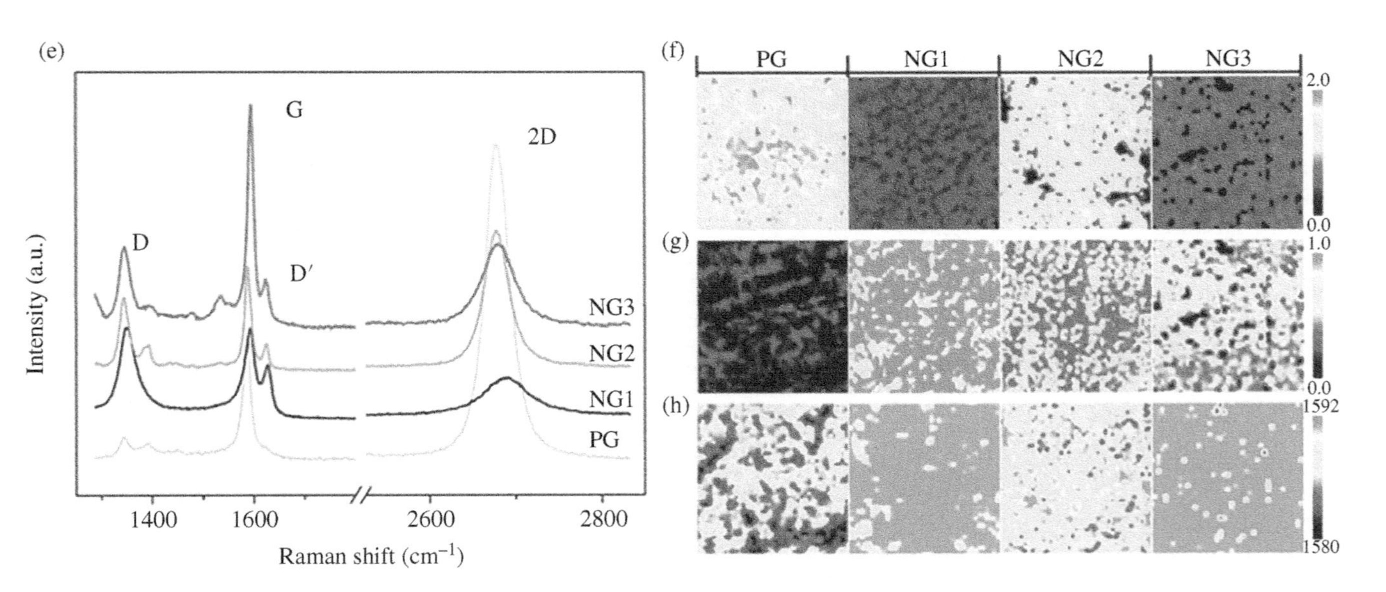

Figure 17.3 (Continued)

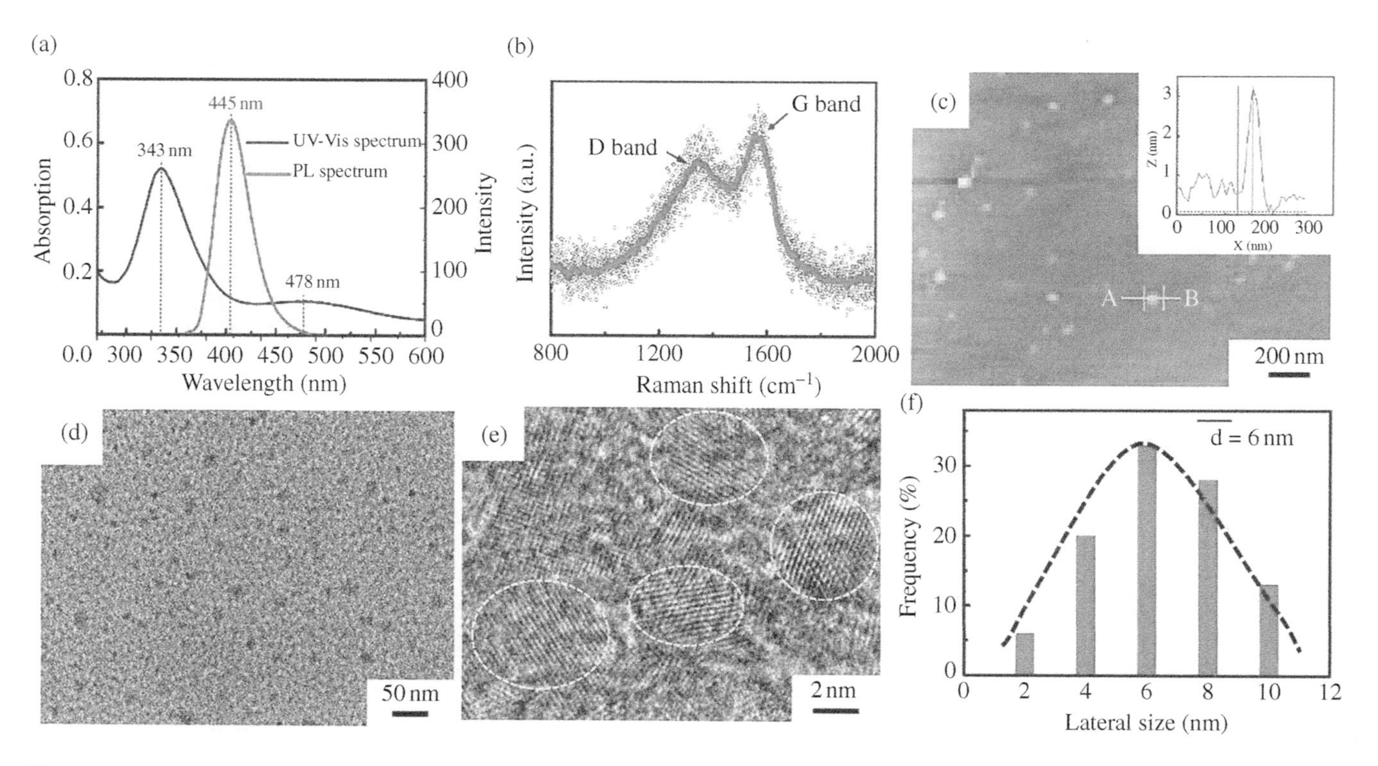

Figure 17.4 The fabricated N-doped graphene quantum dots. *Source:* Ref. [22].

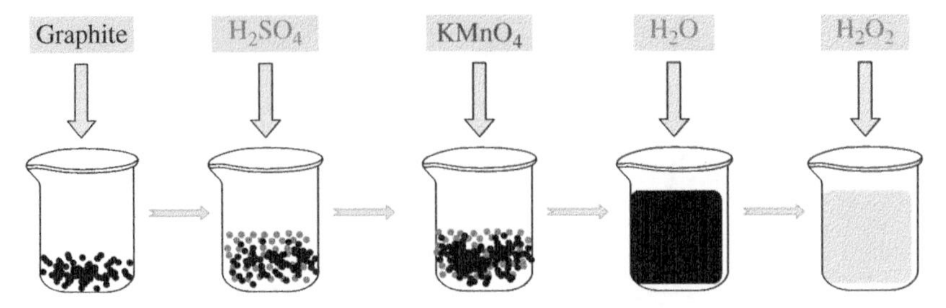

Figure 17.5 The fabricated processes of graphene oxides.

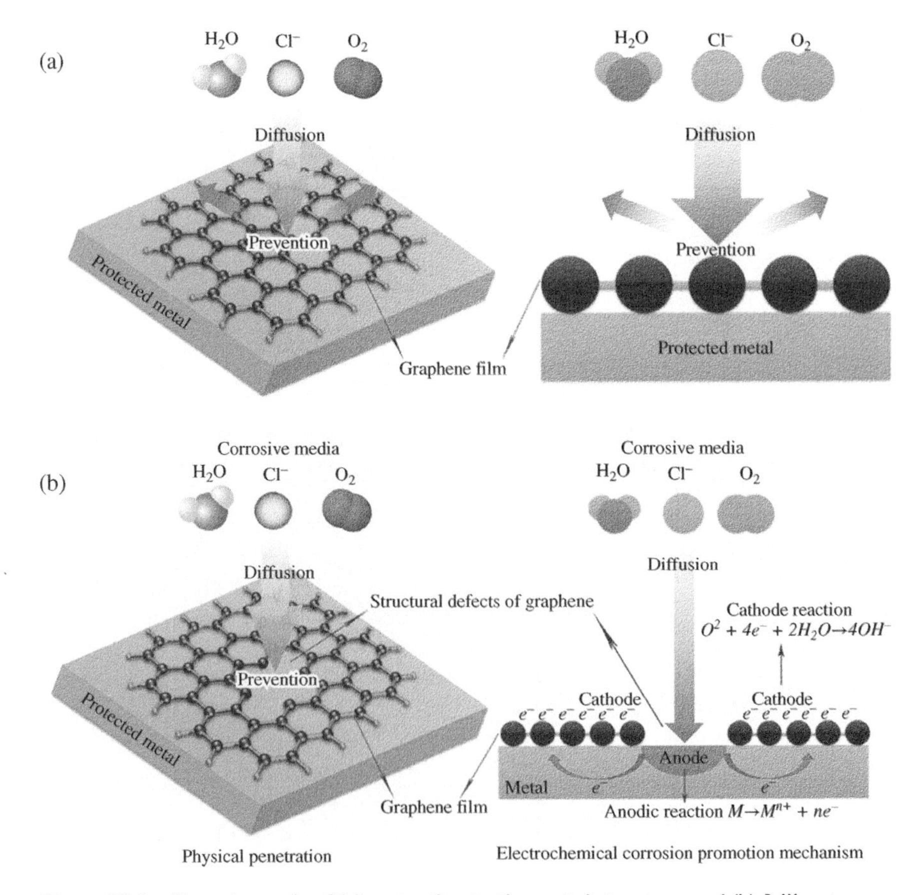

Figure 17.6 The schematic of (a) protecting to the metal structure and (b) failing to metal. *Source:* Ref. [27].

cause of the weak long-term anticorrosion performance of graphene on metals (Figure 17.6b). Besides, it has been observed by many researchers that graphene with excellent conductivity would induce the electrochemical corrosion of metals [39, 40]. In the damaged regions, the lattice defects give potential positions to accelerate corrosion rate for the metal substrates. The metal acting as anode accelerates corrosion rate, and it inevitably forms corrosion microcells between the metal and graphene. Therefore, it is not suitable for corrosion protection in the long term.

Lee and Berman [40] report the impermeability of liquids and gasses into graphene films, due to the thinnest anticorrosion coating for copper and nickel, the existence of defects lead to accelerating corrosion. The analysis indicates that the initial cracks of graphene are the center for iron oxidation and propagation when metals coated with graphene immersing in NaCl. The high concentration of chlorine accelerates metal degradation processes and brings about conspicuous oxidation sites along the defects. Therefore, iron oxidation initially is formed from the cracks of graphene, and then corrosive media slowly propagate into the metal substrates (Figure 17.7).

These different results display that graphene aggravates corrosion rate in the long term, which is attributed to the corrosion mechanism of graphene. The perfect graphene possesses excellent physical barrier, but the defective graphene accelerates the corrosion rate of metal substrates through inducing a micro-galvanic corrosion around the metal substrate. The corrosion inhibition mechanism of graphene and N-doped graphene is illustrated in Figure 17.8.

Reducing the conductivity of graphene and as the corrosion inhibitor in organic coatings would be a better method to improve the anticorrosion properties for graphene.

On the one hand, incorporation of N into graphene nanosheets is considered as an effective way to deduce the conductivity of graphene and inhibits the galvanic reactions between graphene and Cu substrates as declared by Ren et al. [21]. Meanwhile, they report that N-doped graphene as barrier coatings can effectively stand up to ingression of the atomic oxygen [41]. Jiang et al. report that N-doped graphene QDs/polymethyltrimethoxysilane composite coating on AZ91D Mg alloy exhibits excellent corrosion resistance [22]. Additionally, there are other approaches as well for improving anticorrosive coatings. For example, graphene coatings significantly decrease corrosion rate by incorporating Ni, Cu, or alloys. Also, organic substances with different functional groups can improve the dispersion and compatibility between graphene and coatings.

Chen et al. [13] confirmed that incorporating a little of well-dispersed graphene nanosheets could be utilized as protecting additives in epoxy coatings to enhance

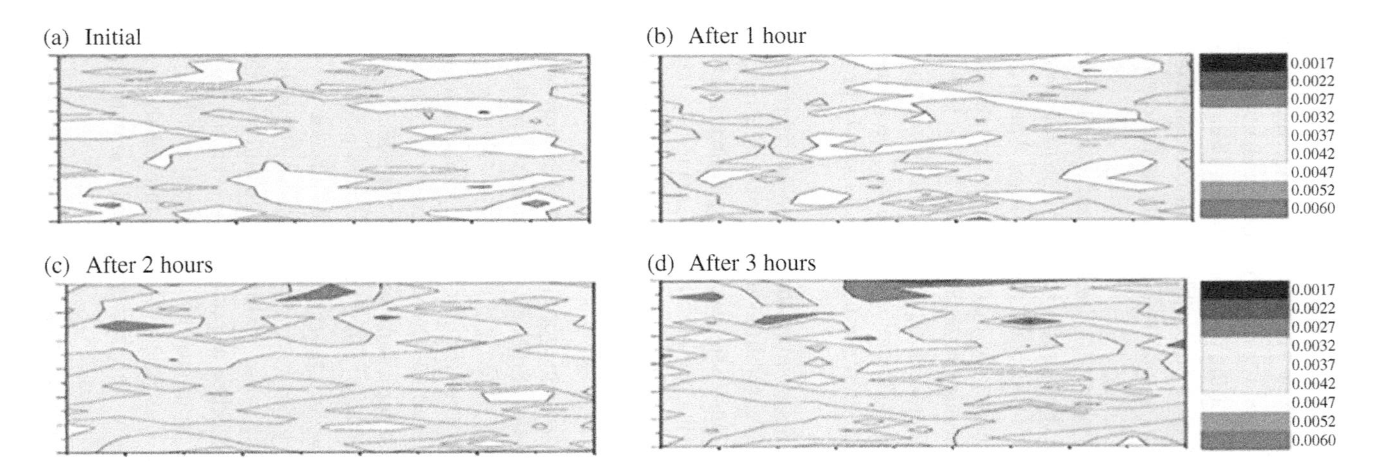

Figure 17.7 The changes of electrochemical potential of the tested samples in 3 wt.% sodium chloride solution. *Source:* Ref. [40].

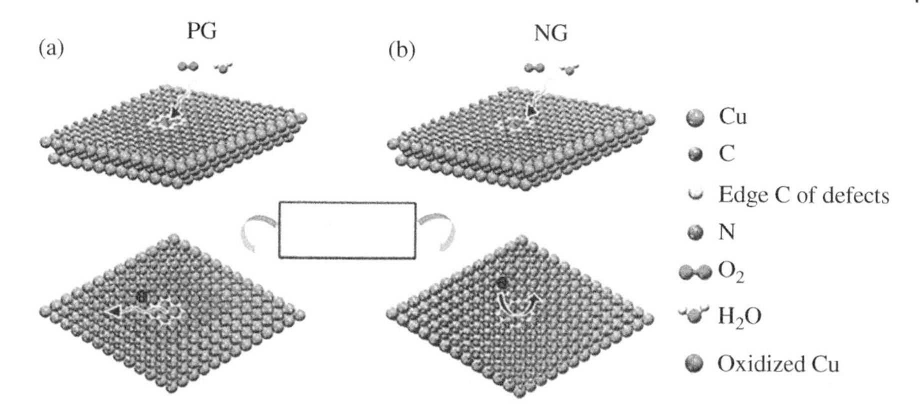

Figure 17.8 The schematic diagrams of corrosion mechanism for graphene and N-doped graphene. *Source:* Ref. [21].

anticorrosion performances and wear resistance properties. They choose poly(2-butylaniline) in terms of non-covalent π–π interactions, showing the superior dispersibility of graphene after modification of poly(2-butylaniline). It is a common concept that graphene is well dispersed in the coatings and can improve the tortuosity of diffusion pathways for corrosive substances. Jena et al. [42] show that two-order decrease in corrosive current-density when he develops a ternary composite layer including graphene and oxide-chitosan-silver on the metal base of Cu-Ni alloy. Both graphene oxide and chitosan play an important role in anticorrosive. As a functional group, chitosan can well prevent the galvanic coupling of graphene oxide and retard the diffusion of corrosive ions into substrate (Figure 17.9).

Nan et al. also report attapulgite as an inorganic nanoparticle bonded with graphene by hydrogen bonding, and results indicate that attapulgite–graphene composites significantly improve anticorrosive performance of waterborne-epoxy coating in long term, the synthetic route could be followed in Figure 17.10. As we know, there are much more hydroxyl groups on the edges. Therefore, the incorporation of attapulgite can obviously enhance the graphene compatibility with water. The synergistic actions of ATP nanofibers and graphene nanosheets exhibit enhanced resistant behavior against the penetration of water, O_2, and corrosion medium [43].

Based on the above analysis, concordant concepts have been adopted by researchers. We can obtain excellent corrosion resistance of graphene in long-term corrosive tests through modifying non-covalent π–π bonds, functional groups, or N doping. The method of different modifiability can enhance infiltration propagation distances, reduce the conductivity of graphene, and provide efficient barriers for corrosive media.

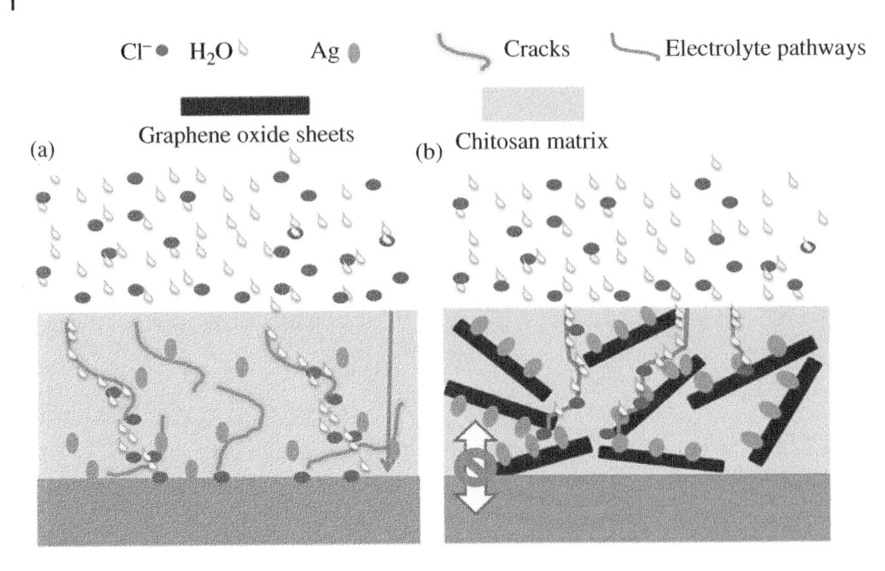

Figure 17.9 The mechanism of anticorrosion performance of (a) without graphene oxide, (b) with graphene oxide. *Source:* Ref. [42].

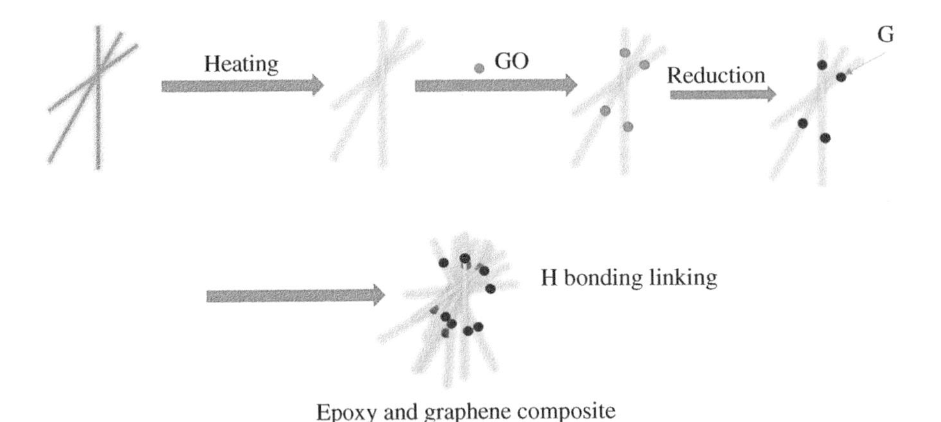

Figure 17.10 Synthetic route of the graphene and waterborne-epoxy coatings.

17.4 The Organic Molecules Modified Graphene as Corrosion Inhibitor

Graphene layers modified by organic molecules is a useful method to enhance the anti-corrosion properties. The hybrid corrosion inhibitor consisting of 1H-benzimidazole and graphene oxide exhibits superior corrosion resistance and self-repair ability for corrosion damage (Figure 17.11) [44]. The phenylenediamine-modified GO provides

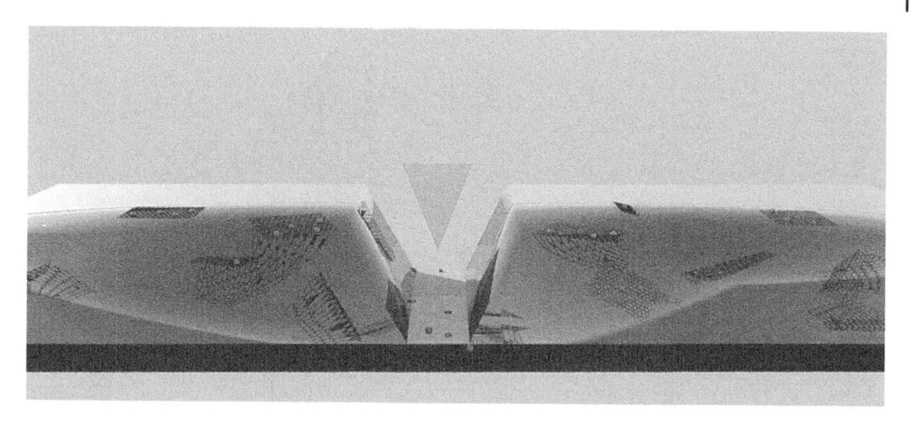

Figure 17.11 The self-repair corrosion-resistant epoxy film. *Source:* Ref. [44].

good corrosion resistance for metals as reported by Hwang et al. [45].Yuan et al. report that polyaniline-modified graphene corrosion inhibitor in epoxy coating could effectively improve the corrosion resistance of Al alloy [46], the corrosive media are effectively inhibited. Graphene tends to link the electronical PANI particles and anchors to the metallic surfaces, resulting in excellent passivation of defect-sites in the composite coating. Therefore, the graphene enhances the chemical activity of PANI to accelerate the passivation capability of the defects. Poly(o-phenylenediamine) modified graphene as the additive in epoxy coatings could effectively improve the protection ability to corrosion, the excellent anticorrosion properties of ECs originate from the synergistic effect in the barrier properties of graphene and the passivated metallic surfaces [47]. Alkyne-chain-modified graphene layer not only could reduce conductivity of reduced graphene oxides but avoid the corrosion-promoting action of carbon steel [48]; the anti-corrosion mechanisms are illustrated in Figure 17.12.

According to the above references, organic molecules modified graphene and reduce its conductivity, which effectively avoid the galvanic reaction between graphene and metal substrates. It effectively inhibits the electron transfer between protective coating and metal substrates.

17.5 The Effect of Dispersion of Graphene in Epoxy Coatings on Corrosion Resistance

The concentration and dispersion of graphene in ECs has a close effect on its anticorrosion performance. Cui et al. report that the graphene oxide and polydopamine (GO-PDA) nanosheet is well dispersed in waterborne-epoxy matrix (Figure 17.13). The electrochemical tests showed that the anticorrosion performance of GO-PDA/EP is remarkably enhanced referee to the blank EP and GO/

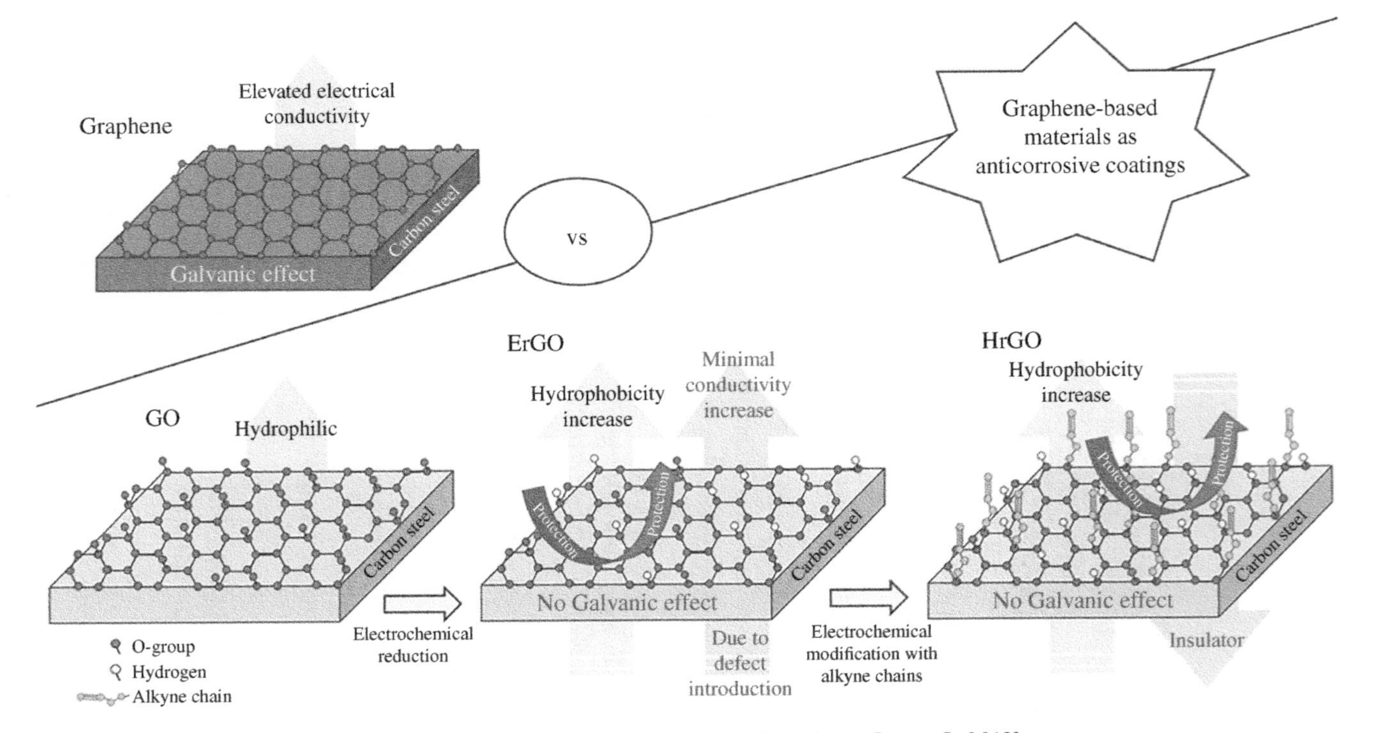

Figure 17.12 The anticorrosion mechanism of alkyne-chain-modified graphene layer. *Source:* Ref. [48].

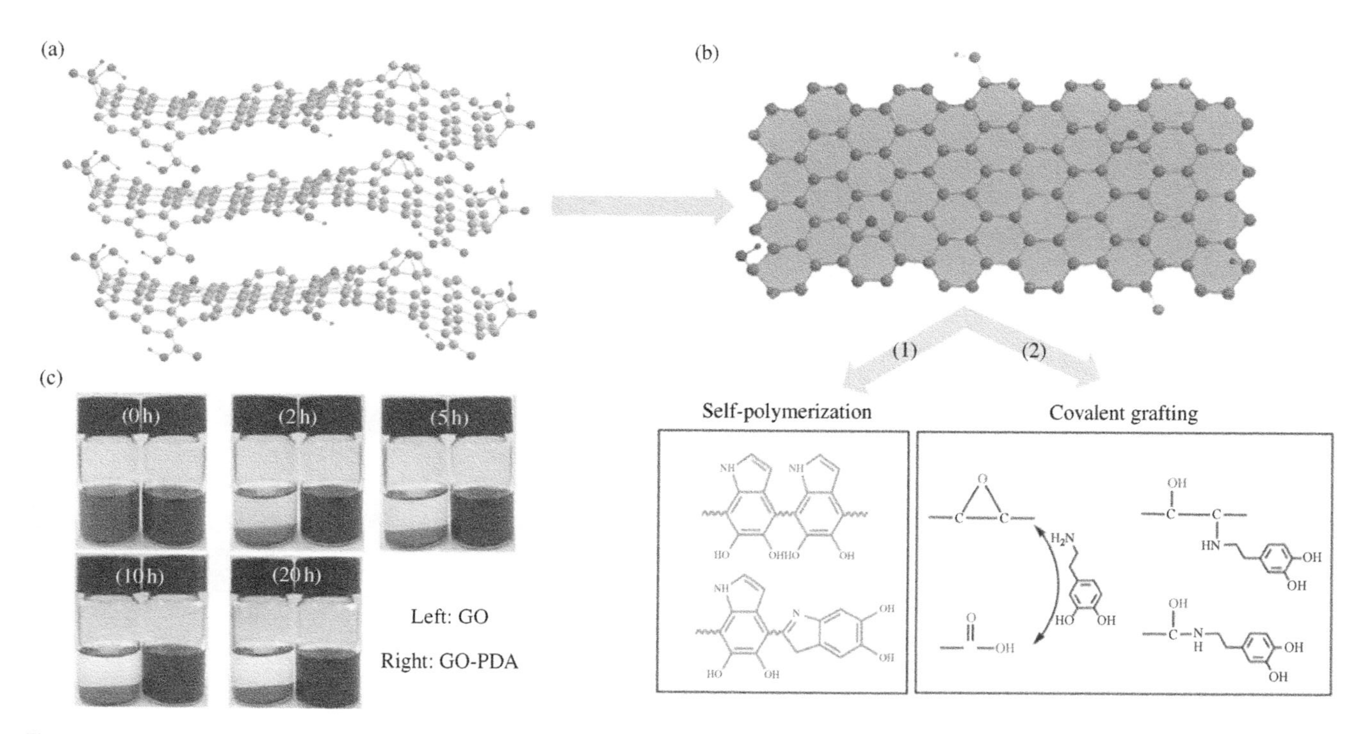

Figure 17.13 The possible reactions of the modification of graphene oxide with PDA. *Source:* Ref. [49].

EP due to the strong bonding between GO-PDA and EP matrix, reinforcing barrier properties originating from GO-PDA [49].

Shang et al. [50] investigate the relationship between the dispersion of graphene and anticorrosion of graphene composite coating. The reduced processes of graphene oxide induced by tea polyphenol (TP) are displayed in Figure 17.14. The good dispersion of graphene corresponds to better corrosion resistance as shown in Figure 17.15. TP-rGO exhibits better dispersibility than the rest kinds of graphite in dimethyl sulfoxide and H_2O. They also report that the better

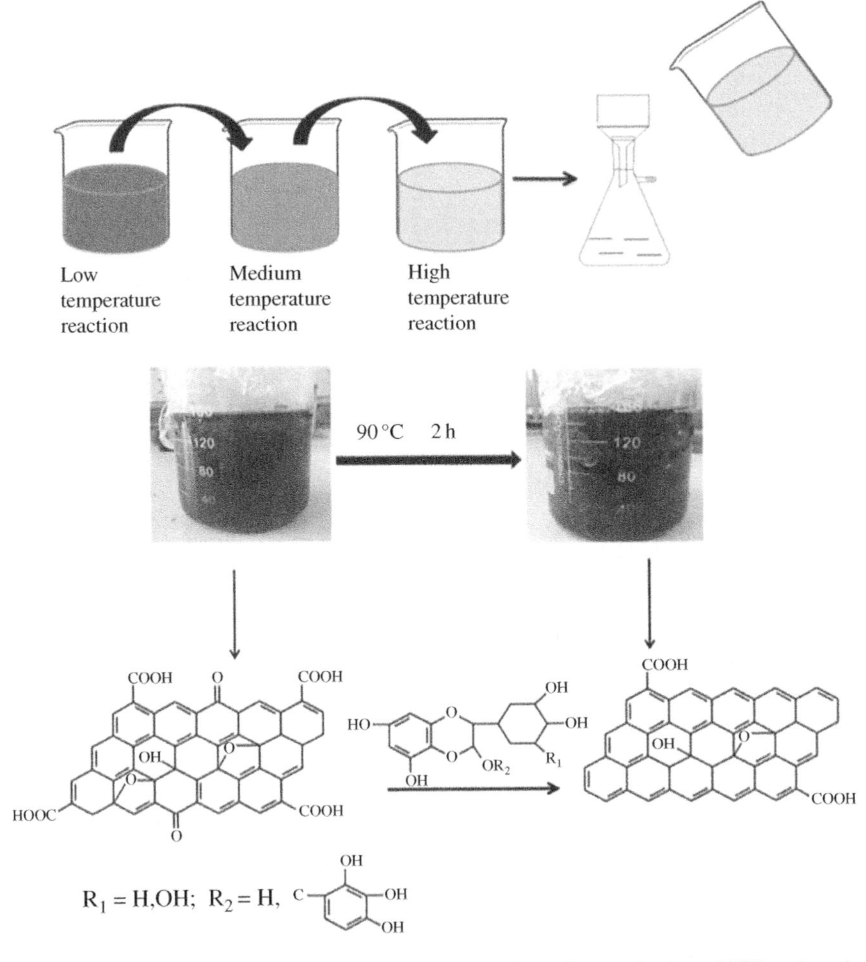

Figure 17.14 Schematic illustration of the preparation of tea polyphenol (TP)-reduced graphene. *Source:* Ref. [50].

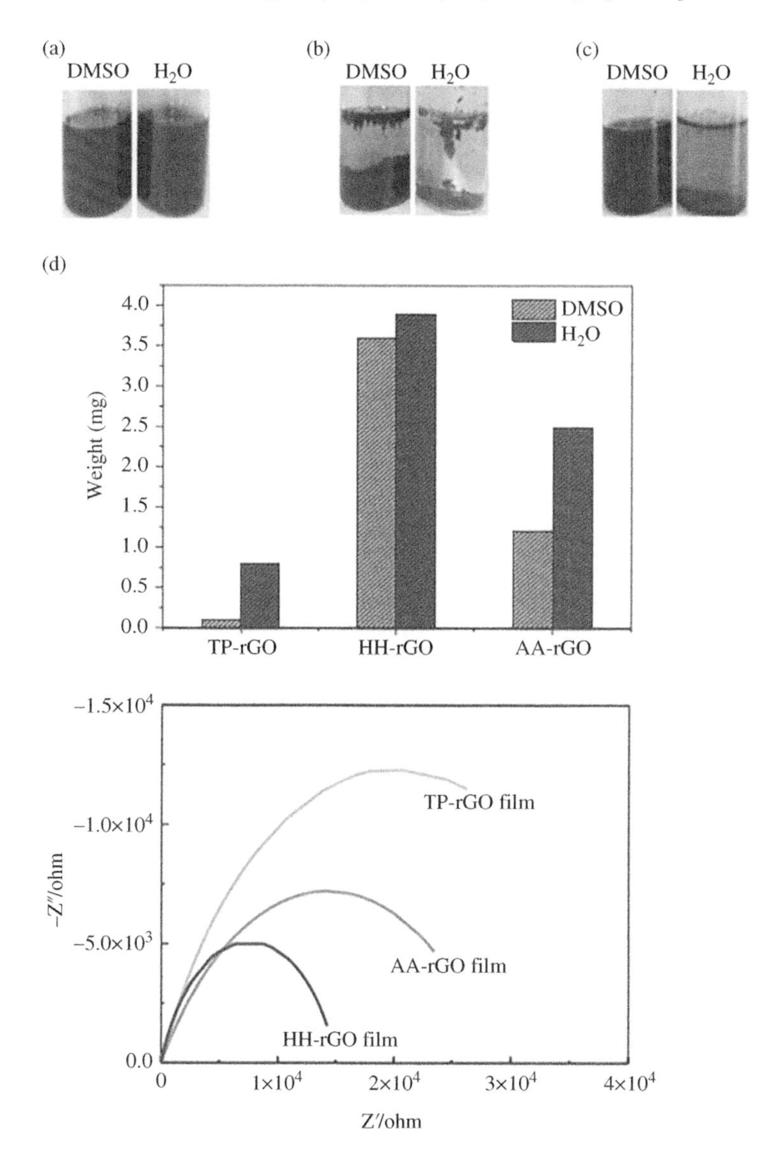

Figure 17.15 The dispersion of graphene and corresponding anticorrosion. *Source:* Ref. [50].

dispersibility of graphite is conducive to enhancing the anticorrosion of the metallic surface.

Chen et al. [51] report that graphic C_3N_4(g-C_3N_4) could enhance the dispersion of graphene in waterborne-epoxy coating (Figure 17.16) and improve the corrosion resistance (Figure 17.17). The results exhibit that the non-covalent π–π bonds are

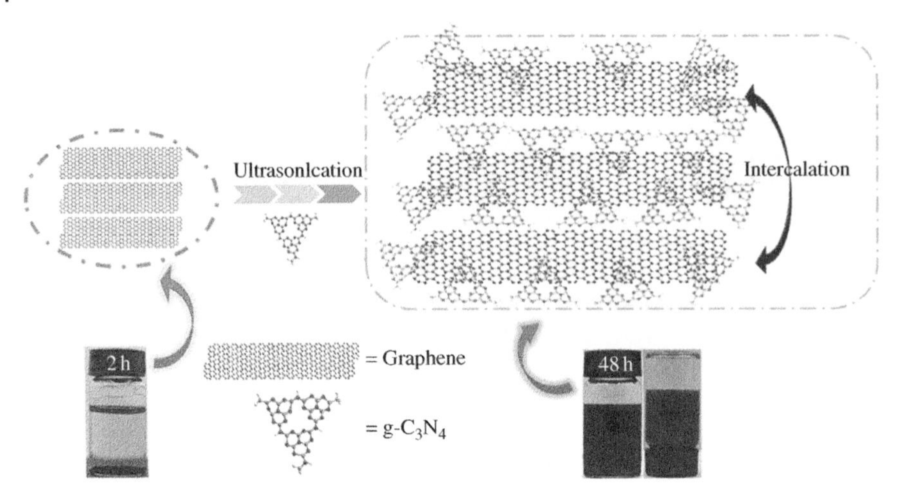

Figure 17.16 Dispersion of graphene in water using g-C_3N_4 as dispersant. *Source:* Ref. [51].

formed between g-C_3N_4 and graphene. Incorporation of g-C_3N_4or Gr/g-C_3N_4 hybrid can effectively improve the anticorrosion performance of epoxy and extend the failure time of the coatings. For comparison, the anticorrosion of Gr/g-C_3N_4/epoxy (0.5 %) is better than that of g-C_3N_4/epoxy (0.5 %) due to the superior barrier performance of Gr.

17.6 Challenges of Graphene

Since graphene is fabricated by Andre Geim and Konstantin Novoselov, 15 years past, only several graphene-based products have reached the market, such as the battery strap, the tennis, the phone touch screens, or the oil drilling mud. All the products show the initial market entry rather than the first full commercial wave of graphene products. In 2025, the current graphene market is estimated to be about US$20 million, showing that until now we are still locating at research and development. The demand of graphene continuously increases, and thus some key techniques should be done to increase production scale and decrease costs, leading to a shift from material sales to the market. In less than 20 years, graphene might become its commercial reality.

17.7 Conclusions and Future Perspectives

In this chapter, we focus on summarizing the latest progress on the synthesis and anticorrosion mechanism of graphene. N-doped graphene and graphene oxides nanosheets generate labyrinth effect to effectively inhibit corrosion. Additionally,

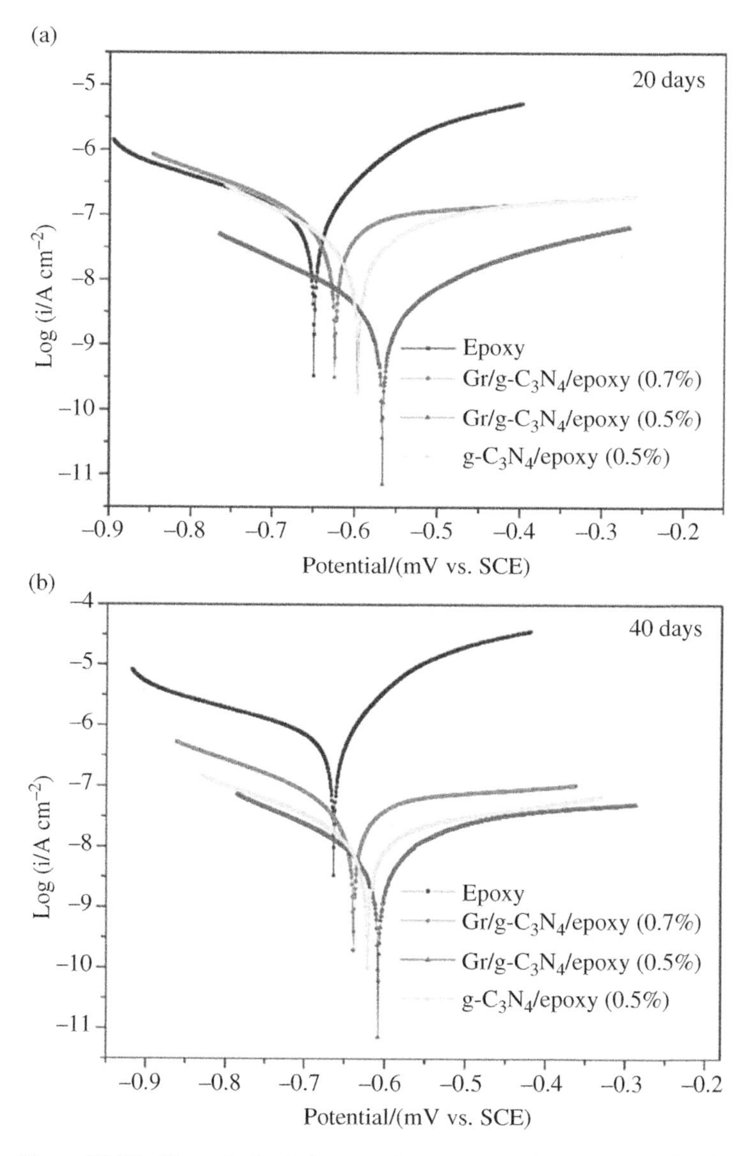

Figure 17.17 The polarization curve of nanocomposites coatings after 20 and 40 days immersion. *Source:* Ref. [51].

the properties of insulating nature and thermal barrier make N-doped graphene and graphene oxides more attractive than graphene in protective coatings. N-doped graphene and graphene oxides is a superior choice for potentially solving micro-galvanic corrosion caused by graphene in the long term.

Powerful progress has been made in the protective coatings. But, the 2D materials' development in coatings is just at the primary stage. We still make efforts to explore significant advantages of graphene or other 2D composite materials so as to decrease the corrosion rate of the metals effectively.

References

1 Das, S.R., Nian, Q., Saei, M. et al. (2015). Single-layer graphene as a barrier layer for intense UV laser-induced damages for silver nanowire network. *ACS Nano* 9: 11121–11133.

2 Novoselov, K.S., Geim, A.K., Morozov, S.V. et al. (2004). Electric field effect in atomically thin carbon films. *Science* 306: 666.

3 Novoselov, K.S., Geim, A.K., Morozov, S.V. et al. (2005). Two-dimensional gas of massless Dirac fermions in graphene. *Nature* 438: 197–200.

4 Zhang, Y., Tan, Y.-W., Stormer, H.L., and Kim, P. (2005). Experimental observation of the quantum Hall effect and Berry's phase in graphene. *Nature* 438: 201–204.

5 Higgins, D., Zamani, P., Yu, A., and Chen, Z. (2016). The application of graphene and its composites in oxygen reduction electrocatalysis: a perspective and review of recent progress. *Energy & Environmental Science* 9: 357–390.

6 Prasai, D., Tuberquia, J.C., Harl, R.R. et al. (2012). Graphene: corrosion-inhibiting coating. *ACS Nano* 6: 1102–1108.

7 Topsakal, M., Şahin, H., and Ciraci, S. (2012). Graphene coatings: an efficient protection from oxidation. *Physical Review B* 85: 155445.

8 Zhao, Y., Xie, Y., Hui, Y.Y. et al. (2013). Highly impermeable and transparent graphene as an ultra-thin protection barrier for Ag thin films. *Journal of Materials Chemistry C* 1: 4956–4961.

9 Li, H., Pu, J., and Zhang, R. (2019). Long-term corrosion protection of Q235 steel by graphene oxide composite coating. *Surface Topography: Metrology and Properties* 7: 045022.

10 Rajitha, K. and Mohana, K.N.S. (2020). Synthesis of graphene oxide-based nanofillers and their influence on the anticorrosion performance of epoxy coating in saline medium. *Diamond and Related Materials* 108: 107974.

11 Huang, H., Tian, Y., Xie, Y. et al. (2020). Modification of graphene oxide with acrylate phosphorus monomer via thiol-Michael addition click reaction to enhance the anti-corrosive performance of waterborne epoxy coatings. *Progress in Organic Coating* 146: 105724.

12 Sari, M.G., Abdolmaleki, M., Rostami, M., and Ramezanzadeh, B. (2020). Nanoclay dispersion and colloidal stability improvement in phenol novolac epoxy composite via graphene oxide for the achievement of superior corrosion protection performance. *Corrosion Science* 173: 108799.

13 Chen, P., Fang, F., Zhang, Z. et al. (2017). Self-assembled graphene film to enable highly conductive and corrosion resistant aluminum bipolar plates in fuel cells. *International Journal of Hydrogen Energy* 42: 12593–12600.

14 Kalisz, M., Grobelny, M., Zdrojek, M. et al. (2015). Determination of structural, mechanical and corrosion properties of titanium alloy surface covered by hybrid system based on graphene monolayer and silicon nitride thin films. *Thin Solid Films* 583: 212–220.

15 Xiao, Y.-K., Ji, W.-F., Chang, K.-S. et al. (2017). Sandwich-structured rGO/PVDF/PU multilayer coatings for anti-corrosion application. *RSC Advances* 7: 33829–33836.

16 Yi, M., Shen, Z., Zhao, X. et al. (2014). Exploring few-layer graphene and graphene oxide as fillers to enhance the oxygen-atom corrosion resistance of composites. *PCCP* 16: 11162–11167.

17 Schriver, M., Regan, W., Gannett, W.J. et al. (2013). Graphene as a long-term metal oxidation barrier: worse than nothing. *ACS Nano* 7: 5763–5768.

18 Zhou, F., Li, Z., Shenoy, G.J. et al. (2013). Enhanced room-temperature corrosion of copper in the presence of graphene. *ACS Nano* 7: 6939–6947.

19 Mahmoudi, M., Farhadian, M., Raeissi, K. et al. (2020). The role of graphene oxide interlayer on corrosion barrier and bioactive properties of electrophoretically deposited ZrO2–10 at. % SiO2 composite coating on 316 L stainless steel. *Materials Science and Engineering: C* 117: 111342.

20 Dehghani, A., Bahlakeh, G., and Ramezanzadeh, B. (2020). Synthesis of a non-hazardous/smart anti-corrosion nano-carrier based on beta-cyclodextrin-zinc acetylacetonate inclusion complex decorated graphene oxide (β-CD-ZnA-MGO). *Journal of Hazardous Materials* 398: 122962.

21 Ren, S., Cui, M., Li, W. et al. (2018). N-doping of graphene: toward long-term corrosion protection of Cu. *Journal of Materials Chemistry A* 6: 24136–24148.

22 Jiang, B.K., Chen, A.Y., Gu, J.F. et al. (2020). Corrosion resistance enhancement of magnesium alloy by N-doped graphene quantum dots and polymethyltrimethoxysilane composite coating. *Carbon* 157: 537–548.

23 Nine, M.J., Cole, M.A., Tran, D.N.H., and Losic, D. (2015). Graphene: a multipurpose material for protective coatings. *Journal of Materials Chemistry A* 3: 12580–12602.

24 András, G. (2018). A review on corrosion protection with single-layer, multilayer, and composites of graphene. *Corrosion Reviews* 36: 155–225.

25 Geim, A.K. and Novoselov, K.S. (2007). The rise of graphene. *Nature Materials* 6: 183–191.

26 Soldano, C., Mahmood, A., and Dujardin, E. (2010). Production, properties and potential of graphene. *Carbon* 48: 2127–2150.

27 Ding, R., Li, W., Wang, X. et al. (2018). A brief review of corrosion protective films and coatings based on graphene and graphene oxide. *Journal of Alloys and Compounds* 764: 1039–1055.

28 Brownson, D.A.C. and Banks, C.E. (2012). The electrochemistry of CVD graphene: progress and prospects. *PCCP.* 14: 8264–8281.
29 Reina, A., Jia, X., Ho, J. et al. (2009). Large area, few-layer graphene films on arbitrary substrates by chemical vapor deposition. *Nano Letters* 9: 30–35.
30 Manna, K., Hsieh, C.-Y., Lo, S.-C. et al. (2016). Graphene and graphene-analogue nanosheets produced by efficient water-assisted liquid exfoliation of layered materials. *Carbon* 105: 551–555.
31 Ciesielski, A. and Samorì, P. (2014). Graphene via sonication assisted liquid-phase exfoliation. *Chemical Society Reviews* 43: 381–398.
32 Treossi, E., Melucci, M., Liscio, A. et al. (2009). High-contrast visualization of graphene oxide on dye-sensitized glass, quartz, and silicon by fluorescence quenching. *Journal of the American Chemical Society* 131: 15576–15577.
33 Liscio, A., Veronese, G.P., Treossi, E. et al. (2011). Charge transport in graphene–polythiophene blends as studied by Kelvin Probe Force Microscopy and transistor characterization. *Journal of Materials Chemistry* 21: 2924–2931.
34 Torrisi, F., Hasan, T., Wu, W. et al. (2012). Inkjet-printed graphene electronics. *ACS Nano* 6: 2992–3006.
35 Coleman, J.N., Lotya, M., O'Neill, A. et al. (2011). Two-dimensional nanosheets produced by liquid exfoliation of layered materials. *Science* 331: 568.
36 Nicolosi, V., Chhowalla, M., Kanatzidis, M.G. et al. (2013). Liquid exfoliation of layered materials. *Science* 340: 1226419.
37 Stankovich, S., Dikin, D.A., Piner, R.D. et al. (2007). Synthesis of graphene-based nanosheets via chemical reduction of exfoliated graphite oxide. *Carbon* 45: 1558–1565.
38 Hou, P.M., Liu, C.B., Wang, X., and Zhao, H.C. (2019). Layer-by-layer self-assembled graphene oxide nanocontainers for active anticorrosion application. *International Journal of Electrochemical Science* 14: 3055–3069.
39 Sun, W., Wang, L., Wu, T. et al. (2015). Inhibited corrosion-promotion activity of graphene encapsulated in nanosized silicon oxide. *Journal of Materials Chemistry A* 3: 16843–16848.
40 Lee, J. and Berman, D. (2018). Inhibitor or promoter: insights on the corrosion evolution in a graphene protected surface. *Carbon* 126: 225–231.
41 Ren, S., Cui, M., Li, Q. et al. (2019). Barrier mechanism of nitrogen-doped graphene against atomic oxygen irradiation. *Applied Surface Science* 479: 669–678.
42 Jena, G., Anandkumar, B., Vanithakumari, S.C. et al. (2020). Graphene oxide-chitosan-silver composite coating on Cu-Ni alloy with enhanced anticorrosive and antibacterial properties suitable for marine applications. *Progress in Organic Coating* 139: 105444.
43 Nan, F., Liu, C., and Pu, J. (2019). Anticorrosive performance of waterborne epoxy coatings containing attapulgite/graphene nanocomposites. *Surface Topography: Metrology and Properties* 7: 024002.

44 Kasaeian, M., Ghasemi, E., Ramezanzadeh, B. et al. (2018). Construction of a highly effective self-repair corrosion-resistant epoxy composite through impregnation of 1H-Benzimidazole corrosion inhibitor modified graphene oxide nanosheets (GO-BIM). *Corrosion Science* 145: 119–134.

45 Hwang, M.-J., Kim, M.-G., Kim, S. et al. (2019). Cathodic electrophoretic deposition (EPD) of phenylenediamine-modified graphene oxide (GO) for anti-corrosion protection of metal surfaces. *Carbon* 142: 68–77.

46 Yuan, T.H., Zhang, Z.H., Li, J. et al. (2019). Corrosion protection of aluminum alloy by epoxy coatings containing polyaniline modified graphene additives. *Materials and Corrosion* 70: 1298–1305.

47 Cui, M., Ren, S., Pu, J. et al. (2019). Poly(o-phenylenediamine) modified graphene toward the reinforcement in corrosion protection of epoxy coatings. *Corrosion Science* 159: 108131.

48 Quezada-Renteria, J.A., Chazaro-Ruiz, L.F., and Rangel-Mendez, J.R. (2020). Poorly conductive electrochemically reduced graphene oxide films modified with alkyne chains to avoid the corrosion-promoting effect of graphene-based materials on carbon steel. *Carbon* 167: 512–522.

49 Cui, M., Ren, S., Zhao, H. et al. (2018). Polydopamine coated graphene oxide for anticorrosive reinforcement of water-borne epoxy coating. *Chemical Engineering Journal* 335: 255–266.

50 Shang, W., Li, J., Rabiei Baboukani, A. et al. (2020). Study on the relationship between graphene dispersion and corrosion resistance of graphene composite film. *Applied Surface Science* 511: 145518.

51 Chen, C., He, Y., Xiao, G. et al. (2020). Graphic C3N4-assisted dispersion of graphene to improve the corrosion resistance of waterborne epoxy coating. *Progress in Organic Coating* 139: 105448.

Part V

Organic Polymers as Corrosion Inhibitors

18

Natural Polymers as Corrosion Inhibitors

Marziya Rizvi

Corrosion Research Laboratory, Department of Mechanical Engineering, Faculty of Engineering, Duzce University, Duzce, Turkey

18.1 An Overview of Natural Polymers

Imagining polymers implies a long chain of repeating units that built up a lot of synthetic substances on daily basis such as plastics. Nature is a huge reservoir of polymers that are harmless and biodegradable (Figure 18.1).

Slightly modifying their chemical structure and applying them even in their crude form appropriately, their immense potential can be exploited. Natural polymers are most abundant in nature as they provide structural strength and contact fluids to the living world. Natural polymers have made their mark in various scientific and industrial processes ranging from being vessels for drug delivery, forming corrosion resistant films on metals, starting materials for fuel cells, forming active surfaces for heterogeneous catalysis, etc.

The lengthy chains of cyclic sugars connected via -O Bridge are diversely found in the living bodies of the organisms, as well as parts of other biological systems. As stated that they have a long chain of cyclic sugars, their molecular lengths and molecular weights are very high. Some very commonly known natural polymers are starch, cellulose, dextrins, chitin/chitosan, and gums. Some natural polymers like chitin can be derived from exoskeletons of arthropods, shells of mollusks, and crustaceans, sometimes even the scales of the fishes. Other major type of natural polymer, which is abundant in nature, is obtained from plants as plant polysaccharides like pectin, arabinogalactans, arabinoxylans, glucomannans, xyloglucans, etc. There are yet other types of natural polymers that are obtained as exudate from bark of the trees like acacia gum, garcinia gum, etc. Sometimes

Organic Corrosion Inhibitors: Synthesis, Characterization, Mechanism, and Applications,
First Edition. Edited by Chandrabhan Verma, Chaudhery Mustansar Hussain, and Eno E. Ebenso.

Natural polymers

Figure 18.1 Sources of naturally occurring natural polymers.

some microorganisms like xanthomonas work upon plant cellulose and convert it into useful substance like xanthan gum. The list of the potentially useful natural polymers is endless. The natural polymers never cease to amaze us as this eco-friendly substance can be crafted into reliable and superior material with a least chemical modification. They are quite inexpensive substances, pose no harm, and are easily accessible. These unmatched properties of this class of compounds has garnered the attention of the scientists all around the world from synthetic compounds to their eco-friendly alternatives.

Some corrosion scientists in the past decades have successfully applied these inexhaustible resources to prevent the metals from corroding in a much natural and humane way by using them as "inhibitors." Inhibitors are the substances that can be introduced to the environment, which is surrounding the metal and causing it to degrade. The role of the corrosion inhibitors is to keep the metals safe from the on-going changes in pH, temperature, composition, and dynamics of the environment. The inhibitors, which occupied the corrosion prevention studied a couple of decades ago, were toxic substances like chromates, arsenate sugars, etc. Not only these compounds were toxic for the workers who handled the industrial equipment where inhibitors were used, they were costly too, and because of their toxic nature, there are laws and regulations to their usage too. On contrary, the natural polymers are natural, nontoxic, biodegradable, inexhaustible/renewable, and easy accessible set of chemical compounds, which the nature offered us. These fascinating compounds inhibit the pure metallic oxidation by offering specific active adsorption sites for the functional molecules that may cause the metal surfaces to otherwise corrode if left to react with the aggressive ions in the corrosion cell. Sometimes the cyclic rings in the long chains trap the corrosion precursors preventing the metallic oxidation. The researchers carried out tremendous lengthy studies on each of the natural polymers in the last two decades. Let us explore some notable natural polymers with the potential to prevent the corrosion of metals.

18.2 Mucilage and Gums from Plants

Bark, stems, and sometimes leaves and seeds of some plants grant viscous exudates and mucilage having properties making them crucial components of adhesives, thickener, binders and stabilizers, micro-encapsulating components of drugs in medicine technology and also corrosion inhibitors of oil and gas industries. These are generally water soluble in nature, which makes them very useful. They tend to bear faint to highly pungent odors.

18.2.1 Guar Gum

Guar beans contain polysaccharides in their endosperm, which is commonly known as guar gum (Figure 18.2). This gum was used as inhibitor in 1 M H_2SO_4 preventing the rusting of carbon steel. The heterocyclic pyran groups in the molecule trapped the corrosion inducing molecules and comprehensively inhibit the corrosion by bonding chemically and physically with the surface moieties responsible for oxidizing the steel surface. The adsorption analysis reveals adsorption of monolayer, Langmuir isotherm being followed by the inhibition process; 1500 ppm of guar gum in 1 M H_2SO_4 inhibited 93.8% of corrosion according to the researchers who have reported this result [1].

18.2.2 Acacia Gum

The acacia trees exude a water-soluble complex mixture of oligosaccharide (Figure 18.3), polysaccharides, and glycoproteins, which is generally called gum Arabic or Acacia gum. The researchers have evaluated its efficiency for multiple substrates like aluminum and mild steel in unit molar of sulfuric acid environment. It effectively protects both the metals but acts more efficiently for preventing the oxidation of aluminum where it acts as physiosorbed inhibitor compared

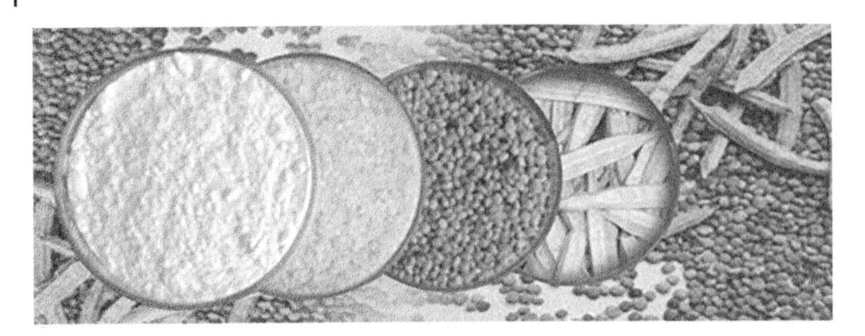

Figure 18.2 Appearance of guar gum.

Figure 18.3 Appearance of acacia gum exudate.

to mild steel where it chemisorbed on the surface. At highest temperature of evaluation, which was 60 °C in case of mild steel, 500 ppm of gum Arabic prevented 37.88% of corrosion. On aluminum substrate, 500 ppm of gum Arabic evaded almost 80% of corrosion at 30 °C [2]. Gum Arabic has also been evaluated for mild steel immersed in HCl by gravimetry, hydrogen evolution analysis, electrochemical analysis, Fourier transform infrared spectroscopy (FTIR), scanning electron microscopy (SEM), and X-ray photoelectron spectroscopic techniques (XPS) [3]. Gum Arabic displayed more synergism in hydrochloric acid compared to sulfuric acid when an external magnetic field is applied. This occurred probably due to changes in interaction mode and oxide layer formed by assistance of a field, which bears superior inhibition toward pitting on re-passivation. Apart from the modification of inhibition by applying a field, the researchers have attempted to synergize the reactions by adding chemicals like surfactants and halides. Thus, SDBS and CTAB were added along with gum Arabic in lesser aggressive solution of 0.1 M H_2SO_4 in a temperature range of 30–60 °C. The response of the metal to addition of surfactants to its environment in this way was inferred using gravimetric analysis, solution analysis, SEM, and atomic force microscopy (AFM) [4]. The surface morphologies displayed a clear increase in the protection offered by gum Arabic after the addition of surfactants. The adsorption studies inferred intra-molecular interaction upon adsorption as Freundlich adsorption isotherm was followed. The researcher had reported 83.36% efficiency for 1000 ppm gum Arabic, which further improved to 90.74% on adding just 1 ppm CTAB at 30 °C. These were some studies carried out with mineral acids in acidic environments. Let us talk about behavior of the same polymer in alkaline environment. The researchers have also conducted the study using gum Arabic as corrosion inhibitor in NAOH at 303 and 313 K using gravimetric studies and hydrogen evolution. Iodide ion was introduced as a synergizing agent to the aggressive solution. The maximum efficiency of 75 % was obtained, and the adsorption isotherm followed was Temkin' [5].

18.2.3 Xanthan Gum

The bacteria Xanthomonas campestris act on plant cellulose and change it to the xanthan gum. This natural polymer is well known to the bakeries and already used as a thickener in food industries. In the recent years, it has caught the attention of the corrosion researchers because of its unique molecular structure (Figure 18.4), water-solubility, inexpensiveness, and easy availability. It was observed that this compound was inhibiting the corrosion of aluminum in 1 M HCl [6]. At 40 °C, xanthan gum was 69.05% efficient in protecting aluminum from 1 M HCl. The adsorption process followed the Temkin and El-Awady adsorption isotherms. Potentiodynamic polarization suggested mixed-type inhibition and cathodic partial inhibitive reaction and the anodic dissolution reaction determined the corrosion rate of the system.

18.2.4 Ficus Gum/Fig Gum

Gum exudates from the various species of fig tree were commonly named as Ficus gum. The gum obtained from African rock fig was used as inhibitor for mild steel corrosion in H_2SO_4 medium [7]. Various corrosion tests have revealed that this gum was 65% efficient at 333 K. The inhibitor protected the metal by chemically adsorbing to the surface of the mild steel. Tannins, glucoproteins, and polysaccharides constitute this gum, which had ensured effective inhibition of corrosion of

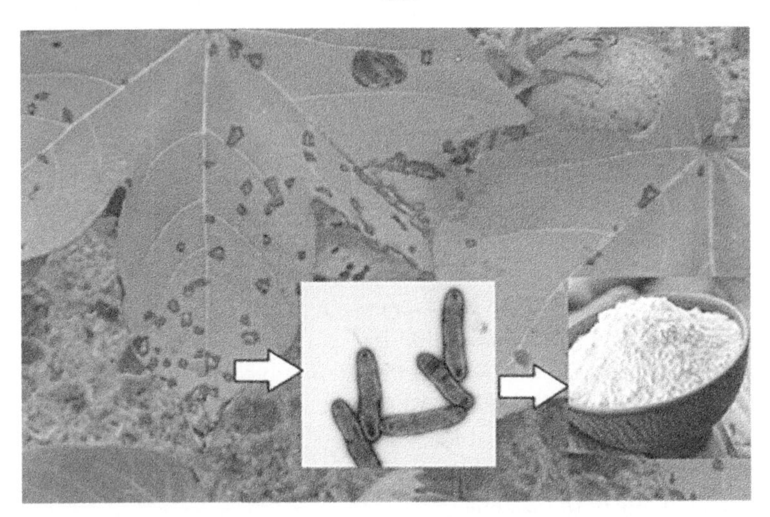

Figure 18.4 Molecular structure of xanthan gum.

surface of the mild steel. With so many components constituting a single gum, there is a chance of formation of multiple layers of adsorption. The researchers have reported Langmuir mode of adsorption where the adsorption is not only exothermic but spontaneous in nature as well as.

Eddy et al. [8] obtained Ficus gum from a different variant of this plant known as Ficus benjamina and used this as inhibitor for aluminum corrosion in 0.1M H_2SO_4 through gravimetric study. Like the previous study, this gum had multicomponent too, and hence formation of multiple layers of adsorption was proposed by the researchers. With multilayered adsorption, it was reported that the process followed Frumkin and Dubinin–Radushkevich adsorption models. A mere 0.5 g/l of this gum successfully inhibited 87% degradation of aluminum at 333K.

18.2.5 Daniella oliveri Gum

Gum derived from the African copaiba balsam tree which is scientifically named as Daniella oliveri. Collected by tapping the bark of this plant, this gum is commonly called hutch gum. This polysaccharide was used to inhibit the corrosion of mild steel in hydrochloric acid environment by the weight loss method and FTIR [9]. It was studied that this gum can inhibit 72.36% mild steel corrosion at a concentration of 0.5g/l at 303K. Endothermic and a spontaneous process was recorded, which supported the mechanism of physical adsorption. It was found that this inhibitor formed a monolayer of molecules on the metal surface following Langmuir adsorption.

18.2.6 Mucilage from Okra Pods

Sometimes a minor modification in the chemistry of the natural substances enhances the characteristics, as well as the properties, making the substance more useful. A research team modified polyacrylamide by grafting it with Okra mucilage, a natural vegetable polysaccharide. This modified mucilage was tested as inhibitor for mild steel corrosion in 0.5 M H_2SO_4 environment using weight loss and electrochemical investigation. A film of polymer was found to adsorb physically on metal substrate obeying the Langmuir adsorption isotherm. The inhibition kept on improving on adding this copolymer to the corrosion cell, and a maximum inhibition was obtained at 100 ppm at 25 °C when this copolymer inhibited almost 97% of corrosion. Thus, we may say that a little modification in the natural structures may render beneficial outcomes when we discuss the natural polymers [10].

18.2.7 Corn Polysaccharide

The polysaccharide from kernels of corn was modified and used as inhibitor of mild steel in 1 M HCl through gravimetric, surface, and electrochemical evaluations [11]. A comprehensive and mixed adsorption was observed following Langmuir adsorption isotherm with maximum efficiency of 91.26% at 313K. The researchers prefer conducting the binding studies on the adsorbed polymer through XPS, UV-visible spectroscopy or FTIR. In this study, the XPS spectra detected the bonded N, C atoms of the heterocycle and the O atom of the hydroxyl group on the metal surface.

18.2.8 Mimosa/Mangrove Tannins

Mangroves are the small shrubs growing in the tropical coastal areas. These interesting plants contain polyphenols called tannins, which impart characteristic colors and odors to these plants. These substances are antifungal in nature, as well as act as natural pesticides against small insects and pests, which may invade this plant. The researchers have incorporated these polyphenols as corrosion inhibitors Along with phosphoric acid, the action of these tannins was tested on pre-rusted steel in 3.5% sodium chloride solution. The efficiencies thus obtained were compared with the tannins from Mimosa [12]. At pH 0.5 and 2.0, inhibition increased but solely when mangrove and mimosa tannins were added, while at pH 5.5 the addition of phosphoric acid gave even higher efficiency of 79% at 30 °C.

18.2.9 Raphia Gum

Gum exudate from Raphia hookeri commonly called ivory coast Raphia palms were also considered potential corrosion inhibitors by some researchers. These gum exudates were used as corrosion inhibitors of aluminum in HCl solution. This gum yielded a moderate corrosion inhibition efficiency of 56.3% at concentration of 500 ppm at 30 °C [13]. The constituent phytochemicals in the exudates adsorbed on the surface of the aluminum metal obeying Temkin adsorption isotherm and kinetic–thermodynamic model of El-Awady et al.

18.2.10 Various Butter-Fruit Tree Gums

African subcontinent pears tree called bush pear and the scientific name of its fruit is Dacryodes edulis. The phytochemical content of this plant inspired the researchers to use it as a corrosion inhibitor of aluminum in 2 M hydrochloric acid environment [14]. It was observed that the inhibitor physiosorbed on the metals surface and yielded an efficiency of 42%. Pachylobus edulis is very closely related to Dacryodes edulis, or maybe it is the same plant. The literature does not have much to say about the similarity of these plants, but both of them are commonly called bush pears or the butter fruits. This bush pear gum was also tested for mild steel in 2 M H_2SO_4 synergized by potassium halides using hydrogen evolution and thermometric methods in the range of 30–60 °C [15]. Synergistic effects increased the inhibition efficiency of the exudates in the presence of potassium halides in the order KI > KBr > KCl. The adsorption of the exudates gum alone and in combination with the potassium halides was approximated by the Temkin adsorption isotherm.

18.2.11 Astragalus/Tragacanth Gum

Astragalus and its various species, which are a set of Middle Eastern legumes, naturally render a ribbon-like dried sap, which is commonly called "goat's thorn" (Greek: tragacanth – goat thorn). The gum refined from this dried sap is called Shiraz gum, elect gum, or dragon gum. Currently, Iran is the largest producer of this gum, which is used for medicines and herbal remedies. It is odorless, viscous, tasteless, yet water-soluble compound, which may have single or multiple polysaccharides. The major part is water-soluble called tragacanthin and the minor part is gel, which swells on proximity of water called bassorin. This substance is generally rained from roots of the plants and dried for usage. The major water-soluble part may be precipitated out, which is mostly comprised of the polysaccharide arabinogalactans. When tested as a corrosion inhibitor of low carbon steel in 1 M HCl, it gives a corrosion inhibition efficiency of 96.35% at 500 ppm concentration [16]. The researchers have studied it extensively using computational methods to judge the exact mechanism of its adsorption. The gravimetric, electrochemical, and the binding studies in collaboration with the computational studies suggest that this polymer with its long chain covers the metal substrate as a monolayer obeying Langmuir adsorption isotherm forming a protective film on the surface of steel. The macromolecular long chain structure of the polysaccharide along with the structure comprising of abundant heteroatom O and functional groups are responsible for efficiently protecting the metallic surface against acidic attack.

18.2.12 Plantago Gum

The species of Plantago, which produce mucilaginous seed coat, are called psyllium. This psyllium is commonly used dietary fiber in many parts of the world. The mucilage extracted from the seed coat is comprised of repeating unit of arabinose and xylose, which are together called as arabixylans. These arabinoxylans are separated from the plant and used as corrosion inhibitor by corrosion researchers. The abundance of this natural polymer, inexpensiveness, and ease of availability prompted the researchers to use this compound as a corrosion inhibitor of low carbon steel in 1 M HCl. At a concentration of 1 g/l in 1 M HCl, it can efficiently inhibit 94% of low carbon steel corrosion [17]. The adsorption of this polysaccharide obeyed Langmuir adsorption isotherm with a comprehensive mode of adsorption involving both physiosorption and chemisorption. But chemisorption was found to play a predominant role in adsorption process.

18.2.13 Cellulose and Its Modifications

A molecule of cellulose consists of innumerable, many a times thousands of C, H, and O atoms (Figure 18.5). It is responsible for imparting the stiffness to the

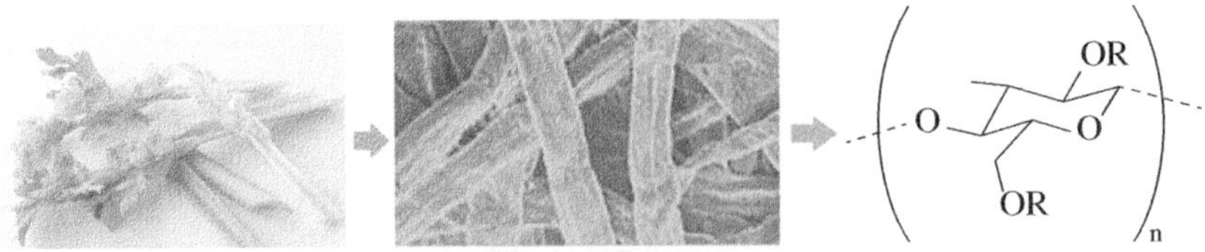

(× 50 magnification)

Figure 18.5 Basic molecular structure of cellulose.

plants. Nondigestible by humans, still it forms an important dietary fiber for us. It is found in many forms and has many variants. For example, most commonly used is carboxymethyl cellulose or CMC.

18.2.13.1 Carboxymethyl Cellulose

The behavior of CMC on mild steel corrosion in 2 M sulfuric acid environment using gravimetric study, hydrogen evolution, and thermometric methods was observed. The adsorption process of CMC obeyed Langmuir and Dubinin–Radushkevich isotherm models. The adsorption studies strongly suggest that CMC physiosorbed on mild steel surface; 500 ppm in corrosive sulfuric acid medium yielded an efficiency of 64.8% at 30 °C [18].

18.2.13.2 Sodium Carboxymethyl Cellulose

With a little further modification to CMC, sodium carboxymethyl cellulose or Na-CMC was used as a corrosion inhibitor of mild steel in 1M HCl solution [19]. The Na-CMC adsorbed on the metal by bridging via the hydroxyl groups and obeyed Langmuir adsorption isotherm. At a weight concentration of 0.04%, an efficiency of 72% was achieved at 298K.

18.2.13.3 Hydroxyethyl Cellulose

Another notable derivative was hydroxyethyl cellulose. Apart from general uses as a thickening, binding and gelling/stabilizing agent hydroxyethyl cellulose or HEC is also employed medicinally for dissolving the drugs in gastrointestinal fluids. Structurally it is similar to CMC except for the presence of the hydroxyethyl group in place of carboxymethyl in the same position. The successful utility of this particular cellulose as a corrosion inhibitor in numerous media is drawn from its uniquely placed functional moieties (-OH, -COOH) on its cellulose backbone, as well as a macro-sized chain molecule, which ensures greater coverage of the metal surface, barring the degrading ions. This interesting variant of cellulose was not only used to protect the metal in neutral saline environment [20], but it was also efficiently utilized as corrosion inhibitor of zinc carbon battery [21]. In both the cases, HEC was more than 90% efficient as an inhibitor.

18.2.13.4 Hydroxypropyl Cellulose

Yet another variant of cellulose, the hydroxypropyl cellulose is very well known to corrosion researchers studying the natural polymers. It was used for the acidic corrosion prevention of cast iron. The investigations were performed by gravimetric and electrochemical testing [22]. Another noteworthy addition to the research work was addition of potassium iodide to the system, which acted antagonistically and synergistically in a simultaneous fashion. Electrochemistry suggested a mixed type of adsorption, and the adsorption analysis suggest its obedience toward Langmuir isotherm. The thermochemical parameters suggested physisorbed film of inhibitors on the metal surface. Cast iron was 89.5% saved by 500 ppm hydroxypropyl cellulose at 298 K.

18.2.13.5 Hydroxypropyl Methyl Cellulose

Another derivative of cellulose, which could be effectively used as corrosion inhibitor of mild steel in sulfuric acid environment, was hydroxypropyl methyl cellulose or HPMC. Investigation techniques comprised of weight analysis, impedance, and polarization calculations [23]. Potassium iodide was added to enhance the corrosion inhibition efficiency. The electrochemical results suggested that HPMS inhibited both cathodic and anodic partial reactions. Quantum chemical descriptors indicate effective adsorption of molecule on the metal surface. The adsorption process follows frendlich adsorption isotherm.

18.2.13.6 Ethyl Hydroxyethyl Cellulose or EHEC

This cellulose derivative was effectively applied as acid corrosion inhibition of mild steel corrosion in 1M H_2SO_4 solution using weight loss calculations, EIS, PDP, and quantum chemical calculation techniques [24]. The IE increased with EHEC concentration and further on addition of KI. The effect of EHEC on corrosion of mild steel was attributed to general adsorption of both protonated and molecular species of the additive on the cathodic and anodic sites. Formation of a chemisorbed film on the mild steel surface was observed. Electrochemical tests showed that EHEC and EHEC+KI were mixed-type inhibitor with predominant cathodic effect.

18.2.14 Starch and Its Derivatives

Starch is a natural macromolecule with multiple glucose units that are connected by glycoside bonds. Normally, starch consists of varying percentages by weight of amylase. When linear and helical the chain is amylose and when branched the chain is amylopectin (Figure 18.6a and b). Starch is the major source of energy for higher animals as it is derived from plant source. They are sometimes processed as simple sugars, thickeners, and glues for general use. Starches have electron rich hydroxyl groups, which coordinate and complete the voids in orbitals of metal substrates. Many researches have used starch as corrosion inhibitors. Among

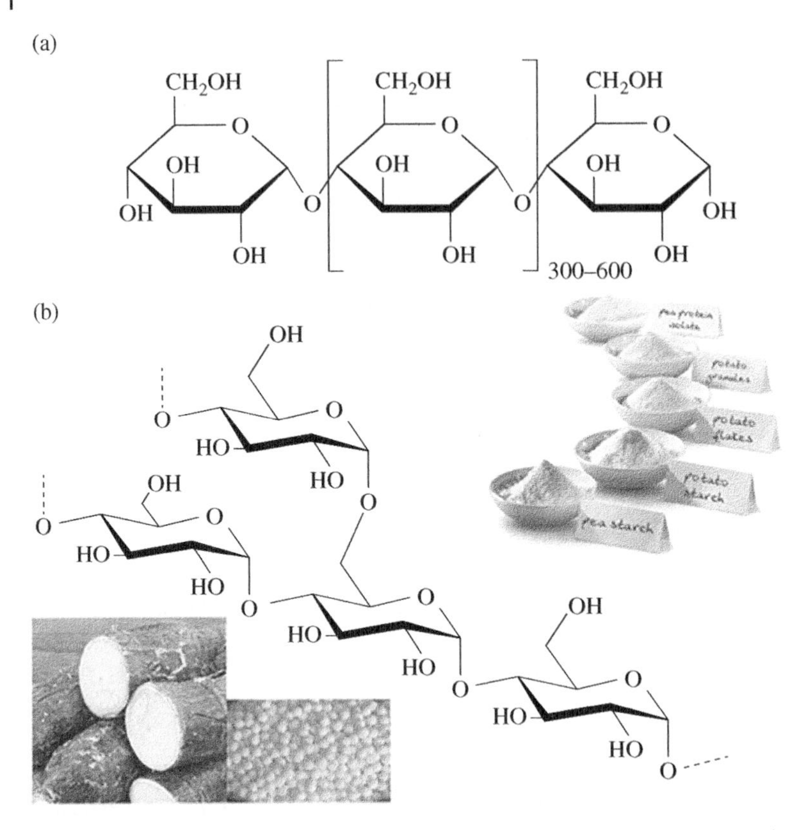

Figure 18.6 Molecular structures of the amylose (a) and amylopectin (b) molecules of starch.

them some notable study was conducted by Mobin et al. [25], using starch for corrosion inhibition of mild steel in H_2SO_4 by using gravimetric and electrochemical measurements. It was observed to inhibit 67% corrosion in 0.1 M HCl at 200 ppm concentration. With synergistic influence of surfactants, its efficiency increased. Similar kind of extensive studies were conducted by other researchers too where modified starches, tapioca starch, and cassava starch were used as corrosion inhibitors [26–29]. In almost all the cases, starch was found to be cathodic inhibitor suppressing the cathodic reactions and yielding high efficiency of protection.

18.2.15 Pectin

Pectins are heteropolysaccharides, which are abundant in cell walls of terrestrial non-woody vegetation. They are easily available commercially as powder or granules as they are already an integral part of food industry. It may be majorly obtained from citrus fruits and apples. The carboxyl and the carboxymethyl groups on its backbone increase its functionality not only as a corrosion inhibitor but also as a scale remover (Figure 18.7).

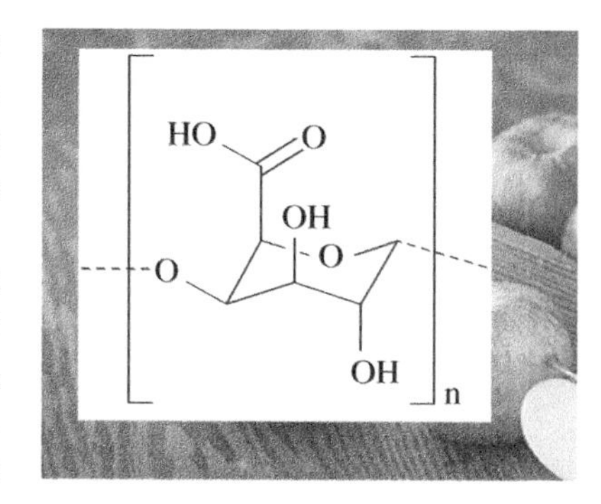

Figure 18.7 Molecular structure of pectate.

Umoren et al. [30] used pectin to protect the X60 pipeline steel from the attack of hydrochloride. It was 98% efficient at 60 °C at a concentration of 1000 ppm. Fares et al. [31] used pectin from citrus fruits as corrosion inhibitor of aluminum in HCl. But the concentration reported for highest obtained IE was very high, i.e. at 8 g/l it might inhibit 91% corrosion. Some researchers have also used lemon peel pectin for mild steel in 1 M HCl [32]. It proved itself to be a mixed-type inhibitor, geometrically blocking the attack of hydrochloride by chemisorbing on the surface. The inhibition efficiency of 2000 ppm pectin at 298 K was observed to be 90.3%, which further increased to 94.2% at 318 K. Grassino et al. studied the effect of tomato pectin [33], while others studied pectin from Opuntia [34] as corrosion inhibitor yielding considerably good efficiencies. Pectates obtained solely from very ripe fruits are already in industrial applications as emulsifying and foaming agents for food and medical. In 2012, a researcher reported corrosion inhibition of pectates for aluminum in 4 M NaOH using gasometric and weight loss methods [35]; 88% of corrosion was evaded when 1.6% pectates were added to the corrosion system. The same researcher reconducted his work this time using sodium pectate for pure aluminum substrate in the same 4 M NaOH. The computed results showed very negligible difference in the results [36].

18.2.16 Chitosan

Chitosan is obtained either from the exoskeletons of marine crustaceans and mollusks or by N-deacetylation of fungal cell-wall chitin. Possessing antibacterial and antifungal properties, it is widely used in cosmetics and skin therapeutics. Chitosan's anticorrosion ability could be drawn from its molecular structure (Figure 18.8), which bears the electron-rich hydroxyl and amino groups in the structure of its molecules capable of bonding to steel surface via coordinate bonds.

Umoren et al. [37] have reported the application of chitosan as a corrosion inhibitor for mild steel in 0.1M HCl. The chemisorption of this polymer was found to accord with Langmuir adsorption isotherm. Applying this polymer to the corrosion system could protect up to 96% of the metal at 60 °C and then drops further increasing the temperature. Increasing the chitosan concentration to

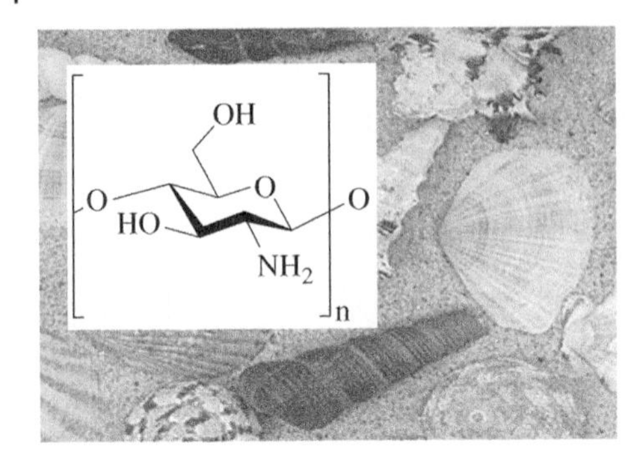

Figure 18.8 Molecular structure of chitosan.

4 μM slightly elevated the IE again. At the same time, El Haddad [38] studied the application of chitosan to protect Cu in 0.5M HCl acid. The weight loss and electrochemical measurement resulted in establishment of chitosan as mixed kind of inhibitor with the maximum efficiency of 93% at 25 °C when 8 μM of inhibitor was added. The quantum computations show that the N and O atoms in the chitosan molecule are the two main active sites that cause it to adsorption on the Cu surface. Cheng et al. [38] have modified this simple polymer as carboxymethyl chitosan-Cu2+(CMCT-Cu2+) mixture for inhibiting mild steel corrosion in 1M HCl and studied the process using gravimetric and electrochemical methods. On addition of CMCT and the mixture of Cu2+ + CMCT, the corrosion effect was controlled. A complex formation occurred between Cu2+ ion and CMCT, which as an inhibitor was much more effective (91.9% IE) when compared to its building constituents. β-Cyclodextrin-modified natural chitosan was used by Liu et al. [39] to inhibit carbon steel corrosion in 0.5M HCl solution. β-CD-chitosan (β- cyclodextrin-modified chitosan) acted as a mixed inhibitor with a maximum inhibition of 96.02% at a small concentration of 230 ppm [40]. Sangeetha et al. [38] synthesized O-fumaryl-chitosan for inhibition of mild steel corrosion in 1M HCl. A corrosion inhibition efficiency of 93.2% was observed at room temperature on adding 500 ppm of inhibitor. There are many other studies where chitosan derivatives were effectively employed as corrosion inhibitors [41–43].

18.2.17 Carrageenan

Carrageenan are a group of gelatinous linear polysaccharides having sulfated β-d-galactose and anhydrous-α-d-galactose backbone (Figure 18.9).

The most common source of carrageenan is seaweeds. Bearing a flexible molecular symmetry and building up unstable helixes makes them gels at room temperature. This gelling capacity makes them suitable as food thickeners and stabilizers. Fares et al. [44] have reported the usage of i-carrageenan for inhibition of aluminum sheets in HCl medium at different concentrations. When pefloxacin mesylate was added as a mediator, it caused an improvement in the magnitude of IE, increasing it from 66.7 to 91.8%. SEM helped in

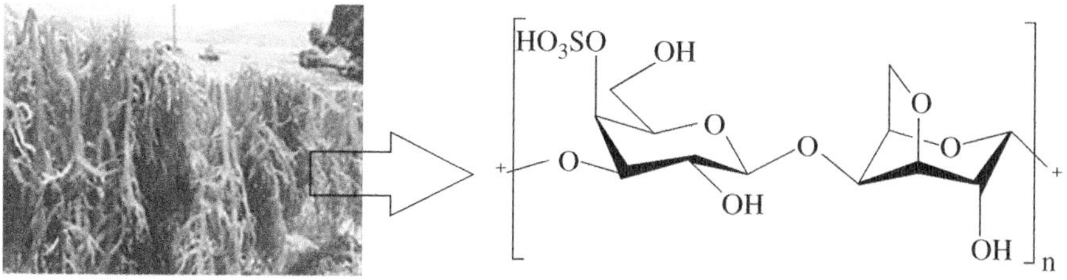

Figure 18.9 Molecular structure of carrageenan.

detection of the inhibitor mediator film on the surface. Zaafarany et al. [45] have also studied all the three available variants of carrageenan, with i, k, λ-carrageenan, as the corrosion inhibitors of low carbon steel in 1M HCl. All these variants were established as being anodic-type inhibitors for steel in the acidic medium studied; 500 ppm concentration of carrageenan displayed efficiencies of 76, 72 and 80%, respectively.

18.2.18 Dextrins

Dextrins are natural polymers having glucose (D) units linked by glycosidic bonds [α-(1→4) or α-(1→6)] shown in Figure 18.10.

They are present in human digestive system as the hydrolysis product of starch upon action of amylases. An alternate way to produce them is synthesis by heat treatment in acidic solution. Numerous dextrins are naturally present like α, β-dextrin, maltodextrin, amylodextrincyclic, and highly branched cyclic dextrin compounds. Researchers have reported dextrin as inhibitor of zinc-plated mild steel in HCl using weight loss and surface morphology tests [46]. In combination with thiourea additive, it has demonstrated improved efficiency protection as examined by SEM/EDX. Many other researches have been performed to evaluate the use of cyclodextrins for different metals and medium. The cyclodextrins have imparted high efficiencies in all cases, thus proving how successfully they inhibit the degradation of metals due to their unique structures and properties [47–50].

18.2.19 Alginates

Alginates are the sugars that have carboxylic acid functional group attached to their molecular structure. Also known as "algins" or "alginic acid," these are linear copolymers having a covalently bonded (1-4)-linked β- d-mannuronate and C-5 epimer β-lguluronate homopolymeric blocks shown in Figure 18.11.

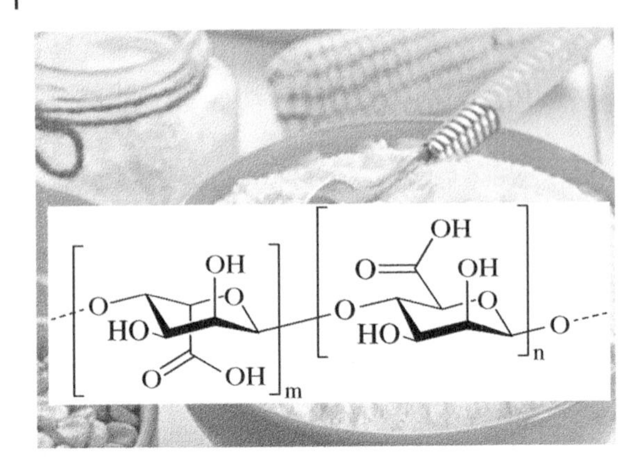

Figure 18.10 Molecular structure of dextrin.

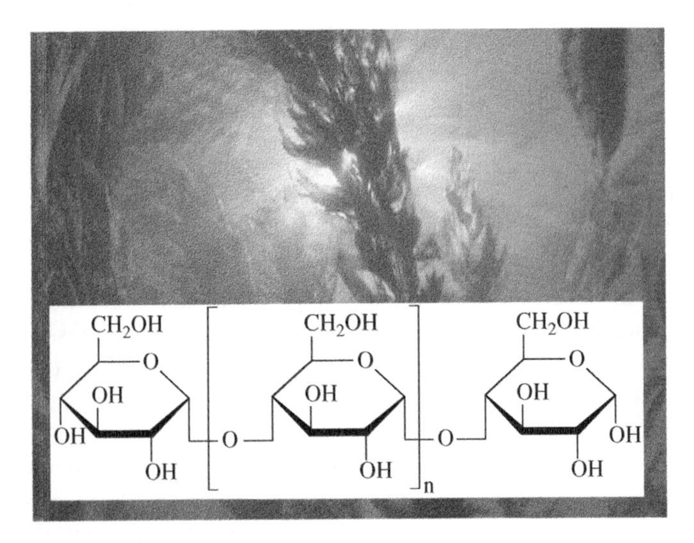

Figure 18.11 Molecular structure of alginate.

This anionic polysaccharide is the main constituent of the hydrocolloids algal cell walls and seaweeds where it attaches with molecular water. Conventional corrosion testing along with surface analysis evaluated the potential of these alginate derivatives as inhibitors for carbon steel rusting in acidic solutions [51]. The results suggested an improvement in efficiency of these inhibitors on the elevation of the solution temperature or their own concentration in the test solution. The electrochemical polarization suggested a mixed-type inhibition with predominant cathodic control. Their adsorption was approximated by Langmuir adsorption isotherm model. It is worthy to mention one of its recently studied derivatives hydroxyl propyl alginate as an inhibitor of mild steel corrosion in 1M HCl at room temperature using chemical and electrochemical techniques. Corrosion inhibition efficiency was observed to improve with the increasing concentration of this compound. A physi-osorbed layer of molecules was observed on the metal surface, which was further assured and confirmed by the results of SEM, AFM, and FTIR [52].

18.3 The Future and Application of Natural Polymers in Corrosion Inhibition Studies

Previous researches clearly indicated that a little modification in the structures of the natural polymers lead to obtaining very high inhibition performances in very aggressive environments having a high molar mineral acid. The only minor short-coming associated with these natural polymers as corrosion inhibitors is that few of them are partially soluble and have a lesser stability when kept for days, i.e. after a set period of time they may not remain functional in highly aggressive solution of 1 or 2 molar mineral acids. The chemical modifications should be aimed to increase the solubility of the natural polymers and increase their stability at high tempera-tures to reduce further the amount and concentration of inhibitor which is required. Considering the industrial and practical applications of corrosion inhibitors, addi-tion of polymers at concentrations ranging from 500 to 1000 ppm, repeatedly, may make the solutions turbid or they may even pose a risk to the efficiency of equip-ment. So the researches that are trending in currently have started application of modified natural polymers, which result in efficient protection of the metal surface at extremely less concentrations. The computational analysis studied the reaction pathways and adsorption mechanisms of these natural polymers when added as inhibitors. Sometimes a single molecule has more than one type of functionality attributed to it. Theoretical calculations are necessary to deduce all the possible points of bonding and the energies involved in the adsorption process. Not only the bonding sites and bonding energies are relevant to the corrosion studies another very important aspect is the spatial orientation of the natural polymer molecules on the metal surfaces. It is worth mentioning that these polymers have a lengthy molecular chains and high molecular weights, which researchers have often consid-ered as one aspect behind the successful application of natural polymers as corro-sion inhibitors. It is very important to study how these molecules align themselves on the metal substrate while being adsorbed on it. Some of the extensively studied natural polymers used as corrosion inhibitors are enlisted in Table 18.1.

The basic trend of using gravimetric, electrochemical, and surface studies pro-vide a clear view of binding occurring on the surface of the metal substrate; how-ever, some of the inhibition may be attributed to the passivation of the metal. Sometimes a researcher keeps on adding the inhibitor to the solution and achieves

Table 18.1 Some natural polymers successfully applied to various corrosion systems.

Natural polymers	Substrate	Medium	Inhibitors' conc (ppm)	Temp. (°C)	Inhibition efficiency IE (%)	References
Guar gum	Carbon steel	1M H_2SO_4 containing NaCl	1500	25	93.88	[1]
Mangrove tannin	Mild steel	3.5% NaCl	3000	25	90	[12]
Tapioca starch	AA6061 alloy	Seawater	1000	25	96	[28]
Iota carrageenan	Aluminum	1 M, 1.5 M, 2 M HCl	1600	40	74.2	[44]
Gum acacia	Mild steel	1 M H_2SO_4	1500	30	91.71	[4]
Pectin	Carbon steel	1 M HCl	2000	45	94.2	[15]
Hydroxypropyl methylcellulose	Aluminum	0.5 M H_2SO_4	2000	30	63.5	[23]
Xanthan gum	A1020 carbon steel	1 M HCl	1000	30	83.17	[53]
Schinopsis lorentzii extract	low carbon steel	1M HCl	2000	29	66	[54]

high values of resistance from the electrochemical instruments, but when the sample surface is studied for the bonded inhibitor, the surface is devoid of any inhibitor molecules, only the passive layer constituents are detected. Referring Pourbaix diagrams do help in setting up a demarcation between the passivation of the surface and inhibition by polymer, but it would be best if all the changes are measured in real time by some precise and accurate electrochemical techniques. Darowicki was the first researcher to successfully apply a real-time corrosion monitoring method in electrochemical analysis [55]. Later many researchers started to apply it in various electro analytical studies. Gerengi and his research group are some of the earliest researchers to work on using a technique that can monitor the corrosion processes occurring in the electrochemical cell in real time [56]. This accurate and precise technique, which is called dynamic electrochemical impedance spectroscopy (DEIS), can detect the changes occurring on the metal sample continuously in real time for a set period of time. With multifunctional structures and lengthy chain, natural polymers can be well studied and evaluated by combined DEIS and theoretical investigations to judge the exact process of inhibition occurring on the metal surfaces.

References

1 Abdallah, M. (2004). Guar gum as corrosion inhibitor for carbon steel in sulfuric acid solutions. *Port. Electrochim. Acta* 22: 161–175.

2 Umoren, S.A. (2008). Inhibition of aluminium and mild steel corrosion in acidic medium using Gum Arabic. *Cellulose* 15: 751–761.

3 Abu-Dalo, M.A., Othman, A.A., and Al-Rawashdeh, N.A.F. (2012). Exudate gum from acacia trees as green corrosion inhibitor for mild steel in acidic media. *Int. J. Electrochem. Sci.* 7: 9303–9324.

4 Mobin, M. and Khan, M.A. (2013). Investigation on the adsorption and corrosion inhibition behaviour of gum acacia and synergistic surfactant additives on mild steel in 1 M H2SO4. *J. Dispers. Sci. Technol.* 34: 1496–1506.

5 Umoren, S.A. (2009). Synergistic Influence of gum arabic and iodide ion on the corrosion inhibition of aluminium in alkaline medium. *Port. Electrochim. Acta* 27: 565–577.

6 Arukalam, I.O., Alaohuru, C.O., Ugbo, C.O. et al. (2014). Effect of Xanthan gum on the Corrosion Protection of Aluminium in HCl medium. *Int. J. Adv. Res. Technol.* 3: 5–15.

7 Ameh, P.O., Magaji, L., and Salihu, T. (2012). Corrosion inhibition and adsorption behaviour for mild steel by Ficusglumosa gum in H2SO4 solution, Afr. *J. Pure Appl. Chem.* 6: 100–106.

8 Eddy, N.O., Ameh, P.O., and Odiongenyi, A.O. (2014). Physicochemical characterization and corrosion inhibition potential of Ficusbenjamina (FB) gum for aluminum in 0.1 M H2SO4. *Port. Electrochim. Acta* 32: 183–197.

9 Eddy, N.O., Odiongenyi, A.O., Ameh, P.O., and Ebenso, E.E. (2012). Corrosion inhibition potential of Daniellaoliverri gum exudate for mild steel in acidic medium. *Int. J. Electrochem. Sci.* 7: 7425–7439.

10 Banerjee, S., Srivastava, V., and Singh, M.M. (2012). Chemically modified natural polysaccharide as green corrosion inhibitor for mild steel in acidic medium. *Corros. Sci.* 59: 35–41.

11 Zhang, H., Wang, D., Wang, F. et al. (2015). Corrosion inhibition of mild steel in hydrochloric acid solution by quaternary ammonium salt derivatives of corn stalk polysaccharide. *Desalination* 372: 57–66.

12 Rahim, A.A., Rocca, E., Steinmetz, E.J., and Kassim, M.J. (2008). Inhibitive action of mangrove tannins and phosphoric acid on pre-rusted steel via electrochemical methods. *Corros. Sci.* 50: 1546–1550.

13 Umoren, S.A., Obot, I.B., Ebenso, E.E., and Obi-Egbedi, N.O. (2009). The Inhibition of aluminium corrosion in hydrochloric acid solution by exudate gum from Raphiahookeri. *Desalination* 247: 561–572.

14 Umoren, S.A., Obot, I.B., Ebenso, E.E., and Obi-Egbedi, N. (2008). Studies on the inhibitive effect of exudate gum from Dacroydesedulis on the acid corrosion of aluminium. *Port. Electrochim. Acta* 26: 199–209.

15 Umoren, S.A. and Ekanem, U.F. (2010). Inhibition of mild steel corrosion in H2SO4 using exudate gum from Pachylobusedulis and synergistic potassium halide additives. *Chem. Eng. Commun.* 197: 1339–1356.
16 Mobin, M., Rizvi, M., Olasunkanmi, L.O., and Ebenso, E.E. (2017). Biopolymer from Tragacanth gum as a green corrosion inhibitor for carbon steel in 1 M HCl solution. *ACS Omega* 2: 3997–4008.
17 Mobin, M. and Rizvi, M. (2017). Polysaccharide from Plantago as a green corrosion inhibitor for carbon steel in 1 M HCl solution. *Carbohydr. Polym.* 160: 172–193.
18 Solomon, M.M., Umoren, S.A., Udosoro, I.I., and Udoh, A.P. (2010). Inhibitive and adsorption behaviour of carboxymethyl cellulose on mild steel corrosion in sulphuric acid solution. *Corros. Sci.* 52: 1317–1325.
19 Bayol, E., Gürten, A.A., Dursun, M., and Kayakırılmaz, K. (2008). Adsorption behavior and inhibition corrosion effect of sodium Carboxymethyl cellulose on mild steel in acidic medium. *Acta Phys. -Chim. Sin.* 24: 2236–2242.
20 El-Haddad, M.N. (2014). Hydroxyethyl cellulose used as an eco-friendly inhibitor for 1018 c-steel corrosion in 3.5% NaCl solution. *Carbohydr. Polym.* 112: 595–602.
21 Deyab, M.A. (2015). Hydroxyethyl cellulose as efficient organic inhibitor of zinc carbon battery corrosion in ammonium chloride solution: electrochemical and surface morphology studies. *J. Power Sources* 280: 190–194.
22 Rajeswari, V., Kesavan, D., Gopiraman, M., and Viswanathamurthi, P. (2013). Physicochemical studies of glucose, gellan gum, and hydroxypropylcelluloseInhibition of cast iron corrosion. *Carbohydr. Polym.* 95: 288–294.
23 Arukalam, I.O. (2014). Durability and synergistic effects of KI on the acid corrosion inhibition of mild steel by hydroxypropyl methylcellulose. *Carbohydr. Polym.* 112: 291–299.
24 Arukalam, I.O., Madu, I.O., Ijomah, N.T. et al. (2014). Acid corrosion inhibition and adsorption behaviour of ethyl hydroxyethyl cellulose on mild steel corrosion. *J. Mater.* 1: 1–11.
25 Mobin, M., Khan, M.A., and Parveen, M. (2011). Inhibition of mild steel corrosion in acidic medium using starch and surfactants additives. *Appl. Polym. Sci.* 121: 1558–1565.
26 Brindha, T., Mallika, J., and Moorthy, V.S. (2015). Synergistic effect between starch and substituted piperidin-4-one on the corrosion inhibition of mild steel in acidic medium. *Mater. Environ. Sci.* 6: 191–200.
27 Bello, M., Ochoa, N., Balsamo, V. et al. (2010). Modified cassava starches as corrosion inhibitors of carbon steel: An electrochemical and morphological approach. *Carbohydr. Polym.* 82: 561–568.
28 Rosliza, R. and Nik, W.B. (2010). Improvement of corrosion resistance of AA6061 alloy by tapioca starch in seawater. *Curr. Appl. Phys.* 10: 221–229.

29 Li, X. and Deng, S. (2015). Cassava starch graft copolymer as an eco-friendly corrosion inhibitor for steel in H2SO4 solution. *Korean J. Chem. Eng.* 32: 2347–2354.

30 Umoren, S.A., Obot, I.B., Madhankumar, A., and Gasem, Z.M. (2015). Performance evaluation of pectin as eco-friendly corrosion inhibitor for X60 pipeline steel in acid medium: experimental and theoretical approaches. *Carbohydr. Polym.* 124: 280–291.

31 Fares, M.M., Maayta, A.K., and Al-Qudah, M.M. (2012). Pectin as promising green corrosion inhibitor of aluminum in hydrochloric acid solution. *Corros. Sci.* 60: 112–117.

32 Fiori-Bimbi, M.V., Alvarez, P.E., Vaca, H., and Gervasi, C.A. (2015). Corrosion inhibition of mild steel in HCl solution by pectin. *Corros. Sci.* 92: 192–199.

33 Grassino, A.N., Halambek, J., Djakovi, S. et al. (2016). Utilization of tomato peel waste from canning factory as a potential source for pectin production and application as tin corrosion inhibitor. *Food Hydrocoll.* 52: 265–274.

34 Saidi, N., Elmsellem, H., Ramdani, M. et al. (2015). Using pectin extract as eco-friendly inhibitor for steel corrosion in 1 M HCl media. *Der Pharm. Chem.* 7: 87–94.

35 Zaafarany, I. (2012). Corrosion inhibition of aluminum in aqueous alkaline solutions by alginate and pectate water-soluble natural polymer anionic polyelectrolytes. *Port. Electrochim. Acta* 30: 419–426.

36 Hassan, R., Zaafarany, I., Gobouri, A., and Takagi, H. (2013). A revisit to the corrosion inhibition of aluminum in aqueous alkaline solutions by water-soluble alginates and pectates as anionic polyelectrolyte inhibitors. *Int. J. Corros.* 30: 419–426.

37 Umoren, S.A., Banera, M.J., Garcia, T.A. et al. (2013). Inhibition of mild steel corrosion in HCl solution using chitosan. *Cellulose* 20: 2529–2545.

38 Cheng, S., Chen, S., Liu, T. et al. (2007). Carboxymethyl chitosan–Cu2+ mixture as an inhibitor used for mild steel in 1.0 M HCl. *Electrochim. Acta* 52: 5932–5938.

39 Liu, Y., Zou, C., Yan, X. et al. (2015). β-Cyclodextrin modified natural chitosan as a green inhibitor for carbon steel in acid solutions. *Ind. Eng. Chem. Res.* 54: 5664–5672.

40 El-Haddad, M.N. (2013). Chitosan as a green inhibitor for copper corrosion in acidic medium. *Int. J. Biol. Macromol.* 55: 142–149.

41 Sangeetha, Y., Meenakshi, S., and Sundaram, C. (2016). Interactions at the mild steel acid solution interface in the presence of O-fumaryl-chitosan: electrochemical and surface studies. *Carbohydr. Polym.* 136: 38–45.

42 Mohamed, R.R. and Fekry, A.M. (2011). Antimicrobial and anticorrosive activity of adsorbents based on chitosan Schiff's base. *Int. J. Electrochem. Sci.* 6: 2488–2508.

43 Li, M.L., Li, R.H., Xu, J. et al. (2014). Thiocarbohydrazidemodified chitosan as anticorrosion and metal ion adsorbent. *J. Appl. Polym. Sci.* 131: 40671–40678.

44 Fares, M.M., Maayta, A.K., and Al-Mustafa, J.A. (2012). Corrosion inhibition of iotacarrageenan natural polymer on aluminum in presence of zwitterions mediator in HCl media. *Corros. Sci.* 65: 223–230.

45 Zaafarany, I. (2006). Inhibition of acidic corrosion of iron by some carrageenan compounds. *Curr. World Environ.* 1: 101–108.

46 Loto, C.A. and Loto, R.T. (2013). Effect of dextrin and thiourea additives on the zinc electroplated mild steel in acid chloride solution. *Int. J. Electrochem. Sci.* 8: 12434–12450.

47 Zou, C., Yan, X., Qin, Y. et al. (2014). Inhibiting evaluation of βCyclodextrin-modified acrylamide polymer on Alloy steel in sulfuric acid solution. *Corros. Sci.* 85: 445–454.

48 Yan, X., Zou, C., and Qin, Y. (2014). A new sight of water-soluble polyacrylamide modified by β-cyclodextrin as corrosion inhibitor for X70 steel. *Starch/Stärke* 66: 968–975.

49 Fan, B., Wei, G., Zhang, Z., and Qiao, N. (2014). Preparation of supramolecular corrosion inhibitor based on hydroxypropyl-bcyclodextrin/octadecylamine and its anticorrosion properties in the simulated condensate water. *Anti-Corros. Methods Mater.* 61: 104–111.

50 Liu, Y., Zou, C., Li, C. et al. (2016). Evaluation of β-cyclodextrin polyethylene glycol as green scale inhibitors for produced-water in shale gas well. *Desalination* 377: 28–33.

51 Tawfik, S.M. (2015). Alginate surfactant derivatives as eco-friendly corrosion inhibitor for carbon steel in acidic environment. *RSC Adv.* 5: 104535–104550.

52 Sangeetha, Y., Meenakshi, S., and Sundaram, C.S. (2016). Investigation of corrosion inhibitory effect of hydroxyl propyl alginate on mild steel in acidic media. *J. Appl. Polym. Sci.* 133: 43004–43010.

53 Mobin, M. and Rizvi, M. (2016). Inhibitory effect of xanthan gum and synergistic surfactant additives for mild steel corrosion in 1 M HCl. *Carbohydr. Polym.* 136: 384–393.

54 Gerengi, H. and Sahin, H.I. (2012). Schinopsis lorentzii extract as a green corrosion inhibitor for low carbon steel in 1 MHCl solution. *Ind. Eng. Chem. Res.* 51: 780–787.

55 Darowicki, K. (2000). Theoretical description of the measuring method of instantaneous impedance spectra. *J. Electroanal. Chem.* 486: 101–105.

56 Gerengi, H., Slepski, P., Ozgan, E., and Kurtay, M. (2015). Investigation of corrosion behavior of 6060 and 6082 aluminum alloys under simulated acid rain conditions. *Mater. Corros.* 66: 233–240.

19

Synthetic Polymers as Corrosion Inhibitors

Megha Basik and Mohammad Mobin

Corrosion Research Laboratory, Department of Applied Chemistry, Faculty of Engineering and Technology, Aligarh Muslim University, Aligarh, Uttar Pradesh, India

19.1 Introduction

Metals strengthened its root as one of the good choices of raw materials used in many industrialized nations. They have a high melting point and boiling point, great tensile strength, good conductivity, and ease of fabrication. Unfortunately, metals get easily corroded in harsh conditions, and the process of deterioration of metals is technically termed as "corrosion." The force that drives the corrosion of metals is the temporary existence of the metal in thermodynamically lesser stable form. This problem adversely affecting the whole world as nearly all materials used in industries and engineering applications are made up of metals and suffers corrosion problem. The undesirable gradual loss of certain valuable and useful metals such as mild steel, aluminum, zinc, copper, etc. when exposed to harsh environments such as acid, base, and salt are the dreadful effect of corrosion [1].

Although materialistic loss is not only the concern but corrosion takes many lives and disastrous conditions if not prevent properly [2]. However, it is very difficult to estimate the actual cost due to corrosion but some leading countries attempted to calculate the damage from corrosion. The first nation to estimate the annual cost of corrosion is the United States. In 2016, a survey conducted by international measures of prevention, application, and economics of corrosion technologies (IMPACT) under NACE estimated the cost of corrosion globally. It was found that about US$ 2.5 trillion was the global cost of corrosion, which was about 3.4% GDP of the nation [3].

Organic Corrosion Inhibitors: Synthesis, Characterization, Mechanism, and Applications,
First Edition. Edited by Chandrabhan Verma, Chaudhery Mustansar Hussain, and Eno E. Ebenso.

The nature of the metal (material composition) and the corrosive environment are the two broad factors to which the rate and extent of corrosion are dependent. So it is important to look into deep for the developments in corrosion mitigation. To prevent metal from corrosion in a corrosive environment is a challenging situation. Although there are several preventive methods to control metal degradation such as cathodic protection [4], anodic protection [5], application of coatings [6], proper material selection, and use of inhibitors [7]. Among them, the use of inhibitors is the best preventive technique in which modification in the environment can be done. This technique is the most practical, effective, and approachable due to its ease of usage for corrosion prevention. Traditionally, compounds containing heteroatoms such as O, N, and S act as good corrosion inhibitors but several research articles describe that the structural aspect of inhibitors is one of the important factors that decide the adsorption process and route of mechanism.

Adsorption of inhibitor on metal surface depends on many physicochemical properties such as active functional moieties present in its structure, steric factors, the electron density of donor atoms, aromaticity, and so on [8]. The selection of proper corrosion inhibitor is a serious concern. Earlier, according to the suitability of inorganic and organic compounds, they were generally used as corrosion inhibitors. Therefore, chromium compounds and lanthanide salts were widely used as effective corrosion inhibitors, but they were prohibited and banned in industrial applications due to their high toxicity and adverse effect on health in the long run [9, 10]. Later on, organic inhibitors came into trends because of their good solubility in various aqueous mediums, effectiveness, stability, and ease of adsorption on the metallic surface [11, 12]. But due to difficult synthesis routes and unreasonable prices, they have been criticized severely. Thus, the search for a replacement of these organic and inorganic corrosion inhibitors has begun. Alternatively, polymers (both natural and synthetic) seek the attention and gained the title of perfect replacement of toxic inhibitors and found to be effective in retarding corrosion. Polymers are large macromolecules having multiple binding sites. They are in trend because of their ease of availability, economic feasibility, excellent performance to withstand, effectiveness, environmental friendliness, inherent stability, and multiple adsorption centers [13]. They have various anchoring functional groups through which they easily get adsorbed on the substrate surface to be protected. It should also be noted that the capacity or potential of polymers as corrosion inhibitors are mainly dependent on the chemical composition of the polymers. The presence of heteroatoms such as oxygen, sulfur, and nitrogen enhances the electron density on polymers and intensifies their corrosion mitigation ability. Thus, the topic focuses on the recent reports on the utilization of synthetic polymers as corrosion inhibitors.

19.2 General Mechanism of Polymers as Corrosion Inhibitors

Polymeric compounds have large carbon chains with multiple bonds and functional moieties attached to their molecular structure. The functional moieties have various heteroatoms such as oxygen, sulfur, and nitrogen, which act as an active center. These atoms are electron-rich in nature as lone pair are present and thus donation of electrons to metal takes place from these active atoms. They easily get adsorbed by interacting with metal surface and thus displace the water molecules forming a barrier layer. The availability of lone pair of electrons facilitates the transfer of electrons from inhibitor molecule to metal surface forming a coordinate covalent bond. Electron density and polarizability of donor group are the factors that decide the strength of the bond formed. The presence of functional group (-NH_2, -NO_2, -CHO, -COOH, etc.) on the C skeleton of polymers enhances the electron density and thus rate of either cathodic or anodic reaction or both decreases. Polymers can easily form large complexes with the metal ions present on surface through their available functional groups or moieties. A large polymer chain covers a large surface area by blanketing the surface of the metal to be protected from the corrosive environment. The inhibitive power of polymers is mainly related to the presence of conjugate bond or cyclic rings and the presence of heteroatoms, which majorly act as an active center for adsorption.

19.3 Corrosion Inhibitors - Synthetic Polymers

Synthetic polymers are synthesized to substitute the natural polymer as they lack in various properties. They are introduced to enhance the qualities of corrosion inhibitors, which could not be achieved by the natural polymer. They are gaining attention due to their inherent stability and cost-effectiveness. The presence of the polar functional group fulfills the requirement as a metal corrosion inhibitor [14]. Literature reveals that a variety of synthetic polymers was studied as corrosion inhibitors, and their performance to inhibit corrosion was mentioned in Table 19.1. The inhibitive effect of polyacrylic acid was tested for aluminum in an alkaline medium. The corrosion mitigation effect of polyacrylic acid was investigated by using gravimetric and electrochemical techniques. The maximum inhibition efficiency (IE) of 97% obtained at 20×10^{-8} M concentration [15]. The performance of polyvinylpyrrolidone on stainless steel corrosion has been tested, and it is found that the inhibitor effectively retard the corrosion to 96% at 0.005 mol/l at 25 °C [16]. Polyethylene glycol methyl ether successfully retard the acid initiated (sulfuric acid) mild steel corrosion. Electrochemical analysis tells that inhibitors behave mixed type providing 90% IE at 308 K temperature [17]. Corrosion inhibition of poly(diphenylamine) was studied for iron surface in 0.5 M sulfuric acid medium by using PDP, LP, and EIS techniques. It was found that synthesized polymer effectively retard corrosion to 96% at minimal concentration of 10 ppm, which could not be achieved by its monomer at 1000 ppm (75 %) [18]. The inhibitive performance of polyvinylpyrrolidone and polyethyleneimine was studied by gravimetric and polarization techniques for low carbon steel in an aqueous phosphoric acid solution. The inhibitors effectively retards the anodic and cathodic reactions [19]. Another combination of two polymers have proven to be an effective corrosion inhibitor. Doped poly(styrenesulfonic acid) has been successfully synthesized and tested to retard corrosion in acidic phase. The corrosion mitigation efficacy was investigated by WL measurement, PDP, hydrogen permeation measurements and AC impedance measurements. The inhibitor effectively retards the corrosion to 85.9% at very low concentration of 70 ppm. The polymeric inhibitor predominantly showing anodic behavior and follows Temkin's adsorption isotherm [20]. Polymeric amines have also been studied as an effective corrosion inhibitor. Polyaniline has been widely used from decades to resist corrosion by coating application. Now, the performance of water-soluble polyaniline has been tested as corrosion inhibitor by using PDP, LP, and EIS in 0.5 M H_2SO_4 for iron substrate. It was found that inhibitor successfully retard the corrosion to 84% at 100 ppm concentration, which could not be achieved by its monomeric part [21].

Synergistic behavior has been proven to enhance the inhibitive performance. Addition of small quantity of certain compound (halides, metal cations, and surfactants) can boost up the inhibitive performance to another level at low concentration. The effect of cerium ions for corrosion retarding performance of polyaniline has been studied for iron surface in 0.5 M H_2SO_4 solution. The IE of polyaniline increased from 53 to 89% at 10 ppm in the presence of cerium ions [22].

Sometimes it is difficult to achieve all the desired properties from a single polymeric compound, so some selected polymers have been synthesized and fused together to form a new class called "terpolymer," which can withstand all the required conditions and have excellent inhibition performance. Verma et al. (2013) successfully synthesized three terpolymers that were synthesized from polycondensation of resorcinol, formaldehyde, and diaminoethane known as TER-1, whereas remaining terpolymers were synthesized by the reaction of resorcinol, formaldehyde, urea (TER-2), and thiourea (TER-3). Their inhibitive action was monitored by gravimetric, PDP, LP, and electrochemical impedance spectroscopy (EIS) techniques. The studied inhibitor effectively retard the corrosion up to 90–94% at 50 ppm concentration [23]. Polymeric compound BFP was synthesized by the condensation reaction of bisphenol-A, formaldehyde, and piperazine in an alkaline medium (Figure 19.1).

The inhibitive effect of the synthesized polymer was examined by gravimetric and electrochemical measurements. The inhibitor effectively retard the corrosion

Table 19.1 List of synthetic polymers used as corrosion inhibitor.

Details		References
a) Poly(p-toluidine) b) Substrate: iron; solution: 1 M HCl c) Electrochemical impedance, linear polarization, Tafel polarization techniques d) SEM and FTIR analysis e) 50–500 ppm concentration	The maximum IE of 94% at 500 ppm	[31]
a) Poly(aminoquinone) b) Substrate: iron ; solution: 0.5 M H_2SO_4 c) Electrochemical impedance and potentiodynamic polarization technique d) SEM and FTIR analysis e) 10–100 ppm concentration	IE of 96% at 100 ppm	[32]
a) Poly(4-vinylpyridine-poly(3-oxide-ethylene) tosyle) b) Substrate: iron; solution: 1 M H_2SO_4 c) Gravimetric, EIS, and PDP d) 10^{-10} M to 2.5×10^{-8} M concentration	The excellent IE of 100% obtained at 2.5×10^{-8} M	[33]
a) Poly(p-anisidine) b) Substrate: iron; solution: 1 M HCl c) Electrochemical impedance and potentiodynamic polarization technique d) 10–100 ppm concentration	IE of 94.4% observed at 100 ppm	[34]
a) Poly p-aminobenzoic acid b) Substrate: iron; solution: 1 mol/l HCl c) Electrochemical impedance and potentiodynamic polarization technique d) $50–500\times10^{-6}$ concentration	Maximum IE of 94.8% observed at 500×10^{-6}	[35]
a) Polyvinyl alcohol b) Substrate: Carbon steel; Solution: 60 ppm Cl^- c) Weight loss measurement d) FTIR analysis e) 25–150 ppm concentration.	IE of 93% achieved at 100 ppm	[13]

(*Continued*)

Table 19.1 (Continued)

Details		References
a) Polyacrylamide b) Substrate: iron; solution: 0.5 M H_2SO_4 c) PDP and EIS studies d) Atomic force microscopy e) 1×10^{-8} to 1×10^{-6} concentration	Inhibition performance of 76% observed at 1×10^{-6}	[36]
a) Polyethyleneimine b) Substrate: iron; solution: 0.5 M sulfuric acid c) Potentiodynamic polarization and electrochemical analysis d) SEM and XPS analysis e) 3–5 mg/l concentration	93.8% IE observed at 5 mg/l	[37]
a) Polyanthranilic acid b) Substrate: mild steel; solution: 0.5 M HCl c) Gravimetric, electrochemical impedance spectroscopy, and polarization measurement d) Atomic force microscopy (AFM) e) 10–80 ppm concentration	Maximum IE of 94.7% observed at 60 ppm	[38]
a) Polyaspartic acid b) Substrate: mild steel; solution: 0.5 M H_2SO_4 c) Weight loss and electrochemical d) Measurements e) SEM, XPS, and FTIR f) 0.1–2 g/l concentration	88% IE obtained at 2 g/l concentration	[39]
a) Polyacrylamide b) Substrate: iron; solution: 3 M HCl c) Weight loss and electrochemical measurements d) 0.5–2 ppm concentration	94% IE	[40]
a) Polyacrylamide b) Substrate: aluminum; solution: 1 M HCl c) Gravimetric, thermometric, and hydrogen evolution techniques d) 1×10^{-5} M to 1×10^{-4} M	The maximum IE of 90% observed at 1×10^{-4} M	[41]

Table 19.1 (Continued)

Details		References
a) Poly(o-phenylenediamine) b) Substrate: mild steel; solution: 1 M HCl c) Weight loss and PDP studies. d) 2.5–15 ppm concentration.	95% IE observed at 15 ppm	[42]
a) Polyvinyl alcohol b) Substrate: mild steel; solution: 0.5 M HCl c) EIS, LPR, and PDP techniques d) SEM e) 50–1000 ppm concentration	IE of 79% was obtained at 1000 ppm	[43]
a) Polyvinyl pyrrolidone b) Substrate: carbon steel; solution: 0.1 NaCl c) Weight loss and electrochemical d) methods e) SEM analysis f) 500–3000 ppm concentration	72.97% efficiency obtained at 3000 ppm	[44]
a) PEG and PVP b) Substrate: mild steel; solution: 0.5 N sulfuric acid and 0.5 N hydrochloric acid c) PDP, LP, EIS d) SEM and AFM analysis e) 10–200 ppm concentration	Maximum IE of 99% (PVP) and 98% (PEG) was obtained at 200 ppm	[45]
a) Poly(4-vinylpyridine-hexadecyl bromide) b) Substrate: mild steel; solution: 1 HCl c) Gravimetric and electrochemical measurement d) 5–300 ppm concentration	95% at 300 mg/l	[46]
a) Polyethylene glycol b) Substrate: aluminum; solution: 1 HCl c) Weight and polarization studies d) DFT and MD simulation e) 0.3×10^{-3} to 15×10^{-3} M	Maximum IE of 66% observed at 15×10^{-3} M	[47]

(Continued)

Table 19.1 (Continued)

Details		References
a) Polypropylene glycol b) Substrate: mild steel; solution: 0.5 M H_2SO_4 c) Gravimetric and electrochemical measurements d) SEM and contact angle analysis e) 50–1000 ppm	IE of 83% obtained at 1000 ppm	[48]
a) Polyvinyl pyrrolidone (PVP) b) Substrate: mild steel; solution: 1 M H_2SO_4 c) Gravimetric and hydrogen evolution methods d) 2×10^{-5} M to 1×10^{-4} M concentration.	67.16% obtained at 1×10^{-4} M	[49]
a) PASP, PESA, and PAPEMP b) Substrate: carbon steel; solution: cooling water c) Weight loss and electrochemical measurements d) SEM, EDX, and AFM analysis e) 10–200 ppm concentration.	PASP = 79.9% at 200 ppm PAPEMP = 78.6% at 50 ppm PESA= 67.1% at 200 ppm	[50]
a) Poly (methacrylic acid) (PMAA) b) Substrate: mild steel; solution: 0.5 M H_2SO_4 c) Gravimetric method and electrochemical measurements d) SEM and contact angle analysis e) (c) 10–1000 ppm concentration	Protection efficiency of 61.6% observed at 1000 ppm	[51]
a) Poly(aniline-formaldehyde) b) Substrate: mild steel; solution: 1 N HCl c) PDP, LP, EIS, and WL measurements d) AFM analysis e) 1–10 ppm	The highest resistance (98.75%) was achieved at 10 ppm concentration	[52]
a) Poly(aniline-co-4-amino-3-hydroxy-naphthalene-1-sulfonic acid) b) Substrate: iron; solution: 1 M HCl c) Tafel extrapolation method and EIS d) 10–70 ppm concentration	The maximum IE of 90% at 70 ppm	[53]

Table 19.1 (Continued)

Details		References
a) Poly(aniline-co-o-toluidine) b) Substrate: carbon steel; solution: 3% NaCl c) Electrochemical measurements d) FTIR, XRD, and UV-Vis spectrophotometry e) 10–100 ppm concentration	IE of 74.5% observed at 100 ppm concentration	[54]
a) Poly(vinylpyrrolidone–methylaniline) b) Substrate: mild steel; solution: 1 M HCl c) Gravimetric and electrochemical measurements d) SEM and EDAX analysis e) 100–2000 ppm concentration	Efficiency of 87% achieved at 2000 ppm concentration	[55]
a) Poly(aniline-formaldehyde), poly(o-toluidene–formaldehyde) and poly(p-chloroaniline–formaldehyde) b) Substrate: mild steel; solution: 1 N HCl c) WL measurements, LP, TP, and EIS d) AFM analysis e) 1.0–15.0 mg/l concentration	Poly(aniline-formaldehyde)= 97.23% Poly(o-toluidene-formaldehyde) = 98.46% Poly(p-chloroaniline-formaldehyde) = 98.96 %	[56]
a) Poly(acrylamide-co-4-vinylpyridine) b) Substrate: mild steel; solution: 1 M H_2SO_4 c) WL measurements, EIS and PDP techniques d) 0.1–100 mg/l	61% efficiency observed at 100 ppm	[57]
a) Poly(vinyl alcohol-o-methoxyaniline) b) Substrate: mild steel; solution: 1 M HCl c) Gravimetric and electrochemical measurements d) SEM analysis e) 100–2000 ppm	91% effectiveness observed at 2000 parts per million concentration	[58]

(Continued)

Table 19.1 (Continued)

Details		References
a) Poly(vinyl alcohol–proline) (PVAP) b) Substrate: mild steel; solution: 1 M HCl c) Gravimetric and electrochemical measurements d) FTIR, SEM-EDX, and XRD e) 0.15–0.60% PVAP	Maximum IE of 94.82% observed at 0.60% concentration of PVAP	[59]
a) Poly(vinyl alcohol-threonine) (PVAT) b) Substrate: mild steel; solution: 1 M HCl c) Gravimetric, LPR, PDP, and EIS d) 0.15–0.60% PVAT	IE of 96.58% observed at 0.60% concentration of PVAT	[60]

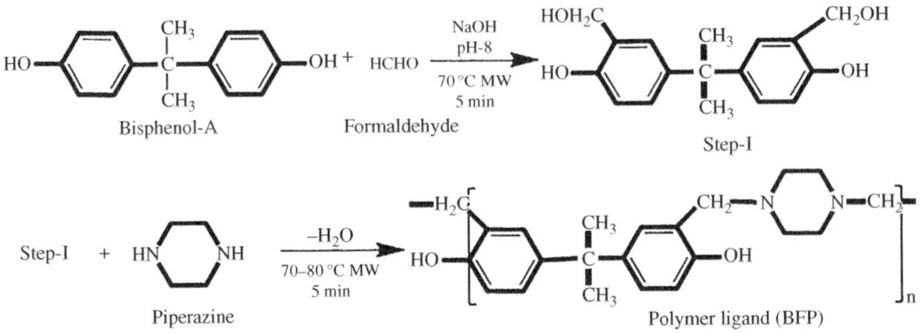

Figure 19.1 Synthesis of BFP polymer.

to 96% at lowest concentration of 75 ppm [24]. The pictorial representation of studied polymer on mild steel surface shown in Figure 19.2.

Protective effect of copolymers, namely, poly(vinyl caprolactone-co-vinyl pyridine) and poly(vinyl imidazol-co-vinyl pyridine), was investigated by different gravimetric and electrochemical techniques for carbon steel in the phosphoric acid medium [25]. Ali and Saeed (2001) and Ali et al. (2012) successfully produced various N, N-diallyl compounds from 1,6-hexanediamine and 1,12-dodecanediamine. The synthesized compound undergoes cyclopolymerization and found to be effective for restricting mild steel corrosion in HCl, H_2SO_4, and NaCl medium [26, 27]. The corrosion mitigation performance of synthesized water-soluble triblock copolymers, PDEA-PDMA-PMEMA and PDPA-PDMA-PMEMA of different molecular weights on mild steel corrosion in 0.5 M HCl medium. The synthesized block

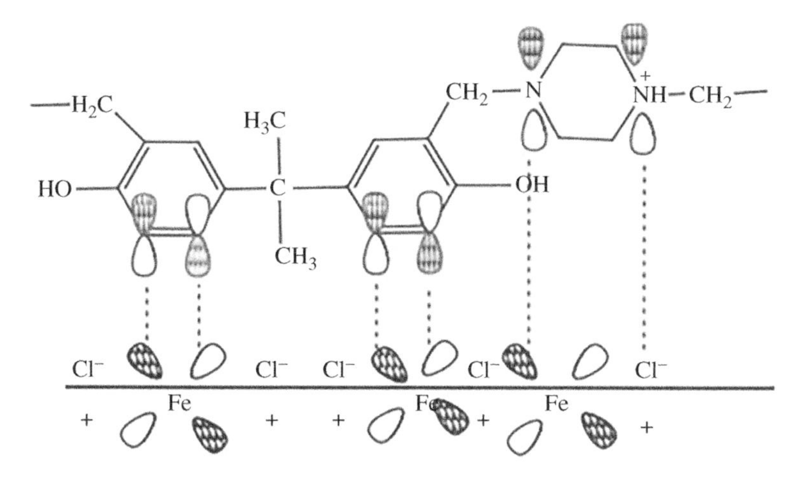

Figure 19.2 Pictorial representation of mechanism of adsorption of BFP polymer on MS surface.

copolymers (Figure 19.3) were investigated by using EIS, PDP and LP methods. The studies reveal that the inhibitive efficiency depends mainly on nature of the substituent present in the molecular structure of the inhibitor and on the molecular weight of the inhibitor [28].

Barak et al. [29] synthesized the terpolymer using poly(ethylene oxide) and polystyrene block copolymers. The inhibitive performance of water-soluble polystyrene-*b*-poly(ethylene oxide)-*b*-polystyrene triblock copolymers was studied in 1 M H_2SO_4 on metal surface using potentiodynamic polarization technique. The inhibitor effectively retards the corrosion up to 90–95% at 800–1600 ppm concentration. Studies shows that the inhibitor behaves as mixed type and depend upon both concentration and temperature.

Water soluble SAP was synthesized by using 2, 2′-benzidinedisulfonic acid and isophthaloyl chloride, and its inhibitive performance has been tested on copper in 1 M HCl solution. The synthesized inhibitor was examined by electrochemical measurements and surface morphological studies. The inhibitor was found to effectively mitigate corrosion to 92% at 500 ppm concentration. The potentiodynamic results show that the inhibitor act as mixed type and obeys Langmuir adsorption isotherm [30]. The synthesis route of the polymer was shown in Figure 19.4.

19.4 Conclusion

The variety of synthetic polymers behaves as efficient corrosion inhibitor in a different medium for different substrates and for different operating conditions. It is found that IE was greatly affected by the molecular weight of the polymer,

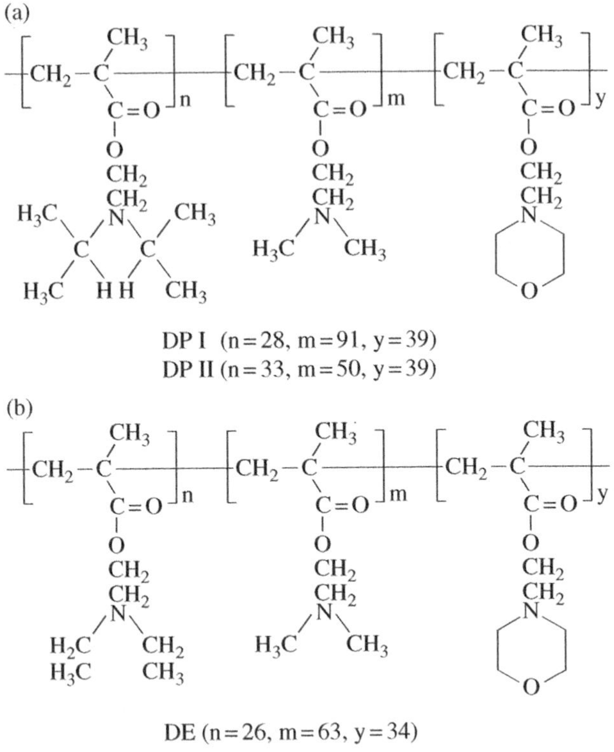

(a)

DP I (n=28, m=91, y=39)
DP II (n=33, m=50, y=39)

(b)

DE (n=26, m=63, y=34)

Figure 19.3 Molecular structures of (a) PDPA-PDMA-PMEMA and (b) PDEA-PDMA-PMEMA triblock copolymers synthesized by using group transfer polymerization.

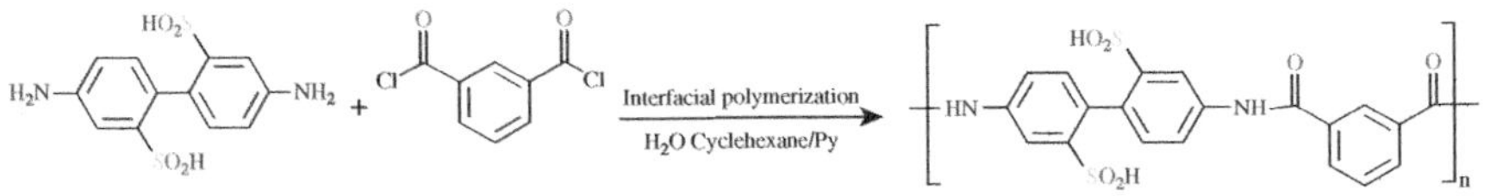

Interfacial polymerization
H_2O Cyclehexane/Py

Figure 19.4 The synthesis route of polymer.

the concentration of polymer used, the structure of the inhibitor, and operating temperature. The mechanism of polymeric inhibition was principally by adsorption of inhibitor molecule through multiple active centers on a metal surface by displacing the water molecule available on its surface. Copolymerization enhances the properties of a polymer, which could not be achieved by a single polymer alone. It enhances the solubility, as well as the corrosion inhibition performance, of the polymer. Literature also reveals the extraordinary IE of synthetic polymers compared with their monomer against corrosion. Thus, the aim was to gather knowledge regarding synthetic polymers as effective corrosion inhibitors, and also to open new frontier research in the field of corrosion by these polymers.

Useful Links

https://link.springer.com/article/10.1186/2228-5547-4-2

https://www.tandfonline.com/doi/abs/10.1080/00986445.2014.934448#:~:text=Several%20works%20have%20been%20reported,metals%20in%20various%20corrosive%20environments.&text=The%20solution%20pH%2C%20concentration%2C%20exposure,their%20role%20in%20inhibition%20performance.

https://www.researchgate.net/publication/265844821_Recent_Developments_on_the_Use_of_Polymers_as_Corrosion_Inhibitors_-_A_Review

https://ieeexplore.ieee.org/document/8340423

References

1 Srimathi, M., Rajalakshmi, R., and Subhashini, S. (2014). *Arab. J. Chem.* https://doi.org/10.1016/j.arabjc.2010.11.013.
2 Mobin, M., Basik, M., and Aslam, J. (2019). *Meas. J. Int. Meas. Confed.* 134: 595.
3 Shekari, E., Khan, F., and Ahmed, S. (2017). *Int. J. Press. Vessel. Pip.* https://doi.org/10.1016/j.ijpvp.2017.08.005.
4 Kim, D.K., Muralidharan, S., Ha, T.H. et al. (2006). *Electrochim. Acta* https://doi.org/10.1016/j.electacta.2006.01.054.
5 Cecchetto, L., Delabouglise, D., and Petit, J.P. (2007). *Electrochim. Acta* https://doi.org/10.1016/j.electacta.2006.10.009.
6 Aslam, J., Mobin, M., Aslam, R., and Ansar, F. (2020). *J. Adhes. Sci. Technol.* https://doi.org/10.1080/01694243.2019.1676599.
7 Mobin, M., Ahmad, I., Basik, M. et al. (2020). *Sustain. Chem. Pharm.* https://doi.org/10.1016/j.scp.2020.100337.
8 Jayalakshmi, M. and Muralidharan, V.S. (1998). *Indian J. Chem. Technol.*
9 Twite, R.L. and Bierwagen, G.P. (1998). *Prog. Org. Coat.* 33: 91.
10 Bernal, S., Botana, F.J., Calvino, J.J. et al. (1995). *J. Alloys Compd.* https://doi.org/10.1016/0925-8388(94)07135-7.
11 El-Maksoud, S.A. (2008). *Int. J. Electrochem. Sci.*
12 Loto, R.T., Loto, C.A., and Popoola, A.P.I. (2012). *J. Mater. Environ. Sci.*
13 Rajendran, S., Sridevi, S.P., Anthony, N. et al. (2005). *Anti-Corrosion Methods Mater.* https://doi.org/10.1108/00035590510584816.

14 Fathima Sabirneeza, A.A., Geethanjali, R., and Subhashini, S. (2015). *Chem. Eng. Commun.* https://doi.org/10.1080/00986445.2014.934448.

15 Amin, M.A., El-Rehim, S.S.A., El-Sherbini, E.E.F. et al. (2009). *Corros. Sci.* https://doi.org/10.1016/j.corsci.2008.12.008.

16 Khaled, M. (2010). *Arab. J. Sci. Eng.*

17 Dubey, A.K. and Singh, G. (2007). *Port. Electrochim. Acta* https://doi.org/10.4152/pea.200702221.

18 Jeyaprabha, C., Sathiyanarayanan, S., Phani, K.L.N., and Venkatachari, G. (2005). *J. Electroanal. Chem.* https://doi.org/10.1016/j.jelechem.2005.08.017.

19 Jianguo, Y., Lin, W., Otieno-Alego, V., and Schweinsberg, D.P. (1995). *Corros. Sci.* https://doi.org/10.1016/0010-938X(95)00008-8.

20 Manickavasagam, R., Karthik, K.J., Paramasivam, M., and Iyer, S.V. (2002). *Anti-Corrosion Methods Mater.* https://doi.org/10.1108/00035590210413566.

21 Prabha, C.J., Sadagopan, S., and Venkatachari, G. (2006). *J. Appl. Polym. Sci.* 101: 2144.

22 Jeyaprabha, C., Sathiyanarayanan, S., and Venkatachari, G. (2006). *Appl. Surf. Sci.* https://doi.org/10.1016/j.apsusc.2005.12.081.

23 Verma, C.B., Quraishi, M.A., and Ebenso, E.E. (2013). *Int. J. Electrochem. Sci.*

24 Singh, P., Quraishi, M.A., and Ebenso, E.E. (2013). *Int. J. Electrochem. Sci.*

25 Benabdellah, M., Ousslim, A., Hammouti, B. et al. (2007). *J. Appl. Electrochem.* https://doi.org/10.1007/s10800-007-9317-1.

26 Ali, S.A. and Saeed, M.T. (2001). *Polymer (Guildf).* https://doi.org/10.1016/S0032-3861(00)00665-0.

27 Ali, S.A., Saeed, M.T., and El-Sharif, A.M.Z. (2012). *Polym. Eng. Sci.* https://doi.org/10.1002/pen.23224.

28 Yurt, A., Bütün, V., and Duran, B. (2007). *Mater. Chem. Phys.* 105: 114.

29 Barak, A., Das, P.J., Vashisht, H., and Kumar, S. (2014). *Int. J. Sci. Res. Publ.*

30 Farahati, R., Ghaffarinejad, A., Rezania, H.J. et al. (2019). *Colloids Surfaces A Physicochem. Eng. Asp.* 123626: 578.

31 Kumar, H. and Yadav, V. (2020). *Chem. Data Collect.* 29: 100500.

32 Jeyaprabha, C., Sathiyanarayanan, S., Phani, K.L.N., and Venkatachari, G. (2005). *Appl. Surf. Sci.* https://doi.org/10.1016/j.apsusc.2005.01.098.

33 Chetouani, A., Medjahed, K., Sid-Lakhdar, K.E. et al. (2004). *Corros. Sci.* https://doi.org/10.1016/j.corsci.2004.01.020.

34 Manivel, P. and Venkatachari, G. (2005). *Corros. Sci. Technol.* 4: 51.

35 Manivel, P. and Venkatachari, G. (2006). *J.Mater.Sci.Technol.* 22: 301.

36 Umoren, S.A., Li, Y., and Wang, F.H. (2010). *Corros. Sci.* https://doi.org/10.1016/j.corsci.2010.01.026.

37 Zhang, X., Wu, X., Li, J. et al. (2009). *Mater. Sci. Forum.*

38 Shukla, S.K., Quraishi, M.A., and Prakash, R. (2008). *Corros. Sci.* https://doi.org/10.1016/j.corsci.2008.07.025.

39 Qian, B., Wang, J., Zheng, M., and Hou, B. (2013). *Corros. Sci.* https://doi.org/10.1016/j.corsci.2013.06.001.

40 DRAGICA, C., MAJA, C., and Grchev, T. (2007). *J. Serbian Chem. Soc.*: 72. https://doi.org/10.2298/JSC0707687C.

41 Umoren, S. and Solomon, M. (2010). *Arab. J. Sci. Eng.* 35: 115.

42 Abd El Rehim, S.S., Sayyah, S.M., El-Deeb, M.M. et al. (2010). *Mater. Chem. Phys.* https://doi.org/10.1016/j.matchemphys.2010.02.069.

43 Umoren, S. and Gasem, Z. (2014). *J. Dispers. Sci. Technol.* https://doi.org/10.1080/01932691.2013.833481.

44 Juhaiman, L.A.A., Mustafa, A.A., and Mekhamer, W.K. (2013). *Anti-Corrosion Methods Mater.* https://doi.org/10.1108/00035591311287429.

45 John, S., Kuruvilla, M., and Joseph, A. (2013). *Res. Chem. Intermed.* https://doi.org/10.1007/s11164-012-0675-x.

46 Belkaid, S., Tebbji, K., Mansri, A. et al. (2012). *Res. Chem. Intermed.* https://doi.org/10.1007/s11164-012-0547-4.

47 Awad, M.K., Metwally, M.S., Soliman, S.A. et al. (2014). *J. Ind. Eng. Chem.* https://doi.org/10.1016/j.jiec.2013.06.009.

48 Solomon, M.M. and Umoren, S.A. (2015). *J. Environ. Chem. Eng.* https://doi.org/10.1016/j.jece.2015.05.018.

49 Umoren, S.A., Eduok, U.M., and Oguzie, E.E. (2007). *Port. Electrochim. Acta* https://doi.org/10.4152/pea.200806533.

50 He, C., Tian, Z., Zhang, B. et al. (2015). *Ind. Eng. Chem. Res.* https://doi.org/10.1021/ie504616z.

51 Solomon, M.M. and Umoren, S.A. (2015). *J. Adhes. Sci. Technol.* https://doi.org/10.1080/01694243.2015.1017436.

52 Quraishi, M.A. and Shukla, S.K. (2009). *Mater. Chem. Phys.* https://doi.org/10.1016/j.matchemphys.2008.08.028.

53 Bhandari, H., Choudhary, V., and Dhawan, S.K. (2011). *Synth. Met.* https://doi.org/10.1016/j.synthmet.2011.01.026.

54 Benchikh, A., Aitout, R., Makhloufi, L. et al. (2009). *Desalination* https://doi.org/10.1016/j.desal.2008.10.024.

55 Karthikaiselvi, R. and Subhashini, S. (2012). *Arab. J. Chem.* https://doi.org/10.1016/j.arabjc.2012.10.024.

56 Shukla, S.K. and Quraishi, M.A. (2012). *J. Appl. Polym. Sci.* https://doi.org/10.1002/app.35668.

57 Mansri, A., Bouras, B., Hammouti, B. et al. (2013). *Res. Chem. Intermed.*

58 Karthikaiselvi, R. and Subhashini, S. (2014). *J. Assoc. Arab Univ. Basic Appl. Sci* https://doi.org/10.1016/j.jaubas.2013.06.002.

59 Sabirneeza, A.A.F. and Subhashini, S. (2014). *Int. J. Ind. Chem.* https://doi.org/10.1007/s40090-014-0022-8.

60 Subhashini, S. and Sabirneeza, A. (2011). *E. J. Chem.* 8: 671–679.

20

Epoxy Resins and Their Nanocomposites as Anticorrosive Materials

Omar Dagdag[1], Rajesh Haldhar[2], Eno E. Ebenso[3], Chandrabhan Verma[4], A. El Harfi[5], and M. El Gouri[1]

[1] *Laboratory of Industrial Technologies and Services (LITS), Department of Process Engineering, Height School of Technology, Sidi Mohammed Ben Abdallah University, Fez, Morocco*
[2] *School of Chemical Engineering, Yeungnam University, Gyeongsan, South Korea*
[3] *Institute for Nanotechnology and Water Sustainability, College of Science, Engineering and Technology, University of South Africa, Johannesburg, South Africa*
[4] *Interdisciplinary Research Center for Advanced Materials, King Fahd University of Petroleum and Minerals, Dhahran, Saudi Arabia*
[5] *Laboratory of Advanced Materials and Process Engineering, Department of Chemistry, Faculty of Sciences, Ibn Tofaïl University, Kenitra, Morocco*

20.1 Introduction

Epoxy resins (ERs) are unique class of material with large number of industrial applications. Among the ERs, those made from condensation of bisphenol A and epichlorohydrin received the most attention due to their superior mechanical, rheological, and anticorrosive properties. Therefore, they have been widely used in industry in applications such as adhesives, coatings, laminates, encapsulating materials, electronics, and in making composite [1–4]. The synthetic methods and characterization procedures of bisphenol A and epichlorohydrin type of resins were covered thoroughly in the literature [5–10]. Aside from above, ERs are profoundly compressive materials that have fantastic consumption obstruction, high rigidity, protection from actual maltreatments, and prevalent weakness strength properties [11].

Organic Corrosion Inhibitors: Synthesis, Characterization, Mechanism, and Applications,
First Edition. Edited by Chandrabhan Verma, Chaudhery Mustansar Hussain, and Eno E. Ebenso.

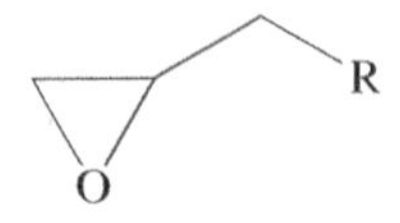

Figure 20.1 Epoxy cycle found in polyepoxides.

In this chapter, we present some synthesized ERs conducted in recent years by different researchers in our laboratory to try to highlight as corrosion inhibitors in aqueous, as well as in coating phase. The term "epoxy resins" designates a wide variety of prepolymers containing one or more epoxy groups (or oxirane) (Figure 20.1) [12, 13].

These resins are most often prepared from epichlorohydrin in two stages:

- **First step:** condensation of epichlorohydrin and formation of α-chlorohydrins (Figure 20.2).
- **Second step:** dehydrohalogenation of α-chlorohydrins and regeneration of epoxy cycles by the action of an alkali metal (Figure 20.3).

ERs were discovered almost simultaneously by the Swiss P. Castan (1939) and by the American S.O. Greenlee (1939).

There are several epoxy prepolymers on the market (bisphenol formaldehyde, phenol novolaks, cresol-novolak, etc.), but DGEBA is the most widespread epoxy prepolymer, with production representing 95% of the world tonnage of epoxy prepolymers [14].

20.2 Characteristic Properties of Epoxy Resins

ERs are appreciated for their unique properties such as their low weight, their resistance to corrosion, or their adhesive character [15]. These polymers also have good mechanical properties in terms of traction, bending, or compression even if their impact resistance constitutes their main weak point.

Other advantages such as their low coefficient of expansion (30 à 60×10^{-6}/K depending on whether the resin is loaded or not), their low thermal conductivity (≈0.2 $W.m^{-1}/K$), and their high resistance to humidity are also worth noting. These properties, associated with a high glass transition temperature (T_g) allowing them to be maintained at high temperature, are all reasons why polyepoxides are

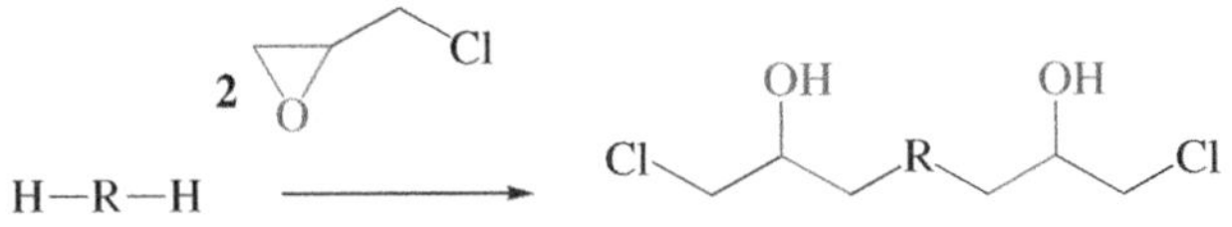

Figure 20.2 Formation of α-chlorohydrins.

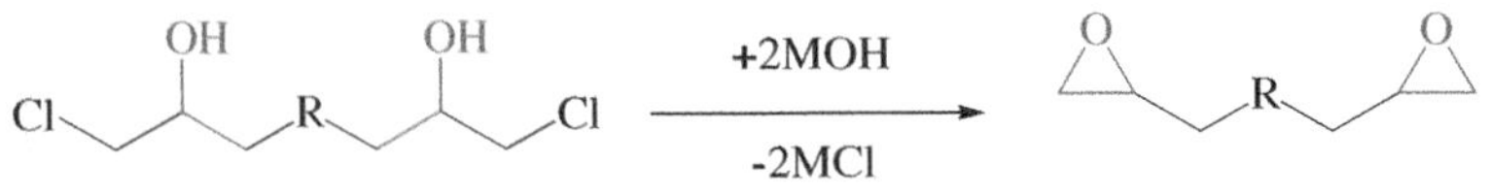

Figure 20.3 Formation of epoxy rings.

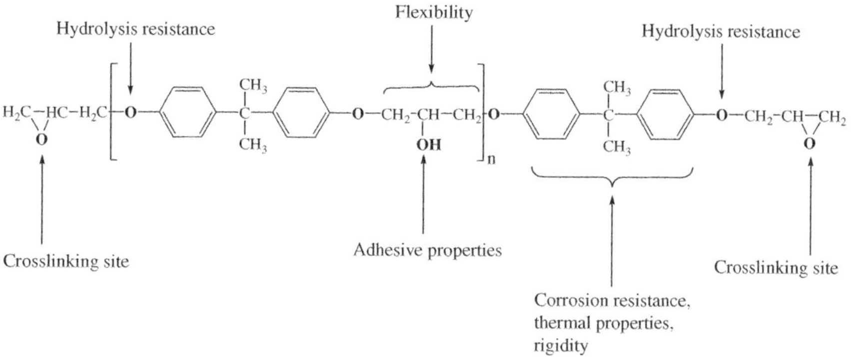

Figure 20.4 Structural formula of the DGEBA – relationship between structure and properties.

so widely used in fields as varied as the aeronautics and automotive industries, resistant coating, sport, or the electrical and electronic industries.

To better understand the origin of the properties of a polyepoxide, one must look at its chemical structure. If we take the example of a resin based on bisphenol A diglycidyl ether (DGEBA) (Figure 20.4), it is observed that each chemical group has an influence on the final properties of the cross-linked material. Ether bridges provide good resistance to hydrolysis so that hydrolysis can only be carried out completely under specific conditions [16]. The alcohol functions exhibit adhesive properties in particular because of their labile hydrogen. The flexibility of the material should be associated with the aliphatic parts, while the aromatic parts are responsible for good corrosion resistance, as well as the thermal and mechanical properties of the resin.

Finally, the cross-linking that makes it possible to obtain a three-dimensional network is provided via the oxirane groups. This cross-linking is traditionally carried out using cross-linking agents commonly called hardeners, although some work has reported homopolymerization of the resin [17, 18]. Its structural formula is presented in Figure 20.4 (where *n* represents the degree of polymerization).

20.3 Main Commercial Epoxy Resins and Their Syntheses

20.3.1 Bisphenol A Diglycidyl Ether (DGEBA)

The first epoxy prepolymer to appear on the market was DGEBA. Its structural formula is presented in Figure 20.5.

The properties of DGEBA gum rely upon the quantity of rehashing units. Low subatomic weight particles will in general be fluid and higher subatomic weight atoms will in general be more thick fluids or solids [19, 20]. Yang et al. [21]

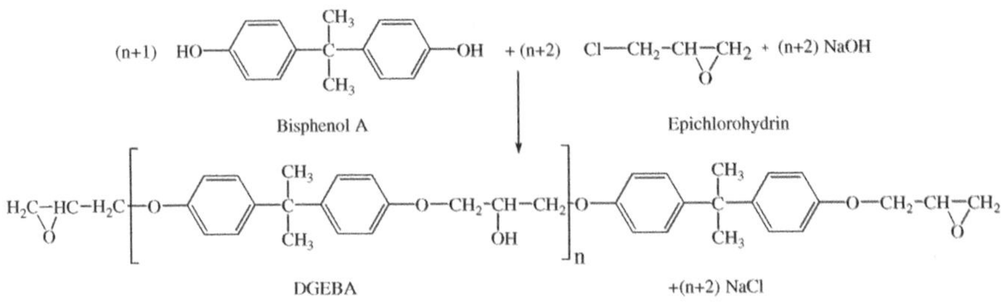

Bisphenol A + Epichlorohydrin → DGEBA + (n+2) NaCl

Figure 20.5 Synthesis of bisphenol A diglycidyl ether (DGEBA).

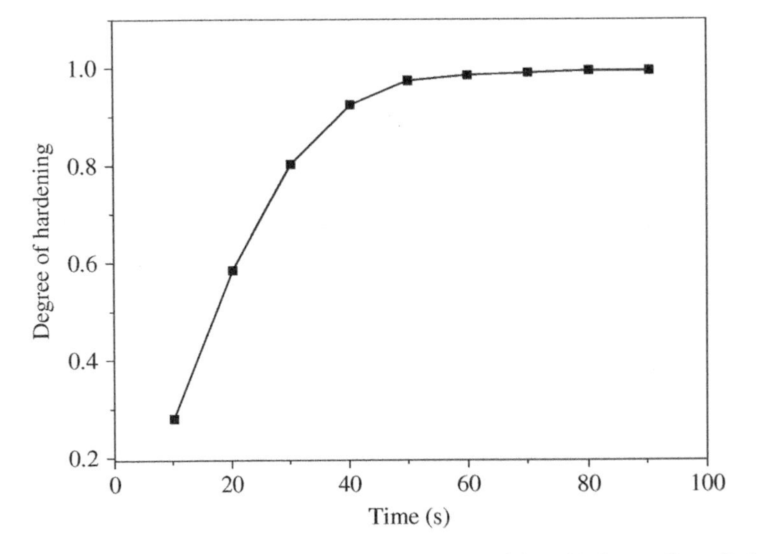

Figure 20.6 Degree of hardening as a function of time (s). *Source:* From Ref. [21].

combined a low thickness epoxy sap by response of polyethylene glycol and polyepoxide (DGEBA). The polyepoxide restored utilizing cationic photo-initiator under UV light and the level of fix of epoxy tar was over 90% in 40s, as appeared in Figure 20.6.

Czub [22] synthesized ERs of high molecular weight from modified natural oils and ERs based on bisphenol A. The ERs obtained are very viscous liquids. Wu et al. [23] synthesized two-step liquefied bamboo bisphenol A copolymer ERs. The curing process of the epoxy copolymer resin can occur at room temperature after addition of triethylenetetramine, the curing being an exothermic reaction.

20.3.2 Cycloaliphatic Epoxy Resins

3,4-Epoxy cyclohexane 3′, 4′ epoxy cyclohexylmethyl (CAE) cycloaliphatic ER is synthesized by reacting 3′-cyclohexenyl methyl 3-cyclohexene carboxylate with peracetic acid. Figure 20.7 shows the compound structure of CAE. This epoxy pitch has an aliphatic spine and a completely immersed subatomic structure, which adds to its great UV solidness, great climate opposition, great warm strength, and phenomenal electrical properties. These properties are essential for gums used to manufacture underlying segments requiring application in high temperature conditions [24, 25].

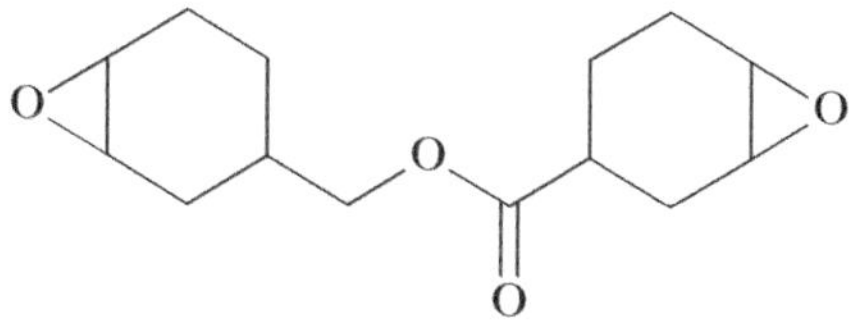

Figure 20.7 Chemical structure of CAE.

Tao et al. [26] orchestrated imide ring and silica containing siloxane, 1, 3-bis [3-(4,5-epoxy-1, 2, 3, 6-tetrahydrophthalimido) propyl] tetramethyldisiloxane (BISE) by a two-venture strategy. Figure 20.8 shows the synthetic structure of BISE.

Completely restored BISE epoxy pitch has great warm strength and a moderately low glass change temperature (T_g) contrasted with industrially accessible CAE. In any case, Gao et al. [27] have likewise incorporated straightforward cycloaliphatic epoxy-silicone saps by a two-venture response course for use in bundling optoelectronic gadgets. In correlation with CAE, the relieved cycloaliphatic epoxy-silicone gums showed better warm solidness, lower water assimilation, and higher UV/warm opposition.

20.3.3 Trifunctional Epoxy Resins

Tetrafunctional epoxy gums are integrated by responding 1, 3-diaminobenzene or 4, 4′- aminodiphenylmethane with epichlorohydrin. Figure 20.9 shows the substance structures of these epoxy tars.

These epoxy tars have high epoxy usefulness and high cross-link densities also, and hence, they are utilized in applications where high temperature obstruction is required. Restored epoxy saps have astounding compound obstruction, high modulus, great UV hindering impact, and great warm dependability [28–30].

Aouf et al. [31] have blended a multifunctional epoxy tar. The relieved epoxy tar showed a higher cross-linking thickness and higher carbon yield than the restored DGEBA under similar conditions.

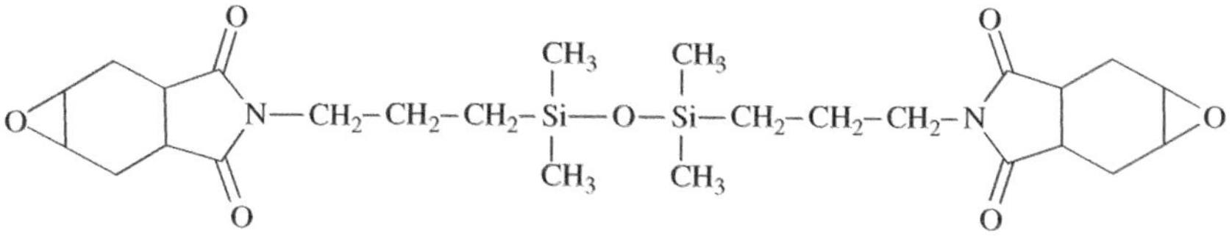

Figure 20.8 Chemical structure of BISE.

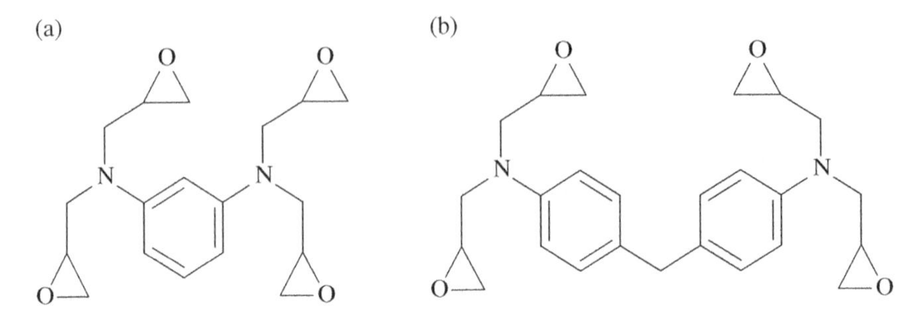

Figure 20.9 Chemical structures of tetra-functional epoxy resins.

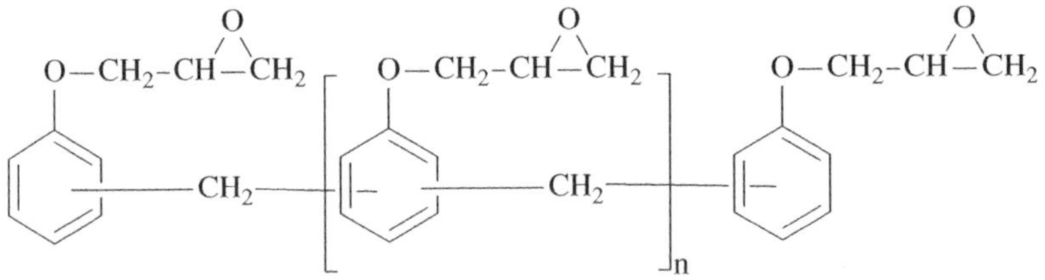

Figure 20.10 Chemical structure of epoxy novolac resins.

20.3.4 Phenol-Novolac Epoxy Resins

Epoxy novolak tars are glycidyl ethers of phenolic novolak tars that have been incorporated by responding phenolic novolak gum with epichlorohydrin. Figure 20.10 shows the compound structure of epoxy novolac gums. The numerous epoxy bunches in epoxy novolak saps add to their high cross-linking densities, bringing about incredible warm, substance, and dissolvable obstruction properties [32].

Lin et al. [33] have orchestrated multifunctional epoxy pitches from an expansion response of a phosphorus item, a monofunctional phosphinate, 9,10-Dihydro-9 oxa-10-phosphaphenanthrene 10-oxide (DOPO), and epoxy novolak tar. Because of its P-H bond, which can bond with epoxy pitch, this synthetic alteration makes epoxy gum fire resistant ordinarily.

20.3.5 Epoxy Resins Containing Fluorine

The utilization of fluorinated monomers to alter epoxy tars is of interest on the grounds that the presence of fluorine improves extraordinary attributes, for example outstanding synthetic opposition, low coefficient of contact, low dielectric steady, low water assimilation, and expanded use temperature [12, 34–36]. Park et al. [37] incorporated a DGEBA epoxy sap containing a CF3 gathering, the substance structure of which is appeared in Figure 20.11.

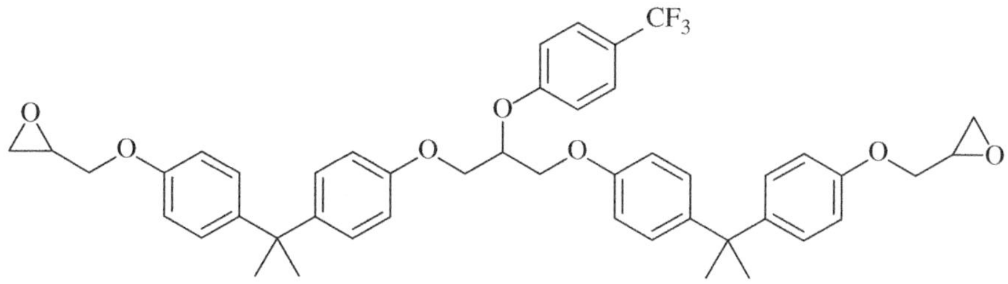

Figure 20.11 Chemical structure of DGEBA resin containing CF_3 groups.

20.3.6 Epoxy Resins Containing Phosphorus

Phosphorus mixes could give high fire retardancy to epoxy tars by repressing gas stage fire and enhancing coals in the dense stage. They have additionally been found to produce less poisonous gas and smoke than mixes containing incandescent light. The joining of covalently bound phosphorus into epoxy pitches could be accomplished utilizing phosphorus-containing oxirane mixes [38–41].

Liu et al. [39] incorporated epoxy saps containing phosphorus for naturally agreeable reusable electronic bundling materials, as appeared in Figure 20.12. The restored epoxy saps were straightforward and shown a high T_g of 227 °C and a high mechanical modulus.

Wang et al. [40] then have arranged epoxy tars containing phosphorus for use as fire-resistant materials. The exploratory outcomes showed that the synergistic impacts coming about because of the blend of the epoxy sap containing phosphorus and the hardener containing nitrogen give an incredible improvement in fire-resistant conduct. Enfin, Liu et coll. [41] have orchestrated hexabis (4-hydroxymethylenephenoxy) cyclotriphosphazene (PN-OH), the last will respond with DGEBA to at long last acquire cyclophosphazene containing the

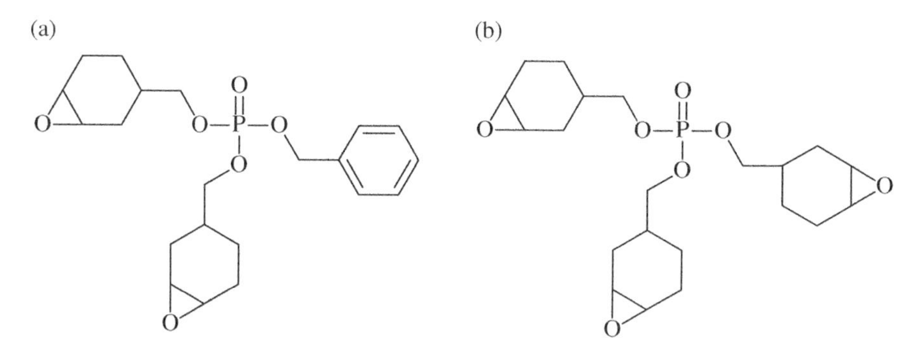

Figure 20.12 Chemical structures of epoxy resins containing phosphorus.

epoxy gathering (PN-EP). The gum in this way incorporated was solidified by sweet-smelling diamine hardeners. Thermogravimetric investigation of arranged examples shows better warm solidness at raised temperature and decay and incredible fire opposition (Figure 20.13).

20.3.7 Epoxy Resins Containing Silicon

Silicon is viewed as a biological fire resistant in light of the fact that it has a less harming sway on the climate than existing materials. Epoxy saps containing silicon can be incorporated by the accompanying two methodologies. One technique is to bring siloxanes into the epoxy aggravates utilizing hydrosilylation responses. Another strategy utilizes transetherification between the alkoxylsilane and glycidol or the build-up of hydroxyl ended siloxane with epoxy pitches or epichlorohydrin. These epoxy gums have focal points of both epoxy saps and silicone gums [28, 42–46]. Mercado et al. [43] integrated epoxy pitches containing silicon, and their compound structures are appeared in Figure 20.14. Relieved epoxy tars have a moderate T_g and a high limiting oxygen index esteem.

Liu et al. [44] incorporated two novel cycloaliphatic epoxy gums containing silicon for electronic bundling applications (Figure 20.15). Restored epoxy tars show high T_g, great warm soundness, and great mechanical properties.

Park et al. [45] incorporated epoxy pitches containing silicon (DGEBA-Si) by expansion of DGEBA with dichlorodiphenylsilane utilizing triphenylphosphine as an impetus. The relieved epoxy gum had a lower T_g and improved mechanical properties than the unadulterated DGEBA epoxy pitch.

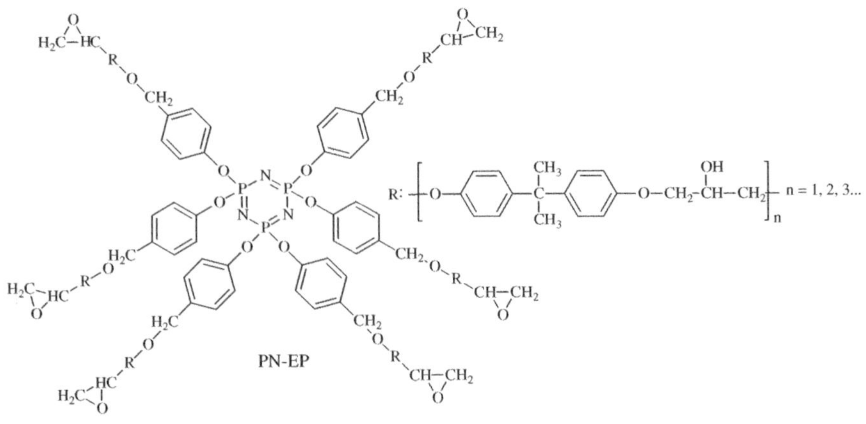

Figure 20.13 Chemical structure of PN-EP.

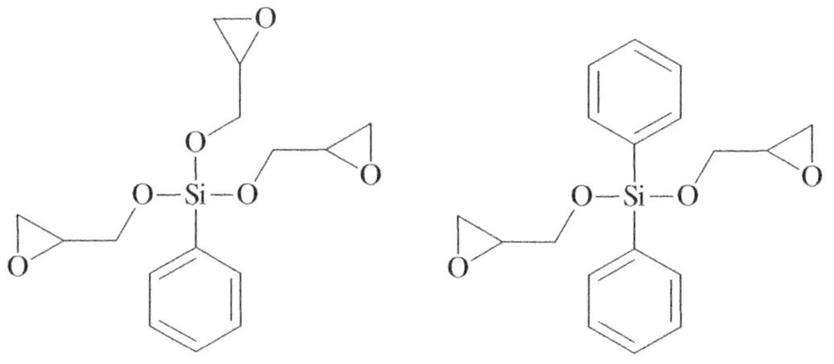

Figure 20.14 Chemical structures of epoxy resins containing silicon.

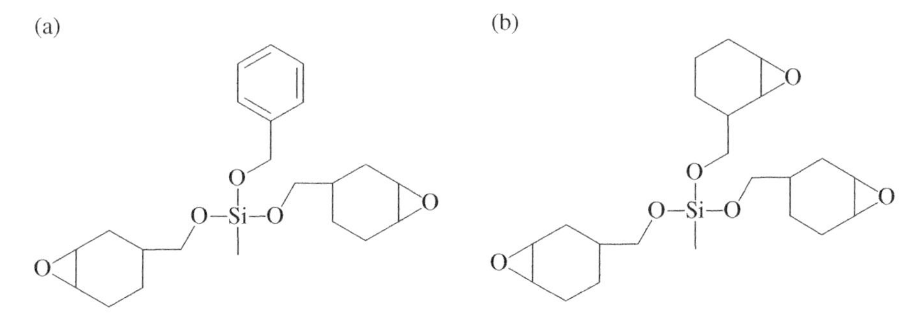

Figure 20.15 Chemical structures of epoxy resins containing silicon.

20.4 Reaction Mechanism of Epoxy/Amine Systems

The amine functions of the cross-linking agent can react with two epoxy groups. Their reactivity depends essentially on their basicity. This is because an aliphatic amine is much more reactive than an aromatic amine. First, the primary amine reacts with an epoxy group, creating a secondary amine, which in turn can react with another epoxy group (Figure 20.16).

The reactivity of primary and secondary amines to epoxies cannot be differentiated due to the observation of a unique activation energy (E_a) and reaction enthalpy (ΔH_a) [47]. The reactivity ratio between these two reactions (k_2/k_1) varies between 0.1 and 1 [15]. The reaction mechanism accepted so far in the literature (Figure 20.16) involves reactive complexes [48] in which a hydrogen bridge is formed between the nitrogen atom of the amine and the oxygen atom of the epoxy group.

However, the actual form of these complexes is still debated [47]. These complexes perform two functions, on the one hand, the hydrogen bond weakens the C—O bond of the epoxy group, the carbon therefore becomes more electrophilic,

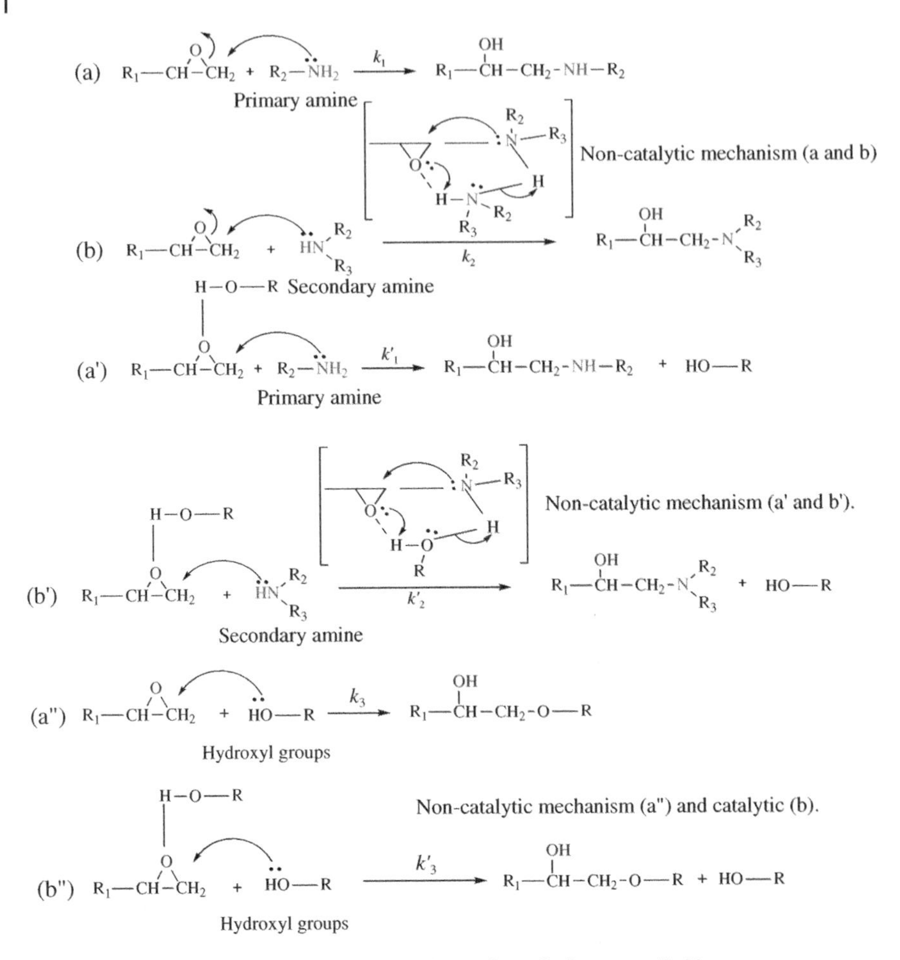

Figure 20.16 Main chemical reactions taking place during cross-linking.

thus facilitating the nucleophilic attack by the nitrogen atom [48]. The formation of such a complex allows the nucleophile to be close enough to the electrophile for enough time for the reaction to take place [48]. A reaction following such a mechanism is considered non-catalyzed [49]. During the reaction between the amine (primary or secondary) and the epoxy group, a hydroxyl function (secondary alcohol) is also formed. These hydroxyl functions have a catalytic effect on the epoxide–amine reaction. In fact, the presence of hydrogen bonds with the oxygen atoms of epoxides facilitates nucleophilic attack [48, 50–53]. In this case, a self-catalyzed reaction is then mentioned because it is the groups resulting from the reaction, which provide the catalysis.

However, the hydroxyl functions are also likely to initiate an etherification reaction with the epoxy functions. This is then the homopolymerization of the ER. This last reaction is in competition with the two preceding ones [15]. It is catalyzed by tertiary amines formed during cross-linking. This reaction has only been highlighted very little because of the special conditions it requires [54]. It can take place in the case of aromatic amines [54] due to the low reactivity of secondary amines compared to primary amines [15]. However, it does not intervene below 200 °C [47]. The mechanism proposed to explain the action of tertiary amine breaks down into three stages: initiation, propagation, and termination [53].

20.5 Applications of Epoxy Resins

Epoxy polymers are important thermosetting materials and can be used in the form of varnishes, but also of paints, adding colorants, additives, or pigments, for their exceptional physical, mechanical, or chemical properties. They can be obtained by adding fillers, in particular zinc phosphates, iron oxides, or metal powder. Finally, the hardeners used in the polymerization of epoxy prepolymers are of great importance. They are the ones who determine how and in which application the finished product is best used.

20.5.1 Epoxy Resins as Aqueous Phase Corrosion Inhibitors

One option in contrast to conventional consumption inhibitors is the utilization of polymer composites [55]. Because of their enormous subatomic size, they give powerful surface inclusion and decreased consumption even at generally low fixations [56]. In such manner, a few classes of regular and engineered polymers are utilized as successful consumption inhibitors [57]. Polar useful gatherings of polymer structure edifices with metal outlines because of coordination bonds. Polymers can be successfully utilized as materials for against consumption coatings; nonetheless, their utilization as watery erosion inhibitors is restricted because of their restricted dissolvability in polar electrolytes [58]. Subsequently, a few endeavors have been made to expand the dissolvability of polymers in the polar electrolyte media. One of the fundamental endeavors is to add polar substances to the polymer chain. Polar (hydrophilic) bunches increment their solvency and go about as adsorption focuses [57]. Another elective strategy is to utilizable mindboggling polymeric materials, for example macromolecules, for example epoxy saps and oligomers, as consumption inhibitors [59]. Macromolecules can be utilized in watery conditions just as covering conditions. Numerous oligomers and macromolecules utilize watery erosion inhibitors. This audit is an assortment of a portion of the key distributed reports on epoxides as erosion inhibitors. Apparently, this will be the primary thorough and itemized report on epoxy gums as consumption inhibitors.

Albeit most epoxy pitches are polymer, they have restricted solvency in polar electrolyte media, including HCl and NaCl arrangements. In any case, a few, which have a generally little atomic size and/or some fringe polar practical gatherings, show great solvency in these electrolytic media [60]. In an acidic climate, an epoxy ring can go through a ring opening response to shape an open chain structure as demonstrated in Figure 20.17 [61].

In our previous work [5, 6, 9, 62–69], we have prepared other ERs. A portion of the principle investigates the antimicrobial movement of epoxy pitches are appeared in Table 20.1. The anticorrosive movement of all mixes was assessed by electrochemical strategies. This investigation shows that these mixes give great CS security against consumption restraint in 1 M HCl. The after effect of the electrochemical test shows that these mixes have an erosion hindering impact and the repressing proficiency increments with expanding inoculums focus. Electrochemical outcomes show that ERs go about as generally great inhibitors of carbon steel with a normal substance of 1 M HCl, and their viability is in the request for: ER3 (98,1%) > ER5 (97,3%) > ER2 (96,5%) = ER11 (96,5 %) > ER4 (95,8%) > ER1 (95,6%) > ER7 (95,4%)> ER10 (95,0%) > ER6 (94,3%) > ER9 (92,9%) > ER8 (91,7%) (Table 20.1).

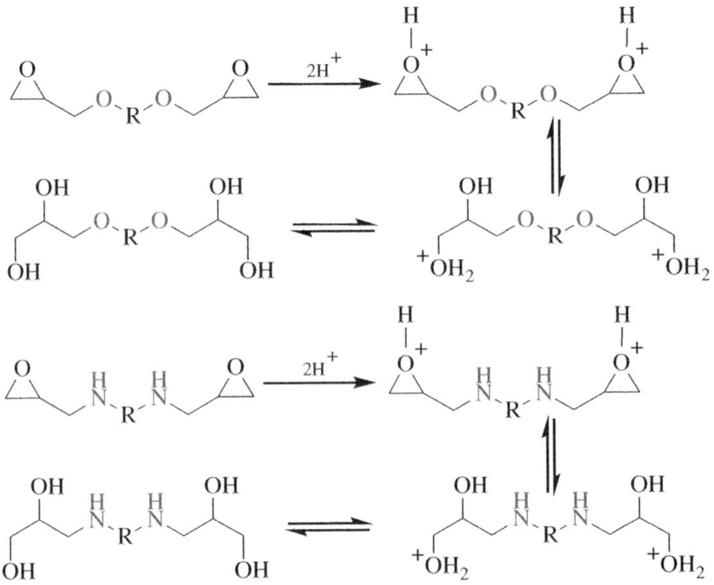

Figure 20.17 Mechanism of ring opening reaction of epoxy resins in acid solution.

Table 20.1 Abbreviations, nature of metals, electrolytes, adsorption behavior, optimum concentration, and maximum protection efficiency of some common epoxy resins evaluated as aqueous phase corrosion inhibitors.

Abbreviation	Nature of metal & electrolyte	Nature of adsorption	%IE & optimum conc.	References
ER1	CS/1M HCl	Anodic type	95.6 at 10^{-3}	[5]
ER2	CS/1M HCl	Cathodic type	96.5 at 10^{-3}	[5]
ER3	CS/1M HCl	Anodic type	98.1 at 10^{-3}	[6]
ER4	CS/1M HCl	Anodic type	95.8 at 10^{-3}	[6]
ER5	CS/1M HCl	Mixed type	97.3 at 10^{-3}	[9]
ER6	CS/1M HCl	Mixed type	94.3 at 10^{-3}	[62]
ER7	CS/1M HCl	Mixed type	95.4 at 10^{-3}	[62]
ER8	CS/1M HCl	Mixed type	91.7 at 10^{-3}	[63]
ER9	CS/1M HCl	Mixed type	92.9 at 10^{-3}	[63]
ER10	CS/1M HCl	Anodic type	95.0 at 10^{-3}	[64]
ER11	CS/1M HCl	Mixed type	96.5 at 10^{-3}	[7]
ER12	CS/1M HCl	Cathodic type	91.3 at 10^{-3}	[8]
ER13	CS/1M HCl	Mixed type	94.2 at 10^{-3}	[65]
ER14	CS/1M HCl	Mixed type	91.0 at 10^{-3}	[69]
ER15	E24 CS/1M HCl	—	88.0 at 10^{-3}	[70]
ER16	CS/1M HCl	Mixed type	95.02 at 10^{-3}	[71]
ER17	CS/1M HCl	Cathodic type	94.0 at 10^{-3}	[72, 73]
ER18	CS/1M HCl	Cathodic type	95.0 at 10^{-3}	[72, 73]
ER19	CS/1M HCl	Mixed type	94.4 at 10^{-3}	[72, 73]
ER20	CS/1M HCl	Mixed type	94.3 at 10^{-3}	[72, 73]
ER21	CS/1M HCl	Cathodic type	93.2 at 10^{-3}	[74]
ER22	CS/1M HCl	Cathodic type	93.5 at 10^{-3}	[74]

In our previous work [5, 6], we have synthesized four ER-based hydroxyl (ER1, ER2, ER3, and ER4). Their structures are shown in Figure 20.18. The corrosion prevention activity of four compounds was evaluated by electrochemical methods. This study showed that these compounds provide good CS protection against corrosion inhibition in 1 M HCl. The result of the electrochemical study shows that these four compounds have an extraordinary inhibitory effect on corrosion, and the inhibitory activity increases with increasing concentration of inhibitor. The highest inhibition efficiency at 10^{-3} M (optimal concentration) was 98.1, 96.5,

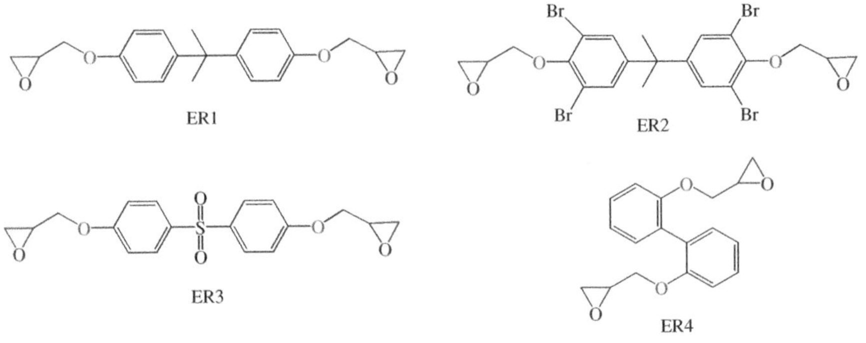

Figure 20.18 Chemical structures of epoxy resins containing diols.

95.8, and 95.6% for ER3, ER2, ER4, and ER1, respectively (Table 20.1). High inhibition efficiencies are attributed to their association with larger molecular sizes and the presence of dense conjugation in the form of two aromatic rings and heteroatoms. Electrochemical results showed that ER1 and ER2 mainly act as anodic corrosion inhibitors and cathodic corrosion inhibitors, respectively. Electrochemical results show that ER3 and ER4 act as anodic and intermediate corrosion inhibitors. Thermodynamic studies have shown that ERs themselves are adsorbed using the chemisorption mechanisms.

In our past work [9, 62–64], we incorporated six amines dependent on epoxy tars (ER5, ER6, ER7, ER8, ER9, ER10, ER11, and ER12). Its construction is appeared in Figure 20.19.

The antimicrobial action of six mixes was assessed by electrochemical techniques. This examination demonstrated that these mixes give great CS assurance

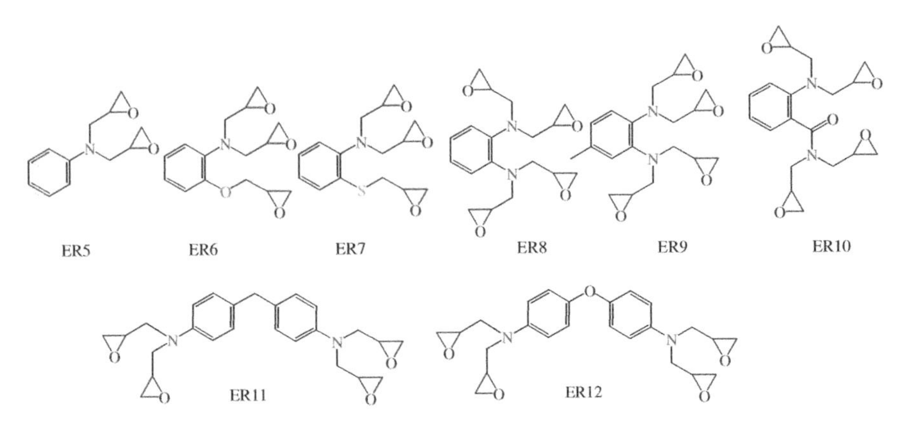

Figure 20.19 Chemical structures of epoxy resins containing amines.

against erosion in 1 M HCl. The consequence of an electrochemical investigation shows that these six mixes have incredible enemy of consumption activity and the inhibitory action increments with the centralization of the inoculators. Electrochemical outcomes show that ERs act as reasonably great inhibitors of carbon steel in a 1 M HCl climate, and their efficiencies are in the request for ER5 (97.3%) > ER7 (95.4%) > ER10 (95.0%) > ER6 (94.3 %) > ER9. (92.9%) > ER8 (91.7%) (Table 20.1). A plain polarization study demonstrated that ER5 goes about as a blended inhibitor with generally cathodic movement, though ER6, ER7, and ER10 are blended inhibitors and ER8 and ER9 are blended inhibitors with minimal anodic strength.

The anticorrosive properties of ER11 and ER12 mixes were assessed for the erosion of CS in 1 M HCl arrangement by electrochemical strategies. Electrochemical examinations show that ER11 and ER12 work as rationale inhibitors that are legitimately acceptable at 1 M HCl and show the most noteworthy adequacy for ER11 and ER12 at 10^{-3} M with 96.5 and 91.3%, individually (Table 20.1). The PDP study indicated that ER11 shows minimal anodic predominance and ER12 acts essentially as a cathode-type inhibitor.

Hsissou et al. [65–69] have arranged other phosphorus-based epoxy gums ER13 and ER14. Its construction is appeared in Figure 20.20. The antimicrobial movement of four mixes was assessed by electrochemical strategies. This investigation demonstrated that these mixes give great CS insurance against erosion in 1 M HCl for mixes (ER13 and ER14). Electrochemical outcomes show that ERs (ER13 and ER14) go about as CS inhibitors, respectively, in 1 M HCl, and their movement is in the request: ER13 (94.18%) > ER14 (91%) (Table 20.1).

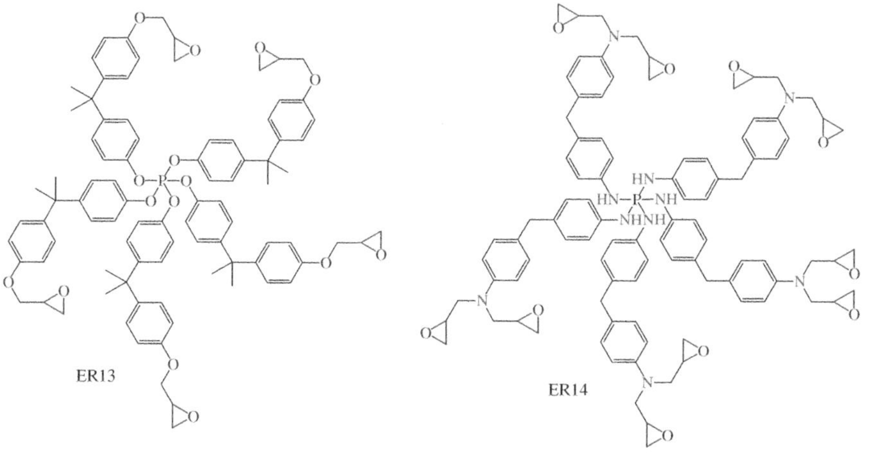

Figure 20.20 Chemical structures of epoxy resins containing phosphorous.

Hsissou et al. [70, 71] reported the preparation of the ERs (ER15 and ER16) synthesized and 2, 4, 6-trichloro-1,3, 5-triazine was evaluated and investigated as corrosion inhibitor of steel in 1 M HCl solution. Their structures are shown in Figure 20.21.

Electrochemical studies demonstrate that ER15 and ER16 act as sensibly good inhibitors for steel in 1 M HCl medium and showed highest efficiency of as high as 88 and 95,02% at 10^{-3} M for the ER15 and ER16 compounds, respectively.

Rbaa et al. [72–74] reported the preparation of the ERs (ER17, ER18, ER19, ER20, ER21, and ER22) and glucose subsidiaries are planned as a characteristic calming specialist for CS in the 1M HCl pathway. The designs are appeared in Figure 20.22. Detestable security measures for the four mixes were surveyed by electrical methods. This investigation shows that these mixes give great assurance to CS from rot as 1 M HCl, and the defensive impact up to the ideal measure of 95, 94.4, 94.3, and 94%, 93, 48, and 93.20 individually for ER18, ER19, ER20, ER17, ER22, and ER21 arrangements. The after effects of the PDP show that the four authorizations of ER17, ER18, ER21, and ER22 are destructive to cathode-type variations and ER19 and ER20 are held on the iron as a blend. An exceptionally negative measure of Gibbs energy estimation proportions (44.41–48 kJ/mol) indicated that these mixes were unequivocally connected with the iron surface.

20.5.2 Epoxy Resins as Coating Phase Corrosion Inhibitors

The expression "polymer" is utilized to depict macromolecules portrayed by numerous rehashes of at least one unit (monomer (s)) joined to one another in adequate amount to give a bunch of properties [75]. Based on their root, they are gathered into regular and engineered [75]. In the class of common polymers, we discover cellulose, starch, gum arabic, dextran, chitosan, and so on Manufactured polymers have three subdivisions: thermoplastics (e.g. polyethylene, polyvinyl chloride, and so forth),

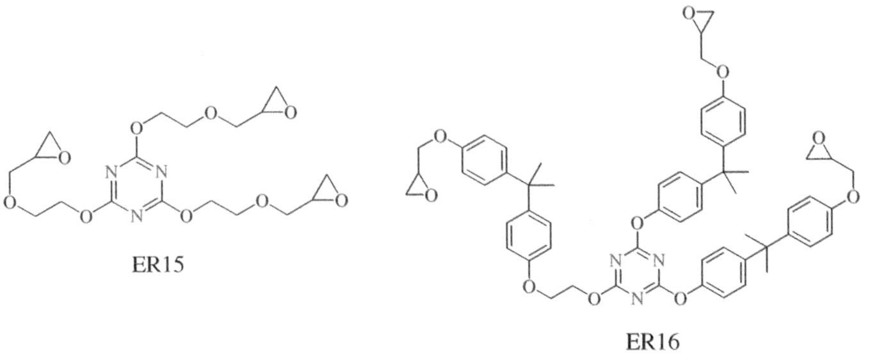

Figure 20.21 Chemical structures of epoxy resins containing 2, 4, 6-trichloro-1,3, 5-triazine.

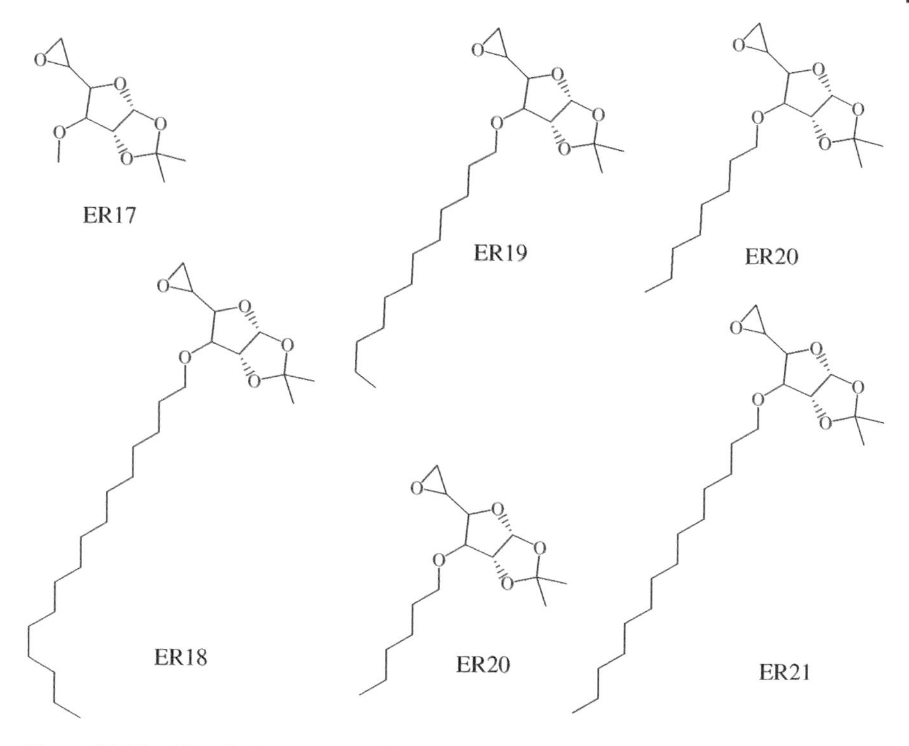

Figure 20.22 Chemical structures of epoxy resins containing glucose derivatives.

thermosets (e.g. epoxy tar, melamine, urea formaldehyde, and so on), and elastomers (e.g. normal elastic). Polymers have profited by a wide scope of uses, including car, aviation, development, toy producing, stains, boat bodies, pastes; and so on. They have appealing highlights that present as key possibility for the interfacial adjustment against metal erosion. This incorporates accessibility, natural neighborliness, cost adequacy, characteristic strength, and numerous adsorption places [75].

20.5.3 Composites of Epoxy Resins as Corrosion Inhibitors

A composite can be characterized as a material that contains at least two synthetically particular segments, isolated by an unmistakable interface, and with special properties [76]. The principle contrast among composites and nanocomposites is the high surface zone to volume proportion of nanoparticles [77]. Because of this high surface territory to volume proportion, a nanocomposite shaped with similar constituents as that of a composite would have qualities just unique in relation to those of the composite.

An assortment of polymers [78–81], including epoxy pitches [82–84], polyesters [85], and polyurethanes [86, 87], are referred to fill in as defensive coatings for metal substrates. They have great grip and adaptability properties contrasted with conductive polymer coatings, yet come up short on the intrinsic conductivity and mechanical properties of conductive polymers [88]. To improve their properties and increment their defensive adequacy, compositing is considered an applied designing methodology [89]. Most compositing is finished with inorganic fillers [90]. Inorganic fillers give coatings a bunch of properties, for example great perspective proportion, solidness, consistency, and adaptability while the host network shields the fillers from unfavorable natural conditions and keeps up their situation all through the lattice [91, 92]. Regularly utilized inorganic fillers incorporate SiO_2 [93–95], TiO_2 [96–98], graphene oxide [99–101], Zn particles [102], dirt [103], and Al_2O_3 [104, 105]. The huge surface territory and smooth non-permeable surfaces of the fillers advance contact between the fillers and the polymer framework. Xia et al. [106] found that epoxy networks where the application in a gamma radiation climate is fundamentally restricted because of antagonistic impacts of debasement could be valuable if the graphene oxide having a critical revolutionary forager conduct is consolidated in epoxy pitch. Alhumade et al. [91] saw that polyetherimide–graphene composite coatings gave long haul assurance of the Cu surface presented to the NaCl climate. A few ordinary epoxy and polymer coatings strengthened with inorganic fillers have been appeared to display predominant erosion opposition execution [84]. Nonetheless, the homogeneous scattering of inorganic fillers in the polymer grid is a genuine test as the high surface energy particles will in general agglomerate and agglomeration causes' helpless collaboration [107, 108]. All things considered, a homogeneous scattering can be gotten by changing the outside of the fillers, which can be physical or substance [76].

The actual alteration of the surface has to do with the utilization of surfactants, which could prompt particular adsorption of polar gatherings on a superficial level at high energy of charges by means of electrostatic cooperation [109], while compound change depends on the covalent connection of a modifier to keep away from desorption because of the filling surface. In electro polymerization, Grari et al. [107] had proposed the utilization of low recurrence (20 kHz) ultrasonic illumination as a method for acquiring an appropriate homogeneous scattering.

20.5.4 Nanocomposites of Epoxy Resins as Corrosion Inhibitors

Propelled ordinarily, ace scientific expert with staggering abilities (bones, for instance, are made [74] progressive nanocomposites produced using clay tablets and natural binders) [110], researchers had the option to fuse inorganic materials at the nanoscale into a polymer grid [111–113]. It is fascinating to take note of that the presentation of a limited quantity of nanoparticles in the polymer frameworks offers the chance of fundamentally improving the properties of the polymer and can likewise give new attributes to the polymer. By definition, a nanocomposite is a composite of which one of the parts has a measurement short of what one micron (by and large <0.1 μm, is 100 nm) [77].

A few strategies are accessible for the blend of polymer nanocomposites and the decision of the amalgamation strategy relies upon the focus on the field of utilization [77]. For application in the field of corrosion protection, techniques, for example wet substance decrease [114], sonochemistry [115], microemulsion [116], photochemical decrease [117], laser intervened [118], aqueous [119], microwave helped [120], just as the biogenic strategy [121, 122] have been utilized for the blend of polymer nanocomposites. These techniques can be assembled into two primary gatherings: ex situ and in situ strategies. The ex situ strategy incorporates all the strategies by which the nanoparticles are shaped external to the host polymer. Pre-assembled nanoparticles are straightforwardly joined into the host polymer grid [123]. The benefits of ex situ strategies incorporate the following: (i) they permit exact power over the size, shape and molecule thickness [77], (ii) reasonable for enormous scope modern creation [77]. The test with ex situ procedures is the means by which to homogeneously scatter nanoparticles in the polymer network and how to keep up long haul dependability against accumulation [77]. In situ techniques, as the name recommends, includes the age of nanoparticles straightforwardly inside the polymer grid [77]. The procedures are basic and frequently require a solitary pot produce [77]. The development of nanoparticles inside the host framework forestalls molecule agglomeration and still keeps up great spatial dissemination. No big surprise in situ methods are exceptionally preferred by consumption researchers [121–123]. In any case, the disadvantage of in situ methods is that un-reacted educts could impact the properties of the readied nanocomposite [77].

Organic coatings form a class, artificially safe, utilized in defensive applications coatings is one of the approaches to battle erosion [124, 125]. By and large, epoxy-based coatings are among the most widely recognized mechanical polymers that have been applied to shield different metals from consumption in forceful conditions [126]. Epoxy coatings have been utilized as a primary or designing cement for the development of airplanes, autos, and so on. The great substance obstruction, great grip to the basic metal surfaces, remarkable warm soundness, great mechanical properties, and electrical protecting properties make epoxy saps an ideal covering material for different applications [78, 79].

For this reason, epoxy coatings decrease the consumption pace of the metals by going about as a viable actual boundary between the metals and destructive climate. In any case, inferable from wear and scraped area, epoxy-based coatings likewise neglect to offer long haul consumption insurance. During the restoring cycle of epoxy coatings, the shrinkage of epoxy sap happens and retains water, air from the climate, and this would thusly make microspores in the epoxy coatings. The pores in the covering encourage the dissemination of destructive species, for example oxygen, water, and particles somewhat at metal/epoxy interface and start the erosion cycle and exhaust the covering [80–82].

Nonetheless, practically speaking, all natural coatings are porous to destructive species, for example oxygen, water, and particles somewhat [132, 133]. Over the most recent couple of years, investigation of the exhibition of nanocomposite coatings has been in the focal point of consideration of natural covering researchers because of their advantageous properties [134, 135]. Different nanoparticles have, consequently, been utilized as fortifications to improve coatings' presentation in the destructive conditions, for example ZnO [136], ZrO_2 [137], CeO_2 [138], $CaCO_3$ [139], Fe_2O_3 [140], SiO_2 [141], and TiO_2 [142]. They have been introduced into coatings to investigate their impacts on defensive properties. Summary of some epoxy-based anticorosive coatings is given in Table 20.2.

Table 20.2 Influence of external inorganic and organic additives on anticorrosive coating effect of different ERs.

Metal and electrolyte	Additives	Anticorrosive coatings/ percentage of additive	Inhibition efficiency or R_{ct} (Ωcm^2)
E24 carbon steel/3.5% NaCl	Natural phosphate (NP)	E1 (TGEEBA/MDA)	96%
		E2 (TGEEBA/MDA/5%NP)	99%
E24 carbon steel/3.5% NaCl	Natural phosphate (NP)	E1 (HGEMDA/MDA)	94%
		E2 (TGEEBA/MDA/5%NP)	96.5%
Carbon steel/3% NaCl	Zinc phosphate (ZPH)	DGEDDS/MDA	$R_{ct} = 26971$
		DGEDDS/MDA/5wt%ZPH	$R_{ct} = 59837$
Carbon steel/3% NaCl	Zinc phosphate (ZP)	ER-MDA	$R_{ct} = 17209$
		ER-MDA-ZP	$R_{ct} = 44968$
Carbon steel/3.5% NaCl	Molybdenum disulfide (MoS_2)	EP	$i_{corr} = 8.908 \times 10^{-7}$
		MoS2/EP	$i_{corr} = 7.437 \times 10^{-9}$
		MoS2@PDA/EP	$i_{corr} = 5.838 \times 10^{-10}$ (μAcm^{-2})
Carbon steel/3.5% NaCl	Nanospheres (MCNSs)	Blank	$i_{corr} = 11.4$ (μAcm^{-2})
		WE@MCNS	$i_{corr} = 0.83$ (μAcm^{-2})
N80 steel/10% NaCl	Reduced graphene oxide (RGO)	ER	$i_{corr} = 49.44$ (μAcm^{-2})
		ER-0.4wt.%RGO	$i_{corr} = 0.137$ (μAcm^{-2})
Carbon steel/3.5% NaCl	GO-BMIM-Cl	ER	$i_{corr} = 9.3 \pm 0.3$ (μAcm^{-2})
		GO-BMIM/ER	$i_{corr} = 2.1 \pm 0.6$ (μAcm^{-2})
Carbon steel/3.5% NaCl	OapPOSS		

Source: From Ref. [83]. © Elsevier.

20.6 Conclusion

Epoxy pitches speak to an uncommon gathering of profoundly receptive prepolymers or polymers that contain epoxide gatherings. Their subatomic structures contain a few fringe polar practical gatherings including hydroxyl (-OH), amino (-NH_2) gatherings, and amide (-$CONH_2$) through which they can get effortlessly adsorbed over the metallic surface and go about as successful anticorrosive materials in covering just as arrangement stage. A few ERs in unadulterated and relieved structures have been utilized as hostile to destructive covering materials particularly for CS in 1M HCl medium. Electrochemical (EIS and PDP strategies) exhibitions indicated that a large portion of the ERs went about as interface type and blended sort anticorrosive materials. The anticorrosive impact of the ER coatings can additionally be improved by adding natural and inorganic added substances. These added substances block the surface micropores present in ER coatings through which destructive species can infiltrate or diffuse and euphoria the covering structures.

Abbreviations

ER1 4,4′-isopropylidenediphenol oxirane
ER2 4,4′-isopropylidene tetrabromodiphenol oxirane
ER3 2,2′-(((sulfonylbis(4,1-phenylene)) bis(oxy))bis(methylene)) bis(oxirane)
ER4 2,2′-bis(oxiran-2-ylmethoxy)-1,1′-biphenyl
ER5 Diglycidyl amino benzene
ER6 2-(oxiran-2-yl-methoxy)-N,N-bis(oxiran-2-yl-methyl)aniline
ER7 N, N-bis(oxiran-2-ylmethyl)-2-((oxiran-2-ylmethyl) thio)aniline
ER8 N^1,N^1,N^2,N^2-tetrakis (oxiran-2-ylmethyl)bbenzene-1,2-diamine
ER9 4-methyl- N^1,N^1,N^2,N^2-tetrakis (oxiran-2-ylmethyl) benzene-1,2-diamine
ER10 Tetraglycidyl-1,2-aminobenzamide
ER11 4, 4′-(ethane-1, 2-diyl) bis (*N, N*-bis (oxiran-2-ylmethyl) aniline)
ER12 4,4′-oxybis(*N,N*-bis(oxiran-2-ylmethyl)aniline)
ER13 Pentaglycidyl ether pentabisphenol A of phosphorus
ER14 Decaglycidyl phosphorus penta methylene dianiline
ER15 Triglycidyl ether of triethoxytriazine
ER16 2,4,6-tris(4-(2-(4-(2-methoxy oxiran) phenyl) propan-2-yl) phenoxy)-1,3,5-triazine
ER17 [(5,6-anhydro-3-*O*-methyl-1,2-*O*-isopropyli-dene-α-D-glucofuran-ose)
ER18 5,6-anhydro-3-*O*-hexadecyl-1,2-*O*-isopropylidene-α-D-glucofuranose
ER19 5,6-anhydro-3-*O*-dodecyl-1,2-*O*-isopropylidene-α-D-glucofuranose
ER20 5,6-Anhydro-3-*O*-hexyl-1,2-*O*-isopropylidene-α-D-glucofuranose
ER21 2-dimethyl-6-(octyloxy)-5-(oxiran-2-yl)tetrahydrofuro[2, 3-d][1, 3]dioxole
ER22 2,2-dimethyl-5-((S)-oxiran-2-yl)-6-(tetradecyloxy)tetrahydrofuro[2, 3-d][1, 3]dioxole

References

1 Hsissou, R., Bekhta, A., Dagdag, O. et al. (2020). Rheological properties of composite polymers and hybrid nanocomposites. *Heliyon* 6: e04187.

2 Dagdag, O., El Gouri, M., El Mansouri, A. et al. (2020). Rheological and electrical study of a composite material based on an epoxy polymer containing cyclotriphosphazene. *Polymers* 12: 921.

3 Dagdag, O., Hsissou, R., El Harfi, A. et al. (2020). Fabrication of polymer based epoxy resin as effective anti-corrosive coating for steel: computational modeling reinforced experimental studies. *Surfaces and Interfaces* 18: 100454.

4 Dagdag, O., Berisha, A., Safi, Z. et al. (2020). Highly durable macromolecular epoxy resin as anticorrosive coating material for carbon steel in 3% NaCl: computational supported experimental studies. *Journal of Applied Polymer Science* 137: 49003.

5 Dagdag, O., Safi, Z., Qiang, Y. et al. (2020). Synthesis of macromolecular aromatic epoxy resins as anticorrosive materials: computational modeling reinforced experimental studies. *ACS Omega* 5: 3151–3164.

6 Dagdag, O., Safi, Z., Erramli, H. et al. (2019). Anticorrosive property of heterocyclic based epoxy resins on carbon steel corrosion in acidic medium: electrochemical, surface morphology, DFT and Monte Carlo simulation studies. *Journal of Molecular Liquids* 287: 110977.

7 Dagdag, O., Safi, Z., Erramli, H. et al. (2020). Epoxy prepolymer as a novel anti-corrosive material for carbon steel in acidic solution: electrochemical, surface and computational studies. *Materials Today Communications* 22: 100800.

8 Dagdag, O., Safi, Z., Wazzan, N. et al. (2020). Highly functionalized epoxy macromolecule as an anti-corrosive material for carbon steel: computational (DFT, MDS), surface (SEM-EDS) and electrochemical (OCP, PDP, EIS) studies. *Journal of Molecular Liquids* 302: 112535.

9 Hsissou, R., Benhiba, F., Dagdag, O. et al. (2020). Development and potential performance of prepolymer in corrosion inhibition for carbon steel in 1.0 M HCl: outlooks from experimental and computational investigations. *Journal of Colloid and Interface Science* 574: 43–60.

10 Dagdag, O., El Harfi, A., Safi, Z. et al. (2020). Cyclotriphosphazene based dendrimeric epoxy resin as an anti-corrosive material for copper in 3% NaCl: experimental and computational demonstrations. *Journal of Molecular Liquids* 308: 113020.

11 Singla, M. and Chawla, V. (2010). Mechanical properties of epoxy resin–fly ash composite. *Journal of Minerals and Materials Characterization and Engineering* 9: 199–210.

12 Jin, F.-L., Li, X., and Park, S.-J. (2015). Synthesis and application of epoxy resins: a review. *Journal of Industrial and Engineering Chemistry* 29: 1–11.

13 Park, S.J. and Jin, F.L. (2011). Epoxy resins: fluorine systems. *Wiley Encyclopedia of Composites*: 1–6.

14 Bardonnet, P. (1992). Monographies. Résines époxydes composants et propriétés. *Journal of Techniques de l'ingénieur. Plastiques et composites* 3: A3465. 1-A3465.

15 Barrere, C. and Dal Maso, F. (1997). Résines époxy réticulées par des polyamines: structure et propriétés. *Revue de l'institut français du pétrole* 52: 317–335.

16 Fromonteil, C., Bardelle, P., and Cansell, F. (2000). Hydrolysis and oxidation of an epoxy resin in sub-and supercritical water. *Industrial and Engineering Chemistry Research* 39: 922–925.

17 Prolongo, S.G., del Rosario, G., and Ureña, A. (2006). Comparative study on the adhesive properties of different epoxy resins. *International Journal of Adhesion and Adhesives* 26: 125–132.

18 Matuszczak, S. and Feast, W.J. (2000). An approach to fluorinated surface coatings via photoinitiated cationic cross-linking of mixed epoxy and fluoroepoxy systems. *Journal of Fluorine Chemistry* 102: 269–277.

19 Jin, F.-L., Ma, C.-J., and Park, S.-J. (2011). Thermal and mechanical interfacial properties of epoxy composites based on functionalized carbon nanotubes. *Materials Science and Engineering A* 528: 8517–8522.

20 Jin, F.L. and Park, S.J. (2006). Thermal properties and toughness performance of hyperbranched-polyimide-modified epoxy resins. *Journal of Polymer Science Part B: Polymer Physics* 44: 3348–3356.

21 Yang, C. and Yang, Z.G. (2013). Synthesis of low viscosity, fast UV curing solder resist based on epoxy resin for ink-jet printing. *Journal of Applied Polymer Science* 129: 187–192.

22 Czub, P. (2009). Synthesis of high-molecular-weight epoxy resins from modified natural oils and Bisphenol A or BisphenolA-based epoxy resins. *Polymers for Advanced Technologies* 20: 194–208.

23 Wu, C.C. and Lee, W.J. (2010). Synthesis and properties of copolymer epoxy resins prepared from copolymerization of bisphenol A, epichlorohydrin, and liquefied Dendrocalamus latiflorus. *Journal of Applied Polymer Science* 116: 2065–2073.

24 Yoo, M.J., Kim, S.H., Park, S.D. et al. (2010). Investigation of curing kinetics of various cycloaliphatic epoxy resins using dynamic thermal analysis. *European Polymer Journal* 46: 1158–1162.

25 Liu, W. and Wang, Z. (2011). Silicon-containing cycloaliphatic epoxy resins with systematically varied functionalities: synthesis and structure/property relationships. *Macromolecular Chemistry and Physics* 212: 926–936.

26 Tao, Z., Yang, S., Chen, J., and Fan, L. (2007). Synthesis and characterization of imide ring and siloxane-containing cycloaliphatic epoxy resins. *European Polymer Journal* 43: 1470–1479.

27 Gao, N., Liu, W., Yan, Z., and Wang, Z. (2013). Synthesis and properties of transparent cycloaliphatic epoxy–silicone resins for opto-electronic devices packaging. *Optical Materials* 35: 567–575.

28 Park, S.-J., Jin, F.-L., and Lee, J.-R. (2004). Thermal and mechanical properties of tetrafunctional epoxy resin toughened with epoxidized soybean oil. *Materials Science and Engineering A* 374: 109–114.

29 Chen, Y., Yang, L., Wu, J. et al. (2013). Thermal and mechanical properties of epoxy resin toughened with epoxidized soybean oil. *Journal of Thermal Analysis and Calorimetry* 113: 939–945.

30 Lee, M.C., Ho, T.H., and Wang, C.S. (1996). Synthesis of tetrafunctional epoxy resins and their modification with polydimethylsiloxane for electronic application. *Journal of Applied Polymer Science* 62: 217–225.

31 Aouf, C., Nouailhas, H., Fache, M. et al. (2013). Multi-functionalization of gallic acid. Synthesis of a novel bio-based epoxy resin. *European Polymer Journal* 49: 1185–1195.

32 Guo, B., Jia, D., Fu, W., and Qiu, Q. (2003). Hygrothermal stability of dicyanate-novolac epoxy resin blends. *Polymer Degradation and Stability* 79: 521–528.

33 Lin, C.-H. and Wang, C.-S. (2001). Novel phosphorus-containing epoxy resins Part I. Synthesis and properties. *Polymer* 42: 1869–1878.

34 Sangermano, M., Bongiovanni, R., Priola, A., and Pospiech, D. (2005). Fluorinated alcohols as surface-active agents in cationic photopolymerization of epoxy monomers. *Journal of Polymer Science Part A: Polymer Chemistry* 43: 4144–4150.

35 Lee, J., Jin, F., Park, S., and Park, J. (2004). Study of new fluorine-containing epoxy resin for low dielectric constant. *Surface and Coating Technology* 180: 650–654.

36 Gupta, D., Gutch, P., Lal, G. et al. (2004). Fluorinated epoxy resins-based sorbent coating materials for quartz piezoelectric crystal detector. *Defence Science Journal* 54: 229.

37 Park, S.-J., Jin, F.-L., and Shin, J.-S. (2005). Physicochemical and mechanical interfacial properties of trifluorometryl groups containing epoxy resin cured with amine. *Materials Science and Engineering A* 390: 240–245.

38 Zhang, W., Li, X., and Yang, R. (2011). Pyrolysis and fire behaviour of epoxy resin composites based on a phosphorus-containing polyhedral oligomeric silsesquioxane (DOPO-POSS). *Polymer Degradation and Stability* 96: 1821–1832.

39 Liu, W., Wang, Z., Xiong, L., and Zhao, L. (2010). Phosphorus-containing liquid cycloaliphatic epoxy resins for reworkable environment-friendly electronic packaging materials. *Polymer* 51: 4776–4783.

40 Wang, X. and Zhang, Q. (2004). Synthesis, characterization, and cure properties of phosphorus-containing epoxy resins for flame retardance. *European Polymer Journal* 40: 385–395.

41 Liu, R. and Wang, X. (2009). Synthesis, characterization, thermal properties and flame retardancy of a novel nonflammable phosphazene-based epoxy resin. *Polymer Degradation and Stability* 94: 617–624.

42 Yang, X., Huang, W., and Yu, Y. (2011). Synthesis, characterization, and properties of silicone–epoxy resins. *Journal of Applied Polymer Science* 120: 1216–1224.
43 Mercado, L., Galia, M., Reina, J., and Stability (2006). Silicon-containing flame retardant epoxy resins: Synthesis, characterization and properties. *Polymer Degradation and Stability* 91: 2588–2594.
44 Liu, W., Wang, Z., Chen, Z. et al. (2012). Synthesis and properties of two novel silicon-containing cycloaliphatic epoxy resins for electronic packaging application. *Polymers for Advanced Technologies* 23: 367–374.
45 Park, S.-J., Jin, F.-L., and Lee, J.-R. (2005). Synthesis and characterization of a novel silicon-containing epoxy resin. *Macromolecular Research* 13: 8–13.
46 Mendels, D.-A. *Coupling of Physical Aging and Internal Stresses in Thermoset Based Multimaterial Systems*. EPFL2001.
47 Rozenberg, B.A. (1986). Kinetics, thermodynamics and mechanism of reactions of epoxy oligomers with amines. *Epoxy Resins and Composites* II: 113–165.
48 Vinnik, R. and Roznyatovsky, V. (2003). Kinetic method by using calorimetry to mechanism of epoxy-amine cure reaction. *Journal of Thermal Analysis and Calorimetry* 74: 29.
49 Shechter, L., Wynstra, J., and Kurkjy, R.P. (1956). Glycidyl ether reactions with amines. *Industrial and Engineering Chemistry* 48: 94–97.
50 Flammersheim, H. (1998). Kinetics and mechanism of the epoxide–amine polyaddition. *Thermochimica Acta* 310: 153–159.
51 Smith, I.T. (1961). The mechanism of the crosslinking of epoxide resins by amines. *Polymer* 2: 95–108.
52 Mezzenga, R., Boogh, L., Månson, J.-A.E., and Pettersson, B. (2000). Effects of the branching architecture on the reactivity of epoxy– amine groups. *Macromolecules* 33: 4373–4379.
53 Enikolopiyan, N. (1977). New aspects of the nucleophilic opening of epoxide rings. *Pure and Applied Chemistry* 48: 317–328.
54 Ramos, J.A., Pagani, N., Riccardi, C.C. et al. (2005). Cure kinetics and shrinkage model for epoxy-amine systems. *Polymer* 46: 3323–3328.
55 Umoren, S.A. and Eduok, U.M. (2016). Application of carbohydrate polymers as corrosion inhibitors for metal substrates in different media: a review. *Carbohydrate Polymers* 140: 314–341.
56 Umoren, S., Ogbobe, O., Ebenso, E., and Ekpe, U. (2006). Effect of halide ions on the corrosion inhibition of mild steel in acidic medium using polyvinyl alcohol. *Pigment & Resin Technology* 35: 284–292.
57 Umoren, S. (2009). Polymers as corrosion inhibitors for metals in different media-A review. *The Open Corrosion Journal* 2.
58 Quraishi, M. and Shukla, S.K. (2009). Poly (aniline-formaldehyde): a new and effective corrosion inhibitor for mild steel in hydrochloric acid. *Materials Chemistry and Physics* 113: 685–689.

59 Chauhan, D.S., Ansari, K., Sorour, A. et al. (2018). Thiosemicarbazide and thiocarbohydrazide functionalized chitosan as ecofriendly corrosion inhibitors for carbon steel in hydrochloric acid solution. *International Journal of Biological Macromolecules* 107: 1747–1757.

60 Yarovsky, I. and Evans, E. (2002). Computer simulation of structure and properties of crosslinked polymers: application to epoxy resins. *Polymer* 43: 963–969.

61 Jacobsen, E.N., Kakiuchi, F., Konsler, R.G. et al. (1997). Enantioselective catalytic ring opening of epoxides with carboxylic acids. *Tetrahedron Letters* 38: 773–776.

62 Dagdag, O., Safi, Z., Erramli, H. et al. (2019). Adsorption and anticorrosive behavior of aromatic epoxy monomers on carbon steel corrosion in acidic solution: computational studies and sustained experimental studies. *RSC Advances* 9: 14782–14796.

63 Dagdag, O., Safi, Z., Hsissou, R. et al. (2019). Epoxy pre-polymers as new and effective materials for corrosion inhibition of carbon steel in acidic medium: computational and experimental studies. *Scientific Reports* 9: 1–14.

64 Dagdag, O., El Harfi, A., Cherkaoui, O. et al. (2019). Rheological, electrochemical, surface, DFT and molecular dynamics simulation studies on the anticorrosive properties of new epoxy monomer compound for steel in 1 M HCl solution. *RSC Advances* 9: 4454–4462.

65 Hsissou, R., Abbout, S., Seghiri, R. et al. (2020). Evaluation of corrosion inhibition performance of phosphorus polymer for carbon steel in [1 M] HCl: computational studies (DFT, MC and MD simulations). *Journal of Materials Research and Technology* 9: 2691–2703.

66 Dagdag, O., El Harfi, A., Essamri, A. et al. (2018). Phosphorous-based epoxy resin composition as an effective anticorrosive coating for steel. *International Journal of Industrial Chemistry* 9: 231–240.

67 Dagdag, O., El Harfi, A., El Gana, L. et al. (2020). Designing of phosphorous based highly functional dendrimeric macromolecular resin as an effective coating material for carbon steel in NaCl: computational and experimental studies. *Journal of Applied Polymer Science* 138: 49673.

68 Galai, M., El Gouri, M., Dagdag, O. et al. (2016). New hexa propylene glycol cyclotiphosphazene as efficient organic inhibitor of carbon steel corrosion in hydrochloric acid medium. *Journal of Materials and Environmental Science* 7: 1562–1575.

69 Hsissou, R., Dagdag, O., Berradi, M. et al. (2019). Development rheological and anti-corrosion property of epoxy polymer and its composite. *Heliyon* 5: e02789.

70 Hsissou, R., Dagdag, O., Abbout, S. et al. (2019). Novel derivative epoxy resin TGETET as a corrosion inhibition of E24 carbon steel in 1.0 M HCl solution. experimental and computational (DFT and MD simulations) methods. *Journal of Molecular Liquids* 2: 182–192.

71 R. Hsissou, S. Abbout, Z. Safi, F. Benhiba, N. Wazzan, L. Guo, *et al.*, "Synthesis and anticorrosive properties of epoxy polymer for CS in [1 M] HCl solution: electrochemical, AFM, DFT and MD simulations," *Construction and Building Materials,* p. 121454, 2020, 270.

72 Rbaa, M., Benhiba, F., Dohare, P. et al. (2020). Synthesis of new epoxy glucose derivatives as a non-toxic corrosion inhibitors for carbon steel in molar HCl: experimental, DFT and MD simulation. *Chemical Data Collections* 27: 100394.

73 Rbaa, M., Dohare, P., Berisha, A. et al. (2020). New Epoxy sugar based glucose derivatives as eco friendly corrosion inhibitors for the carbon steel in 1.0 M HCl: experimental and theoretical investigations. *Journal of Alloys and Compounds* 833: 154949.

74 Koulou, A., Rbaa, M., Errahmany, N. et al. (2020). Synthesis of new epoxy glucose derivatives as inhibitor for mild steel corrosion in 1.0 M HCl, Experimental study: Part-1. *Moroccan Journal of Chemistry* 8 (4): 775–787.

75 Umoren, S. and Solomon, M. (2014). Recent developments on the use of polymers as corrosion inhibitors-a review. *The Open Materials Science Journal* 8.

76 Ruslantsev, A., Portnova, Y.M., and Tairova, L. et al. (2016). Analysis of mechanical properties anisotropy of nanomodified carbon fibre-reinforced woven composites. *IOP Conference Series: Materials Science and Engineering*, p. 012003.

77 Solomon, M.M., Gerengi, H., Umoren, S.A. et al. (2018). Gum Arabic-silver nanoparticles composite as a green anticorrosive formulation for steel corrosion in strong acid media. *Carbohydrate Polymers* 181: 43–55.

78 Hang, T.T.X., Truc, T.A., Nam, T.H. et al. (2007). Corrosion protection of carbon steel by an epoxy resin containing organically modified clay. *Surface and Coatings Technology* 201: 7408–7415.

79 Dagdag, O., Guo, L., Safi, Z. et al. (2020). Epoxy resin and TiO2 composite as anticorrosive material for carbon steel in 3% NaCl medium: experimental and computational studies. *Journal of Molecular Liquids* 317: 114249.

80 Rathish, R.J., Dorothy, R., Joany, R., and Pandiarajan, M. (2013). Corrosion resistance of nanoparticle-incorporated nano coatings. *European Chemical Bulletin* 2: 965–970.

81 Ghanbari, A. and Attar, M. (2015). A study on the anticorrosion performance of epoxy nanocomposite coatings containing epoxy-silane treated nano-silica on mild steel substrate. *Journal of Industrial and Engineering Chemistry* 23: 145–153.

82 Dagdag, O., Hsissou, R., El Harfi, A. et al. (2020). Development and anti-corrosion performance of polymeric epoxy resin and their zinc phosphate composite on 15CDV6 Steel in 3wt% NaCl: experimental and computational studies. *Journal of Bio-and Tribo-Corrosion* 6: 1–9.

83 Verma, C., Olasunkanmi, L.O., Akpan, E.D. et al. (2020). Epoxy resins as anticorrosive polymeric materials: a review. *Reactive and Functional Polymers* 156: 104741.

84 Suleiman, R., Dafalla, H., and El Ali, B. (2015). Novel hybrid epoxy silicone materials as efficient anticorrosive coatings for mild steel. *RSC Advances* 5: 39155–39167.

85 Bahlakeh, G., Ramezanzadeh, B., and Ramezanzadeh, M. (2017). Corrosion protective and adhesion properties of a melamine-cured polyester coating applied on steel substrate treated by a nanostructure cerium–lanthanum film. *Journal of the Taiwan Institute of Chemical Engineers* 81: 419–434.

86 Mo, M., Zhao, W., Chen, Z. et al. (2015). Excellent tribological and anti-corrosion performance of polyurethane composite coatings reinforced with functionalized graphene and graphene oxide nanosheets. *Rsc Advances* 5: 56486–56497.

87 Gurunathan, T., Rao, C.R., Narayan, R., and Raju, K. (2013). Synthesis, characterization and corrosion evaluation on new cationomeric polyurethane water dispersions and their polyaniline composites. *Progress in Organic Coatings* 76: 639–647.

88 Conradi, M., Kocijan, A., Zorko, M., and Jerman, I. (2012). Effect of silica/PVC composite coatings on steel-substrate corrosion protection. *Progress in Organic Coatings* 75: 392–397.

89 Riaz, U., Ashraf, S., and Ahmad, S. (2007). High performance corrosion protective DGEBA/polypyrrole composite coatings. *Progress in Organic Coatings* 59: 138–145.

90 Pagotto, J.F., Recio, F.J., Motheo, A.d.J., and Herrasti, P. (2016). Multilayers of PAni/n-TiO2 and PAni on carbon steel and welded carbon steel for corrosion protection. *Surface and Coatings Technology* 289: 23–28.

91 Alhumade, H., Abdala, A., Yu, A. et al. (2016). Corrosion inhibition of copper in sodium chloride solution using polyetherimide/graphene composites. *The Canadian Journal of Chemical Engineering* 94: 896–904.

92 Ruhi, G., Bhandari, H., and Dhawan, S.K. (2014). Designing of corrosion resistant epoxy coatings embedded with polypyrrole/SiO2 composite. *Progress in Organic Coatings* 77: 1484–1498.

93 Chen, X., Shen, K., and Zhang, J. (2010). Preparation and anticorrosion properties of polyaniline-SiO2-containing coating on Mg-Li alloy. *Pigment and Resin Technology* 39: 322–326.

94 Sambyal, P., Ruhi, G., Dhawan, R., and Dhawan, S.K. (2016). Designing of smart coatings of conducting polymer poly (aniline-co-phenetidine)/SiO2 composites for corrosion protection in marine environment. *Surface and Coatings Technology* 303: 362–371.

95 Ruhi, G., Modi, O., and Dhawan, S. (2015). Chitosan-polypyrrole-SiO2 composite coatings with advanced anticorrosive properties. *Synthetic Metals* 200: 24–39.

96 Sathiyanarayanan, S., Azim, S.S., and Venkatachari, G. (2007). Corrosion protection of magnesium ZM 21 alloy with polyaniline–TiO2 composite containing coatings. *Progress in Organic Coatings* 59: 291–296.

97 Li, Z., Ma, L., Gan, M. et al. (2013). Synthesis and anticorrosion performance of poly (2, 3-dimethylaniline)–TiO2 composite. *Progress in Organic Coatings* 76: 1161–1167.

98 Sathiyanarayanan, S., Azim, S.S., and Venkatachari, G. (2007). Preparation of polyaniline–TiO2 composite and its comparative corrosion protection performance with polyaniline. *Synthetic Metals* 157: 205–213.

99 Yu, Z., Di, H., Ma, Y. et al. (2015). Fabrication of graphene oxide–alumina hybrids to reinforce the anti-corrosion performance of composite epoxy coatings. *Applied Surface Science* 351: 986–996.

100 Sari, M.G., Shamshiri, M., and Ramezanzadeh, B. (2017). Fabricating an epoxy composite coating with enhanced corrosion resistance through impregnation of functionalized graphene oxide-co-montmorillonite Nanoplatelet. *Corrosion Science* 129: 38–53.

101 Huang, C., Li, C., and Shi, G. (2012). Graphene based catalysts. *Energy & Environmental Science* 5: 8848–8868.

102 Xi, Z., Tan, C., Xu, L. et al. (2015). Preparation of novel functional Mg/O/PCL/ZnO composite biomaterials and their corrosion resistance. *Applied Surface Science* 351: 410–415.

103 Singh-Beemat, J. and Iroh, J. (2012). Characterization of corrosion resistant clay/epoxy ester composite coatings and thin films. *Progress in Organic Coatings* 74: 173–180.

104 Thenmozhi, G., Arockiasamy, P., and Mohanraj, G. (2014). Evaluation of corrosion inhibition of mild steel: chemically polymerized PpAP/Al2O3 composite in the presence of anionic surfactants. *Portugaliae Electrochimica Acta* 32: 417–429.

105 Zhang, D. (2006). Preparation of core–shell structured alumina–polyaniline particles and their application for corrosion protection. *Journal of Applied Polymer Science* 101: 4372–4377.

106 Xia, W., Xue, H., Wang, J. et al. (2016). Functionlized graphene serving as free radical scavenger and corrosion protection in gamma-irradiated epoxy composites. *Carbon* 101: 315–323.

107 Grari, O., Taouil, A.E., Dhouibi, L. et al. (2015). Multilayered polypyrrole–SiO2 composite coatings for functionalization of stainless steel: characterization and corrosion protection behavior. *Progress in Organic Coatings* 88: 48–53.

108 Rong, M., Zhang, M., and Ruan, W. (2006). Surface modification of nanoscale fillers for improving properties of polymer nanocomposites: a review. *Materials Science and Technology* 22: 787–796.

109 Ruslantsev, A., Portnova, Y.M., Tairova, L., and Dumansky, A. (2016). Analysis of mechanical properties anisotropy of nanomodified carbon fibre-reinforced woven composites. IOP Conference Series: Materials Science and Engineering, p. 012003.

110 Sanchez, C., Arribart, H., and Guille, M.M.G. (2005). Biomimetism and bioinspiration as tools for the design of innovative materials and systems. *Nature Materials* 4: 277–288.

111 Seidi, F., Jenjob, R., and Crespy, D. (2018). Designing smart polymer conjugates for controlled release of payloads. *Chemical Reviews* 118: 3965–4036.

112 Alkbir, M., Sapuan, S., Nuraini, A., and Ishak, M. (2016). Fibre properties and crashworthiness parameters of natural fibre-reinforced composite structure: a literature review. *Composite Structures* 148: 59–73.

113 Umoren, S.A. and Madhankumar, A. (2016). Effect of addition of CeO2 nanoparticles to pectin as inhibitor of X60 steel corrosion in HCl medium. *Journal of Molecular Liquids* 224: 72–82.

114 Trung, V.Q., Van Hoan, P., Phung, D.Q. et al. (2014). Double corrosion protection mechanism of molybdate-doped polypyrrole/montmorillonite nanocomposites. *Journal of Experimental Nanoscience* 9: 282–292.

115 Talebi, J., Halladj, R., and Askari, S. (2010). Sonochemical synthesis of silver nanoparticles in Y-zeolite substrate. *Journal of Materials Science* 45: 3318–3324.

116 Li, J.L., An, X.Q., and Zhu, Y.Y. (2012). Controllable synthesis and characterization of highly fluorescent silver nanoparticles. *Journal of Nanoparticle Research* 14: 1325.

117 Maretti, L., Billone, P.S., Liu, Y., and Scaiano, J.C. (2009). Facile photochemical synthesis and characterization of highly fluorescent silver nanoparticles. *Journal of the American Chemical Society* 131: 13972–13980.

118 Ossi, P.M., Agarwal, N.R., Fazio, E. et al. (2014). Laser-mediated nanoparticle synthesis and self-assembling. In: *Lasers in Materials Science*, 175–212. Springer.

119 Li, Y.-f., Gan, W.-p., Jian, Z. et al. (2015). Hydrothermal synthesis of silver nanoparticles in Arabic gum aqueous solutions. *Transactions of Nonferrous Metals Society of China* 25: 2081–2086.

120 Sreeram, K.J., Nidhin, M., and Nair, B.U. (2008). Microwave assisted template synthesis of silver nanoparticles. *Bulletin of Materials Science* 31: 937–942.

121 Solomon, M.M. and Umoren, S.A. (2016). In-situ preparation, characterization and anticorrosion property of polypropylene glycol/silver nanoparticles composite for mild steel corrosion in acid solution. *Journal of Colloid and Interface Science* 462: 29–41.

122 Solomon, M., Umoren, S., and Abai, E. (2015). Poly (methacrylic acid)/silver nanoparticles composites: in-situ preparation, characterization and anticorrosion property for mild steel in H2SO4 solution. *Journal of Molecular Liquids* 212: 340–351.

123 Salehoon, E., Ahmadi, S.J., Razavi, S.M., and Parvin, N. (2017). Thermal and corrosion resistance properties of unsaturated polyester/clay nanocomposites and the effect of electron beam irradiation. *Polymer Bulletin* 74: 1629–1647.

124 Yuan, X., Yue, Z., Liu, Z. et al. (2016). Comparison of the failure mechanisms of silicone-epoxy hybrid coatings on type A3 mild steel and 2024 Al-alloy. *Progress in Organic Coatings* 90: 101–113.

125 Verma, C., Olasunkanmi, L.O., Akpan, E.D. et al. (2020). Epoxy resins as anticorrosive polymeric materials: a review. *Reactive and Functional Polymers* 156: 104741.

126 Charitha, B.P., Chenan, A., and Rao, P. (2017). Enhancement of surface coating characteristics of epoxy resin by dextran: an electrochemical approach. *Industrial & Engineering Chemistry Research* 56: 1137–1147.

127 Hang, T.T.X., Truc, T.A., Nam, T.H. et al. (2007). Corrosion protection of carbon steel by an epoxy resin containing organically modified clay. *Surface and Coatings Technology* 201: 7408–7415.

128 Dagdag, O., Guo, L., Safi, Z. et al. (2020). Epoxy resin and TiO2 composite as anticorrosive material for carbon steel in 3% NaCl medium: experimental and computational studies. *Journal of Molecular Liquids* 317: 114249.

129 Rathish, R.J., Dorothy, R., Joany, R., and Pandiarajan, M. (2013). Corrosion resistance of nanoparticle-incorporated nano coatings. *European Chemical Bulletin* 2: 965–970.

130 Ghanbari, A. and Attar, M. (2015). A study on the anticorrosion performance of epoxy nanocomposite coatings containing epoxy-silane treated nano-silica on mild steel substrate. *Journal of Industrial and Engineering Chemistry* 23: 145–153.

131 Dagdag, O., Hsissou, R., El Harfi, A. et al. (2020). Development and anti-corrosion performance of polymeric epoxy resin and their zinc phosphate composite on 15CDV6 steel in 3wt% NaCl: experimental and computational studies. *Journal of Bio-and Tribo-Corrosion* 6: 1–9.

132 González-García, Y., González, S., and Souto, R. (2007). Electrochemical and structural properties of a polyurethane coating on steel substrates for corrosion protection. *Corrosion Science* 49: 3514–3526.

133 Dagdag, O., El Harfi, A., El Gana, L. et al. (2020). Designing of phosphorous based highly functional dendrimeric macromolecular resin as an effective coating material for carbon steel in NaCl: computational and experimental studies. *Journal of Applied Polymer Science* 138: 49673.

134 Sangermano, M., Roppolo, I., Shan, G., and Andrews, M.P. (2009). Nanocomposite epoxy coatings containing rare earth ion-doped

LaF3 nanoparticles: film preparation and characterization. *Progress in Organic Coatings* 65: 431–434.

135 Dagdag, O., Hsissou, R., El Harfi, A. et al. (2020). Epoxy resins and their zinc composites as novel anti-corrosive materials for copper in 3% sodium chloride solution: experimental and computational studies. *Journal of Molecular Liquids* 315: 113757.

136 Ramezanzadeh, B. and Attar, M. (2011). Studying the effects of micro and nano sized ZnO particles on the corrosion resistance and deterioration behavior of an epoxy-polyamide coating on hot-dip galvanized steel. *Progress in Organic Coatings* 71: 314–328.

137 Xavier, J.R. and Nallaiyan, R. (2016). Application of EIS and SECM studies for investigation of anticorrosion properties of epoxy coatings containing ZrO 2 nanoparticles on mild steel in 3.5% NaCl solution. *Journal of Failure Analysis and Prevention* 16: 1082–1091.

138 Schem, M., Schmidt, T., Gerwann, J. et al. (2009). CeO2-filled sol–gel coatings for corrosion protection of AA2024-T3 aluminium alloy. *Corrosion Science* 51: 2304–2315.

139 Yu, H., Wang, L., Shi, Q. et al. (2006). Study on nano-CaCO3 modified epoxy powder coatings. *Progress in Organic coatings* 55: 296–300.

140 Dhoke, S.K., Sinha, T.J.M., and Khanna, A. (2009). Effect of nano-Al 2 O 3 particles on the corrosion behavior of alkyd based waterborne coatings. *Journal of Coatings Technology and Research* 6: 353–368.

141 Dolatzadeh, F., Moradian, S., and Jalili, M.M. (2011). Influence of various surface treated silica nanoparticles on the electrochemical properties of SiO2/ polyurethane nanocoatings. *Corrosion Science* 53: 4248–4257.

142 Dong, Y., Zhang, Q., Su, X., and Zhou, Q. (2013). Preparation and investigation of the protective properties of bipolar coatings. *Progress in Organic Coatings* 76: 662–669.

Index

Organic Corrosion Inhibitors: Synthesis, Characterization, Mechanism, and Applications, First Edition. Edited by Chandrabhan Verma, Chaudhery Mustansar Hussain, and Eno E. Ebenso.

b

d

f

g

h

i

k

l

V

X

Printed and bound by CPI Group (UK) Ltd, Croydon, CR0 4YY

07/02/2023

03189351-0001